5/78

QL
737
C2.2
F6.94

THE WILD CANIDS

BEHAVIORAL SCIENCE SERIES

Being and Becoming Human; Essays on the Biogram, *Earl W. Count*
Imprinting, Early Experience and the Developmental Psychobiology of Attachment, *Eckhard H. Hess*
Motivation of Human and Animal Behavior: An Ethological View, *Konrad Lorenz and Paul Leyhausen*
The Cheetah: The Biology, Ecology, and Behavior of an Endangered Species, *Randall L. Eaton*
The Wild Canids: Their Systematics, Behavioral Ecology and Evolution, *M. W. Fox, Editor*

THE WILD CANIDS

Their Systematics, Behavioral Ecology and Evolution

edited by

M. W. Fox, Ph.D., B. Vet. Med., M.R.C.V.S.
Department of Psychology, Washington University,
St. Louis, Missouri

FOREWORD BY KONRAD LORENZ

BEHAVIORAL SCIENCE SERIES

VNR VAN NOSTRAND REINHOLD COMPANY
New York / Cincinnati / Toronto / London / Melbourne

Van Nostrand Reinhold Company Regional Offices:
New York Cincinnati Chicago Millbrae Dallas

Van Nostrand Reinhold Company International Offices:
London Toronto Melbourne

Copyright © 1975 by Litton Educational Publishing, Inc.

Library of Congress Catalog Card Number: 74-23194
ISBN: 0-442-22430-3

All rights reserved. No part of this work covered by the copyright hereon may be reproduced or used in any form or by any means — graphic, electronic, or mechanical, including photocopying, recording, taping, or information storage and retrieval systems — without permission of the publisher.

Manufactured in the United States of America

Published by Van Nostrand Reinhold Company
450 West 33rd Street, New York, N.Y. 10001

Published simultaneously in Canada by Van Nostrand Reinhold Ltd.

15 14 13 12 11 10 9 8 7 6 5 4 3 2 1

Library of Congress Cataloging in Publication Data

Fox, Michael W 1937-
The wild canids.

(Behavorial science series)
Bibliography: p.
Includes index.
1. Canidae. I. Title.
QL737.C22F694 599'.7442 74-23194
ISBN 0-442-22430-3

BEHAVIORAL SCIENCE SERIES

The Van Nostrand Reinhold Behavioral Science Series will publish a broad range of books on animal and human behavior from an ethological perspective. Although presently observable behavior is the focus of this series, the development of behavior in individuals, as well as the evolutionary history in various species, will also be considered. It is felt that such an holistic approach is needed to come to a fuller understanding of behavior in general. This series is a contribution toward this goal.

Erich Klinghammer, Consulting Editor
Purdue University

FOREWORD

KONRAD Z. LORENZ

I am in some doubt as to whether I am qualified to write a foreword to a book such as this, a book which tries to supply comprehensive information on the systematics, the behavioral ecology, and the evolution of such a large and diversified group of animals as the Canidae are. It is, however, essential to the very nature of this book as a pioneer effort to bring order into a vast field of knowledge that, besides the compilation of facts, speculations are legitimate which, as inspired guesses, may direct future research by calling attention to that which we do not know and which we badly want to know. In fact, many of the contributors are particularly good at doing just this. If I myself venture to voice some opinions of my own and even to offer some criticism, my contributions are to be taken as this kind of hint. I emphasize the speculative nature of all I am going to say. All of it is based on the rather uncertain ground of taxonomic and behavioral Gestalt perception.

My factual knowledge about Canidae is confined to what I learned in spending a life in close company of domestic dogs, in keeping one dingo, and in closely following the work of two near friends of mine who are specialists on canid behavior. The first is Alfred Seitz, who studies golden jackals, coyotes, wolves, racoonlike dogs, and many domestic breeds, also making a special study of the fertility of coyote—jackal hybrids. The second is Eric Zimen, who for years has been studying wolves and wolf–poodle hybrids and who is maintaining a pack of wolves under near-natural conditions in the Bavarian National Park.

I am guilty of writing a popular book on domestic dogs and also of having propounded an erroneous hypothesis: I had inherited, from my teacher Oskar Heinroth, the assumption that the bulk of domestic dog races are descended from the golden jackal, *Canis aureus*. It is good strategy to have one's errors corrected by one's own pupils and it may well have been my suggestion that caused Alfred Seitz, during the time of his directorship of the Nuremberg Zoo, to conduct a thorough investigation of the ethology, mainly of the communicating behavior, of some species which could be regarded as potential ancestors of the domestic dog. The result clearly contradicted Heinroth's and my assumption. The golden jackal, like the coyote, howls with a falling pitch, beginning with a high sharp "yip, yip, yip" and then dropping into a melancholy-sounding decrescendo while, conversely, practically all breeds of domestic dogs howl exactly like wolves, beginning softly at a low pitch and rising gradually in pitch as well as in loudness. The only highly interesting exception is represented by the basenji, of which some individuals howl in a definitely jackallike manner. There is one other peculiarity of this race to which I found reference in this book: according to Nunez and his co-workers (1970) the thyroid activity of this race is significantly higher than in other domestic dogs. It would be interesting to know whether this is also the case in golden jackals.

Although the golden jackal has to be excluded from being a possible ancestor of most domestic breeds, I still maintain that the races which I originally have called "lupus dogs," e.g., chow, husky, Greenland dogs etc., have a wild ancestor different from that of most or all other breeds. I believe that they are closely related to timber wolves, though I can give no other reason for thinking so except that they look like timber wolves and other breeds do not. As *Canis aureus* is out of consideration as an ancestor of all these dogs, my next best bet is *Canis pallipes,* now regarded as subspecific to *Canis lupus*. Mendelsohn (personal communication), who has been keeping a captive pack of *pallipes* in Jerusalem and who is also familiar with Finnish wolves as well as with timber wolves, is of the opinion that the rank order in a *pallipes* pack is less rigid and cruel and there is much less stress among the members of a captive pack than there is in that of any other subspecies. Also, *pallipes* is a scavenger who is apt to follow the big feline carnivores and is therefore eligible as a dog ancestor on the basis of my pet hypothesis, which assumes that dogs were domesticated without the means of keeping them in captivity.

I was interested to find support in the opinion of Clutton-Brock who regards *Canis lupus pallipes* as the ancestor of the dingo, and the dingo, of course, is a domestic dog gone feral, and not so feral as that. If N.W.G. Macintosh has seen fit to add the word "an enigma" to the title of his excellent contribution on the dingo, I think that his own findings have solved much of that riddle. A carnivore which allows itself to be fondled, even if it is hardly ever fed by the man fondling it, and which even sleeps in a huddle with him must have gone through a long, long history of symbiosis with man, even if the mutual advantages at present seem to be small. There is no other way in which sufficient trust could have developed between aborigines and dingoes, and a lot of trust is indeed needed to sleep together in a huddle! The fact that the aborigines and their "warm blanket" dingoes do not hunt together does not seem to me particularly significant: the men of most cultures in which pariah dogs are tolerated do not do so either. Otto Koenig, who lately has studied pariah dogs in Turkey and made some most instructive films of them, told me a very curious thing. The pariahs round Istanbul often live in close vicinity of domestic stock; chickens, goats, and sheep range free and unprotected, but they never touch them, even when very hungry. Even bitches with litters refrain from eating small chicks and newly born sheep. The survival value of this inhibition is obvious, as, in an area closely populated by humans and domestic animals, the dogs would soon be exterminated if they ever attacked livestock. The question, however, is how they learned not to do it. It is important to state in this connection, that unlike dingoes, Turkish pariah dogs prove to be completely amenable to normally "civilized" house-dog education.

Another point in this book about which my taxonomic Gestalt perception made me doubtful is the "lumping" of the three genera *Lycaon, Cuon,* and *Speothos* into the subfamily of Simnocyoninae, Zittel 1893. Although I know these animals only by a rather superficial zoo observation, I cannot bring myself to believe that these rather peculiar genera are much more closely related to each other than any of them is to all the rest of the Canidae.

On the other hand, I felt quite happy about missing, in a book on Canidae, the genus *Nyctereutes* (except in one list). I am acquainted with this animal and its behavior because A. Seitz had and bred it for some generations and I always felt that it was a dog-like racoon rather than a racoonlike dog.

Some puzzlement was also aroused in me by N.W.G. Macintosh's statement . . . "the animals which superficially look somewhat like dingoes were either dholes (*Cuon alpinus),* or jackals (*Canis aureus*), or pariahs (*Canis indicus*) in decreasing order of similar appearance and behavior." I assume that "decreasing" slipped in, instead of "increasing." There are pariahs which are externally indistinguishable from dingoes, so are all their expression movements and it is only on attempting to teach them to let small domestic animals alone that one would note a difference.

It is concerning this difference that I want to add a few of my own observations to the foreword of this book. They are meant only as a possible pointer to future research. The problems of domestication have been an obsession with me for many years. On the one hand I am convinced that man owes the life-long persistence of his constitutive curiosity and explorative playfulness to a partial neoteny which is indubitably a consequence of domestication. In a curiously analogous manner does the domestic dog owe its permanent attachment to its master to a behavioral neoteny which prevents it from ever wanting to become a pack leader. On the other hand, domestication is apt to cause an alarming disintegration of valuable behavioral traits and an equally alarming exaggeration of less desirable ones. The hypertrophy of eating and copulating are as characteristic of the domestic pig as they are of modern man. Anyone doubting this need only to observe people in a public swimming pool and note the increasing incidence of much-too-fat boys and young men, or to have a look at some modern illustrated magazines or films. The revulsion I felt and still feel against these phenomena which, I believe, are seriously threatening our culture, has misled me in regard to the difference between the domestic dog and its wild ancestor. Although I was always aware that the "vulgarisation" — as Julian Huxley calls it — of most domestic animals is caused by the selection pressure exerted by man's need of animals which are easy to breed and to fatten, and though I always emphasized that two domestic animals, the horse and the dog, were exempt from this degrading trend of domestication, I never realized the consequence of the very particular sort of selection pressure to which the domestic dog has been exposed for nearly 10,000 years. The romantic enthusiasm I had felt as a boy for Jack London's *White Fang* may also have contributed to make me believe that a tame purebred timber wolf would be the highest ideal of a dog to which a man could aspire.

This is an error. A good pig is one that eat and breeds indiscriminately and is easy to fatten. A "good" dog, on the other hand, is one that is faithful and friendly, that defends its master courageously at the danger of its own life, and also one that is intelligent and easy to teach. In other words, domestic dogs have been selected by man for properties which in human society are regarded as virtues. It is, therefore, not astonishing at all if this selection pressure acting through so long a time has made the domestic dog into something which is different from and in some way superior to its wild ancestor. If we feel that there are some specifically human traits in our dogs, this need not be due to an uncritical and erroneous projection of human properties into our dog but to a perfectly natural and understandable consequence of selection: the domestic dog really has been "humanized" in regard to certain of its faculties.

To love one's brother as one does oneself is one of the most beautiful commands of Christianity, though there are few men and women able to live up to it. A faithful dog, however, loves its master much more than it loves itself and certainly more than its master ever can be able to love it back. There certainly is no creature in the world in which "bond behavior," in other words personal friendship, has become an equally powerful motivation as it has in dogs.

Another specifically human trait found in no animals except in dogs is the faculty of having a "bad conscience." Some humanists will be up in arms against this statement, contending that "conscience" presupposes rational insight into the consequences of one's own acts and, therewith, conscious morality. I don't mean this. I believe that there is an unreflected feeling of displeasure, closely akin to guilt, in a dog which has committed an infraction on one of its instinctive or conditioned inhibitions. I have seen dogs who, on having bitten their human friend by mistake, have gone into the deepest dejection bordering on neurosis. In puppies of my own I often diagnosed the fact that they had done something forbidden on seeing their obviously "guilty" behavior. It was highly characteristic of my tame dingo that this particular faculty was obviously missing.

These two quasi-human properties of domestic dogs may have the most interesting consequences for the structure of their society. I was made to realize this by an incident which I observed lately. Claudia, the seven year old alpha-bitch of my little pack of chow hybrids of different composition, had just had her first serious fight with Babette, who, at the age of just over one year had outgrown the immunity to attack enjoyed by puppies. The outcome of the fight was inevitable, as the older bitch was heavier than the younger by at least a third. After the fight Claudia, to my utmost surprise, did not behave in the least as the victor in a dog fight usually does, but showed all the symptoms of bad conscience exactly as if she had, by mistake, bitten a human friend. She approached Babette with unmistakable submissive behavior. When the latter was lying at rest in a narrow passage, Claudia hardly dared to squeeze past until Babette got up and, again to my amazement, performed the ritual of anal control in the role of a rank-superior dog. Then they smelled each other's behind and gradually began to wag tails exactly as strangers do in effecting a "peace treaty." During the next days when Babette had fully recovered from her shock, she made the utmost of the indulgence caused in Claudia by her "bad conscience." She pulled her about in play in a most uninhibited manner, and, in pursuit games took predominantly the role of the pursuer. She even mounted Claudia in the manner demonstrating superiority, as described by Schenkel, which I had never observed her doing before the fight. No clear rank order has been developed between these two bitches during the four months that have elapsed since then. Claudia does, from time to time, lose her patience and administers a sound beating to Babette, thoroughly intimidating her for the moment, though never drawing blood. However, she never follows up her victory by any further persecution as a she-wolf would, and so Babette's subjugation never lasts longer than a few days and the relationship between the two bitches remains essentially unchanged.

These few observations raise a lot of questions. What would be the structure of a canine society composed of dogs "humanized" in the way my old Claudia is? Would such a society be able to function at all or would it be subject to grave malfunctions? Perhaps it would be, as it is well known that domestic dogs kept together in large numbers are quite likely to kill pack members in sudden excesses of mass fighting. This has repeatedly been recorded of beagles. These, however, are a very old and highly bred race and it is altogether possible that in more primitive dogs, such as my externally dingolike chow hybrids, killing inhibitions effective in the wolf pack might still be present and prevent murder while a "humane" bad conscience could soften the cruelty of rank order. These are things which I badly want to know and which I intend to investigate a bit in my retirement in Altenberg.

I have ventured to include these few chance observations in the foreword to this book because they represent just one more of the many hints it contains about "what should be done." In fact I regard it as one of its great merits that, besides the encyclopedic knowledge compiled in it, it also contains so many indications to lines of promising research still open to us. The hope is well justified that this book will attract the attention of many biologists, ecologists, and ethologists to a great group of mammals, the study of which cannot fail to become more and more important as it progresses. Our ecological and ethological knowledge of the Canidae is particularly important because of the many analogies which obviously exist between the structures and dynamics of canid and primate societies — including human society. Distribution of food gained by one individual between the members of a social group as well as a division of labor between individuals which become more and more different in the process is a phenomenon which, except for some slight beginnings found in the chimpanzee, does not occur in any other vertebrate except in man and in some Canidae. Nor are there many species in existence in which a nonsexual bond of personal friendship plays an equally important part as it does in some Canidae and in *Homo sapiens*.

KONRAD Z. LORENZ
Altenberg

PREFACE

Comparative studies of behavior within a particular family or order such as the Anatidae (ducks and geese) and primates, as exemplified by Johnsgard (1965) and Crook (1970), constitute a valuable approach to understanding the evolution of behavior. By comparing the development of behavior in a closely related species and after following their life history under often very different ecologies, insight may be gained as to their evolution, taxonomic relationships and *socio-ecological* fitness or adaptation. This latter point has been elegantly considered in Crook's review of the ecology and social organization of primates; he and also Kruuk (1972a) emphasize that comparative studies of the same species (or of races) adapted behaviorally to different environments can give very clear indications of contemporary evolutionary changes. Thus, the study of intraspecific variations in relation to ecology involves a recognition and understanding of the interaction of many variables. When assembled, these interacting variables form a matrix of intrinsic complexity within which the organism interacts with its socio-environmental milieu (see Chapter 30). Crook stresses the importance of such socio-ecological research as a new frontier for the ecologist, ethologist and anthropologist.

Such comparative studies in behavioral ecology therefore go far beyond earlier behavior studies in which the comparative analysis of action patterns or displays were used to determine taxonomic relationships between closely related species (e.g., Lorenz, 1941), and indirectly give some intimation as to the evolution of behavior. Given the added dimensions of an interdisciplinary approach as urged by Crook, a comparative study of inter- and intraspecific similarities and differences in social ecology would give a more complete picture of the evolutionary process. A single channel analysis of intra- and interspecific variations in displays can lead to circular arguments about homology, analogy, and evolutionary convergence or parallelism. A multichannel analysis of the intrinsic complexities of various socio-ecological systems or life styles in closely related (or the same) species under similar (or different) habitats includes the following parameters:

(a) availability, distribution, and seasonal variations in food supply;
(b) timing of reproduction in relation to (a);

(c) changes in social relationships and group organization in relation to (a) and (b);
(d) social dynamics in relation to division of labor (e.g., hunting), care of young, defense of territory, control intraspecific aggression via dominance hierarchy;
(e) analysis of social relationships (and their change due to death, emigration, or birth of offspring) between members of same and opposite sex, parents and young, and members of different age and social rank;
(f) determination of socio-ecological population regulating mechanisms;
(g) elucidation of socialization processes, including parent–infant and infant–infant allegiance and kinship, and philopatry or attachment to a particular locale (i.e., "locality imprinting");
(h) intra- and interspecific analysis of communication, including auditory and visual signals and displays, contact behaviors (grooming and licking), and scent marking;
(i) changes in communication with age, and in relation to social rank and season;
(j) correlation of morphology with social behavior, age, and rank;
(k) determination of relationships of the species in question with sympatric species — prey, predators, commensals, parasites, etc.

In this book some of the above parameters will be examined in the Canidae which as a family provides an excellent model for comparative studies, for its members range from the relatively solitary red fox to the gregarious wolf. Our knowledge on these matters is far from complete and indeed very little is known about the development and life habits of many canids. There is a dire need for more field and laboratory work on the ecology and ethology of the Canidae, and it is hoped that this book will not only be a useful reference source, but will also stimulate more research on these animals in the near future. Such research might lead us to a better understanding of the intrinsic complexities of socio-ecological systems, and consequently provide us with the knowledge essential for future conservation programs and the improvement of maintenance of such species in captivity.

CONTENTS

D. HIDEN RAMSEY LIBRARY
U.N.C. AT ASHEVILLE
ASHEVILLE, N. C. 28804

THE WILD CANIDS

PART

I

TAXONOMIC AND MORPHOLOGICAL STUDIES

This section contains a selection of papers which deal with various traditional and new methods relevant to taxonomy and systematics. Morphological variations may be correlated with ecology and these variations together with more recent "evolutionary" changes following hybridization of various canid species in the wild will be discussed in this section.

The studies described in this section provide a framework for ethological studies (see Parts III and V) where a knowledge of taxonomic relationships can facilitate comparative analysis at the behavioral level.

1
DISTRIBUTION AND TAXONOMY OF THE CANIDAE

HOWARD J. STAINS
Department of Zoology
Southern Illinois University
Carbondale, Illinois

EVOLUTION AND GEOLOGICAL HISTORY OF THE CANIDS

The terrestrial order Carnivora has long been subdivided into two infraorders: the Aeluroidea and the Arctoidea (Romer, 1966) or two superfamilies, Canoidea and Feloidea (Simpson, 1945; Stains, 1967). The family Canidae (together with the Procyonidae, Ursidae, and Mustelidae) is placed in the Arctoidea (Canoidea).

Although there are a few exceptions, the arctoids (canoids) differ from the aeluroids (feloids) in lacking retractile claws, having tympanic bullae formed from the tympanic bone only, and having a canal for the carotid artery which lies beneath the bullae.

Two of the most primitive canids are *Hesperocyon,* from the Lower Oligocene of North America, and *Cynodictis,* from the Upper Eocene of Europe and Asia and the Oligocene of Europe. These primitive canids resemble, in some respects, weasels and civets as the skeleton indicates that the body was elongate and the limbs short. This is due to the origin of all carnivores from the Miacidae, the members of which had this basic structure. *Hesperocyon* and *Cynodictis* were primitive enough to be ancestral to many of the terrestrial carnivores besides the canids. According to Matthew (1930), members of the subfamily Caninae probably evolved through forms like *Nothocyon, Cynodesmus,* and *Tomarctus,* and members of the subfamily Simocyoninae may have originated from *Daphoenus.*

Canids appear in the Eocene. Romer (1966) lists 5 genera in the Eocene, 19 in the Oligocene (3 genera held over from the Eocene), 42 in the Miocene (9 of which appeared in the Oligocene), 20 in the Pliocene (12 of which were present in the Miocene and 1 in the Oligocene), and 14 in the Pleistocene (4 remaining from the Pliocene and 1 from the Miocene). There are 15 genera today comprising some 35 natural species (Fiennes and Fiennes, 1971, are ultraconservative and list only 5 genera). Of these 15 genera, *Vulpes* appeared in the Miocene, *Canis* and *Nyctereutes* in the Pliocene, and *Alopex, Chrysocyon, Cuon, Dusicyon, Lycaon, Otocyon, Speothos,* and *Urocyon* in the Pleistocene. There is no fossil record for *Atelocynus, Cerdocyon, Dasycyon,* or *Fennecus*. From the fossil record, the open plains of North America seem to have been the chief center for the evolution and dispersal of the canids.

Trends revealed in the fossil record for the canids are: development of a carnassial tooth, ossification of the tympanic bullae, consolidation of the scaphoid and lunar bones of the foot, and a reduction from 5 spreading toes to 4 compact toes on both feet with a vestigial pollex (fifth toe) on the front foot.

BASIC CANID CHARACTERISTICS

In general, canids differ from most other arctoids in possessing adaptations for running: semirigid, elongate legs ending in 4 well-developed toes placed close together and tipped with blunt, nonretractile claws which are nearly straight, with the reduced fifth toe (pollex) forming the dew claw, and a locked radius and ulna which prevent rotation of the front leg. Thus, canids are cursorial digitigrades, running on their toes or the small pad under their toes. The shearing carnassials are well-developed with some adaptation for grinding as well. The basic dental formula for canids is $I\,\frac{3}{3}$, $C\,\frac{1}{1}$, $P\,\frac{4}{4}$, $M\,\frac{2}{3}$: a total of 42 teeth. This tooth series includes unspecialized but relatively large incisors, long and powerful canines, sharp premolars, well-developed sectorial carnassials (P 4/M 1), and crushing molars. The elongate skull, long nose, and powerful cheek muscles are adaptations for seizing, biting, and holding prey. Other characteristics are: large bullae with longitudinal septa, a zygomatic process that projects strongly outward, elevated cheeks, smooth tongue, no entepicondylar foramen of the humerus, a small, cartilaginous clavicle, a grooved and well-developed baculum, a caecum coiled into an S-like form, and a complex cerebral cortex indicating these carnivores have a high level of intelligence. Exceptions to the above general characteristics will be mentioned under each genus.

Canids range in size from the 3 pound (1.5 kg) fennec to the 175 pound (80 kg) timber wolf, and most have a long, bushy tail with a scent gland dorsally near the base. The ears are pointed and erect; in some desert forms the ears are large, in the Arctic fox and bush dog, the ears are small. Most canids are brown or gray in color, the Arctic fox is blue or white and the African (Cape) hunting dog is blotched. Gestation usually lasts 63 days but varies from 49 in *Fennecus* to 80 in *Lycaon*. Most canids are sexually mature when 1 year old but the larger forms, such as the timber wolf, may be 3 years before attaining sexual maturity. One litter a year is usual. Mammae range from 6 to 16 in number and litter size ranges from 2 to 13. In general, the eyes open in 2 weeks and offspring nurse for 4 to 6 weeks. The sense of smell is keen as is the sense of hearing; sight is less important although movement is instantly perceived (Fiennes and Fiennes, 1971; Stains, 1967). Canids which roam in packs (such as the wolf, African hunting dog, and the dhole) tend to be strictly carnivorous while the remaining canids, which tend to hunt alone, are omnivorous, feeding on a variety of foods.

In the following discussion, the reader is instructed to consult the "Distribution Map Section" as each species is presented.

GENERA AND SPECIES OF CANIDS

SUBFAMILY CANINAE Gill, 1872. Talonid of lower carnassial with a basin.

Canis Linnaeus, 1758. The general characteristics listed previously for the canids are possessed by members of this genus. Other characteristics are: upper incisors prominently lobed, outer pair of upper incisors extending beyond middle of canines, canines short and heavy, upper carnassials highly developed and equal or greater in length than the two upper molars, occipital shield nearly triangular with apex not depressed, sagittal crest a simple ridge with no tendency to show a flattened lyrate area, postorbital process thick and convex dorsally, subcaudal glands (if present) do not produce strong odors, pupils of eye round in strong light, mammae numbering 8 to 10, and tail bushy, broadest in middle and shorter than half length of body. Characteristics that distinguish *Canis* from *Alopex* and *Vulpes* include a markedly elevated frontal region, a postorbital process that is convex from above, an apex of the lower canine that does not reach the upper edge of the alveolus of the upper canine when the jaw is closed, a maximum diameter of the upper canine that is ⅔ the external length of the upper carnassial, tympanic bullae that are not parallel to each other on the anterointernal side, and a facial region that is relatively short and broad. The genus *Canis* is represented by 7 natural wild species.

Canis adustus Sundevall, 1847. The side-striped jackal occurs in central Africa from about 15° N latitude to 30° S latitude. They are absent along the Atlantic coast area from southern Sierra Leone to central French Equatorial Africa, northern Belgian Congo, and southwest Africa. This jackal has a white-tipped tail and a pair of light and dark stripes on each side from the shoulder to the hip. The ears are short, thick, upright and pointed; not as long as other jackals. The cry is a series of yaps quite different from the howling of the black-backed jackal (Smithers, 1966a; Dorst, 1970). Monotypic (not divided into subspecies).

Canis aureus Linnaeus, 1758. The golden jackal is the most northern of the jackals. If the Asiatic and African forms are the same, then the Asiatic or golden jackal range occurring in northern Africa (south to the northern range of the side-striped jackal) presumably extends across the Arabian area through India. This patternless, pale, reddish-brown jackal has a black-tipped tail about ⅓ of body length, is more completely nocturnal than other jackals, and perhaps is more wary. Approximately 12 subspecies are recognized.

Canis mesomelas Schreber, 1775. Black-backed jackals occur in the eastern and southern parts of Africa. Characteristics of the black-backed jackal, which separate it from other members of the genus *Canis,* are the black-speckled-with-white mantle extending from the neck to the tail, the long, triangular foxlike ears, and the black-tipped rufous tail. This jackal is the most diurnal of the 3 species called jackals. Three subspecies are recognized.

Canis dingo. See *Canis familiaris.*

Canis familiaris Linnaeus, 1758. The domestic dog is thought by many to have been derived by man from the wolf. Since domestic dogs have been successfully crossed with the wolf, coyote, and jackal, it is reasonable to consider the various domestic dogs having been derived from different wild canids in different regions of the world (Beddard, 1902; Fiennes and Fiennes, 1971). Some domestic dogs have a vestigial fifth toe on the rear foot. *Canis dingo* and *Canis hallstromi* are considered by most mammalogists to be canids carried by natives to Australia and Papua; thus, they are not natural environmentally created species but man-produced breeds which do not deserve species status. *Canis familiaris* is therefore not a natural species but one created which can be used as a category for those dogs developed by man for man. This canid, however, is the type species of the genus *Canis*.

Canis hallstromi Troughton, 1957. See *Canis familiaris*.

Canis latrans Say, 1823. The coyote, in size, is the jackal of North America, occurring from Costa Rica north through southern and western Canada to Alaska, but most common in the plains area. In food habits, the coyote is quite different from the jackals as it is more of a predator than a scavenger. Slightly larger and heavier than the jackals, the coyote is the smallest of the 3 species of the genus *Canis* in North America. The coyote and the jackal (*C. aureus*) have been crossed and backcrossed by man. To separate the coyote from the other forms in North America, size of nosepad, diameter of head, diameter of canine, width of zygomata, and other size differences have been used. Approximately 19 subspecies are recognized.

Canis lupus Linnaeus, 1758. The gray or timber wolf is a northern holarctic circumpolar species which has been largely exterminated over most of its previous range. This is the largest living canid, weighing from 60 to 175 pounds (27 to 80 kg). Thus, most characteristics used to separate this canid from others involve size. Differences between the gray and red wolves involve, other than size, color of pelage and the absence of a conspicuous cingulum on the outer edge of the first upper molar in the gray wolf. Approximately 38 subspecies are recognized. As with most subspecies of canids, differences are largely based on color and size.

Canis niger (Bartram), 1791. The red wolf is intermediate in size between the coyote and gray wolf, slightly closer to the coyote. Considerable debate exists at present as to the taxonomic status of the red wolf. Numerous opinions exist: proper species, subspecies of wolf, subspecies of coyote, coyote–dog cross, coyote–wolf cross, or even dog–wolf cross (McCarley, 1962; Nowak, 1970). Other than being of larger size, the red wolf supposedly differs from the coyote in having a coarser pelage, a jugal more deeply inserted on maxilla, occipital condyles extending further transversely, nearly equal posterior and anterior external cusps of the first molar, and being a pacer when running rather than a loper (Grossenheider, personal communication). Other than being of smaller size, the red wolf supposedly differs from the gray wolf in being browner in color, having molariform teeth with more deeply cleft crowns, and a first upper molar with a more prominent cingulum and a posterioexternal cusp which is distinctly smaller than the anteroexternal cusp. The arrangement of the cerebellum indicates the red wolf to be a distinct species related to the gray wolf (Atkins and Dillon, 1971). Three subspecies of the red wolf have been recognized. This canid is an endangered species with but a few remaining along the Gulf Coast of North America (Figure 1-1).

Figure 1-1 The so-called red wolves *(Canis niger)*. Photo: M. W. Fox.

Canis simensis Rüppell, 1869. The Abyssinian wolf or simenian fox is today an extremely rare canid occurring in two adjacent locations in Ethiopia. This canid resembles a long-legged dog with a foxlike head. Rufous in color above and white below holds true for the body as well as the tail, although the tip of the tail is blackish. Across the white chest are two successive rufous collars. Monotypic.

Alopex lagopus Linnaeus, 1758. Arctic foxes live farther north than most terrestrial mammals. Like the timber wolf, the Arctic fox is holarctic circumpolar, largely occurring near the coast from Norway, east through North America, to Greenland. This is the only canid that has well-established colonies with 2 color phases: a white, which becomes brownish-gray in summer, and a blue phase. Characteristics of this fox are small, rounded ears, long hairs on the soles of the feet, a relatively broad rostrum, closely spaced teeth on the lower jaw, elongated pupils, and a tail more than half the body length. This genus perhaps has 2 litters a year at times. They resemble *Vulpes* and have been placed in this genus by some taxonomists or are regarded as a subgenus of *Vulpes* by others. It differs from *Vulpes* in that the distance from the posterior edge of infraorbital foramen to the posterior alveolus of the canine is not more than equal to the width of the skull over the canines, the length of the hard palate is less than 3 times the width between anterior edges of the third upper molar, the frontal region is markedly elevated, the postorbital process is flat above, and the apex of the lower canine reaches, but does not extend beyond the edge of the upper canine alveolus when the jaw is closed. There are 9 recognized subspecies.

Figure 1-2 Small-eared dog *(Atelocynus microtis)*. Photo: M. W. Fox.

Atelocynus microtis (Sclater), 1882. The small-eared dog is found in the Amazon basin of Brazil, in parts of Ecuador, Peru, Columbia, and probably Venezuela. The ears are relatively shorter and rounder than those of other species of wild dogs. It is one of the darkest of the South American canids. The thickly-haired tail touches the ground when hanging. Unlike other canids, the small-eared dog moves like a cat. The eyes glow in dim light and the male has anal glands that emit a strong, musky odor (Walker, 1964). This canid has a straight or faintly curved caecum in the form of a simple cylinder, heavy teeth which are considerably longer than those of other South American canids except *Chrysocyon,* a very high masseteric ridge of jugal, a well-developed sagittal crest with no suggestion of a parietal expansion, and a much enlarged second lower molar. Monotypic. (Figure 1-2).

Cerdocyon thous (Linnaeus), 1766. The crab-eating fox occurs in the tropical and subtropical forests and grasslands from Columbia and Venezuela to northern Argentina and Paraguay. This genus has features of the feet, skull, and teeth which differ from other genera in South America: very large digital and plantal pads on the feet, interdigital webs which are barely extensible, short claws, snout progressively narrow toward front and a little shorter than that of *Dusicyon*, a high, convex forehead, a hardly pronounced postorbital constriction, a masseteric ridge of the molar that is near the lower edge, a condyle of the mandible that is drawn upwards at a very broad vertical angle, short canines, thick premolars, and ears and tail that have dark tips. A good deal of color variation can be found even in the same locality. There are approximately 7 recognized subspecies. (Figure 1-3)

Figure 1-3 Crab-eating "fox" *(Cerdocyon thous).* Photo: M. W. Fox.

Chrysocyon brachyurus (Illiger), 1815. The maned wolf occurs on the pampas and along edges of swamps from northeastern Brazil south to northern Argentina. This canid resembles a red fox on stilts as the animal is only 125 cm in length (head plus body) and yet 75 cm high at the shoulder. Such long legs enable the animal to run swiftly. The tail is short (half as long as the shoulder height), the lower part of the leg is darker than the body (almost black), the sagittal crest is developed as a ridge, the front of the orbit is almost level with the rear border of the

Figure 1-4 Maned wolf *(Chrysocyon brachyurus)*. Photo: Eva Rappaport.

first upper molar, the frontals are wide and flattened, the caecum of *Chrysocyon,* like *Atelocynus,* is a simple cylinder, the pupils of the eye are round, and an erectile mane runs along the back from the top of the shoulders and neck. Monotypic. (Figure 1-4)

Dasycyon hagenbecki Krumbiegel, 1949. The Andean wolf was originally named in 1949 as *Oreocyon.* Known from one specimen, skin only, the Andean wolf is believed restricted to the high Andes of southern South America. This canid is larger than the maned wolf, with comparatively shorter legs, longer hair with a thick, felty light-gray underfur, short, rounded ears, and a dark brown coloration which is brighter on the abdomen. A study of the hair (Dieterlen, 1954) indicates a closer relationship to the shepherd dog than the maned wolf. Monotypic.

Dusicyon H. Smith, 1839. This South American canid resembles a small coyote but differs in the moderate flattened sagittal area. These animals, in general, have long snouts, large canines, and laterally compressed premolars. They differ from *Cerdocyon* in having small digital and plantar pads on the feet, extensible interdigital webs, long slender claws, low forehead, more sectorial molars with a masseteric ridge near the middle, and a mandible with the condyle drawn backward at a slender angle. There are at least 6 recognized species in this group though 1 of these is extinct (Osgood, 1934; Cabrera, 1931).

Dusicyon australis (Kerr), 1792. The Falkland "wolf" (the genotype of *Dusicyon*) occurred on the Falkland Islands and was exterminated around 1876. It differed from the living species in having a shorter tail which was white-tipped, a larger snout, a fourth upper premolar with a protocone drawn backwards and facing the middle of the main cusp, a first lower molar with the metaconid placed conspicuously low (almost at the level of the paraconid), large carnassial molars, and reduced tubercular molars. Two subspecies are recognized.

Dusicyon culpaeus (Molina), 1782. These South American foxes are perhaps best known as the culpeos. They are found in the Andes from Ecuador and Peru south to Tierra del Fuego. There is considerable transverse extension of the upper molars, the postorbital constriction of the frontals is more pronounced than *D. gymnocercus,* the tubercular molars are larger than *D. gymnocercus,* and the temporal crests remain separate. About 6 subspecies are recognized.

Dusicyon griseus (Gray), 1836. The Chico gray fox has a distribution similar to that of *D. culpaeus* from Ecuador south to southern Chile except that this fox occurs at lower elevations. Included in *D. griseus* is the Chiloe Island species, *D. fulvipes*. Postorbital constriction of frontals are more pronounced and the tubercular molars are larger than that of *D. gymnocercus*. Approximately 7 subspecies are recognized.

Dusicyon gymnocercus (Fisher), 1814. The pampas fox is found in Paraguay and southeastern Brazil south through the pampas region of Argentina to the river Negro. The skull is more compressed behind the postorbital apophysis and minute constriction of the frontals is evident, unlike *D. griseus*. Frequently, an extensive sagittal crest is present. Two subspecies are recognized.

Dusicyon sechurae (Thomas), 1900. The Sechura fox is restricted to the arid desert coastal zone of northwestern Peru and southwestern Ecuador. This species is similar to *D. vetulus* in molar index, muzzle shortness, in possessing a less transverse extension of molars, and in having an oblique third upper molar overlapping the second internally. Monotypic.

Dusicyon vetulus (Lund), 1842. The small-toothed dog occurs in southcentral Brazil. This is one of the more distinct species of this genus as the muzzle is short, the upper carnassials are small in proportion to the molars, the tubercular molars are proportionately larger than any other living canid, the lower first molar is about the same length as the combined length of the second and third lower molars, the upper molars are subquadrate with very little extension transversely, and they have a lyriform sagittal area. Monotypic.

Fennecus zerda Zimmermann, 1780. The fennec is confined to northern Africa, occurring over the whole of the Sahara Desert. This 3 pound (1.5 kg) canid resembles a small *Vulpes* with enormous bullae and ears. This smallest and palest of any of the foxes has feet with densely-furred soles and a tail with a blackish-brown tip and a black patch located dorsally near the base. Monotypic.

Nyctereutes procyonoides Gray, 1834. The raccoon dog is a native of eastern Siberia, Japan, Manchuria, China and northern Indochina. This dog has been widely introduced into a number of European countries: European Russia, Poland, Rumania, Sweden, and Finland.* This canid has short ears, shorter legs and tail than foxes, and a large dark spot beneath and behind the eyes on the cheek resembling the mask of the raccoon. The lower jaw has a peculiar steplike subangular pro-

*This species is now feral in many of these European countries — Ed.

cess for the insertion of the digastric muscle, a situation similar to that found in *Otocyon* and *Urocyon* but different from *Canis, Vulpes, Alopex,* and *Fennecus*. *Nyctereutes* differs from *Alopex, Canis,* and *Vulpes* in having a very high angular process (not low and bent forward), a skull that rests on the entire jaw (not just the posterior part), nasal bones whose midsections are slightly depressed (not marked), a distance from the anterior of the premaxillary to the end of the postorbital process almost equal to the distance from the postorbital process to the foramen magnum (not longer), and a posterior edge of the hard palate that extends beyond the posterior edge of the last molar (Ognev, 1931; Novikov, 1956). Some have thought that the raccoon dog resembles the raccoon in a number of characteristics and habits (Frechkop, 1959), but most mammalogists think the relation is canid.* Five subspecies are recognized. (Figure 1-5)

Figure 1-5 Raccoonlike dog *(Nyctereutes procyonoides)*. Photo: M. W. Fox.

Otocyon megalotis Desmarest, 1822. The bat-eared fox occurs in southwestern and eastcentral Africa. A weak bunodont dentition, procumbent, long incisors, large auditory bullae, an extension of the hard palate beyond the molars, extra molars in both jaws (3–4/4–5) for a total

*Although it is the only canid that reportedly may hibernate — Ed.

of 46–50 teeth, wide diastema between the inner and outer upper incisors, and first and second molars of similar size and shape make a unique combination found in no other canid. The bat-eared fox has very large black-tipped ears, black feet, and short legs. Some (Smithers, 1966a; Dorst, 1970) have likened this fox to a jackal because of its build, odor, and breeding habits, but the peculiar steplike subangular process on the lower jaw for insertion of the digastric muscle, the roughened temporal area of the skull, and the heavy lyrate sagittal ridge are features found in the gray fox, *Urocyon* (Guilday, 1962). Once considered to belong to a distinct subfamily by itself largely because of the extra molars, the bat-eared fox is now placed in the Caninae with most of the canids (Van Valen, 1964). There are 1 or 2 subspecies.

Urocyon cinereoargenteus (Schreber), 1775. Gray foxes are found from extreme northwestern South America through most of Central America, and from the eastern United States west to the central area of the prairie states and north to Canada in the east. The combination of a step on the ventral border of the dentary, the broad lyrate sagittal ridge, the roughened temporal area, the mane of stiff, black-tipped hairs on top of the tail, the large dorsal tail gland, and the 6 mammae will distinguish this fox from other canids. The similarity to *Otocyon* and *Nyctereutes* is mentioned above. An island species, *U. littoralis,* is smaller in size and, by the standards of some mammalogists, should not be regarded as a distinct species. Sixteen subspecies of continental gray foxes and 5 subspecies of the island gray fox are presently recognized.

Urocyon littoralis (Baird), 1858. See *U. cinereoargenteus.*

Vulpes Bowdich, 1821. *Vulpes* has the widest distribution of any canid. Comparisons with *Canis, Alopex,* and *Nyctereutes* are made under those genera. The upper part of the body is a vivid red, the ears long and pointed, the limbs relatively long and slender, the postorbital process thin and concave dorsally, the muzzle sharp and elongate, the toe pads naked with the remainder of the sole densely covered with hair, the pupils elongate and elliptical, the subcaudal glands often producing a strong pungent odor, the tail long and bushy. *Vulpes* usually has 6 mammae. Approximately 11 species.

Vulpes bengalensis Shaw, 1800. The Bengal fox is found in India, Assam, and Nepal. The tail is black-tipped. Some have regarded this fox as a subspecies of *V. vulpes,* others as a subspecies of *V. corsac.* Monotypic.

Vulpes cana Blanford, 1877. Blanford's or the hoary fox is found in southwestern Russian Turkestan, Afghanistan, northeastern Iran, and Baluchistan. This fox, like *V. bengalensis,* has a weakly black-tipped tail but is smaller in size. The dark fur is very thick with a dark middorsal line. The chin and lower lip is a deep brown — almost black, the upper canines are elongate and reach the lower edge of lower jaw, the nasal bones are of a narrow, wedgelike form, and the central pairs of the incisors are awllike, without additional lateral denticles. Monotypic.

Vulpes chama A. Smith, 1834. The cape fox, silver-backed fox, or asse is found in southern Africa from Southern Rhodesia and southern Angola south. The tip of the tail of this larger species of fox is black and the ears elongate, more than 8.5 cm. The back is silvery-gray in color and there is little or no unpleasant smell to this animal. Monotypic.

Vulpes corsac Linnaeus, 1768. Corsac foxes occur from southeastern Russia, Russian Turkestan, and Chinese Turkestan through Mongolia and Transbaikalia to northern Manchuria and northern Afghanistan. This medium-sized, rusty-colored fox with a gray tone is not as big as *V. ferrilata,* with a smaller skull and bullae and a nose which is not especially elongate or narrow, but larger than *V. cana.* The tail and ears are smaller than those of *V. ruppelli, chama, bengalensis, cana,* or *pallida.* The chin and lower lip are white. There are 3 subspecies.

Vulpes ferrilata Hodgson, 1842. The Tibetan sand fox is found in Tibet and Nepal. This species may be a subspecies of *V. vulpes.* The ears are the same general color as the body, the tail and ears are short compared to *V. ruppelli, chama, bengalensis, cana,* or *pallida,* the skull, bullae, and upper canines are larger than *V. corsac,* and the muzzle is longer and narrower. Monotypic.

Vulpes fulva (Desmarest), 1820. See *Vulpes vulpes.*

Vulpes macrotis Merriam, 1818. The kit fox is found in southwestern United States and northwestern Mexico. It differs from *V. velox* in having longer ears, larger bullae, a narrow skull, and being slightly smaller. There are 8 described subspecies. (Figure 1-6)

Vulpes pallida (Cretzschmar), 1826. This pale fawn-colored fox occurs in the plains and desert areas of northern Africa from the Red Sea to the Atlantic Ocean. The fur is thin and short, the black tip of the tail sharply contrasting, and the ears are proportionately small (although still large) and rounded at the tips. Monotypic.

Figure 1-6 Kit fox *(Vulpes macrotis)*. Photo: M. W. Fox.

Vulpes ruppelli Schinz, 1825. The sand or Ruppell's fox occurs in a number of separated populations in northeastern Nigeria and northern Cameroons, northeastern French Equatorial Africa and northwestern Anglo-Egyptian Sudan, several places in northern Anglo-Egyptian Sudan, and in British Somaliland. Some have considered this fox to be a subspecies of *V. corsac.* The tip of the long tail is white or in some cases the entire tail is whitish and long as compared to *V. ferrilata* or *corsac.* The ears are large, the back rufous, the sides gray, the belly white, and a black patch on the muzzle extends toward the eyes. Ruppell's fox resembles a large fennec with rather short legs. Six subspecies are recognized.

Vulpes velox (Say), 1823. The swift fox occurs north and east of *V. macrotis* from New Mexico and Texas to Canada. Considered a subspecies of *V. macrotis* by some, see comparisons under *V. macrotis*. Two subspecies. (Figure 1-7)

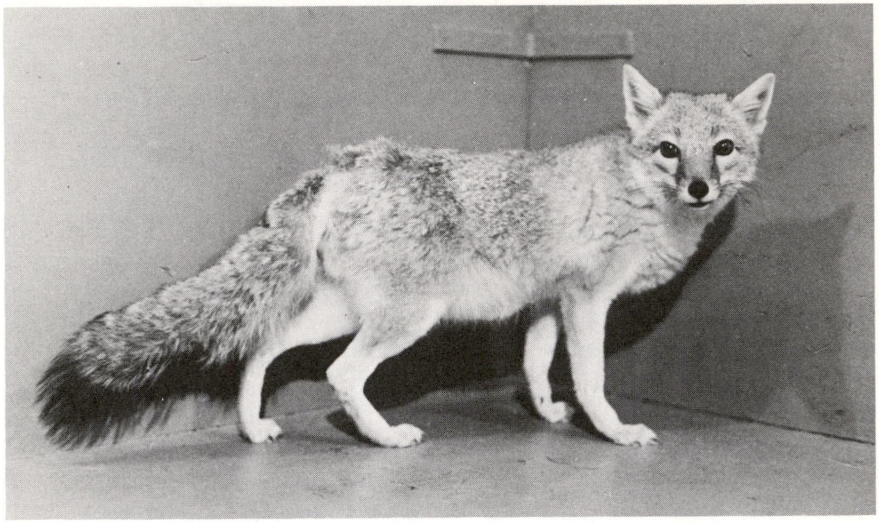

Figure 1-7 Swift fox *(Vulpes velox)*. Photo: M. W. Fox.

Vulpes vulpes Linnaeus, 1758. The red fox is circumpolar occurring through much of northern Asia, the Arabian peninsula, Turkestan, the Iranian plateau, northern India, Europe, the coastal regions of northern Africa south to northeastern Anglo-Egyptian Sudan, and North America (except for part of the central plains area) south to central Texas. The back of the ears of this species is black or dark brown, contrasting sharply with the color of the head and nape. This is the largest of the foxes. Various color phases such as the silver and cross-foxes occur in nature. When *V. fulva* of North America is included in this species, approximately 48 subspecies have been described giving this canid the largest taxonomic complex and perhaps the most extensive distribution. In addition, this species has been introduced into Australia and some Pacific islands.

SUBFAMILY SIMNOCYONINAE Zittel, 1893. Members of this subfamily lack an internal tubercal of the crested talonid on the lower carnassial and, in some genera, the metaconid of the same tooth is absent. The dental formula ranges from 1–2/1–2 molars. This group is more specialized than the Caninae and was once more widely distributed than today.

DISTRIBUTION OF SPECIES OF CANIDS

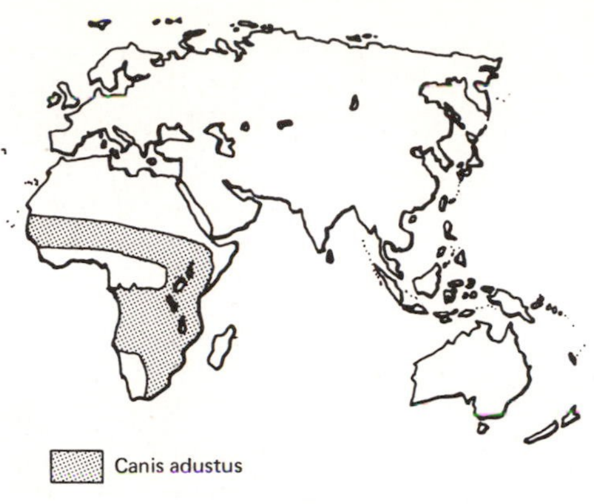

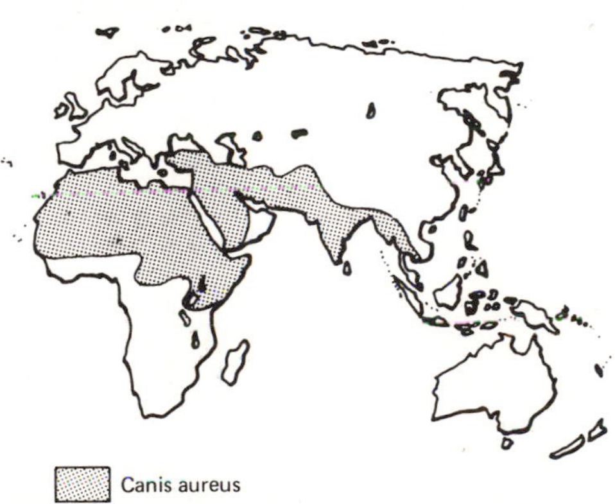

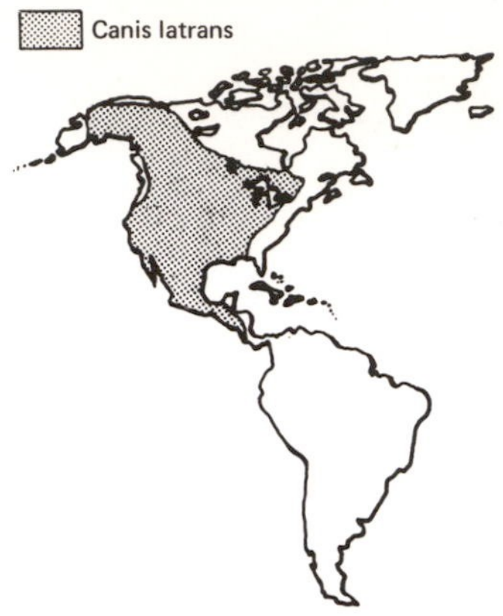

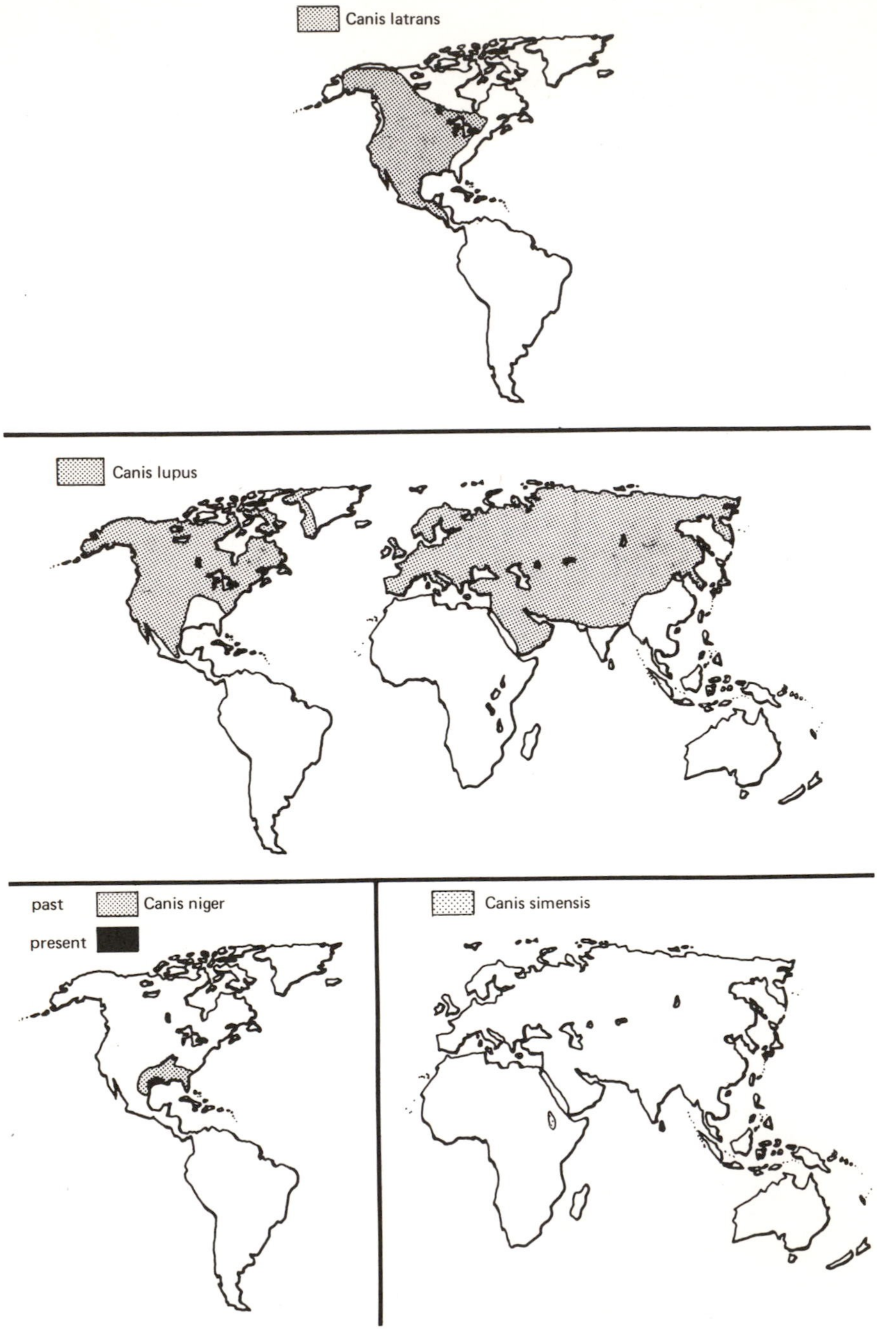
Canis latrans
Canis lupus
past
Canis niger
present
Canis simensis

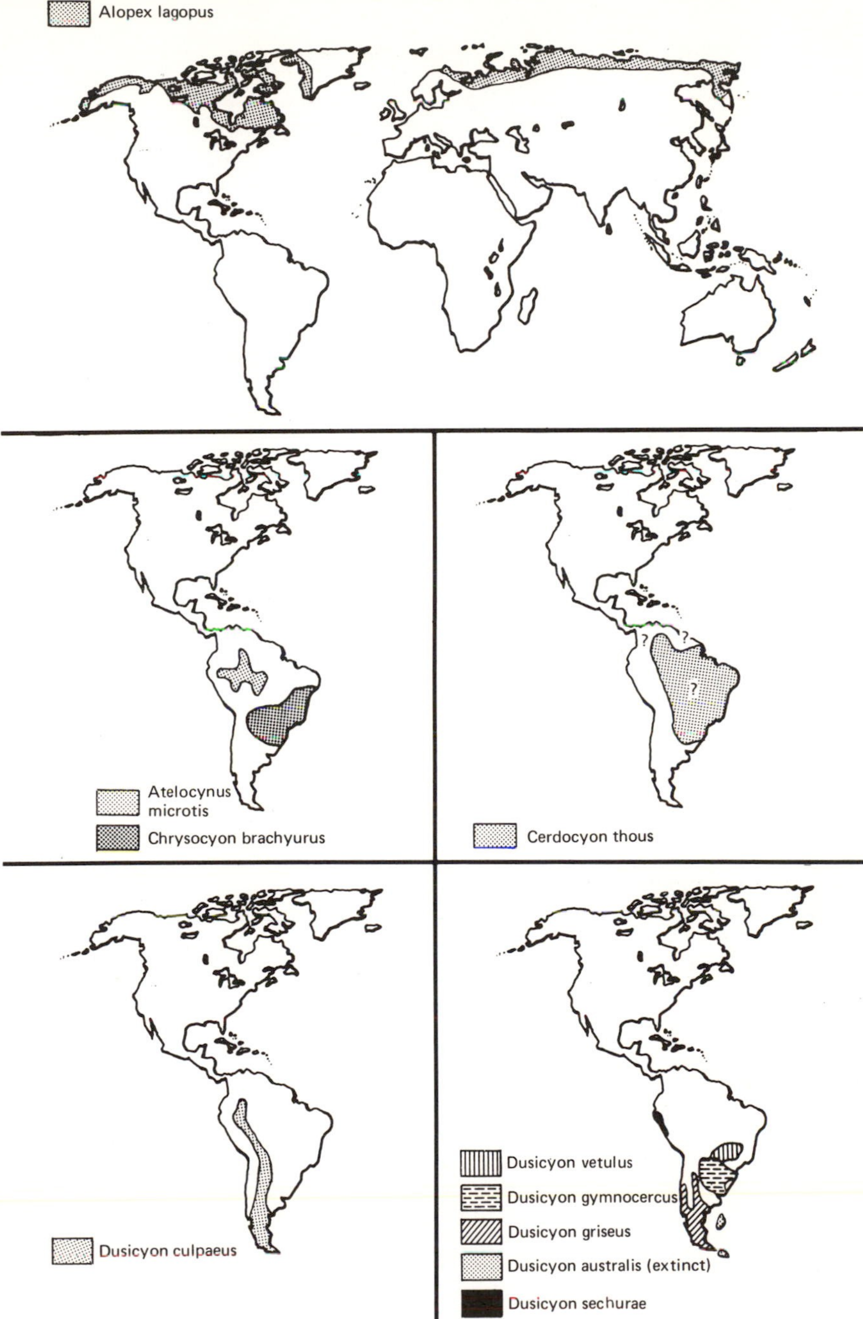
Alopex lagopus
Atelocynus microtis
Chrysocyon brachyurus
Cerdocyon thous
?
?
?
Dusicyon culpaeus
Dusicyon vetulus
Dusicyon gymnocercus
Dusicyon griseus
Dusicyon australis (extinct)
Dusicyon sechurae

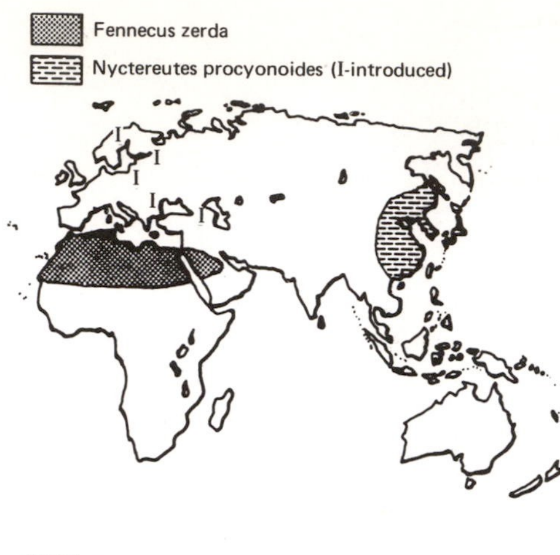
Fennecus zerda
Nyctereutes procyonoides (I-introduced)

Otocyon megalotis

Urocyon cinereoargenteus

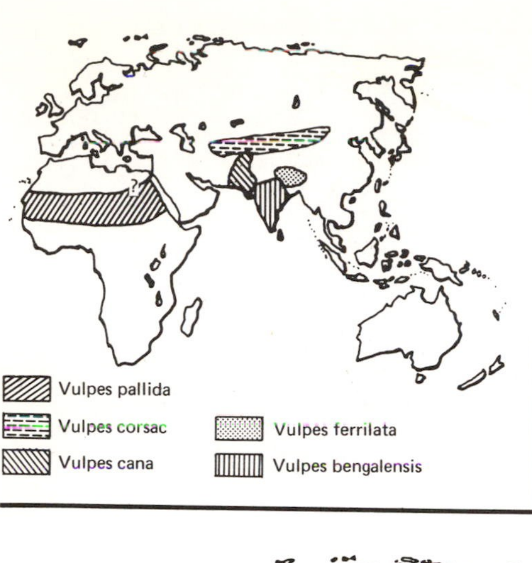
?
Vulpes pallida
Vulpes corsac
Vulpes cana
Vulpes ferrilata
Vulpes bengalensis

Vulpes velox
Vulpes macrotis

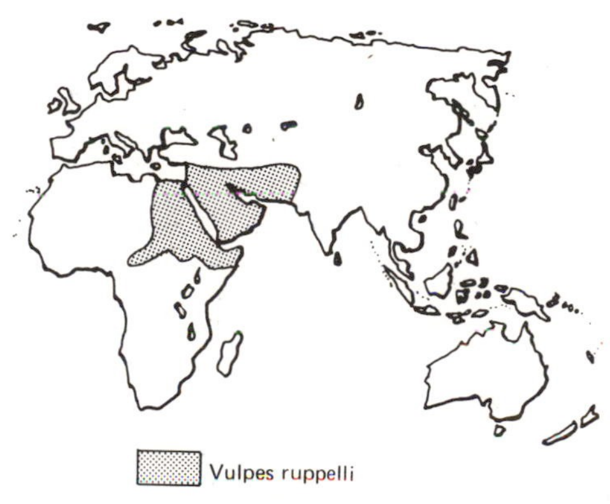
Vulpes ruppelli

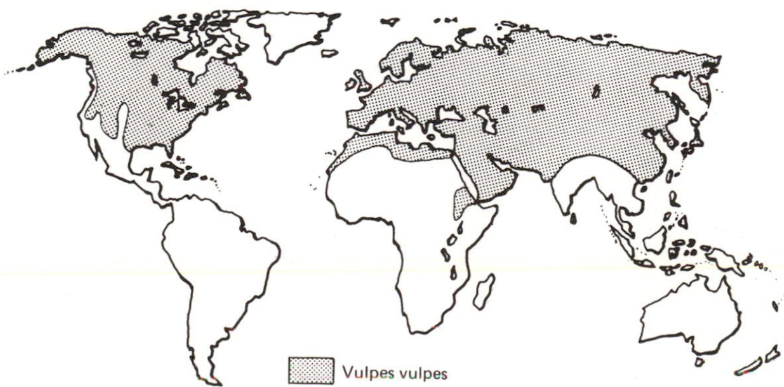
Vulpes vulpes

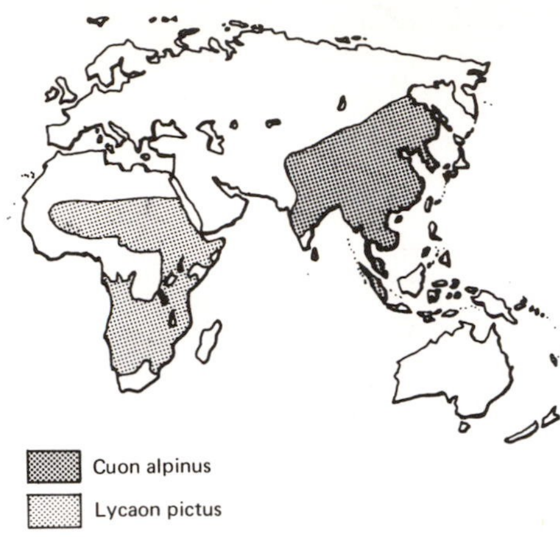

Speothos venaticus

Cuon alpinus Pallas, 1811. The dhole is found from the Altai and mountains of central Asia to the Far East, and south through forested regions to southern India and Java. This large brownish or reddish-yellow canid resembles *Canis* in general external features but has more rounded ears, 6 to 8 pairs of mammae, a markedly short facial region, 2 molars above and below, the first lower molar with only 1 (not 2 or 3) small denticle behind the main one and a slightly developed internal lobe, and a body with a white ventral surface. There are 10 subspecies.

Lycaon pictus Temminck, 1820. The African hunting dog is found from about the 20° N latitude south to the Union of South Africa. It is absent from the western coast of Africa above central French Equatorial Africa and from most of the Belgian Congo. The color of this dog is extremely variable, even in the same pack, being mottled in black, yellow, and white with a blackish skin. Other characteristics are long, broad and rounded ears, long and slender toes with only 4 toes on both the front and rear feet, and an offensive, very powerful musky odor. It also has long, thin legs, a powerful skull with a strong sagittal crest, a well-developed clavicle, and ⅔ molars. This canid resembles a big dog with a massive head and slender body. Four subspecies are recognized.

Figure 1-8 Bush dogs *(Speothos venaticus)*. Photo: M. W. Fox.

Speothos venaticus (Lund), 1842. The bush dog or vinegar fox is found from Panama south to northeastern Peru, Paraguay to Mato Grosso, and through most of central and northern Brazil to the interior of the Guianas. Features of *Speothos* are: skull and mandible short and robust, nose very short, palate wide, occipitals very angular, sagittal crest low and extensive, postorbital apophysis farther forward than in other canids, upper and lower second molars rudimentary, upper first molar always present with internal part simpler than in other canids (more restricted with fewer number of cuspids), lower first molar lacking metaconid and entoconid (thus molar is reduced to one sectorial plate with a paraconid, protoconid and hypoconid), molars 1–2/2, short legs, ears and tail, and tail not bushy. Three subspecies. (Figure 1-8)

2 MOLECULAR APPROACHES TO TAXONOMIC PROBLEMS IN THE CANIDAE*

U. S. SEAL
Veterans Administration Hospital and
Department of Biochemistry, University of Minnesota
Minneapolis, Minnesota

Biochemical studies of taxonomic or systematic problems are based upon the concept that protein sequences are maps of genetic DNA sequences. Comparisons of homologous proteins thus provide a comparison of homologous genes. Relationships are then defined in terms of sequence similarity. Protein sequences provide the direct evidence for such comparisons. However, these are limited by the time required for completion of sequence analyses. Indirect methods are based upon the assumption that they can detect differences between proteins in some proportion to sequence differences. These include quantitative immunological techniques and mobility differences on electrophoresis of proteins and enzymes. Used with proper care, both groups of indirect techniques have proven powerful tools for analysis of problems ranging from the population to the ordinal level. Application of these tools to the Carnivora has been limited to family comparisons. There have been no published studies directed primarily at the Canidae. Recent interest in the genus *Canis* led to our beginning such a study. The available literature has been examined to identify possibly useful information from studies on the domestic dog and to summarize the comparative data.

*Supported in part by USPHS grant 5 RO1 AM11376-11 and USDI-FWS-BSFW contract 14 16 0008 548.

STUDIES INCLUDING THE DOMESTIC DOG BUT NOT DEMONSTRATING POLYMORPHISMS OR SPECIES DIFFERENCES IN THE CANIDAE

The sequence of elephant seal, *Mirounga leonina,* heart cytochrome *c* differs from the dog protein in one amino acid at position 100 where isoleucine replaces the lysine found in dog. No other carnivore sequences have been reported, but this example strongly suggests that cytochrome *c* will be of no value for phylogenetic problems within this order or in the Canidae (Augusteyn, *et al.,* 1972). This also would have been predicted from the limited number of differences found in comparing the available sequences of cytochrome *c* from mammals.

A survey of erythrocyte glucose-6-phosphate dehydrogenase activity in 28 vertebrate species included 3 dogs with identical 1-band patterns and 1 maned wolf (*Chrysocyon brachyurus*) with a faster 1-band pattern. Two cats and a jaguar (*Panthera onca*) had differing 1-band patterns of slower mobility (Nobrega, *et al.,* 1970). Dog erythrocytes (3 specimens) predominantly contained a single carbonic anhydrase corresponding electrophoretically to component *B* of humans and rabbits. Antisera to human component *B* crossreacted more strongly with the dog enzymes than the goat, sheep, or cow preparations. No other carnivores were included (Funakoshi and Deutsch, 1971). Measurement of canine erythrocyte catalase activity in 83 beagles and 5 of other breeds yielded values in the range of 0–43 perborate units/ml blood. The values for other mammalian species (man and laboratory rodents) ranged from 100–500. Thus, the dog was considered to range from acatalasemic to hypocatalasemic. Tissues from 4 dogs contained the enzyme in "normal" amounts (Feinstein, *et al.,* 1968). The erythrocytes of the polar bear, dog, seal, and cat are of low potassium and high sodium content in contrast to species from other mammalian orders (Manery, *et al.,* 1966). We have examined members of all carnivore families and found similar results (Seal, unpublished).

Canine plasminogen purified (to immunochemical purity) from a single donor and homogeneous in the analytical ultracentrifuge exhibited 4 active bands on starch gel electrophoresis. No evidence was collected as to possible polymorphism (Heberlein and Barnhart, 1968). Canine haptoglobin was isolated from plasma of a post-turpentine stimulated animal by ion-exchange chromatography. The amino acid and carbohydrate composition were determined and a molecular weight value of 81,000 was obtained. This is similar to the size of the human haptoglobin 1-1 form (Dobryszycka, *et al.,* 1969). Canine serum haptoglobin was purified and found to be a tetramer (similar to the human and swine protein) of molecular weight 81,000. It is composed of 2 α chains, molecular weight 10,000, and 2 β chains, molecular weight about 30,000. The β chains contain the hemoglobin binding site. Canine haptoglobin forms intermediate complexes with hemoglobin which exhibit an intermediate electrophoretic mobility (Shim, *et al.,* 1971). Thus, for electrophoretic identification of haptoglobins and possible polymorphisms, it is important to use an excess of hemoglobin or else spurious multiple bands may be seen. We have verified this possibility experimentally with dog and bear serum samples.

Dog plasma was shown to contain transcobalamin II, a vitamin B_{12} binding protein, similar in properties to the human protein and crossreactive with antisera to the human protein (Rappazzo and Hall, 1972). A rabbit antiserum to human serum inhibitor of C1 esterase crossreacted with serum containing similar inhibitory activity from other primates but not with other mammals including dog (Donaldson and Pensky, 1970). Antisera to canine fibrinogen, complement C'3, and immunoglobulins have been prepared and used for study of interstitial nephritis in dogs (Krohn *et al.,* 1971).

An antiserum to canine growth hormone crossreacted with the growth hormone of sheep, cow, rat, and pig but not with human or "simian" growth hormones. No other carnivores were examined (Tsushima, *et al.,* 1971). Antisera to human placental lactogen crossreacted weakly with comparable extracts from dog placenta. This suggests the possibility that a placental lactogen may be found in pregnant dog serum (and other carnivores) as well (Gusdon, *et al.,* 1970). Goat antisera to hog pancreatic amylase were examined for their crossreactivity with the amylases of pancreas, saliva, serum, liver, and kidney in 8 species including a dog and a cat. No salivary amylase could be demonstrated in these carnivores, but crossreactivity was evident in the other tissues (McGeachlin, *et al.,* 1966).

STUDIES DEMONSTRATING POLYMORPHISMS OR SPECIES DIFFERENCES IN THE CANIDAE

Measurements of thyroid function in 5 breeds of dogs — beagle, sheltie, cocker, terrier, and basenji — documented significantly greater thyroidal activity in the basenji (Nunez, *et al.,* 1970). This perpetually active state of the basenji thyroid was reflected in structural modifications of the thyroid follicular cells (Nunez, *et al.,* 1972). Year-round studies of serum thyroxine in the wolf in wild populations indicates seasonal variations in activity (Seal, unpublished), but actual measurements of thyroid activity have not been reported for a wild canid species.

Canine blood typing based upon extensive studies of transfusion reactions and experimental studies have documented the presence of polymorphic blood grouping systems in the dog (Bowdler, *et al.*, 1971; Swisher and Young, 1961; Swisher, *et al.*, 1962). However, I have found no references including other canids in such studies. Smith and McKibbin (1972) describe separation by thin layer chromatography of the fucose-containing glycolipids of dog small intestine mucosa into two types. Type I containing galactosamine was found in 5 animals, and Type II containing no galactosamine was found in 5 animals. These complex lipids may be related to blood group activity.

Six antigenically distinct immunoglobulins have been found in canine serum including 4 groups of IgG, an IgA, and an IgM (Johnson and Vauhan, 1967). These have been quantitated in canine serum, colostrum, milk, and intestinal secretions with some evidence for concentration differences between purebred strains and mongrels (Reynolds and Johnson, 1970). There were no data on other canid species. Immunological studies have demonstrated crossreactivity between equine, canine, and human immunoglobulins of the same class (Allen and Johnson, 1972). Serum antitrypsin is present in dogs at concentrations of 100 ± 30 mg/ml, similar to most other mammals (Ihrig, *et al.*, 1971). Two separate trypsin-binding α-2 macroglobulins were isolated and characterized. Antisera prepared in rabbits were specific for each protein; however, antisera to human α-2 macroglobulin cross-reacted with both proteins, indicating a common structure. Both these proteins appear to be multimorphs since no variants were observed in 15 animals. There were no family data. Both had the same sedimentation coefficient (Ohlsson, 1971).

Most earlier reports agree that dog hemolysates contain a single hemoglobin component (Gratzer and Allison, 1960). I reported in 1969 on hemoglobins in the Carnivora and included information on 99 hemoglobin samples from 11 genera and 17 species in the Canidae. All exhibited identical electrophoretic mobility on polyacrylamide gel. Since that time, the list has expanded to 320 samples from 12 genera and 19 species plus 2 subspecies of *Canis familiaris* (Table 2-1) with the same results. Thus, we found no evidence of electrophoretic polymorphism or differences between genera in this family. Neither we nor LeCrone (1970) found evidence for a fetal hemoglobin in *Canis familiaris* (Seal, 1969).

However, crystallographic studies by Reichert and Brown (1909) and microcomplement fixation studies by Sarich (1972) have indicated that structural heterogeneity is present in these hemoglobins. This demonstrates the greater sensitivity of the immunochemical methods to sequence differences and suggests that the mutations accepted in the

TABLE 2-1 Tabulation of Blood Samples from the Canidae

Species	Specimens
Canis lupus	87
Canis rufus	35
Canis latrans	65
Canis aureus	9
Canis familiaris (domestic)	40
Canis familiaris (dingo)	4
Canis familiaris (Hallstroms)	4
Canis mesomelas	2
Alopex lagopus	2
Vulpes velox	1
Vulpes vulpes	35
Vulpes pallida	2
Fennecus zerda	5
Urocyon cinereoargenteus	7
Nyctereutes procyonoides	7
Atelocynus microtus	1
Cerdocyon thous	2
Speothos venaticus	1
Lycaon pictus	5
Otocyon megalotis	3
Chyrsocyon brachyurus	3

hemoglobins have been conservative with respect to charge. We have pursued these comparisons using automated sequencing methodology and have found differences between dog, black bear, and sea lion within the first 30 residues in both the α and β chains. The complete sequences of dog α and β chains have been reported by Jones, *et al.* (1970). The information is too limited as yet to permit firm interpretations in terms of family relationships (Table 2-2). However, these studies are being extended to include all families of the Canoidea and Pinnipedia as well as other genera of Canidae and species of *Canis*.

Tetrazolium oxidase is polymorphic in canine erythrocytes (Baur and Schorr, 1969). Two alleles were identified in a survey of 189 animals. Type A had a frequence of 0.94. The patterns of 2 coyotes and a coyote–dog hybrid were identical with canine oxidase Type A and, hence, this enzyme does not provide a marker for distinguishing members of these 2 canine species. Kaminski and Balbierz (1965) described protein and esterase patterns on starch gel for 81 *Alopex lagopus*, 26 *Vulpes vulpes* (22 silver and 4 of the platinum variety), and 17 *Canis familiaris* samples. There were no photographs and no specific identification of the polymorphic bands observed. Multiple

TABLE 2-2 Partial Sequences of Hemoglobin Chains from Dog *(Canis familiaris)*, Back Bear *(Ursus americanus)*, and Sea Lion *(Zalophus californianus)*

Alpha Chain[b]

Dog[a]	Val	Leu	Ser	Pro	Ala	Asp	Lys	Thr	Asn	Ile	Lys	Ser	Thr	Trp	Asp	Lys	Ile	Gly	Gly	His	Ala	Gly	Asp	Tyr	Gly	Gly	Glu	Ala	Leu	Asp
Bear	—	—	—	—	—	—	—	—	—	Val	—	Ala	—	—	—	—	—	—	Ala	—	—	—	Glu	—	—	—	—	—	—	—
Sea Lion	—	—	—	—	—	—	—	—	—	Val	—	Thr	—	—		—	—		—	—	—	—	Glu	—	—	—	—	—	—	Glu

Beta Chain[b]

Dog[a]	Val	His	Leu	Thr	Ala	Glu	Glu	Lys	Ser	Leu	Val	Ser	Gly	Leu	Trp	Gly	Lys	Val	Asn	Val	Asp	Glu	Val	Gly	Gly	Glu	Ala	Leu	Gly	Arg
Bear	—	—	—	—	Gly	—	—	—	—	—	—	Thr	—	—	—	—	—	—	—	—	—	—	—	—	—	—	—	—	—	—
Sea Lion	—	—	—	—	—	Asp	—	—	Ala	Ala	—	Thr	Thr	—	—	—	—	—	—	—	—	—	—	—	—	—	—	—	—	—

[a]The dog sequences are from Jones, *et al.* (1972).
[b]The dashes indicate the residue was idential with that found in the dog. Differences are listed. Uncertain residues are left blank.

intermediate bands are described which are difficult to interpret from the account.

Braend (1966) examined samples from 65 dogs representing 25 breeds by starch gel electrophoresis at *p*H 8.1. Identical patterns were observed in 61 specimens consisting of 3 bands. One specimen had 3 additional faster bands. Three other specimens, 2 from related animals, exhibited slower moving bands in addition to the common pattern. These results suggest that the 3-banded pattern is controlled by a single locus and that the variants represent codominant alleles. Stevens and Townsley (1970) surveyed 248 dogs of 19 breeds, including 37 families of 7 mating types producing 124 offspring by starch gel electrophoresis at *p*H 7.1. A 3-banded pattern was observed in homozygotes. One group of fast variants exhibited 5 to 6 bands in the heterozygotes. A single fast homozygote had a 3-band pattern compatible with control of all 3 bands by a single locus and codominant alleles. However, the bulk of the variation was in the middle intense band (see Figure 2-1) which showed a polymorphism in 15 breeds independently of the weaker 1st and 3rd bands of the pattern. The lower *p*H of this electrophoresis procedure perhaps allowed recogni-

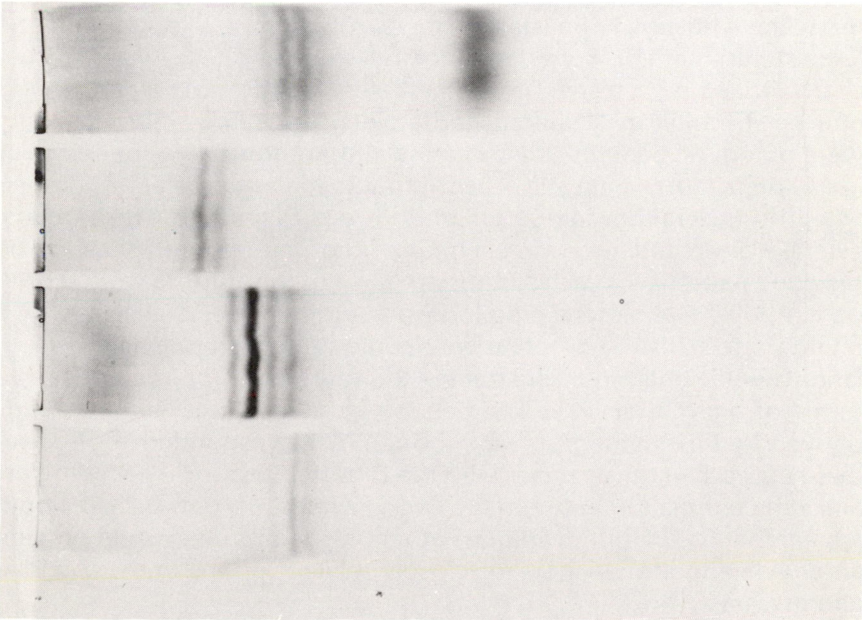

Figure 2-1 Acrylamide gel electrophoresis of rivanol-treated serum samples to demonstrate multiple transferrin bands of Canidae. The species from top to bottom are *Ursus americanus, Canis familiaris, Nyctereutes procyonoides,* and *Otocyon megalotis.*

tion of the common polymorphism in these samples which had not been observed by Braend (1966) in his material. The results were supported by consistent family data. The two types of polymorphism suggest independent control of 2 parts of the transferrin structure by 2 loci. None of these authors added Fe to their samples. Apotransferrin and Fe-saturated transferrin have slightly different mobilities and, hence, can confuse a pattern. Masson and Heremans (1971) demonstrated that dog milk lacks both transferrin and lactoferrin but contains more than 5 μg/ml of Fe. It would be of interest to examine milk from other carnivores.

Thrombin acting upon fibrinogen produces 3 distinct peptides: *A*, *B*, and *C*. They vary in length from 16 to 22 amino acids. The primary sequence of peptide *A* has been reported for many artiodactyls and 7 carnivores including dog, fox, badger, mink, brown bear, gray seal, and cat. Peptide *B* sequences were reported for dog, fox, and cat (Osbahr, *et al.*, 1964; Blomback and Blomback, 1968). Peptide *A* is identical in dog and fox, whereas they differ at 1 position in peptide *B*. Thus, these peptides are likely to be of little value for species comparisons within genera in the Canidae.

Using starch gel electrophoresis of canine heparinized plasma and different substrates, Ecobichon (1970) found 10 distinct esterase bands including 4 butyrylcholinesterase, an estrolytically active albumin, and 5 diesterase and/or arylesterase active bands. These were further characterized with respect to size classes, kinetics (*p*H and K_m), and effects of inhibitors. Differences between dogs in staining intensity were noted. An earlier study by Leone and Anthony (1966) of esterases in 110 dogs representing 40 breeds combined immunoelectrophoresis using 10 antisera to whole serum of different dog breeds with aromatic esterase and cholinesterase staining of the immunoprecipitates. The resulting patterns for each serum with each of the antisera were scored on a 4-point scale for staining intensity and the results analyzed by a multivariate statistical program on a computer which calculated correlation coefficients and constructed dendrograms for the breeds in terms of enzyme activity. The geneologic derivations of the breeds derived by this method provide about 65% correspondence of placement for each of the enzymes with the reported historical geneologies. Since the correlation between the two enzymes was only 0.23, it would appear that use of this technique with multiple enzymes would provide an effective tool of analysis for closely related groups in the Canidae and in *Canis*.

Immunological techniques, whether serum flocculation (Nuttall, 1904), turbidity (Leone and Wiens, 1956; Pauly and Wolfe, 1957–8; Paulk, 1962), or microcomplement fixation (Seal, *et al.*, 1971; Sarich,

TABLE 2-3 Immunological Comparison of Serum Albumins from Canidae Species to Dog Albumin with Antiserum 80

Species	Name	I.D.
Canis familiaris	Domestic dog	1.00
Canis lupus	Timber wolf	1.00
Canis latrans	Coyote	1.00
Canis rufus	Red wolf	1.05
Canis aureus	Golden jackal	1.05
Lycaon pictus	Cape hunting dog	1.05
Alopex lagopus	Arctic fox	1.08
Vulpes pallida	Pale fox	1.12
Otocyon megalotus	Bat-eared fox	1.13

TABLE 2-4 Immunological Comparison of Serum Albumins from 17 Canidae Species to Red Fox Albumin

Species	Name	I.D.
Vulpes vulpes	Red fox[a]	1.00
Vulpes pallida	Pale fox	1.05
Vulpes velox	Swift fox	1.09
Fennecus zerda	Fennec	1.07
Canis lupus	Timber wolf	1.16
Canis rufus	Red wolf	1.15
Canis latrans	Coyote	1.13
Canis aureus	Golden jackal	1.15
Canis mesomelas	Black-backed jackal	1.18
Canis familiaris	Domestic dog	1.20
Alopex lagopus	Arctic fox	1.22
Urocyon cinereoargenteus	Gray fox	1.18
Chrysocyon brachyurus	Maned wolf	1.20
Speothos venaticus	Bush dog	1.20
Atelocynus microtis	Small-eared dog	1.28
Lycaon pictus	Cape hunting dog	1.27
Otocyon megalotis	Bat-eared fox	1.28
Nyctereutes procyonoides	Raccoon-like dog	1.58

[a]Red fox albumin antiserum No. 84 was prepared in a rabbit. The 50% MC'F titer was 1/22,000 at a serum dilution of 1/30,000 or 1 μg/ml of red fox albumin.

1969a and 1969b) have been used primarily to study family relationships in the Carnivora and Pinnipedia. Leone and Wiens (1956) prepared rabbit antisera to whole serum from wolf, coyote, and dog and made the possible comparisons. Single samples from each species were used so variance estimates are not possible. They concluded that the dog and coyote are more closely related to each other than to the wolf. Pauly and Wolfe (1957–8), using an anti-wolf serum, compared wolf (100%), dog (88%), and red fox (50%), and again differentiated dog and wolf but did not make the coyote comparison. Our antisera to black bear serum albumin differentiated *Canis* from 4 other genera of the Canidae examined. Direct comparisons within the Canidae have been made by microcomplement fixation using rabbit antisera to *Canis familiaris* and *Vulpes vulpes* serum albumins (Tables 2-3 and 2-4). The results indicate that the spread of values within the family is not great, discrimination among members of the genus *Canis* is not significant, and that studies with antisera to albumin from each of the genera will be necessary to allow cluster analysis for construction of a significant dendrogram for the genera of this family. Interfamily relationships within the order have been considered in the above papers and constitute a separate problem.

Comparison of the genera of Canidae can be accomplished with a number of proteins as shown in Figures 2-1, 2-2 and 2-3, illustrating transferrins, serum proteins, and esterases on polyacrylamide gel elec-

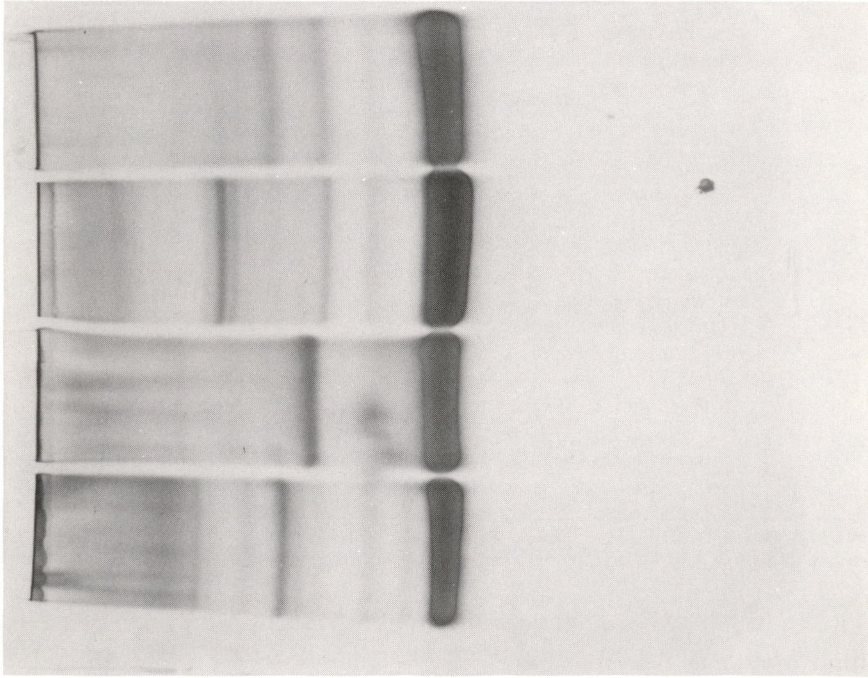

Figure 2-2 Serum protein electrophoresis patterns (from top to bottom) of samples from *Urocyon cinereoargenteus*, *Vulpes vulpes*, *Speothos venaticus*, and *Atelocynus microtus*.

Figure 2-3 Serum esterase patterns (from top to bottom) of samples from *Urocyon cinereoargenteus, Vulpes vulpes, Speothos venaticus,* and *Atelocynus microtus.*

trophoresis. However, differences between species of *Canis* are not readily evident (Figures 2-4 and 2-5). We have done similar studies on red cell hemolysates and soluble protein extracts of heart, kidney, liver, and skeletal muscle and have found no reliable difference between *Canis familiaris, Canis lupus, Canis latrans,* and *Canis rufus.* It has been possible to distinguish *Canis aureus* and *Canis mesomelas* from each other and from the other species of *Canis.* These studies have included some 30 loci as represented by different proteins and enzymes. These surveys are also notable for the paucity of polymorphisms so far observed in contrast to similar studies in other mammalian orders. We have found a similar lack of polymorphism in black, brown, and polar bears and in several Antarctic seal species as well (Seal, *et al.,* 1970; Seal, *et al.,* 1971). Studies of milk proteins in the bears (Jenness, *et al.,* 1972) have demonstrated several differences between closely related species in this family and suggest the value of a similar study of *Canis* species.

Khan, *et al.* (1973) have reported on 15 enzymes representing 21 loci in the leucocytes and erythrocytes from 92 unrelated mongrel dogs. They found polymorphism at the 3 loci phosphoglucomutase$_3$ (PGM$_3$), dimeric-tetrazolium oxidase (D-TO), and PGM_2. Three apparent heterozygotes of PGM_2 were observed. The observations on PGM_3 and D-TO were interpreted in terms of 2 codominant alleles for each enzyme. The gene frequencies found were PGM_3 1, 0.7167; PGM_3 2, 0.2778, and PGM_3 3, 0.0056. The gene frequencies for D-TO were D-TO 1, 0.8750 and D-TO 2, 0.1250. No variants were found at 14 loci and a single variant was observed for glucose-6-phosphate dehydrogenase, PGM_1, cytoplasmic glutamic oxaloacetic transaminase, and mitochondrial malate dehydrogenase.

Figure 2-4 Serum protein electrophoresis patterns (from top to bottom) of samples from *Canis familiaris* (Hallstroms), *Canis lupus, Canis aureus,* and *Canis latrans.*

Cumulatively, these studies indicate that the molecular techniques can easily contribute information on generic relationships within the Canidae and on the relationships of families within the Carnivora and Pinnipedia. However, the identification of protein polymorphisms within species and of protein genetic differences between species of the genus *Canis* has had only limited success. The available data suggest a substantially lower incidence of polymorphic loci in these species (Canoidea and Pinnipedia) than has been encountered in other mammalian orders.

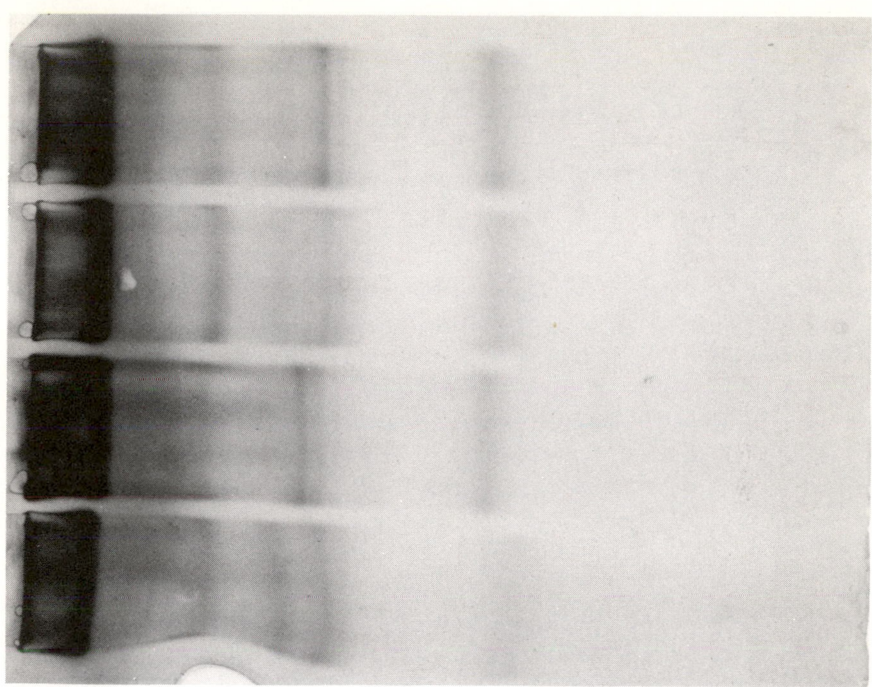

Figure 2-5 Serum esterase patterns (from top to bottom) of samples from *Canis familiaris* (Hallstroms), *Canis lupus, Canis aureus,* and *Canis latrans.*

3
THE CHROMOSOMES OF THE CANIDAE

A. B. CHIARELLI
Institute of Anthropology
University of Turin, Italy
and
Department of Anthropology
University of Toronto
Toronto, Ontario, Canada

A karyological approach to the taxonomy of the Canidae and the origin of the domestic dog seems especially relevant at this time. In many different animal groups, recently acquired knowledge on chromosome number and morphology has furnished important data necessary for an attempt to reconstruct their phyletic evolution and has provided us with one more criterion for taxonomic organization.

In spite of the fact that there is much new material on the chromosomes of the Canidae now available, to my knowledge no synthesis of this data has yet been attempted, apart from the more general review of Wurster and Benirschke (1968) for the entire group of Carnivora.

Knowledge of chromosomes is important to the study of phyletic evolution and taxonomy because these structures are the direct carriers of genetic information which has a very strict and stable organization. For each species of Eukaryota, the structural affinity between homologous chromosomes is consistently controlled by their pairing at meiosis in each individual of a chromosome population. Incomplete chromosome pairing at meiosis suggests the presence of structural changes in the organization of the genetic information which is linearly distributed along the chromosomes. If such structural changes are not restricted and limited, they are no longer compatible with the functional organization of the genetic information. The result of this is the inability of homologous chromosomes to pair at meiosis and consequently, the spoiling of the meiotic process.

In all higher organisms, homologous chromosome pairing at meiosis plays the role of a filter through which only a functionally patterned genetic system can pass. Furthermore, in native populations meiosis represents a barrier which prevents exchange between diverging genetic systems. The constancy of the features of the karyotype for each species (which we can appreciate at the level of the entire chromosome) is therefore a consequence of the meiotic filter. On the other hand, the existing differences between karyotypes of different species are due not only to meiosis but also to the process of natural selection, operating on the variation existent in the somatic chromosomes of the germinal cell line.

By studying and hypothesizing the possible steps by which the chromosomes of two related species differentiated, one can construct a hypothetical phyletic line connecting them and establish their probable taxonomic affinities. However, before one can establish chromosome identity between individuals of different species, it is important to be sure of their homologous genetic content. The only control one can have (as it occurs normally during meiosis) is the observation of the behavior of the chromosomes during meiosis (especially diakinesis) in individuals derived from crossbreeding two related species (F_1). In cytogenetic research, therefore, hybridization data deserve important consideration.

In the present paper, the differences in chromosome number and morphology of the different Canidae will be analytically considered and interpreted at the phylogenetic and taxonomic level, keeping in mind hybridization information. Afterwards, some consideration will be given to the origin of the karyotype of the domestic dog.

THE SOMATIC CHROMOSOMES OF THE CANIDAE

The taxonomic nomenclature followed here is the same as that used by H. J. Stains also in this volume.

A synthesis of the available information for the chromosomes of the species which have been studied to date is presented below. The calculation of the fundamental number (N.F.) is made by counting metacentrics and submetacentrics as 2, and acrocentrics and subacrocentrics as l in a female karyotype.

Alopex lagopus Linn (Figure 3-1a) $2n = 48–50$. N.F. 88 (Karyological information obtained from Gustavsson and Sundt, 1965). There are 24 pairs of meta or submetacentric chromosomes. The X chromosome is of medium size and is submetacentric. It is the smallest of the set. Cases of intraspecific polymorphism of the centric fusion type with chromosome numbers of 48, 49, 50, and 52 have been observed.

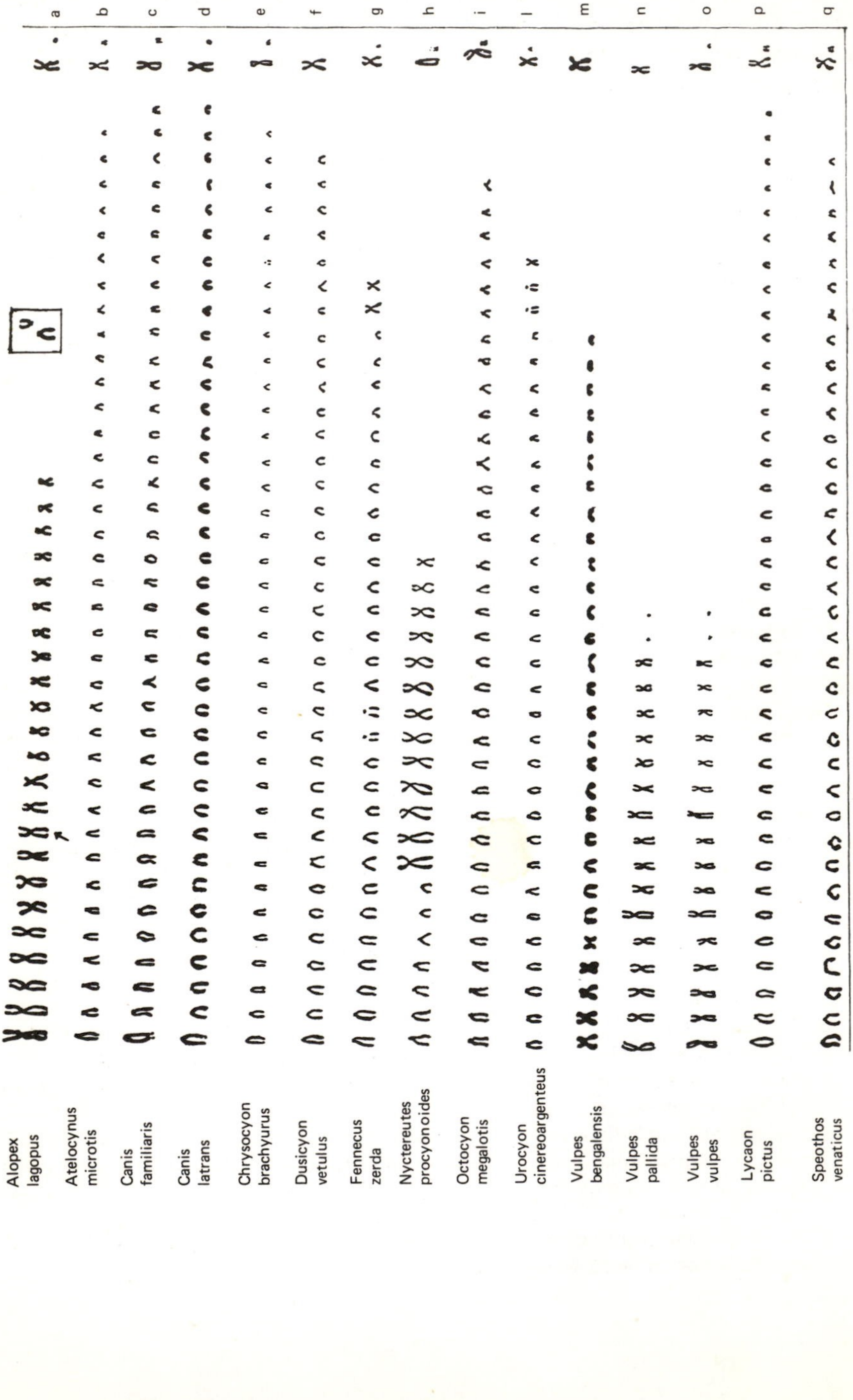

Alopex lagopus a
Atelocynus microtis b
Canis familiaris c
Canis latrans d
Chrysocyon brachyurus e
Dusicyon vetulus f
Fennecus zerda g
Nyctereutes procyonoides h
Octocyon megalotis i
Urocyon cinereoargenteus l
Vulpes bengalensis m
Vulpes pallida n
Vulpes vulpes o
Lycaon pictus p
Speothos venaticus q

1 10 20 30 xy

Atelocynus microtis Sclater (Figure 3-1b) $2n = 76$. N.F. 78 (Karyological information obtained from Wurster and Benirschke, 1968). There are 36 pairs of acro- or subacrocentrics and 1 pair of large submetacentrics. The latter are supposedly the X chromosomes, which would be in keeping with the other canids, but this is not yet proven, as the 1 individual studied was female.

Canis familiaris Linn (Figure 3-1c) $2n = 78$. N.F. 80 (Karyological information obtained from Chiarelli, 1965). The autosomal complement is 38 acrocentric chromosomes in decreasing order of size. The X is the largest submetacentric chromosome and the Y is the smallest meta- or submetacentric.

Canis aureus Linn $2n = 78$. N.F. 80 (Karyological information from Ranjini, 1966). The karyotype appears to be the same as that of *Canis familiaris*.

Figure 3-1 The haploid set of somatic chromosomes of different species of Canidae. Each chromosome set has been prepared by selecting one homologue from the reconstruction in pairs presented by the author quoted in the text. The chromosome order used by the original author has been maintained. The difference in dimension found in each set is due to the different original enlargement of the karyotype used and is not related to a real difference in chromosome dimension between the species. In the karyotype of *Alopex lagopus* the individual with 49 chromosomes studied by Gustavsson (1964) has been reported. The arrow indicates the chromosome homologue resulting from the centric fusion of 2 acrocentrics which are presented in the square box in the same line and which are to be considered as homologues of the previous one.

Important features to be noted in the different karyotypes are:

1) The exact morphological resemblance of the somatic chromosomes of the different species of the genus *Canis* (only two are presented here: *Canis familiaris* and *Canis latrans*). No noticeable differences have been observed in the other species studied.
2) The similarity of the karyotype of *Lycaon* with *Canis*.
3) The close resemblance of the karyotypes of *Chrysocyon, Atelocynus, Dusicyon,* and *Speothos* with the karyotype of *Canis*.
4) The similar number of arms in *Vulpes bengalensis* and *Vulpes vulpes,* suggesting the hypothesis that the two species differentiated by centric fusion mutations.
5) The two chromosomes with an achromatic region near the centromere in *Fennecus* (13–14) and in *Urocyon* (31–32).
6) The presence of 2 metacentric autosomal pairs in *Fennecus* (the last two) with 32 pairs of chromosomes and the presence of only 1 pair of metacentrics in *Urocyon* but 33 pairs of chromosomes, suggesting the hypothesis that the two species differ from one another by 1 centric fusion mutation.
7) The strict resemblance of the X chromosomes of all the species presented here with the exception of *Nyctereutes procyonoides*.

Canis latrans Say (Figure 3-1d) $2n = 78$. N.F. 80 (from Hsu and Benirschke, 1967). The karyotype appears to be identical to that of *Canis familiaris*.

Canis lupus Linn $2n = 78$. N.F. 80 (Karyological information obtained from Hungerford and Snyder, 1966). The karyotype appears to be the same as that of *Canis familiaris*.

Canis niger Batrum $2n = 78$. N.F. 80 (Karyological information obtained from Hsu and Arrighi, 1966). The karyotype appears identical to that of *Canis familiaris*.

Chrysocyon brachyurus Illiger (Figure 3-1e) $2n = 76$. N.F. 78 (Karyological information obtained from Newnham and Davidson, 1966). The autosomal complement consists of 37 acrocentric chromosomes. One autosomal pair appears to be incompletely acrocentric and one of the smallest pairs seems to have an achromatic region near the centromere. The X is a large submetacentric chromosome and the Y a small acrocentric.

Dusicyon vetulus Lund (Figure 3-1f) $2n = 74$. N.F. 76 (Karyological information obtained from Wurster and Benirschke, 1968). There are 36 pairs of acrocentrics and a large submetacentric. Since the individual studied was a female, the identification of the X chromosome is uncertain, but judging from other members of Canidae, it is presumed to be the metacentric pair.

Fennecus zerda Zimm (Figure 3-1g) $2n = 64$. N.F. 70 (Karyological information obtained from Matthey, 1954 and from Wurster and Benirschke, 1968). The autosomal complement consists of 2 pairs of meta- and submetacentric chromosomes and 29 pairs of acrocentrics. Two pairs of medium-sized acrocentrics possess an achromatic region in the long arm next to the centromere. The X chromosome is a large submetacentric and the Y is a very tiny chromosome.

Nyctereutes procyonoides Gray (Figure 3-1h) $2n = 42$. N.F. 68 (Karyological information obtained from Minouchi, 1929, revised by Todd, 1969). The autosomal complement consists of 7 pairs of acrocentric chromosomes and 13 submetacentric or metacentric chromosomes. The X chromosome is the largest acrocentric. The Y is a small satellited acrocentric and one of the smallest of the submetacentric autosomes. The satellite in the Y chromosome is located on the short arms.

Octocyon megalotis Desmaret (Figure 3-1i) $2n = 72$. N.F. 80 (?). (Karyological information obtained from Hsu and Benirschke, 1969). The autosomal complement consists of 35 pairs, almost all of which are acrocentric. The number of metacentric and submetacentric chromosomes seems undefined as it appears to be different in different animals studied.

Urocyon cinereoargenteus Schreber (Figure 3-1l) $2n = 66$. N.F. 70 (Karyological information obtained from Wurster and Benirschke, 1968). The autosomal complement consists of 1 pair of medium-sized metacentrics and 31 pairs of size-graded acro- or subacrocentric chromosomes. Two pairs of medium-sized acrocentrics possess achromatic regions in the long arm next to the centromere. The X chromosome is the largest element and it is metacentric; the Y is the smallest with the position of the centromere undefined.

Vulpes bengalensis Shaw (Figure 3-1m)$2n = 60$. N.F. 72 (Karyological information obtained from Ranjini, 1966; revised by Srivastava and Bhatnagar, 1969). The karyotype is composed of 5 pairs of metacentric or submetacentric chromosomes and 24 pairs of acrocentric or subacrocentric chromosomes. The X chromosome is metacentric and the Y is a small acrocentric.

Vulpes vulpes Linn (Figure 3-1o) $2n = 34–38$. N.F. 72 (Karyological information obtained from Chiarelli, 1965, from Omodeo and Renzoni, 1965, and from Lin, 1972). The karyotype is composed of 32 metacentric or submetacentric chromosomes and of 1 to 4 (occasionally) minute chromosomes. The latter could be interpreted perhaps as satellites broken off from the chromosomes to which they were originally attached. The study on North American *Vulpes vulpes* by C. C. Lin shows a variation in the number of minute chromosomes from 0–5, which can also be found in different cells of the same animal. The X chromosome is metacentric; the Y is a small acrocentric chromosome.

Vulpes pallida Cretzshmar (Figure 3-1n) $2n = 38$. N.F. 72 (Karyological information obtained from Chiarelli — unpublished data). The karyotype appears to be almost identical to that of *Vulpes vulpes;* however, more complete data is needed for absolute certainty, as there is some doubt about the taxonomy of the individuals studied. Moreover, no data are available on the sex chromosomes, as the individual examined was a female.

Table 3-1 Karyological information and taxonomic organization of Canidae

Taxa	2n	N.F.	References
Caninae			
Alopex lagopus Linn.	48–52	88	Andres, 1938; Lande, 1960; Wipf, *et al.*, 1949; Wurster, *et al.*, 1968; Gustavsson, *et al.*, 1965
Atelocynus microtis Schlater	74–76	76	Hsu, *et al.*, 1970; Wurster, 1969
Canis adustus			
Canis aureus Linn.	78	80	Ranjini, 1966
Canis familiaris Linn.	78	80	Ahamed, 1941; Awa, *et al.*, 1959; Barberis, *et al.*, 1964; Basrur, *et al.*, 1966; Borgaonkar, *et al.*, 1968; Brown *et al.*, 1966; Chiarelli, 1965 1966, 1972; Ford, 1965, 1969; Gustavsson, 1964; 1965; Hsu, *et al.*, 1967; Minouchi, 1928, 1929; Moore, *et al.*, 1963; Painter, 1925; Rath, 1894; Reither, *et al.*, 1963; Srivastava, 1969; Valenti, 1965; Makino, 1949; Maxione, 1918; Newnham, *et al.*, 1966
Canis latrans Say.	78	80	Benirschke, *et al.*, 1965; Hsu, *et al.*, 1967; Hungerford, *et al.*, 1966
Canis lupus Linn.	78	80	Hungerford, *et al.*, 1966
Canis mesomelas			
Canis niger Bartram	78	80	Hsu, *et al.*, 1966
Canis simensis			
Cerdocyon thous			
Chrysocyon brachyurus Illig.	76	78	Newnham, *et al.*, 1966
Dasycyon hagenbecki			
Dusycyon australis			
Dusycyon culpaeus			
Dusycyon griseus			
Dusycyon gymnocercus			
Dusycyon sechurae			
Dusycyon vetulus Lund	74	76	Wurster, *et al.*, 1968
Fennecus zerda Zimm.	64	70	Matthey, 1954; Wurster, *et al.*, 1968
Nyctereutes procyonoides	42	68	Minouchi, 1929; Todd, 1969
Otocyon megalotis	72	80	Hsu, *et al.*, 1969
Urocyon cinereoargenteus Schr.	66	70	Hsu, *et al.*, 1970; Wurster, *et al.*, 1968
Vulpes bengalensis Shaw	60	72	Manna, *et al.*, 1965; Ranjini, 1966; Srivastava, *et al.*, 1967; Talukdar, 1966.
Vulpes cana			
Vulpes chama			
Vulpes corsac			
Vulpes ferrilata			
Vulpes macrotis			

Taxa	2n	N.F.	References
Vulpes pallida	38	72	Chiarelli, 1972
Vulpes ruppelli Schinz	40	72	Matthey, 1954
Vulpes velox			
Vulpes vulpes Linn.	34–38	72	Andres, 1938; Bishop, 1942; Buckton, *et al.*, 1971; Gustavsson, 1964, 1965, 1967; Lande, 1958; Makino, 1947, 1951; Moore, *et al.*, 1965; Omodeo, 1965; Wipf, 1942; Wipf, *et al.*, 1949; Wodsedalek, 1931; Lin, 1972; Chiarelli, 1965
Simnocyoninae			
Cuon alpinus			
Lycaon pictus Temm	78	80	Hsu, *et al.*, 1970; Wurster, 1969
Speothos venaticus Lund	74	76	Wurster, 1969

Vulpes ruppelli Schinz $2n = 40$. N.F. 72 (Karyological information obtained from Matthey, 1954). The karyotype has been described as made up of metacentric and submetacentric chromosomes, except for 2 pairs of achromatics and 4 minute chromosomes in the male. The X seems to be a medium-sized metacentric and the Y a very small chromosome.

Lycaon pictus Temminck (Figure 3-1p) $2n = 78$. N.F. 80 (Karyological information obtained from Wurster, 1969). The autosomal complement is made up of 38 acrocentrics arranged in decreasing size. The X chromosome is a large metacentric and the Y a small metacentric.

Speothos venaticus Lund (Figure 3-1q) $2n = 74$. N.F. 76 (Karyological information obtained from Wurster and Benirschke, 1968). The autosomal complement consists of 36 acrocentric chromosomes. The X chromosome is a large metacentric; the Y a small subacrocentric.

In Table 3-1 the living species of Canidae are listed together with references on their karyological data. Unfortunately, information on chromosomes is available only for 19 of the 36 recognized species. The lack of complete data for all the living species makes the interpretive work only tentative. A better understanding of the evolutionary pattern of this group will be possible when the missing data are supplied. In spite of this, however, a preliminary attempt can be made to interpret the phyletic relationships of the different species.

Table 3-1 shows the diploid number ($2n$) of the chromosomes and the so-called fundamental number (N.F.). The latter is based on the assumption that centric fusion (or misdivision of the centromere) is one of the more successful mutations to pass the sieve of meiosis. Such a mutation would not interfere with the structural organization of the genetic information on the chromosomes. The only change which would result would be the reduction or increase in the random distribution of the genetic information in the offspring. The reduction or increase of chromosome units certainly presents an adaptive advantage to an organism, reducing or increasing the potential variability in the population. A relation between chromosome morphology and the chiasma frequencies must, moreover, exist, although at the moment we do not have enough information to define the exact relationship.

If we take into consideration this second type of data (N.F.), the uniformity of the group in Table 3-1 appears to be more clear. The diploid chromosome number varies between 34 (*Vulpes*) and 78 (*Canis*), with a range of 44. If, instead, we consider the number of arms (N.F.), it varies from 88 (*Alopex*) to 68 (*Nyctereutes*); a range of only 20 fundamental units (arms), which in terms of chromosomal mutation, centric fusion, or misjunction of the autosomes are only 10 major events. This range is further reduced if we consider that *Alopex* seems to be an exceptional case, with a karyotype due to some other more intricate chromosomal rearrangement.

The advantage of the use of the fundamental number for the study of this group of mammals is better shown in Table 3-2, where the morphological information on the chromosomes is summarized. The genera, or the karyological species, are listed in order of decreasing number of chromosomes.

From these data it is easy to observe the strict relation between total number of chromosomes and the number of acrocentrics and metacentrics. When the metacentric chromosomes increase, there is always a decrease in the number of acrocentrics.

The table, moreover, shows another type of relationship which is of interest from a taxonomic point of view. The genera *Canis, Lycaon, Chrysocyon, Atelocynus, Dusicyon,* and *Speothos* seem more closely related among themselves than to the others. All the species belonging to this group have only acrocentric autosomes and all have a high chromosome number (from 74 to 78). Also, they appear to share morphologically identical X chromosome and some of the autosomes present similar morphological peculiarities (as n.1.).

The genera *Urocyon* and *Fennecus* also appear to be closely related: they have an identical N.F. and share the same 2 pairs of chromosomes which are marked by an achromatic region near the centromere. In general terms, the chromosomes appear to be very similar. The only difference is that *Fennecus* has 1 more pair of metacentric chromosomes. However, this difference can easily be explained with a centric fusion mechanism.

The two species of *Vulpes* considered also have a different number of chromosomes, but the same fundamental number (36). If the 24 acrocentric chromosomes of *Vulpes bengalensis* are appropriately matched, it is possible to obtain the karyotype of *Vulpes vulpes*.

Alopex, Nyctereutes, and *Otocyon* instead, present karyotypes which are not completely comparable with the others in terms of chromosomal morphology.

TABLE 3-2 Synthesis of the Karyological information actually available on the living Canidae

Genera or Karyological Species	n	H.N.F	M+SM pair	A+SA pair	Peculiar Chromosome Pair	X	Y
Canis	39	40	0	38	0	IM	SM or A
Lycaon	39	40	0	38	0	IM	SM
Chrysocyon	38	39	0	37	0	IM	SA
Atelocynus	37–38	38	0	36	0	IM	?
Dusicyon	37	38	0	36	0	IM	?
Speothos	37	38	0	36	0	IM	SM
Otocyon	36	40	—	—	—	IM	—
Urocyon	33	35	11	29	2	IM	SM
Fennecus	32	35	2	27	2	IM	SM
Vulpes bengalensis	30	36	5	24	—	mM	SA
Vulpes vulpes	18–20	36	16	0	1–3	IM	SA
Alopex	25	44	—	—	—	—	—
Nyctereutes	21	34	13	7	—	IA?	SA

Knowledge of the chromosome number and morphology of the other 3 living species of Canidae (*Cerdocyon thous, Dasycyon hagenbecki*, and *Cuon alpinus*) will be of obvious importance for a more complete interpretation of the chromosomal evolution of the species belonging to this family. However, it seems already clear that the family Canidae, from a karyological viewpoint, could be differentiated into 2 (or maybe even 3) chromosomal subgroups. A similar or identical karyotype, however, does not mean that the species are phylogenetically related. A karyotype similarity or even identity can be reached through both convergent or parallel evolution. It is, in fact, possible that different species shared similar trends in evolution and that the karyotypes differentiated in a parallel way.

THE HYBRIDIZATION INFORMATION

The homology of the karyotype could be tested by the occurrence of hybrids. Two individuals belonging to different species may be forced to reproduce in captivity. There are many examples of individuals from separate species, or even separate genera, interbreeding in captivity, producing hybrids of various degrees of viability and fertility (Gray 1954, 1966). These interspecific or intergeneric hybrids are of enormous interest to prove the genetic homology of their karyotypes, especially when the offspring are completely viable and backcrosses can be produced and studied.

The animals resulting from planned crosses (Seitz, 1959a; 1965) or naturally-occurring crosses (wolves and coyotes in Ontario) between different species of *Canis* show a complete interfecundity, and the fertility of many of these hybrids has been proved at the level of backcrossing. Thus, the real homogeneity of their karyotype seems to be demonstrated. The occasional hybrids occurring between species with different karyotypes are of even greater interest. The data available on these intergeneric hybrids are presented in Table 3-3. Their existence stresses the close karyological and phylogenetic relationship between

Table 3-3 Intergeneric hybrids recorded in Canidae

	Offspring	References
Alopex lagopus X *Vulpes vulpes*	Several, normal,	*American Fur Breeder,* 1929–41; Cole and Shockelford, (1947–48); Starkov, 1940; Wipf, 1969; Karekuledstvo and Zverovodstvo 1949; Green–Armytage, 1941, Gunn, 1946; Leekley, 1943; Boelteger, 1895; Wipf and Shockelford, 1949.
Canis familiaris		
X *Cerdocyon thous*	Only one, normal	Krieg, 1925;
X *Vulpes vulpes*	sterile embryos	Cole and Shockelford, 1947–48, Sipinen, 1949; Autonius, 1951; Maples, 1941; Presner, 1933.
Urocyon cinereoargenteus X *Vulpes vulpes*	Normal sterile	Bezdel, 1944; Cole and Shockelford, 1947–48.

Alopex and *Vulpes,* the affinity of *Canis* with *Cerdocyon thous* (for which we have no karyological data as yet), the affinity of *Urocyon cinereoargenteus* and *Vulpes vulpes* (whose karyotypes are quite different), and a possible affinity (conception with formation of embryos) between *Canis* and *Vulpes vulpes.* The occurrence of these hybrids, however, is almost always occasional and no complete study has been made to ascertain the hybrid karyotypes and the behavior of their chromosomes at meiosis. The only serious attempt at a study of this nature is that of Wipf and Shackleford (1949).

TRENDS IN CHROMOSOME VARIATION IN THE CANIDAE

From the previously discussed data, it appears clear that for the entire group of Canidae there has been a general trend to increase or decrease the number of chromosomes while keeping constant the number of arms (N.F.). However, what direction actually was followed? The ancestral stock which differentiated from the Miacidae during the late Eocene had a) a high number of mostly acrocentric chromosomes as does *Canis*, or b) a low number of mostly metacentric chromosomes as does *Vulpes,* or c) was an intermediate type such as *Otocyon.*

A general observation on the number of chromosomes of the families which are most closely related to the Canidae is unproductive, however. In fact, the Mustelidae have a range in chromosome number from 30 to 64 (with an N.F. ranging from 58 to 96).

The hypothesis of chromosomal evolution through centromere fissioning has no clear-cut proof in natural species; instead, there is abundant evidence that Robertsonian fusion of chromosomes is a spontaneous and frequent occurrence in nature as in the Equidae, Bovidae, Muridae, human translocational aberrations, and in long-term *in vitro* cultures of dog cells (personal observation).

It is, moreover, a general belief among cytotaxonomists that fusions and inversions are the most common events occurring in karyotype evolution and that the ancestral or primitive karyotype of mammals must have possessed mostly acrocentric elements. The ancestral Carnivora (Miacids) likely had a high number of chromosomes (80?). In some of the families derived from them, like the Canidae, little change has taken place despite marked phenotypic evolution, while in other families, like the Felidae, both marked phenotypic and karyotypic differentiation has occurred.

From these data it can be inferred that *Canis*, therefore, has a more generalized type of karyotype, while *Vulpes* has a more specialized one. These data are certainly in agreement with the ecological requirements of the two groups, the *Canis* type being more widespread and in some ways more variable and adaptive, the fox more highly differentiated and perhaps less biologically adaptable.

THE CHROMOSOMES OF THE DOMESTIC DOGS

The differentiation of the domestic dog into many different races and varieties is a major problem in itself. The present great variability in this species is certainly due to artificial and sometimes aesthetic selection carried on by man on a biologically suitable species. However, in spite of this seemingly recent extreme variability in different breeds of domesticated dog, such variations are also well-documented in archaeological remains from prehistoric sites.

Different species of dogs shared with early man a carnivorous, aggressive type of life, as did other members of the Canidae in ancient times. However, later on some stock of Canidae probably adapted to cooperate with the ancient hunters and in later times were accepted in early human settlements, where they scavenged the remains of human foods.

There is therefore the possibility that different human populations domesticated different dog species, and then these different human populations came into contact with the different stocks of domesticated dogs that had been hybridized. This phenomenon of introgressive hybridization can explain the large variation existing among the actual living domestic dogs and their close or distant similarity (as the case may be) with the wolf, the jackal, or even the fox.

The history of dog domestication and the retracing of the biological history of the variation in this animal can therefore provide more information on the development of many human cultures and on human migration. The discovery of some peculiar trait in the chromosomes of an actual living wild species of dog which could perhaps be recognized in some domestic stock would be an important finding for tracing the biological history of the domestic dog.

Unfortunately, microscopic inspection performed up to now on the somatic chromosomes of the different species of *Canis* has revealed no differences in morphology, and similar results are known for the different races of the dog (Chiarelli, 1965 and unpublished data). The more recently available techniques of chromosome banding with fluorescent acridine mustard (Casperson, 1970), NaOH treatment (Schnedl, 1971; Patil, *et al.*, 1971; Arrighi and Hsu, 1971) or trypsin (Chiarelli, *et al.*, 1972; 1973) have not yet been applied to the chromosomes of these animals and the possibility of finding structural differences is therefore still open.

Another approach to the search for heterogeneity in chromosomes is the study of the behavior of the chromosomes when they pair in meiosis. This approach has been undertaken by us on many races of dogs with known reduced fertility and on mongrel dogs with promising results. Several anomalies have been found in both these groups. It is interesting that the parental karyotype in at least one of the mongrels was different because of a translocation (Chiarelli, 1965) and that the reduced fertility of some of the dog races may be due to the reduced possibility of producing viable sperm; the process of diakinesis being rendered too complicated by many chromosome arrangements.

Detailed research on the chromosomes, both somatic and meiotic, on *Canis familiaris,* the domestic species *par excellence,* using selective staining and more elaborate techniques could therefore be valuable in solving anthropological problems, such as those relating to human migration in the Americas (Chiarelli, *et al.,* 1972).

Acknowledgments The present paper is developed under a grant from the Italian National Council of Research and the Mexican CNEN Radiobiology Laboratory. The author is indebted to all the authors mentioned here, from whom he obtained important information.

4

SEXUAL DIMORPHISM AND GEOGRAPHICAL DISTANCE AS FACTORS OF SKULL VARIATION IN THE WOLF *Canis lupus* L.

PIERRE JOLICOEUR
Department of Biology
University of Montreal
Montreal, Quebec, Canada

Because of the complexity and variability of living organisms, multivariate statistical techniques have become an important tool in studies of variation and evolution. In the case of canids, there have been several successful multivariate studies of cranial variation (Giles, 1960; Jolicoeur, 1959; Lawrence and Bossert, 1967, 1969). Multivariate techniques have also been applied with success to other mammals (Rempe and Bühler, 1969), as well as to humans and many other kinds of animals and plants. Blackith and Reyment (1971) have written a good general review of multivariate studies in biology and in other natural sciences. Until now, most biological applications of multivariate analysis have been to morphological (often skeletal) characters, and multivariate analyses of physiological, ecological, and behavioral data would likely be more difficult and, in many cases, require preliminary variate transformations. However, the ultimate approach to any problem would clearly consist of a joint multivariate study of all available types of data. The major aim of the present analysis is to investigate the relationship between sexual dimorphism and geographical distance as factors of skull variation in the wolf *Canis lupus* L.; this aspect was not studied at the multivariate level in the author's previous publication on the subject (Jolicoeur, 1959).

MATERIALS AND METHOD

Because of its particular aim, the present analysis is based mostly on the two areas from which large samples of male and female population data were available: Canada's Northwest Territories between Great Slave Lake and Great Bear Lake, and Manitoba. Many smaller samples from the author's previous study are not used here, either because they were very small, or because sex was undetermined, or because specimens were less fully documented than desirable. However, small samples are included from two distant areas in which wolf populations are already known to be clearly different: the Arctic Archipelago and the Rocky Mountains (north of the Canada–U.S. boundary). The approximate limits of each region and the number of available male and female specimens are listed in Table 4-1. Also excluded from the present data, more rigorously than in the earlier analysis, are several young adult specimens, whose skulls might not have fully reached final proportions. Finally, the number of characters is cut down from the original 12 to 9 here, in part to avoid redundancy and in part because of difficulties in defining and measuring some characters in a perfectly consistent manner. The characters retained in the present analysis are thus the following, all dimensions being expressed in mm:

X_1 = Palatal length
X_2 = Postpalatal length
X_3 = Zygomatic width
X_4 = Palatal width outside the first upper molars
X_5 = Palatal width inside the second upper premolars
X_6 = Width between the postglenoid foramina
X_7 = Interorbital width
X_8 = Least width of the braincase
X_9 = Crown length of the first upper molar

In spite of this drastic amount of culling, the conclusions of the present analysis turn out to agree with those of the earlier study wherever there is overlap.

In the author's earlier study, the statistical analysis was carried out in a highly unified fashion, and separate significance tests were not done

TABLE 4-1 Data used in the present analysis

Region	Approximate limits		Number of specimens	
	Latitude N	Longitude W	Males	Females
Arctic Archipelago	73° to 80°	83° to 121°	10	6
Northwest Territories	62° to 66°	111° to 124°	106	103
Manitoba	53° to 60°	94° to 102°	58	42
Rocky Mountains	49° to 53°	114° to 119°	6	3

for the numerous possible pairs of samples. Although that approach had statistical advantages because it decreased the risk of rejecting the null hypothesis incorrectly, it could leave lingering doubts in the reader's mind about the possibility that the wolves of some regions, like the Northwest Territories and Manitoba, might in fact be similar. Consequently, before engaging in the discriminant analysis described hereafter, several preliminary analyses were made, using Hotelling's T^2 test and Mahalanobis' generalized distance as well as discriminant analysis, in order to check whether the large samples from the Northwest Territories were homogeneous and whether they did differ significantly from those of Manitoba. On the one hand, although the present samples from the Northwest Territories are in fact made up of specimens from three neighboring areas (labeled "D", "E" and "G" in the author's earlier publication), preliminary analyses fail to disclose significant differences between specimens from those three areas: the present large samples from the Northwest Territories can therefore be considered homogeneous. On the other hand, preliminary analyses show Manitoba wolves to differ very conclusively ($P < 0.0005$) from those of the Northwest Territories. The detailed numerical results of these preliminary analyses are not reported here because they are contained implicitly in those of the discriminant analysis hereafter, which is judged to be of more general interest.

The statistical method preferred for the present study is *Discriminant analysis,* also variously known as *Discriminatory analysis, Multiple discriminant analysis, Analysis of dispersion* or, less frequently nowadays, *Canonical analysis* (Seal, 1964). The approximate methods chosen here for testing statistical significance are due mostly to Bartlett (1947, 1965), whose writings unfortunately do not outline explicitly step-by-step procedures. More detailed indications can be found in the books of Rao (1952) and Seal (1964). Morrison's book (1967) is perhaps the best reference for applied multivariate analysis but, regrettably, it does not cover multiple discriminant analysis *per se;* however, Morrison's book does include a treatment of the ordinary discriminant function of

Fisher, obtained in the case of 2 samples, and a detailed and valuable discussion of the multivariate analysis of variance, which is partly related to multiple discriminant analysis.

RESULTS

The averages of the 9 characters for male and female samples from the four regions considered are given in Table 4-2. Examination of these mean values shows that, inasmuch as can be judged from skull dimensions, males are larger than females on the average by approximately 3 to 8%. Joint examination of Tables 4-1 and 4-2 also indicates that, as latitude increases, the wolves of each sex become smaller; this adds an exception to Bergmann's rule. Wolves from different regions do not differ merely in size, however, but also in proportions: although male and female wolves have longer skulls (characters X_1 and X_2) in Manitoba than in the Northwest Territories, for instance, their first upper molars have the same crown length (X_9) in the two regions and the interorbital width (X_7) is smaller in Manitoba.

TABLE 4-2 Averages (mm)

Region	Sex	Characters								
		X_1	X_2	X_3	X_4	X_5	X_6	X_7	X_8	X_9
Arctic Archipelago	Males	115.80	100.80	142.40	81.68	33.53	67.07	45.54	39.99	17.98
	Females	110.83	96.17	137.00	79.23	32.33	64.73	46.18	39.97	17.38
Northwest Territories	Males	120.79	101.78	143.31	80.48	34.36	65.15	48.17	42.15	17.69
	Females	115.04	96.43	135.16	76.40	32.87	62.83	45.05	40.37	16.91
Manitoba	Males	121.31	103.69	143.59	81.50	35.59	65.52	47.78	42.13	17.66
	Females	115.95	98.50	135.50	76.46	33.00	62.82	44.68	40.89	16.95
Rocky Mountains	Males	126.50	108.17	145.17	82.12	33.55	67.28	49.42	43.20	18.18
	Females	117.33	102.67	128.67	75.50	31.10	63.73	43.70	41.43	16.90

The results of the discriminant analysis of the 9 skull dimensions are displayed in Tables 4-3 and 4-4 and in Figures 4-1 and 4-2. Out of the 7 dimensions which could possibly have been *"occupied"* by the dispersion of the 8 sample means about the grand mean (Table 4-3), 4 bring out trends of variation which are highly significant from a statistical viewpoint (discriminant functions Y_1, Y_2, Y_3 and Y_4); the last 3 dimensions fall so far short of statistical significance that they are obviously negligible. Female samples fall clearly on the left and male samples fall clearly on the right side of Figure 4-1: the first discriminant function (Y_1) thus corresponds very closely to sexual dimorphism, and the latter

TABLE 4-3 Discriminant analysis: Bartlett's significance tests

Function	% of Variation	Chi square	Degrees of Freedom	Probability
1st	63.26	452.16	63	< 0.0005
2nd	18.51	199.35	48	< 0.0005
3rd	9.21	103.14	35	< 0.0005
4th	7.26	51.721	24	< 0.001
5th	1.30	10.506	15	> 0.70
6th	0.43	2.738	8	—
7th	0.03	0.176	3	—

constitutes the most important trend of variation between the present groups. Looking again at Figure 4-1, one notes that, although no direct use is made of longitudes and latitudes in the discriminant analysis of skull dimensions, male and female samples fall strictly in the same order according to the second discriminant function (Y_2) as they would according to latitude; this reflects the presence of gradual *geographical variation,* of a *"cline"* in J. S. Huxley's terminology. The double *"polarity"* present in Figure 4-1, of sexual dimorphism and latitudinal variation along the first and second discriminant functions, respectively, is analogous to some of Blackith's pioneering discoveries of double and triple polarities in grasshoppers (reviewed by Blackith in 1965). In contrast with Figure 4-1, Figure 4-2 does not separate the sexes: males and females of each region fall closely together in this scatter diagram of the third and fourth discriminant functions, and the main fact emerging is that, even though Rocky Mountain wolves (represented by black symbols) follow the *cline* brought out by the second discriminant function (in Figure 4-1), they appear to deviate from all others in their own particular direction.

TABLE 4-4 Coefficients of discriminant functions and pooled standard deviations (mm)

Function	Characters								
	X_1	X_2	X_3	X_4	X_5	X_6	X_7	X_8	X_9
1st	.1002	.1101	.0367	.1543	−.0372	−.0390	−.0343	.0250	.1985
2nd	−.2089	−.1414	.1310	.1919	−.2115	.1453	.0525	−.1181	.3675
3rd	−.0711	.2161	−.1276	.1003	−.2723	.2225	−.1405	.0288	.0592
4th	−.1351	.0754	−.0655	.0405	.6204	.0042	−.1475	−.0063	.0214
Pooled standard deviations	3.7022	3.5247	5.0284	2.5973	1.6884	2.3797	2.7580	2.7407	.82752

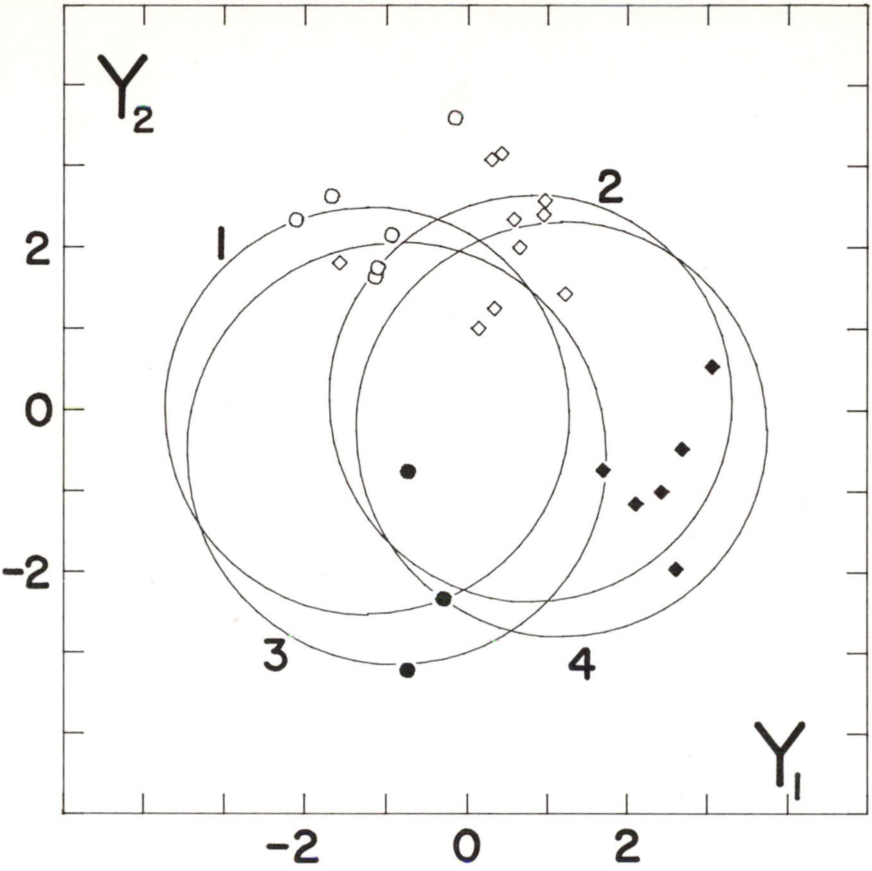

Figure 4-1 Scatter diagram of 1st and 2nd discriminant functions, Y_1 and Y_2, based on 9 skull dimensions. Large samples are outlined by 95% equal-likelihood circles numbered as follows: (1) females, and (2) males from the Northwest Territories; (3) females, and (4) males from Manitoba. Individual specimens of small samples are represented by the following symbols: ○ , females and ◇ , males from the Arctic Archipelago; ● , females and ◆ , males from the Rocky Mountains.

DISCUSSION

Although, as discussed in the author's previous publication (Jolicoeur, 1959), much of this variation might be explained by direct environmental influences on physiology and growth, the general biological context and current experience with variation of this nature make it likely that these size and shape differences are determined at least partly by heredity. It is remarkable in this respect that, once the first discrimi-

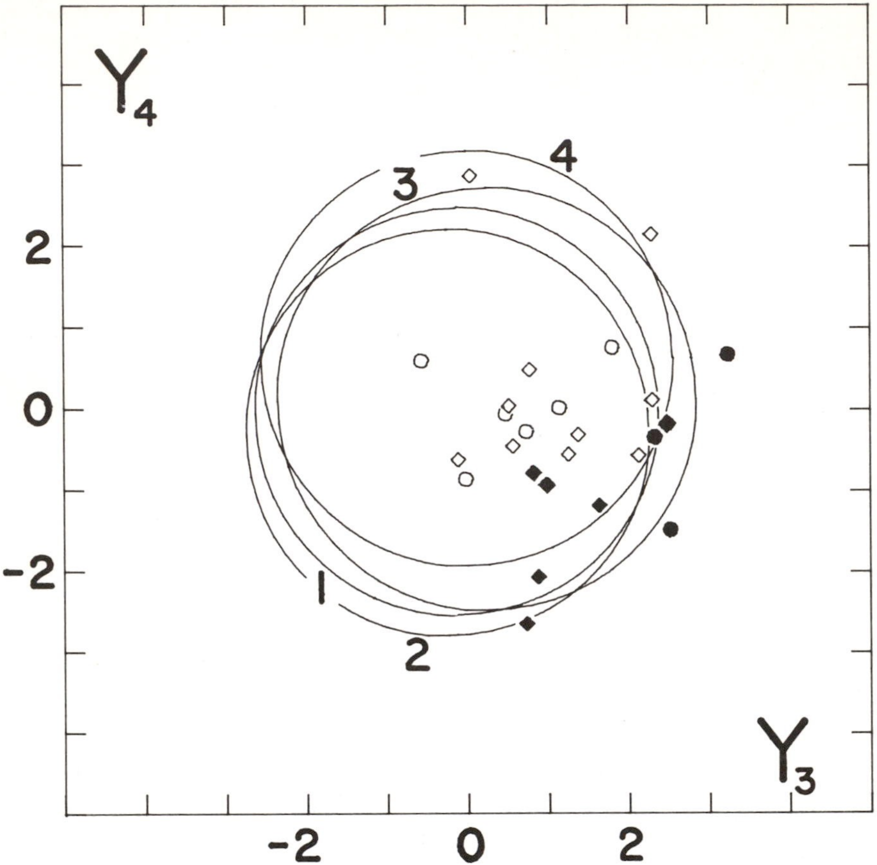

Figure 4-2 Scatter diagram of 3rd and 4th discriminant functions, Y_3 and Y_4, based on 9 skull dimensions. 95% equal-likelihood circles are numbered and individual symbols are defined as in Figure 4-1.

nant function corresponding to sexual dimorphism is removed, the variation between specimens of the four regions considered is still very significantly tridimensional. Thus, even though the specimens compared here come from only four regions between which there are no impassable barriers (the Arctic Archipelago is linked to the mainland by icefields during much of the year) and between which there should be a certain amount of gene flow, the pattern of variation is nevertheless too complicated to be interpreted entirely as the result of simple genetical differentiation over a two-dimensional environment.

A brief speculation is enough to show how little we know about the origin and the biological significance of differences between local populations. For instance, to what extent are presently observed variations remnants from genetic differences established during isolation by Pleistocene glaciers, to what extent are they the result of genetic drift, and to what extent are they still maintained or enhanced by natural selection? Also, though some characters seem rather clearly adaptive, like the long and white pelage of northern wolves, what about other trends of variation, like those of skull dimensions? Is the short face of northern wolves adaptive too (another instance of Allen's rule?), is it merely a case of environmental stunting, or is it a case of environmental stunting that turned out to be adaptive and has become genetically assimilated? The fact that the face of northern wolves is so short that teeth are seriously cramped (Jolicoeur, 1959) suggests that, at least in some cases, variations are not adaptive in every respect. But there are some troubling parallels: the Arctic fox (*Alopex lagopus*) also has a short face in comparison with other foxes. Possibly, adaptations to the arctic are less ancient, less complete, and less harmonious in the wolf, a widely-distributed species, than in the Arctic fox, a specialized resident of the north.

Acknowledgments In closing this reanalysis, the author would like to seize the opportunity to complement the acknowlegments expressed in his earlier publication. If so many excellent and well-documented specimens were collected, it is largely thanks to the professional enthusiasm and efficiency of the following zoologists, then members of the Canadian Wildlife Service and of the Manitoba Game Department: J. Kelsall, D. Flook, and J. E. Bryant. Finally, Dr. I.McT. Cowan of the Department of Zoology of the University of British Columbia supervised the inception and the development of the collection, collected many specimens personally, made the material available for study, and mentored the author throughout the latter's graduate studies in Vancouver, from 1956 to 1958. To these persons, and to all others mentioned in his earlier publication, the author is most grateful.

5 MORPHOLOGICAL AND ECOLOGICAL VARIATION AMONG GRAY WOLVES (*Canis lupus*) OF ONTARIO, CANADA

G. B. KOLENOSKY and R. O. STANDFIELD
Research Branch
Ontario Ministry of Natural Resources
Maple, Ontario, Canada

The Province of Ontario is about 1,068,382 km^2 in area with a climate that varies from humid continental (warm summer) in the vicinity of Lake Erie to subarctic along the shores of Hudson Bay. Rowe (1959) classified three major forest types in Ontario: a small area of deciduous forest in the extreme south, a mixed deciduous–coniferous forest in the vicinity of the Great Lakes, and a boreal forest in the north. A relatively narrow band of almost treeless subarctic tundra occurs along the coasts of Hudson and northern James Bays.

Prior to the beginning of European settlement in the 17th century, the gray wolf (*Canis lupus*) probably inhabited the entire region of Ontario (Bates, 1958). Subsequently, the development of agriculture, with its attendant clearing of forests, destroyed wolf habitats in the southern peninsula and smaller areas in the central and western parts of the province.

Reports of early explorers and settlers lead one to believe that wolves were numerous throughout the province, but there is little of a descriptive nature that is of use in understanding the distributional and ecological situation at that time. About the middle of the 19th century, wolves began to disappear from farming areas along the north shores of the St. Lawrence River and Lakes Ontario and Erie. Agricultural areas are now inhabited by the coyote (*Canis latrans*) which began its invasion into southern and eastcentral Ontario early in the 20th century (Anonymous, 1968).

In his detailed work on the taxonomy and distribution of the gray wolves of North America, Goldman (1944) concluded that almost the whole of Ontario was inhabited by *C. l. lycaon* whose range originally extended deep into the southeastern United States. He also recognized the occurrence of *C. l. hudsonicus* in the northwestern part of the province bordering Hudson Bay and, according to his distributional map, the possible occurrence of *C. l. nubilus* in the parairie outliers in the southwest. His criteria for *C. l. lycaon* were based on a total of 77 specimens. However, only 14 were from Ontario, and all of these originated from the coniferous–deciduous forests on the southern fringes of the boreal forest.

In 1957, an intensive wolf and coyote research program was started in Ontario. Emphasis was given to wolf ecology and distributions, and, as part of these studies, approximately 3000 carcasses or heads of wolves and coyotes were received for postmortem examination. A large volume of data was accumulated on morphological characteristics and these have been analyzed with a view to clarifying the distribution and ecology of races and types of Ontario wolves. Morphological variations among the wolves collected from different regions of Ontario and field observations on behavior indicated that *C. l. lycaon* as recognized by Goldman (1944) did not inhabit the greater part of Ontario. Regional variations in type could possibly be explained by dynamic, long-term associations between wolves and their prey, or by hybridization between wolves and coyotes.

EXISTING SUBSPECIES

On the basis of our studies, we recognize the following subspecies and types of *C. lupus* in Ontario: *C. l. hudsonicus* in the Hudson and James Bays coastal areas, *C. l. lycaon* (Boreal type) in the northern and central boreal forets, *C. l. lycaon* (Algonquin type) in the deciduous-coniferous forest, and a small wolf with some coyotelike characteristics *C. l. lycaon* (Tweed type) along the southern limits of range of the Algonquin type. The main distributions of these and the coyote are shown in Figure 5-1. The terms applied to the various types refer to their approximate centers of origin, and are used throughout as a matter of convenience.

The zone of separation between the Boreal and Algonquin types closely approximates the division between the boreal forest and the coniferous–deciduous forest of the Great Lakes region. Within this zone both types occur. The limits of the Tweed type are not well-defined, but centers of origin are indicated by the triangular symbols. The larger triangle designates the region of most frequent occurrence. Similarly, only confirmed identifications of *C. l. hudsonicus* are shown, but this race probably inhabits most of the Hudson and James Bays coasts (Figure 5-1).

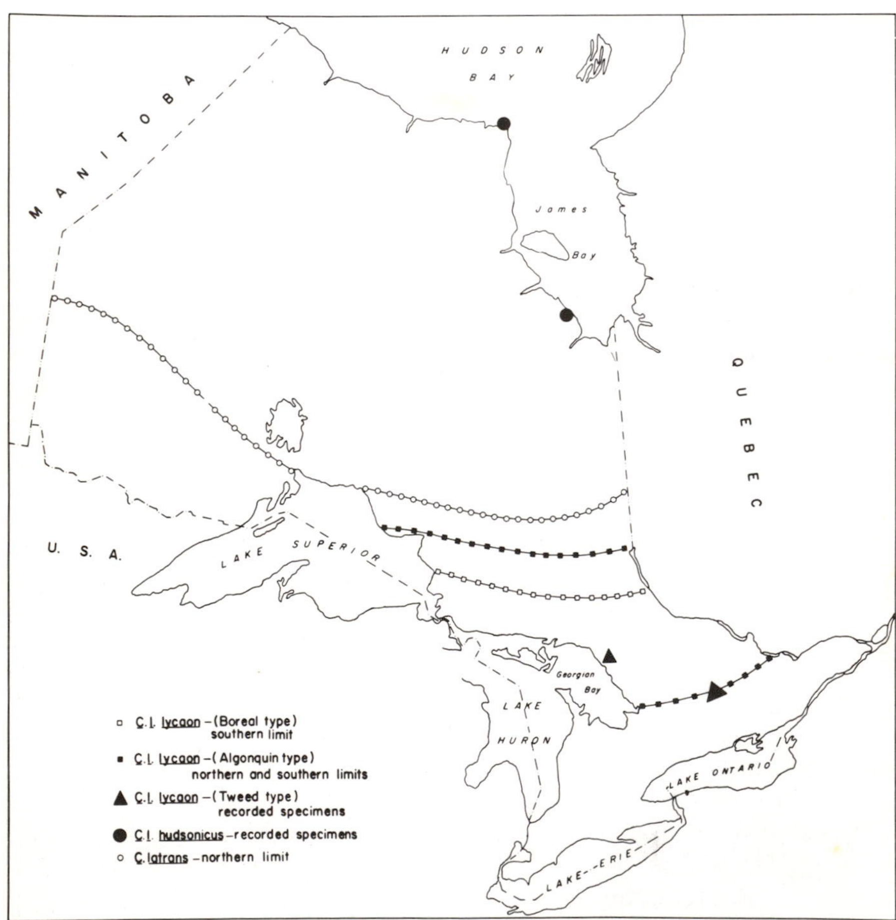

Figure 5-1 Present distributions of races and types of the gray wolf *(Canis lupus)* and coyote *(Canis latrans)* in Ontario.

The Algonquin type is small, slender, with a narrow muzzle, relatively long ears, and is invariably gray-fawn in color. The description given by Goldman (1944) for *C l. lycaon* fits this type admirably. Most individuals of the Boreal type are larger with more massive skulls and proportionately shorter ears. They may vary in color from almost pure white to jet black. The Tweed type is smaller than the Algonquin type but is essentially similar in color. The width of its rostrum is only slightly greater than that of most coyotes but the teeth are almost as large as those of the Algonquin type. This often results in a pronounced overlapping of some of the teeth, especially P^2 and P^3. Suspected hybrid canids from New England possessed similar dental characteristics (Lawrence and Bossert, 1969). A comparison of relative sizes and general characteristics of the skulls of these various types, and that of the coyote are shown in Figure 5-2. Note the slender rostrum of the Algonquin type and the larger rostrum and more massive zygomata of the Boreal type.

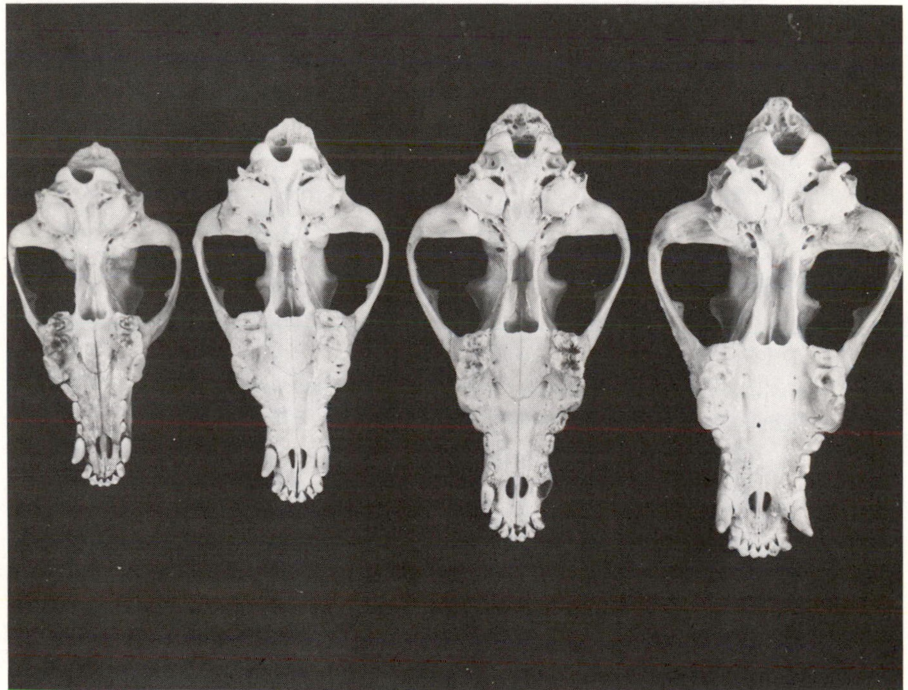

Figure 5-2 Typical adult male skulls (from left to right) of *C. latrans, C. lupus lycaon*

COMPARATIVE STUDIES

As one means of determining the degree of separation between the Algonquin and Boreal types, a series of 105 adult skulls of the former and 122 adult skulls of the latter were compared quantitatively. The sample of skulls of the Tweed type was insufficient to permit a comparable statistical analysis. Only animals that were 1 year of age or older were accepted for analysis in the current studies. Age determination was based on the complete closure of the basisphenoid–basioccipital suture which occurs at approximately 12 to 14 months of age. Known age specimens and specimens aged by the cementum technique were used to test the validity of the method. A test for homogeneity, based on 42 adult skulls of the Algonquin type from Algonquin Provincial Park aged by the cementum technique, revealed no significant differences in skull measurements among the various adult age classes.

The following 27 measurements and 2 ratios were recorded for each specimen. *Skull:* (1) maximum length from premaxilla to occipital condyle, (2) maximum length from premaxilla to posterior of sagittal crest, (3) maximum width of skull across canines, (4) minimum interorbital width, (5) maximum width at postorbital processes, (6) maximum width across zygomatic arches, (7) minimum cranial width at temporal fossa, (8) maximum width of skull at lateral borders of occipital crest, (9) maximum width of long axis of one condyle, (10) maximum width of short axis of one condyle, (11) maximum width across both condyles, (12) maximum skull height from bullae to sagittal crest, (13) ratio of maximum width of skull at protocones of upper carnassials to (3) above. *Mandible:* (14) maximum length from symphysis to angular process, (15) maximum length from symphysis to condyle, (16) maximum height from angular process to coronoid process, (17) maximum height of ramus between P_4 and M_1, (18) maximum width of ramus below metacone of M_1, (19) maximum width of short axis of articular condyle, (20) maximum width of long axis of articular condyle, (21) ratio of maximum length of tooth row from anterior of canine to M_3 at enamel line to (3) above. *Dentition:* (22) maximum bucco-lingual width of P^4 at enamel line, (23) maximum anterior-posterior width of P^4 at enamel line, (24) maximum bucco-lingual width of M^1 at enamel line, (25) maximum anterior-posterior width of M^1 at enamel line, (26) maximum bucco-lingual width of P_4 at enamel line, (27) maximum anterior-posterior width of P_4 at enamel line, (28) maximum bucco-lingual width of M_1 at enamel line, (29) maximum anterior-posterior width of M_1 at enamel line.

Each series of skull measurements of both types was segregated by sex and then analyzed by the procedure of multiple discriminant

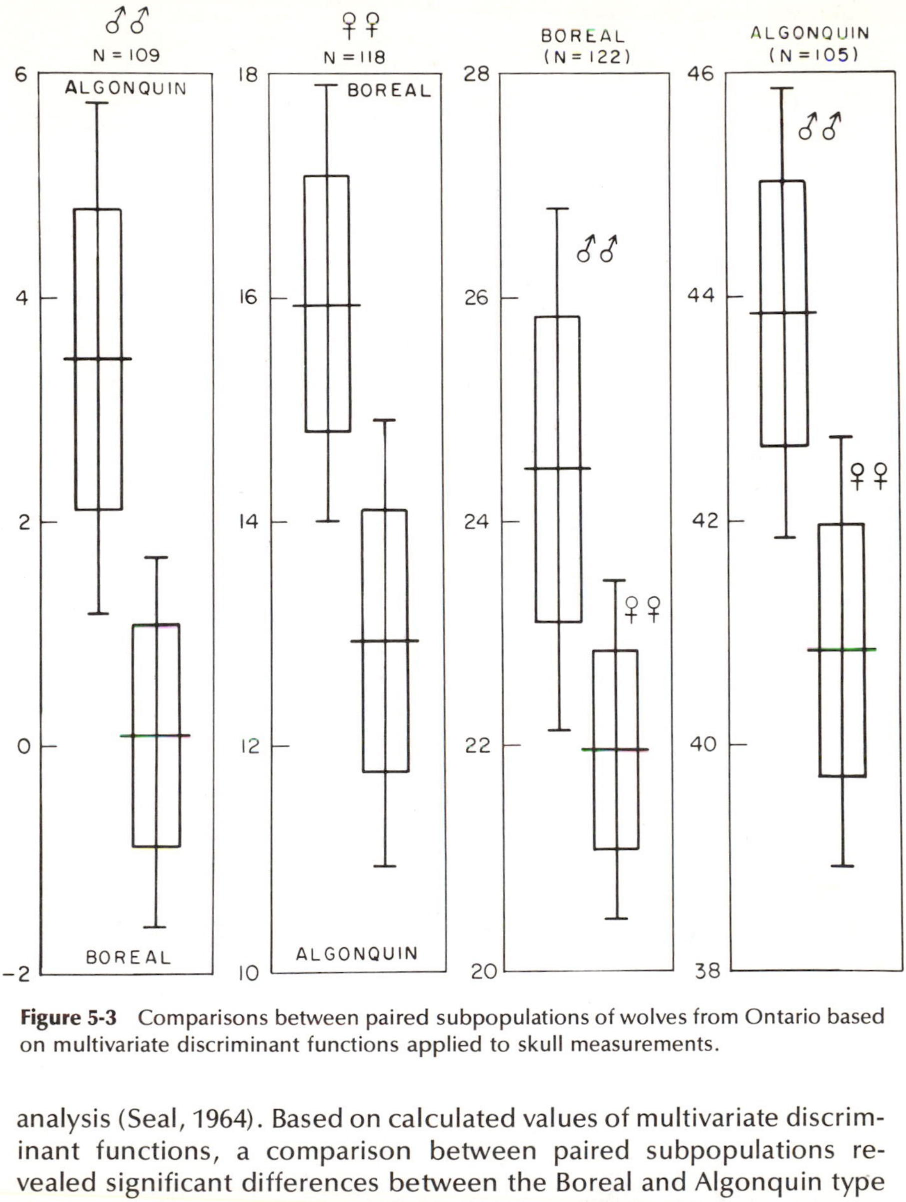

Figure 5-3 Comparisons between paired subpopulations of wolves from Ontario based on multivariate discriminant functions applied to skull measurements.

analysis (Seal, 1964). Based on calculated values of multivariate discriminant functions, a comparison between paired subpopulations revealed significant differences between the Boreal and Algonquin type males ($F = 7.721$, $df = 79$) and females ($F = 7.006$, $df = 88$). The degree of separation is shown graphically in Figure 5-3. The 75% limits represent the mean ± 1.15 standard deviations; the 95% limits represent the mean ± 1.96 standard deviations. The plot indicates that over 75% of males and females of the Boreal type could be completely separated from the males and females of the Algonquin type. These analyses also revealed significant differences between males and females within the Boreal type (F = 5.074, df = 92) and Algonquin type (F = 5.893, df = 75). Similar sexual dimorphism in wolf skulls has been recorded by Jolicoeur (1959) and Rossolimo and Dolgov (1965).

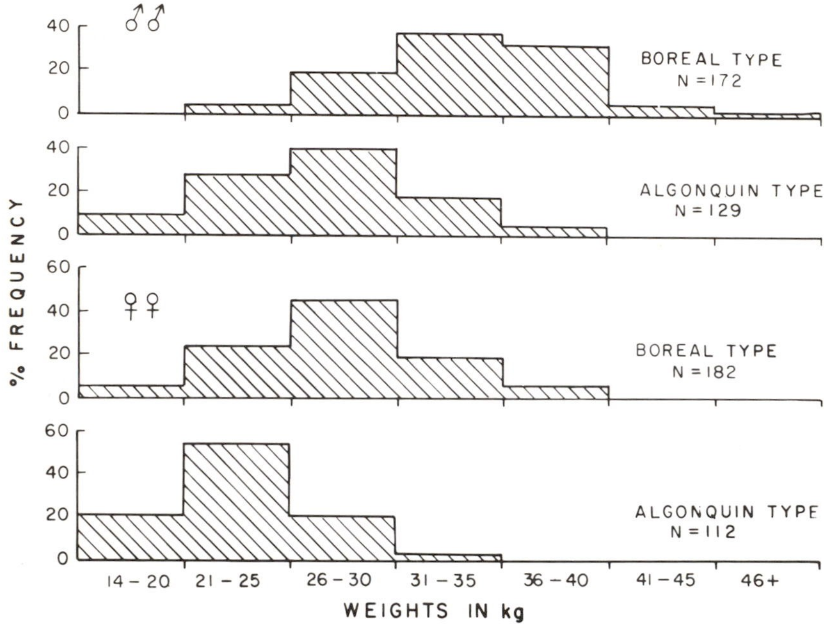

Figure 5-4 Frequency distribution of weights of adult *C. lupus lycaon* (Algonquin type) and *C. l. lycaon* (Boreal type).

In addition to skull measurements, whole weights of 594 adult wolves were available for comparison. Algonquin type males ranged from 17.9 to 38.5 kg and averaged 27.5 kg. Boreal type males were over 25% heavier and ranged from 21.8 to 50.3 kg with an average of 34.5 kg. Over 76.1% of the Boreal type males were greater than 30 kg, compared to 22.5% for the Algonquin type (Figure 5-4). Although weights greater than 79 kg have been reported for North American wolves (Goldman, 1944), wolves from Ontario weighed considerably less. Even in the larger Boreal type only 12.2% were greater than 40 kg and only 2.6% surpassed 45 kg. Boreal type females with a mean weight of 28.2 kg and a range of 14.5 to 39.5 kg were approximately 23.5% heavier than Algonquin type females. The latter averaged 23.5 kg and ranged from 14.7 to 33.8 kg. Almost 76% of the Algonquin type were 25 kg or less; less than 30% of the Boreal type fell in this weight range. Boreal type males were over 22% heavier than females, whereas the weight difference between sexes of the Algonquin type was slightly more than 17%.

A total of 1504 unskinned specimens, adults and sub adults, were placed in one of four color classes during postmortem examination. The classes were designated as black, gray-fawn, cream, and white.

Figure 5-5 Distribution, sample size, and percent occurrence of color classes of *C. lupus lycaon* (Algonquin type) and *C. l. lycaon* (Boreal type). Each circle represents the approximate center of the area of collection.

The Algonquin type was invariably gray-fawn, the color typical of most eastern wolves and coyotes. The Boreal type ranged from black to almost pure white, although the majority were gray-fawn. The incidence of cream-colored individuals was greatest in the northeastern and the northwestern regions of the province. Black individuals were relatively common in the northeast, but occurred less frequently in the northwest (Figure 5-5). Only 4 individuals were classified as white; 2 of these came from the northwest and 2 from the northeast.

DISCUSSION

We conclude that there is convincing evidence that the Boreal type and the Algonquin type are morphologically dissimilar in many respects. Although comparable data are not yet available for a similar assessment of the Tweed type, all evidence points toward its future recognition as a distinctive variety of the gray wolf.

The wolf evolved as a cursorial, pack-hunting predator capable of killing animals much larger than itself. Its lack of specialization and great adaptability enabled it to occupy most of the northern regions of both the Old and New Worlds prior to extensive settlement by man. Goldman (1944) considered it to be a living representative of one of the most successful land mammals that ever evolved. A wide variety of animal species, and occasionally plants, have been recorded in its diet, but it has always been primarily dependent upon ungulates. Generally, species that were most readily available were preyed upon most frequently (Pimlott, 1967). We can postulate, therefore, that the close association between wolves and their ungulate prey probably influenced the ecology and distribution of this predator.

During the beginning of European settlement in the 17th century, white-tailed deer (*Odocoileus virginianus*) were uncommon in the coniferous–deciduous forest zone of Ontario (Dawson, 1963, Cumming and Walden, 1970). Moose (*Alces alces*) were recorded in the extreme western and northeastern sections of the province, but were probably scarce in the interior (Peterson, 1955). The range of the woodland caribou (*Rangifer tarandus*) extended as far south as the upper Great Lakes (Peterson, 1955). With the development of agriculture and extension of logging operations in the middle of the 19th century, the habitat throughout eastern and northcentral Ontario became more suitable for white-tailed deer. Ultimately, the northern fringe of their range extended into the southern boreal forest. During this period, moose were extending their range north of the Great Lakes from the east and west, and woodland caribou all but disappeared from the eastern boreal forests. Within the last two decades there has been a reversal in the distributional patterns of moose and white-tailed deer; the former is increasing in frequency in the coniferous–deciduous forest and the northern limit of the latter has been moving southward. Undoubtedly, the patterns of change of predator–prey relationships proceed rather slowly, but much of the present variability in the gray wolves of eastern Ontario is probably related to changes in ungulate distributions.

The larger Boreal type probably invaded the province from the west and northeast following the retreat of the Wisconsin glacier. Its more massive skull and heavier weight suggest it has always been a predator on the larger ungulates such as moose, caribou, elk (*Cervus canadensis*), and possibly bison (*Bison bison*). Its size and variation in color indicate a closer affinity with the subspecies *C. l. nubilus* or *C. l. hudsonicus* than *C. l. lycaon.* Mech and Frenzel (1971) have suggested that the wolves of Minnesota are remnant populations of *C. l. nubilus*. Morphological differences may exist within the Boreal type at the extremes of its range in Ontario, but these remain to be investigated.

The smaller Algonquin type occurs mainly in the mixed coniferous–deciduous forests and preys principally on deer and smaller mammals (Pimlott, *et al.,* 1969; Kolenosky, 1972) but seldom on moose (Kolenosky, unpublished data). As white-tailed deer extended their range northward throughout the eastern Great Lakes region and into Ontario as far as the boreal forest zone, the Algonquin type probably followed. Therefore, it is possibly a much more recent inhabitant of Ontario than the Boreal type, particularly in the northern regions of the coniferous–deciduous forest.

As shown in Figure 5-1 the ranges of the two types overlap throughout a broad band across eastcentral Ontario, but there is no conclusive evidence of their interbreeding. Wolves taken from this zone can usually be assigned to one type or the other without difficulty by a combination of morphological characteristics. If the gray wolf is as adaptive as we believe, the characteristics which separate the Boreal and Algonquin types must be very deep-rooted. We can only speculate that in the areas where these types are sympatric, highly developed social and behavioral characteristics and adaptations to their prey have maintained an effective barrier against hybridization.

The origin of the Tweed type, a relatively small coyotelike wolf, is still obscure, but evidence is mounting that it originated from hybridization between *C. latrans* and *C. lupus lycaon* of the type described by Goldman (1944). Kolenosky (1971) recorded the successful mating in captivity of a female Algonquin type with a male coyote. Members of the F_1 generation are similar in appearance to many specimens of the Tweed type. A subsequent mating of F_1 siblings has produced a litter of 4 pups. Thus, there is conclusive evidence that hybrids from the coyote and gray wolf (Algonquin type) are fertile. At the time of writing, May, 1972, the F_2 litter is so young that morphological characteristics cannot be assessed.

It is perhaps significant that the Tweed type has occurred in the areas, as shown in Figure 5-1, where ecological niches occupied by coyotes or wolves of the Algonquin type are frequently closely interspersed. The origin of the Tweed type may, indeed, provide the clue as to the identity of the other small wolves in North America. In fact, the existence of this type in the wild and the success of the breeding experiments, described above, lead us to question the origin and identity of *C. l. lycaon*.

The evidence from Ontario, particularly the Great Lakes region, indicates that the morphology and ecology of wolves and the types which occur there are extremely complex and in our estimation very dynamic. Post-glacial origins and movements coupled with drastic changes in habitat and distributions of prey species in the province over recent centuries have left us with a confusing but interesting legacy. The relatively recent invasion of the coyote into the Great Lakes region has made the work of taxonomists even more complex (Lawrence and Bossert, 1967).

In the more northerly regions of Ontario where the habitat has not changed significantly, we have been unable to collect evidence of morphological or ecological instability in gray wolf populations.

All evidence leads us to conclude that human intervention of one kind or another has resulted in great morphological and ecological variations in wolf populations in the Great Lakes region of Ontario. We suggest that comparable changes may be taking place in other regions where major habitat alterations have occurred.

Acknowledgements We wish to acknowledge the value of the stimulating disagreements we have had with many of those who have worked on wolves in Ontario and elsewhere. Dr. C. S. Churcher, University of Toronto, has contributed greatly to our ideas about morphology and distribution. However, ideas would have remained only ideas if Mr. P. H. R. Stepney of the University of Toronto had not undertaken the tedious chore of measuring skeletal material and Dr. F. Raymond, statistician of the Research Branch, Ontario Ministry of Natural Resources, had not designed the program and supervised the analysis of data with great enthusiasm. We also appreciate the talents of our colleagues, the late Mr. G. Cameron (draftsman) and Miss J. Robinson (photographer), for their contributions to our research.

6
RELATIONSHIPS OF NORTH AMERICAN *Canis* SHOWN BY A MULTIPLE CHARACTER ANALYSIS OF SELECTED POPULATIONS

BARBARA LAWRENCE
Curator of Mammals
Museum of Comparative Zoology, Harvard
Cambridge, Massachusetts
and
WILLIAM H. BOSSERT
Gordon McKay Professor of Applied Mathematics and
Member of Department of Biology, Harvard
Cambridge, Massachusetts

Study of the *Canis* which have been moving into the empty predator niche in New England has shown them to be somewhat intermediate in size between wolves and coyotes, and with a blend of characters which is typical of neither. The area is one originally inhabited by *Canis lupus lycaon,* a subspecies which was virtually extinct there by the beginning of this century (Allen, 1942; Silver and Silver, 1969). It is also on the eastern edge of the coyote's expanding range (Young and Jackson, 1951). The question which then arises is whether the New England canids are coyotes which have evolved to occupy a more heavily wooded niche and to prey to a certain extent on deer or whether they are in fact hybrids, and if so, of what ancestry.

Generally speaking, in stable, well-established populations, extreme individuals are accepted as showing a normal variation for the form in question. In a possibly nonhomogeneous or rapidly evolving population, an assumption of identity for atypical individuals based on group membership is no longer valid, and identification must rest on precise diagnostic characters. In the case of wolves and coyotes, where they occur together in relatively undisturbed situations, they are not difficult to tell apart, primarily by size. With drastic changes in habitat or in actual numbers of individuals as has happened in parts of North America, identification becomes more complex; as one species replaces another, hybridization with a resulting blend of characters may occur. If coyotes have evolved to prey on larger animals, an increase in size might make them more wolflike.

METHODS

While size, whether overall or of teeth, is the most obvious character separating wolves and coyotes, a number of reasons limit its diagnostic usefulness for the present study. The following are particularly important. Eastern wolves, *Canis lupus lycaon,* are a small race (Young and Goldman, 1944) whereas eastern races of coyote, *Canis latrans frustror* and *C.l. thamnos* are large (Young and Jackson, 1951; Burt, 1946). Since males are larger than females, unsexed skulls are less distinct than those in which the sex is known. The red wolf, *Canis rufus,* whose range has been described as overlapping with both *frustror* and *lycaon,* is intermediate in size between them. Finally, within limits, closely related subspecies are more likely to be distinguished by differences in size than by pronounced differences in cranial proportions. *Canis familiaris* in size spans the combined range of coyote and wolf and the combination of differences which distinguish particular specimens of dog from the others varies individually.

For all of these reasons, before a systematic evaluation of the New England canids could be undertaken, it was necessary to determine whether, regardless of size or sex, differences of proportions could reliably separate wolves, coyotes, and dogs. In addition, red wolves were also compared with these species for phenotypical clues to how genetically distinct they might be.

Twenty specimens of each species were used. This number was selected as being just large enough to insure against the risk of a large fraction of the specimens being atypical and to give an estimate of population variability sufficient to make accurate species discriminations. The ability of a sample of this size to represent this intrapopulation variability will be demonstrated later. As with many statistical methods, the ability to discriminate populations increases as the square root of the sample sizes, so that increasing the precision of the discrimination by an additional order of magnitude (a decimal place) would have required samples 100 times as large. This was neither required nor justified for our particular study.

Sample selection was biased random to minimize the distance between species, and included both males and females. For wolf, large animals were deliberately avoided, for dog, the most wolflike and coyotelike were chosen, and for coyote, wide geographic distribution within its original range was used to include a variety of cranial types — from long and narrow to short and broad.

Since the desired result is to separate significantly the species in question, selection and numbers of characters to be used as well as method of comparison is critical. The use of too few characters, some of which may be overlapping or which may merely be expressions of an overall difference in size, does not furnish enough information. Ratios, which express two-way differences in proportion remove this difficulty but are liable to misinterpretation when the totality of the difference lies in the combined relationship of 3 or 4 characters. Separating populations by the mean and standard deviations of each of a series of measurements or ratios makes it difficult to determine whether groups of individuals cluster together or how distant such groups may be from other clusters.

Clearly, if the proposed analysis is to be useful as a basis for investigations of the ancestry of unknown animals, then sufficient characters to provide a wide degree of separation are necessary. Characters were not randomly selected. Out of an initial 42 (Lawrence and Bossert, 1967), 24 were tested for diagnostic value in pairwise comparisons. Since, as stated above, we wished to find out whether differences of proportion existed in addition to differences of size, all measurements were related to total length of skull. The mean and standard deviation of these characters, as a fraction of total length, were computed for each of the selected series. The value of each character in distinguishing a pair of species was tested by computing single character distances for the pair, dividing the difference in means for the two populations by the average standard deviation. From this analysis, 9 cranial and 6 tooth measurements were found to be most diagnostic, though no single character was found without overlap between a pair of species, and not all characters were equally important for distinguishing between all pairs.

We therefore turned to multivariate statistical analysis for methods of combining these 15 measurements. The technique of linear discrimination to make pairwise species comparisons was selected because it is fairly straightforward and resulted in a clearcut separation of the species involved. The method is that described by Kendall (1951). Jolicoeur (1959) gives an excellent graphical explanation of the technique. It consists basically of the calculation of a set of weights for each character, the discriminant function coefficients, such that the weighted sum of the characters, the discriminant function value, is most different for individuals of two species being compared.

Our results, not unpredictable, show that *latrans, lupus,* and *familiaris* are clearly distinct and that the two latter species are more similar than either is to the first (Figure 6-1). Discriminant function coefficients for these 3 pairwise discriminations are given in Lawrence and Bossert (1967). A measure of the distance separating the species, the Mahalanobis D^2 statistic (Rao, 1952), is shown in Table 6-1. This statistic is a multivariate generalization of the single character distance used in selecting important characters. It is the square of the difference in mean values of the linear discriminant function separating two species divided by the average variance of the values. All pairwise D^2 distances between the three species were statistically highly significant.

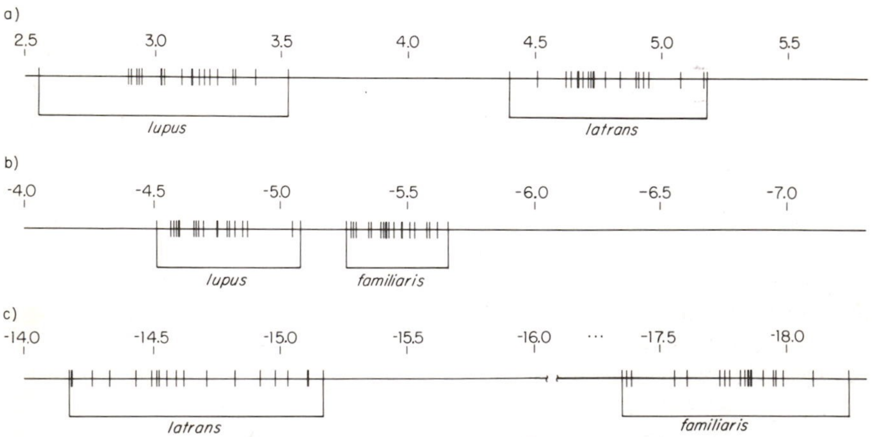

Figure 6-1 Linear discriminant values for pairwise generalized species populations: a) *C. lupus* with *C. latrans,* b) *C. lupus* with *C. familiaris,* c) *C. latrans* with *C. familiaris.*

TABLE 6-1 The generalized distance, D^2, between populations described in the text

	C. familiaris	C. lupus	C. latrans	C. latrans thamnos
C. lupus	27.2	—	—	—
C. latrans	119.9	64.1	—	—
C. latrans thamnos	—	—	6.71	—
New England *Canis*	44.55	29.84	26.83	9.12

A clear view of the degree of separation of the populations achieved by the discriminant functions results from the *a posteriori* identification of the original individual specimens using the functions. For each of the pairwise discriminations, the specimens were assigned to one species tentatively. A final identification was then made by assigning the specimen to that species for which two tentative assignments had been made. For example, if between *latrans* and *lupus* the specimen was assigned to *lupus,* between *latrans* and *familiaris* to *latrans,* and between *lupus* and *familiaris* to *lupus,* then the specimen was identified as *lupus.* In this way all 60 of the specimens were unambiguously and correctly identified; there was no overlap in the values of the various discriminant functions for the populations on which they were based.

An extension of the use of linear discrimination to distinguish between well-defined species is its application to the identification of hybrids. A series of known *C. latrans* × *familiaris* and a second series, the result of variously breeding the hybrids amongst themselves or back to dog, were used. Discriminant function values were calculated on the basis of the pairwise discrimination of *latrans–familiaris.* Both series were found to be intermediate between the parent forms and with a somewhat wider range of variation in discriminant function values than the parent species (Figure 6-2a). One specimen each of the F_1 series is within 3 standard deviations of each parent form; otherwise they cluster around a point midway between the two species. The F_2 series showed a larger proportion falling within 3 standard deviations of one or the other and their mean standard deviation tends a little towards *familiaris.* This suggests that on the basis of multiple characters the technique of linear discrimination can be useful for identifying hybrids between two known ancestors.

Additional comparisons involving a third species, *lupus,* were helpful in defining the characteristics of these known hybrids. Discriminations using the *latrans–lupus* discriminant functions showed that, in spite of the close relation between dogs and wolves, the hybrids as a

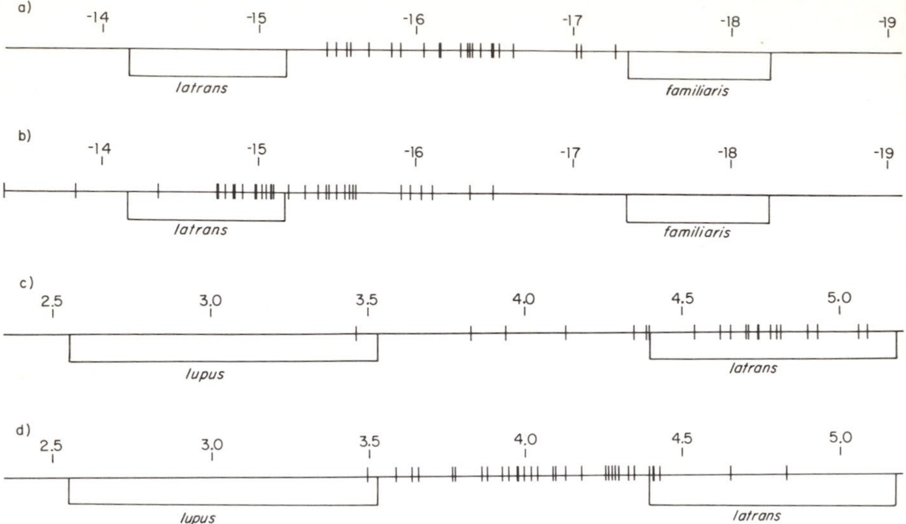

Figure 6-2 Linear discriminant values of known *familiaris* × *latrans* hybrids and of unknown New England *Canis* for the discriminant functions in Figure 6-1: a) known hybrids for *latrans–familiaris* discrimination, b) New England *Canis* for *latrans–familiaris* discrimination, c) known hybrids for *lupus–latrans* discrimination, d) New England *Canis* for *lupus–latrans* discrimination.

group tended to be coyotelike rather than wolflike (Figure 6-2c). When individual specimens are considered, the most doglike of the hybrids fall within the range of *latrans* in this latter discrimination, and the few specimens which are wolflike were exactly intermediate in the *latrans–familiaris* discrimination.

New England *Canis*

Studies of the New England unknowns were based on 22 animals, offspring of wild New Hampshire siblings raised in captivity for behavioral studies (Silver and Silver, 1969) and on 15 wild specimens from New Hampshire, Vermont, and Massachusetts. Discriminant function values calculated on the basis of 3 pairwise discriminations showed the population as a whole to be somewhat intermediate between *latrans* and *familiaris* on the one hand, while there was considerable overlap with *latrans* on the other. Even the most doglike is widely separated from *familiaris*. Compared with the known hybrids, on the *latrans–lupus* discrimination the population is more completely intermediate with less overlap with *latrans* and a number of specimens approach *lupus* closely. Figures 6-2b and 6-2d show these distributions.

The conclusions that can be drawn from these comparisons are that the unknowns differ from all three species and that they resemble coyotes more closely than the known hybrids do. They also are more wolflike. Since dogness in known dog × coyote hybrids seldom shows up as wolfness in a *latrans–lupus* discrimination, it is unlikely that the trend of the unknowns towards *lupus* can be attributed entirely to dogness in this series.

The New England animals are not an isolated population. Coyotes have been expanding their range eastward from the northcentral states for over 50 years. A reported increase in size, as in Michigan (Burt, 1946), or multiplicity of forms of *Canis* as in Ontario (Standfield and Kolenosky, Chapter 5) are apparently closely connected with this extension.

Comparisons between the New England unknowns and these intermediate populations have not been attempted, but linear discriminations of a series from Minnesota of *C. l. thamnos,* the easternmost subspecies of coyote, were made. Using the 3 pairwise discriminations described above, the discriminant function values of 32 *thamnos* were compared, not only with the original species, but also with the known hybrids and the New England animals. While the Minnesota population overlaps strongly with *latrans,* 5 individuals are intermediate towards *lupus* in the *lupus–latrans* discrimination and only 1 towards *familiaris* in the *latrans–familiaris* discrimination.

Furthermore, *thamnos* is intermediate between *latrans* and the New England population and, while it overlaps strongly with the former, it overlaps almost equally with the latter. The trend away from typical coyote and toward both *lupus* and *familiaris,* already apparent in the Minnesota *thamnos,* has progressed considerably farther in the New England population (Figure 6-3). This trend is supported by the D^2 distances shown in Table 6-1.

These comparisons suggest the possibility that the divergence from the typical coyote pattern, culminating in the east is perhaps, in part, caused by some mixing with wolf as well as with dog stocks. Further evidence is provided by the rather high degree of variability found in the two populations. The multicharacter variance, the spread of a population in a multidimensional scatter diagram, was represented by the sum of the principal components of variation (Rao, 1952). In this calculation, the variance along the direction of greatest spread in the diagram is added to the variance along the direction of next greatest spread at right angles to it, and so on until a number of directions equal to the number of characters have been added. This multicharac-

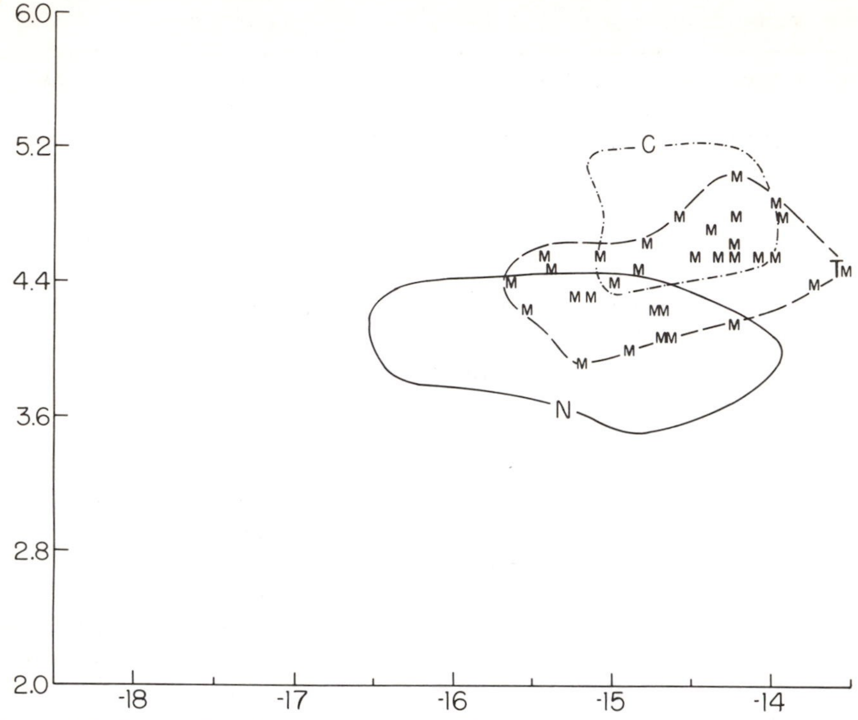

Figure 6-3 Range of the linear discriminant values of *latrans* (dotted), *latrans thamnos* (dashed), and New England *Canis* (line) for *latrans–familiaris* discriminant function (abscissa) and *lupus–latrans* discriminant function (ordinate). Individual values for *latrans thamnos* specimens are shown (M), from Lawrence and Bossert, 1969.

ter variance for the population is shown in Table 6-2. Note that when coyotes are compared with wolves and dogs, the within group variation is relatively small, and the between group distances are relatively large, whereas the reverse is true when *thamnos* or the New England *Canis* are so compared.

In summary then, although the multivariate analysis does not provide definite proof of the genetic composition of the New England population, a number of points may be deduced from it. The differences between the New England population and known dog–coyote hybrids are sufficient to show that the former are not coy-dogs. It establishes that they are in fact predominantly coyote, and that they are not a purely local phenomenon but are instead an extreme example of a progressive change that had already begun in the eastern portion of the coyote's extending range.

TABLE 6-2 Multiple character variability within populations: A measure of the scatter of the cranial proportions in multivariate space

Population	Sum of Principal Axes of Variation ($x\ 10^3$)
C. familiaris	3.40
C. lupus	1.00
C. latrans	0.66
C. latrans thamnos	1.28
New England *Canis*	1.69

Canis rufus

In the southeastern and southcentral parts of the United States, a situation analogous to that in the Northeast exists in that hybridization has been postulated (McCarley, 1962; Paradiso, 1968) in areas where coyotes are replacing wolves. The wolves in question have been described as a distinct species, *Canis rufus,* with a range which, in its easterly and northeasterly part, originally overlapped that of *Canis lupus lycaon*. In the absence of precise comparisons between these species, it has been assumed that the differences separating them are greater than those distinguishing other races of *lupus,* and are in fact large enough to warrant specific separation.

An initial study (Lawrence and Bossert, 1967) of 12 specimens of *Canis rufus* from Florida to eastern Louisiana raises some doubt as to the validity of this assumption. The specimens, here referred to as *floridanus,* which had been collected from well outside of the range of *Canis latrans,* were tentatively identified using the discriminant functions described earlier. All fell within the *lupus* category. They were on the whole less coyotelike and less doglike than the original *lupus* population (Table 6-3). As discussed above, this is not by itself conclu-

TABLE 6-3 The generalized distance, D^2, between populations described in the text

	C. latrans	C. lupus	C. familiaris	C. lupus lycaon
C. lupus	64.1	—	—	—
C. familiaris	119.9	27.2	—	—
C. lupus lycaon	69.5	10.0	66.6	—
C. (rufus) floridanus	116.0	20.3	87.6	56.0

sive because this use of linear discrimination tacitly assumes the specimens to be from one of these groups. To further test the possibility that this population might be a distinct species, the study was expanded to include a series of typical, though northern, *Canis lupus lycaon.* Discriminant function coefficients and D^2 values (generalized distance between populations based on morphological features) were calculated for all pairs of the five populations. Using these and the cluster grouping technique of Rao, the *lycaon* and *floridanus* populations, though fairly distinct, form a cluster with the selected *lupus* population. When the latter is compared with *lycaon* and *floridanus* using the discriminant coefficients that best separate these two, it is found to be intermediate between them. Further, *lycaon* and *floridanus*, though separable, agree in being less coyotelike than the generalized *lupus* (Table 6-3). Of the 8 most significant single characters separating *lupus* and *floridanus,* 1 is found to be more extreme in *lycaon,* and in 5 more, *lycaon* trends away from *lupus* in the direction of *floridanus*.

While the early, eastern populations of *rufus* seem clearly not more than subspecifically distinct from *lupus,* it is by no means correct to extrapolate from this that specimens identified as *rufus* over its entire range described are *lupus*. In the lower Mississippi Valley, a more complex situation exists. Wild *Canis* of some sort still occur there in what was once wolf territory. Presently, a number of investigations (behavioral, ecological, serological) are characterizing these modern populations. Understanding their genetic composition depends also in part on a knowledge of their relationships to the wolves which originally occurred there. For this reason, a comparison of a population from Arkansas called *rufus gregoryi*, with *floridanus, lycaon,* and generalized *lupus* was made. Thirty specimens, collected for the most part in the early 1920s and none later than 1928, from south of the Arkansas River and east of the early range of coyotes (Gipson, 1972) were used. Discriminant function values, D^2 distances, and multicharacter variances were calculated, as described above. The results (Figure 6-4, Table 6-4) show them to differ less from centralized *lupus* than either *floridanus* or *lycaon.* Further, when the three forms are compared with each other, *lycaon* and *floridanus* are seen to be farthest apart with *gregoryi* about equally distant from both. Geographically, this means that the population north of the Great Lakes, the *lycaon* series, is more like the Mississippi Valley population than it is like the southeastern one.

Since there is general agreement that today hybridization is widespread in Arkansas (Gipson, 1972), the homogeneity of this early population was compared with that of *lycaon, floridanus,* and the

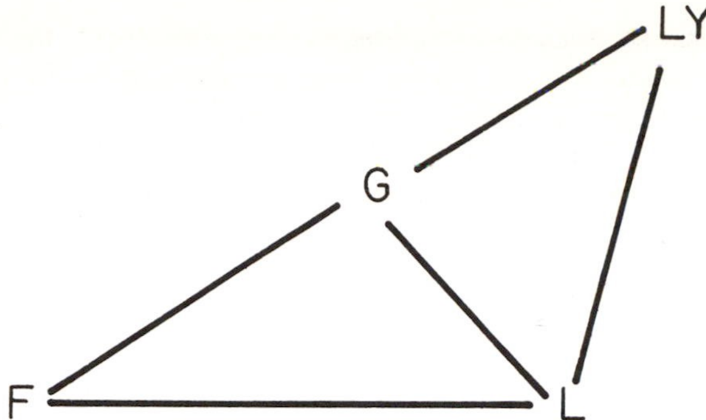

Figure 6-4 Relative distances (square root of D^2 statistic) between generalized *lupus* (L), *l. lycaon* (LY), *r. floridanus* (F), and *r. gregoryi* (G). The ability to express the pairwise species relationships without inconsistencies in a plot such as this is not a necessary property of the D^2 statistic, but was possible in this case because of similarity in variances in the various populations.

generalized *lupus*. Using as our measure the sums of the principal axes of variation, the variance so calculated is seen to be low in all of them (Table 6-5) relative to the New England *Canis* (Table 6-2). It is in no way suggestive of the extensive hybridization there. Also included in Table 6-5 is the multicharacter variance for an additional population of 33 individuals of *gregoryi* from north of the Arkansas River in Arkansas and Missouri. This additional population differs little from that south of the Arkansas River in its D^2 distances to other populations studied and is only slightly more variable. Note that when both populations of *gregoryi* were combined, doubling the sample size, the total variance was only a few percent different from the average of the two subpopulations. This indicates that the smaller sample sizes are adequate in representing this variance for the purposes of this study.

Within this rather homogeneous, *lupus*-like population of *rufus gregoryi*, a few specimens are seen to be coyotelike (Figures 6-5 and 6-6). None is as extreme as 2 Fallsville specimens studied in an earlier

TABLE 6-4 The generalized distance, D^2, between populations described in the text

	C. rufus gregoryi	**C. lupus**	**C. rufus floridanus**
C. lupus	6.5	—	—
C. (rufus) floridanus	13.2	20.3	—
C. lupus lycaon	10.5	10.0	56.0

TABLE 6-5 Multiple character variability within populations described in the text

Population	Sum of Principal Axes of Variation (x 10^3)
C. *(rufus) floridanus*	0.68
C. (rufus) gregoryi (south of Arkansas River)	0.84
C. (rufus) gregoryi (north of Arkansas River)	0.88
C. (rufus) gregoryi (joint population)	0.91
C. lupus lycaon	0.61

paper (Lawrence and Bossert, 1967), but 2 additional skulls from Yell and Saline counties are consistently more coyotelike than the rest of the series. No effort is made here to interpret these extreme individuals, and their inclusion in the series analyzed does not mask the close resemblance of the population as a whole to *lupus*.

While the North American wolves studied cluster in a rather close group, this only sheds light on part of the problem. It is also interesting to consider the relationships of Old World *lupus* with *rufus* as well as with *lupus* of the New World. For this purpose, a series of 11 specimens of the small race, *Canis lupus pallipes,* from India was compared with all the New World series used in the present study. In agreement with the presently accepted relationships of *lupus*, the D^2 distances between *l. pallipes* and both *lycaon* and the generalized *lupus* are small, and in fact the populations are not significantly distinct in this statistic (Table 6-4). It is noteworthy, also, that the Arkansas *gregoryi*, supposedly belonging to a different species, is even less distant from the *lycaon* and the generalized *lupus* than is *pallipes*. Within-group variance in *pallipes* and in the Arkansas series is the same (Table 6-5). In sum, the evidence suggests that these *gregoryi* are no more than subspecifically distinct from *lupus*.

It should be emphasized that multivariate comparisons of this sort produce precise answers for the specific population studied: in this instance, populations which no longer exist. What is true of these *floridanus* and *gregoryi* may or may not be true of other populations elsewhere — western *rufus*, for instance. Nor does it hold that what was true of populations in an area 50 years ago is equally true today. Far from it, where wolf populations have been sharply reduced, coyotes or feral dogs have repeatedly occupied the empty niche. While it is understood that under such circumstances interbreeding may take place, it has been difficult to study the extent of this because the degree of distinctness of the possible parent stocks and the cranial characteristics of hybrid populations needed further investigation.

Reliance on cranial evidence is of course not new. Over the years it has been found suitable and consistent for classifying *Canis* on specific and subspecific levels; the work discussed here is essentially an extension and refinement of such previous investigations. Even the measurements, carefully tested for significance, express characters which for the most part are those recognized and used by earlier workers. Typically, cranial proportions were found to be more significant for separating the forms studied than were tooth dimensions when size was not used as a character. Further, such tooth dimensions as were used, for instance, width of M_2 and of the upper canine, and proportions of P_4, were only of occasional significance. Confirmation of the value of the measurements selected for this analysis lies in the fact that

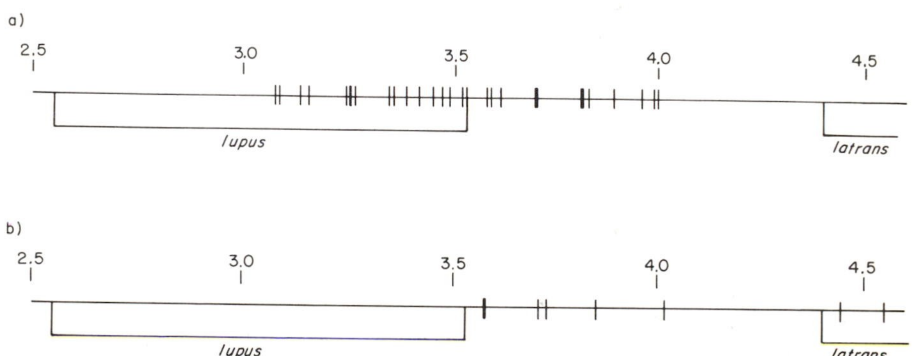

Figure 6-5 Linear discriminant values of *r. gregoryi* for *lupus–latrans* discriminant function: a) population from south of the Arkansas River, b) population from Fallsville, Arkansas.

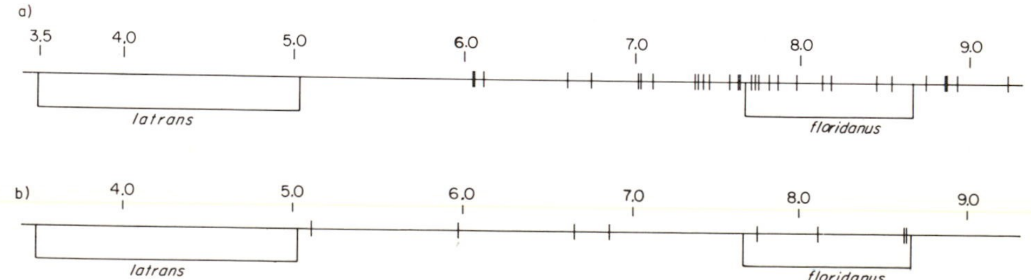

Figure 6-6 Linear discriminant values of *r. gregoryi* for *latrans–r. floridanus* discriminant function: a) population from south of the Arkansas River, b) population from Fallsville, Arkansas.

they express differences which have long been considered diagnostically important for the forms studied.

With the techniques described here, it has been possible to combine a sufficient number of these measurements to eliminate reliance on characters which singly or in ratios are overlapping. It has been possible also both to express relative distances between populations and to evaluate individual specimens in relation to populations whose characteristics have been defined.

Since much of the present ecological, behavioral, and serological work on *Canis* is being carried on in areas which have been to some extent disrupted, interpretation of the results depends heavily on a precise knowledge of the identity of the animals being investigated. Cranial characters, being a common denominator for existing and vanished populations, are the most useful means of relating past and present.

7
THE ORIGIN OF THE DINGO: AN ENIGMA

N. W. G. MACINTOSH
Challis Professor of Anatomy
The University of Sidney
Sidney, Australia

The greatest problem in trying to write about the dingo is that one has no proof of the animal's identification, ancestry, affinity, place of origin, or precise time of arrival in Australia; even the questions of whether it was on arrival a partly-domesticated animal, subsequently becoming feral, or whether it is truly a wild dog, is disputed by many writers, and on almost completely polemical grounds.

There is a most extensive literature directly or indirectly related to the dingo, most of it subjective and legendary, little of it scientific.

It is frustrating to reflect that after 23 years of admittedly sporadic and desultory efforts together with occasional bursts of intensive continuous work, one is not much closer to answering the questions one set out to explore.

At the risk of oversimplification, some attempt will be made to briefly summarize the work of various researchers, especially those who have used the dingo in their comparisons of prehistoric canid remains.

Coon's (1951) Iran canid of 11,480 years B.P. if proved to be dog would be the most ancient so far discovered. The Starr Carr bones of 9500 years B.P. (Degerbøl, 1961) are the most ancient domesticated dogs so far proven.

The dingo (*Canis antarcticus,* Kerr, 1972 — see Iredale, 1947) is the major candidate for close similarity with the largest number of geographically widespread fossils and recent canids, i.e., the pariah, the Tengger dog, *C. f. poutiatini,* the Senckenberg dog, the Starr Carr limb bones (these being like *C. f. poutiatini*), *C. f. matris optimae, C. l.*

pallipes, and perhaps *C. l. hodophylax.* As the dingo and pariah are considered to be so similar, the pariah presumably must have an equal range of similarities, but the dingo is far more commonly mentioned in comparisons. (Figure 7-1).

Figure 7-1 Compare this dingo pup with Zeuner's (1963, Fig. 4:13 p. 91) photo of pariah pup stated to "closely resemble dingoes in many respects." Note the very different fore and hindlimb proportions and shape and the stance. Note also the different shape and placement of the ears, the lesser frontonasal curve in the dingo, and a totally different tail.

As the dingo and *C. f. poutiatini* are considered so very generalized, they must be close at least to the original domestic dog, and Degerbøl (1961) goes so far as to refer to the Senckenberg dog as a European Mesolithic dingo. This would seem to argue a good case for Dahr's (1942) views of a generalized dog, spread Eurasian-wide in early Quaternary times, and descended neither from wolf nor jackal; in other words, a *C. ferus* as preferred by Studer (1901) and Zeuner (1963), but denied by Degerbøl who apparently elects northern wolf ancestry for the Starr Carr and Maglemose dogs. Clutton-Brock (1963) agrees with Degerbøl, but adds that *C. l. pallipes*, the Asiatic wolf, is clearly the ancestor of the dingo and the pariah and that the dingo alone is purely derived from the Indian wolf. Her opinions would appear to be influenced not only by John Hunter and Darwin as she herself says, but also by Degerbøl and by Studer's original division of northern and southern stem groups of canids. From an Australian viewpoint, the most stimulating opinions are those of Studer (1901, 1905), Prashad (1936), and Tichota (1937) on the Tengger dog of Java and Studer's view that it is the ancestral form of the southern stem dogs, the dingo being ancestral to the half wild pariah.

Clearly there are diametrically opposed views and, apart from what appears to be agreement that the dingo is probably the purest and most generalized type of living dog with geographically widespread resemblances, there is certainly no unanimity as to its ancestor. To give a definitive opinion now without a great deal more archaeological revelation, particularly from Asia, and without a much more intense personal examination of all extant canid fossils would be presumptuous.

FOSSILS FROM AUSTRALIA

McCoy (1882) described and figured part of a left mandible and its "first true molar," and part of a right maxilla containing 5 teeth. The mandible had been found in 1840 in Lake Colungulac deposits, 80 miles southwest of Melbourne, together with bones of recent and extinct animals including the first example of *Thylacoleo carnifex* recorded by Owen of the British Museum. Red ferruginous mineralization appeared to be similar on all these bones. The maxilla together with some very juvenile dingo skulls and other animal bones came from the Gisborne bone caverns believed then to predate the basalt flows of Mt. Macedon some 40 miles north of Melbourne; these bones also were thought to be contemporary. McCoy began writing on these in 1857 in the *Argus,* a Melbourne newspaper. The presumptive evidence plus McCoy's observations that the dingo is singularly averse to domestication and to man's society, shows little or no variation over the whole of Australia, and is increasingly numerous towards the desert interior remote from man convinced him that the dingo was an indigenous animal of early post-Tertiary time preceding man's arrival.

From measurements of the teeth and jaw fragments, he claimed the Colungulac specimen to be perceptibly more robust than the modern variety, but the Gisborne specimens identical with living dingoes. He referred also to large carnassial ratios in these fossils.

By courtesy of the Melbourne Museum I have measured these specimens with vernier calipers. The mesio-distal diameters of the

Gisborne teeth are right upper P^2 10.7/mm, P^3 11.9mm, P^4 17.38/mm, M1 10.88/mm, and the Colungulac left lower M^1 23.1 mm, all within the range of modern dingoes. Colungulac Coxiella shells recorded 13,700 ± 250 B.P., but fluorine–phosphate ratios suggested the Colungulac mandible was younger than a female aboriginal skeleton found there, but older than the European advent.

Etheridge (1916, Pl. 10–12) illustrated 3 of 7 *Canis* teeth from the Wellington Caves, N.S.W., which included those earlier written about by Krefft (1865), and said they were "of a dog somewhat superior in size to the warrigal." Masses of extinct marsupial remains have been removed from these caves, and the 7 *Canis* teeth were found among this stored material. My remeasurement of these also puts them within the modern range. Gregory (1906) observed similar association in deposits at Lake Eyre.

Each of these four authors elected to regard the dingo as indigenous. Modern studies have shown that these various deposits cannot be stratigraphically identified; while the dingo and some fossil marsupial remains may be contemporary, the extinction of some of the species occurred much more recently than formerly believed.

McCarthy (1964) recovered teeth fragments from Site 3, Layer D, Section 6, Bondaian Phase, depth 19–24 in., and so appreciably more recent than a date B.P. of 3623 ± 69 for charcoal at a lower level, at his Capertee Valley excavation on the dividing range southwest of Sydney. At his request I reassembled these fragments, producing 9 dingo teeth from a single postmature animal, some complete enough to measure. They are large but within the modern range.

The late Drs. T. D. Campbell and P. Hossfeld excavated a rock shelter at Mt. Burr in South Australia in 1963–4. Their notes are being written up by Mr. R. Edwards. Teeth and jaw fragments claimed from between 8600 ± 300 and 7450 ± 270 B.P., are said to have been identified as dingo.

Dortch and Merrilees (1971) in salvage excavation at Devil's Lair, a limestone cave near Cape Leeuwin, Western Australia, found a single worn canine tooth of *Canis* in disturbed material and there is the possibility of its incorporation from surface litter. If in context, it is not less than 8500 years.

Macintosh (1964) extracted and reassembled an almost complete juvenile male dingo from a paraffin-encased mass of sand 33 × 27 × 13 cm excavated *en bloc* from a stratum between 3170 ± 94 and 2950 ± 91 B.P. of Mulvaney's excavation near a bank of the Murray River in South Australia. The dingo had died curled up on its right side and the stratigraphic evidence indicates it had taken refuge in the shelter during an hiatus in aboriginal occupancy, rather than that it was in any symbiotic or domestic association with the Aborigines.

In 1964, I thought dingo and domestic dog had approximately similar time patterns for tooth eruption although I knew the order of epiphyseal closure was different, occasionally very irregular, mostly earlier in the dingo. Influenced by the teeth, I compromised with an age for the animal at death of 23 to 27 weeks. Its teeth are indistinguishable from and are within the size range of modern dingoes and with which its skeletal indices coincide, but its skeletal frame is absolutely small for such a personal age. The cingulum on the upper first molars is scarcely reduced, certainly not to degree illustrated by Jones (1925, Fig. 233). This led subsequently to a study of tooth eruption in conjunction with detailed epiphyseal closure in our collections, as a result of which I now have to revise the personal age of the Fromm's dingo to an

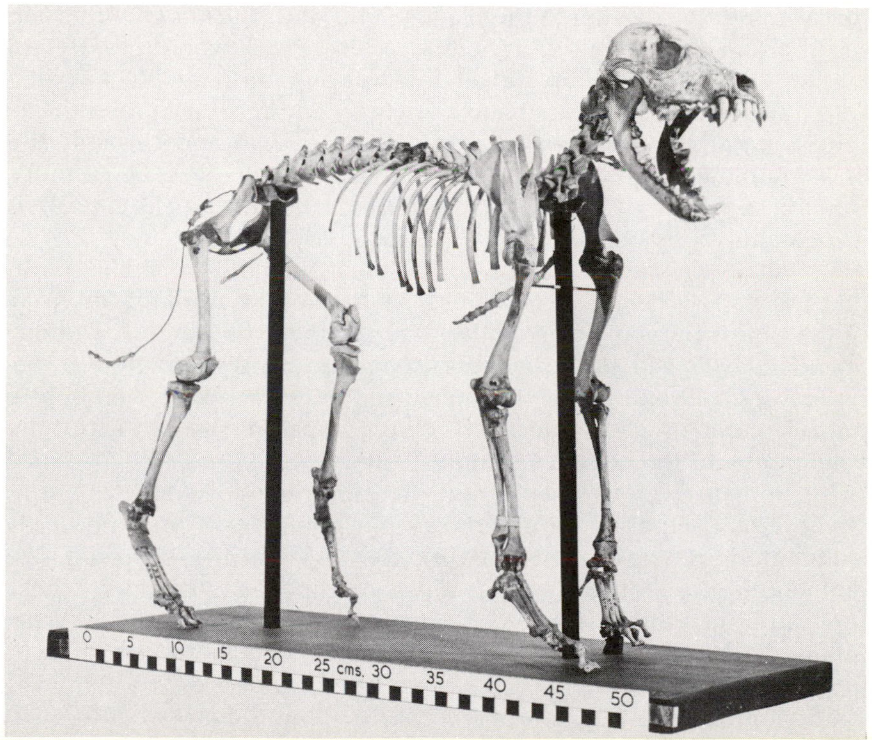

Figure 7-2 3000 year old male dingo. Its personal age at death was 18 weeks. Teeth, cranial characters, postcranial skeletal metrics, and indices are within the ranges of a dingo colony of 5 generations bred at the University of Sydney. This relic proves that the dingo morphological pattern has remained unchanged over 3000 years.

estimate of 18 weeks, which makes its total size for personal age similar to modern dingoes. This relic proves that dingo morphological pattern has remained unchanged for 3000 years (Figure 7-2).

Tate (1952) described 6 dingo crania from far north Queensland. He agrees with Jones (1921, 1925) and Tichota (1937) on the cranial uniformity of the dingo, but like Tichota he disagrees about the value of M^1 cingulum reduction as a dingo criterion.

Longman (1928), referring to 10 dingo crania in the Queensland Museum, agrees with Jones, Tate, and others on the many cranial and dental differentiations of dingo from the red wild dogs (*Cuon*) of Southeast Asia. But he notes that skin of the Indian wild dog from the Nilgiri Hills was not differentiated from other flat skins of dingo by local naturalists. That helps to explain why so many unreliable reports are made about the plastic features of coat color, tail, and ear carriage of canids seen in the Australian bush. Of equal interest is a rather casual reference by Tate to the black tail tip of *C. l. pallipes.* Although this is also a plastic feature, no dingo, even a black whole coat dingo, fails to have a white tail tip and all dingo pups at birth are black coated with a white tail tip and feet and perhaps a white nape of the neck.

In summarizing this section, the earliest proven and certainly the only complete dingo skeleton recovered from Australia is the Fromm's Landing specimen of 3000 years B.P. This male aged 18 weeks at death is inseparable from living dingoes of comparable personal age.

The dingo remains at Devil's Lair (not less than 8600 B.P.) and at Mt. Burr (perhaps 8000 B.P.) if confirmed, are not very short of the antiquity of the oldest proven domesticated dog (Starr Carr 9500 B.P.). Dubiety attends Devil's Lair because there is only a single tooth from a disturbed deposit, and at Mt. Burr because the two workers died before preparing papers and some queries were apparently raised about the stratigraphy of the charcoal samples collected.

The remains from remanié deposits described by McCoy, Krefft, Gregory, and Etheridge are certainly morphologically and metrically within modern dingo ranges and probably less than 2000 years old. The deductions about an indigenous dingo of Pliocene antiquity by these four authors were repeated in many mammalogical texts. What influence this may have exerted on European palaeontologists is an interesting speculation.

Certainly the view of Jones (1921, 1925) about the arrival of the first Aboriginal with his wife and his dog and his dog's wife has been widely accepted, being mentioned by Degerbøl as recently as 1961, and the delightful sketch in Lorenz (1954) of a canid happily gazing ahead from the prow of a paddled canoe is suspiciously similar to the Jones concept. (In passing, my experience is that it is a long time before anything in the nature of a wild dog will adopt that pose in a canoe or raft. It curls up defensively as far from the human occupants as possible, but if it can be placed aboard, and the craft pushed off, it will not jump overboard.)

My attempt to study the dingo began in 1949 solely as an ancillary tool towards the elucidation of the region of origin or racial derivation and affinity of the Aboriginal Australians. Speculation about the latter began very early in the 17th century and an enormous literature has not resulted in definitive conclusions; instead, three major hypotheses and variables of these are ineluctably evocated by their respective adherents.

Stimulated by the logical deductions of Jones (1921), I had the naive thought that if one could accept his postulation that the dingo was brought by sea by the first Aboriginal colonists, a search for dingo affinity and origin might solve both problems. Certainly his identification of the true northern wolf as ancestor of the dingo did not promise very much; one wanted some intermediate similarity to the dingo from regions geographically nearer Australia, i.e., India, Malaya, the Indonesian Islands, and Southeast Asia generally, all in which I had spent some time. The first disappointment was that the animals which superficially looked something like dingoes were either dholes (*Cuon alpinus*) or jackals (*Canis aureus*) or pariahs (*Canis indicus*) in decreasing order of similar appearance and behavior. The final diappointment was the realization that the problem was like that of the Aboriginal Australian, but appreciably more difficult. By that time, however, I had covered a lot of ground, acquired little new knowledge, and had become sold on lesser ambitions about learning the truth of some aspects of the dingo.

I had first looked through the records of the early navigators and explorers. In the main, these men were devoted diarists and dedicated to objective recording. Jones' comment (1921) that the "first white men who came in contact with the dingo remarked that both black dogs and red ones were common" is more than amply borne out, and he could have added white, cream, yellow, tawny, rust, brown, etc. For example, Sir Thomas Mitchell describes a black dingo in the region of the Murray and he was the first white man to traverse the area. Also, a skin in the West Australian Museum in Perth obtained from the western edge of the Dead Heart in 1876 was of a brindle parti-color described by some observers (Figure 7-3).

The next phase was an attempt to gather information from all the daily or weekly newspapers of the major cities and of country towns back to their inception. Library indices helped in this and led me then into pastoral, and stock and station journals. The derived information

Figure 7-3 Pastoralist Adamson and Day, in Gibson Desert W. A. examining the skeleton of a trapped dingo. Some 300 skeletons were collected across the continent. The remains of hide on many of the specimens helped to plot coat color, which matches the environs; i.e., cream, pale yellow, and white on Nullabor Plain, black in shale country, rich red-brown in the eastern section of the mountains, rusty-red in Arnhem Land, and yellowish-brown in the Gibson Desert, where an enclave of dappled and brindle dingoes ("pretty dogs") occurs. All these colors plus very rare piebald and skewbald varieties were noted by early explorers.

about dingoes was so totally contradictory that I drew up a questionnaire and distributed it to graziers, farmers, pastoral committees, vermin boards, and doggers and trappers continent-wide. I should have asked 20 questions instead of 200; nevertheless, some 30% replied. The conflict of opinions was again apparent and it was obvious I would have to look for myself. Obvious starting points would be the areas from which people had generously replied to the questionnaire. The fulminations of Jones (1925) about the impossibility of countering the belief that a pure bred dingo scarcely exists because of crossing with station dogs is just as applicable today. Also, "to say anything in favor of the hated wild dog is treason in Australia" still holds with a majority of outback people who regard it as a "treacherous, intelligent, cunning, ruthless killer" of sheep, calf, or poultry and a few are even prepared to believe dingoes would attack a helpless man. Other creatures, of course, are also victims of unreasoned killing. Hornadge (1972) has just published *"If it Moves, Shoot It,"* an indictment of the slaughter of the kangaroo; but the dingo has infinitely less chance of being taken off the scalp-bounty list, let alone having conservationists pleading his cause.

Meanwhile, however, I had advertised I wanted to purchase dingoes given some evidence of their remote and isolated derivation. From the mountain range west of Gloucester some 170 miles north of Sydney I collected a litter of 1 male and 3 female very young pups captured after their mother and father had been shot. Parents and litter were identical in rich reddish brown whole coat color. From these 4 pups, 5 successive generations totalling 42 dingoes were inbred in captivity at the University of Sydney. (Figure 7-4)

Figure 7-4 Litter of 6 pups. Color at birth is black except for a white tail tip, white feet, and sometimes a white patch on the nape of the neck. Note the pendent bushy tail and the cranial sagittal crest of the mother, as well as her flattened, retracted ears, a response to the stroking of her mandible while facing her. The alleged stigmata of domestication did not appear in any generation of our colony: all were in the upper size range of wild dingoes.

ATTEMPTS AT DINGO DOMESTICATION

The experience of Lorenz (1954) with his dingo, which "harbored the warmest feelings . . . but submission and obedience play no part in these feelings," is totally and uniformly characteristic. While I have

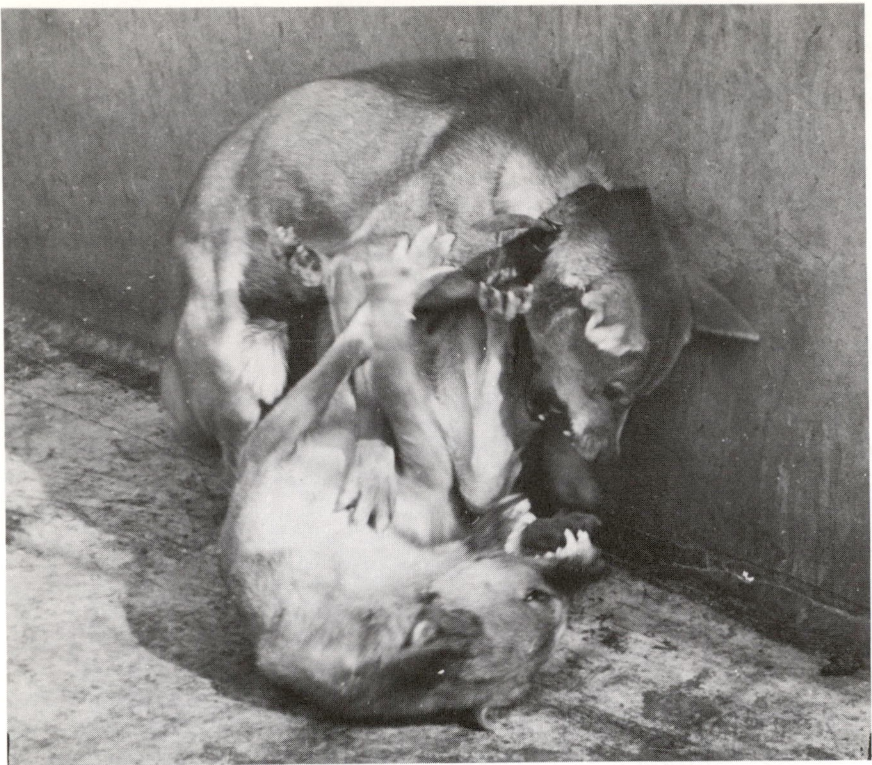

Figure 7-5 Two dingoes of our colony engaged in premating fury. Note the crumpled right ear of the male, damaged in a fight during his youth. There is no anatomical or physiological barrier to crossing with the domestic dog, but depending on relative population density of the two, the dingo exhibits a preference barrier, and the domestic dog, including the Alsatian, is often killed.)

met a number of people who claim and truly believe they have successfully domesticated a nonhybridized dingo, checking out their story and the animal's behavior reveals that they are either self-deluding, or anthropomorphizing which largely amounts to the same thing.

An interesting experiment by kennel clubs in Holland, with some backing from Dr. A. C. van Bemmel, Director of the Rotterdam Zoological Gardens, parceled out individual dingoes to individual families. In August, 1965 I visited the home of Mr. Jon Th. Vis in Enschede who had one of the dingoes. The affection of Jon and his wife for the dingo and the dingo for them was obvious. The physical condition of the animal was excellent, but Jon was very concerned for the animal's nervous state and his own admitted failure to inspire confident acceptance by an animal quite clearly not living in its natural milieu. Talking subsequently with Dr. van Bemmel, it seemed that up to that time, all experiences were like those of Lorenz. Police dog trainers in Sydney some 20 years ago, even with their great experience, patience, and affection for animals, failed to produce anything resembling obedience. (See Figure 7-5)

Many accounts exist of Aboriginal and dingo relationship. Roughly, these are divisible into half alleging Aboriginal use of the dingo as an aid to hunting, and half alleging the uselessness of the dingo for any purpose at all to the Aborigines. Zeuner (1963) accepts one such story of Aborigines tethering a pup and releasing it when it has got used to the human environment. In my experience, I have never found an Aborigine who made such a claim. The dingo would simply take off once released, and return only to steal food.

Gould (1970) in November, 1969 visited a small group of Aborigines in the heart of the Gibson Desert in Western Australia. Most had not previously seen Europeans. Nineteen dingoes were counted in and around the camp. Gould therefore had a unique opportunity to look into the past; equally important, he is biologically and ecologically oriented as well as being an anthropologist and an archaeologist. The dingoes were often fondled but rarely fed, and could therefore only exist by whatever their own personal hunting provided and by invading and stealing anything in the camp. Gould says that in this never ending battle with these "skinniest dogs I have ever seen . . . compulsive cringers and skulkers . . . the dingoes' persistence inevitably won out." The men setting out to hunt constantly drove the dingoes back to camp. As the Aborigines hunt from concealment, the dingoes are a liability. What then for Aborigines and for dingoes was the point of propinquity? Answer: sleeping huddled together for warmth in the near freezing desert nights. When Gould took a flash photo of this huddle, the dingoes made off and did not return for several nights, the Aborigines shivering without their dog blankets.

I accept that account totally. It indicates that the limit of symbiosis is robbing of the camp by the dingo which indeed has to circumvent defenses against its robbery, but it is still free to make off when so inclined. On the other side of the coin of symbiosis, the Aborigines use it as a blanket and an unfed companion.

I think this must have been the level of association of dingo and man when it was brought to Australia. In its entire history, it has never been domesticated and cannot be domesticated. It must have been brought solely as a possession and a companion.

Macintosh (1962) recorded an engraved rock at Mootwingee showing various associations of men including Cult Hero with a minimum of 11 and a maximum of 16 dingoes. Living Aborigines know nothing about rock engraving which appears to have ended at least 1000 years ago while rock painting has persisted almost to the present day, but is now largely supplanted by bark painting. Some Aborigines are on record as providing interpretation of the rock engravings. This has to be a case of grafting the mythical stories onto the existing engravings. So one has to be wary of the Aboriginal interpretations and even more wary of one's own. An Aborigine provided some interpretations of the rock engravings on "Dingo Rock" but had no comment about the animals. The rock appears to exhibit a sequence of diminishing association of man and dingo, but European eyes cannot arrive at the same answers as given by an Aboriginal informant if he can be found. It is mentioned only because of the large number of animals.

Rock engravings of two, more rarely three dingoes hunting kangaroos are common, and similarly in rock paintings. There are none carrying conviction to my eye, or provided by informants, of man and dingo combining in a hunt. Aboriginal galleries are characterized by superimposed paintings. The dingo is invariably missing from the earlier strata and appears only in the more superficial superimpositions. The dingo figures prominently in myths and legends but is usually mentioned in a mythical rather than a pragmatic role. Macintosh (1965b) recorded paintings of male and female horned anthropomorph, and of dingo and echidna in a gallery adjacent to another gallery packed with local fauna paintings, but no dingo. The two galleries were dated from charcoal associated with different ochre colors in the stratified floor layer: the dingo gallery 581 ±120 B.P., the gossip gallery 520 ± 155 B.P.

The only purpose in this digression is to indicate an apparent relatively late advent of the dingo in rock art and the absence of depiction of pragmatic symbiosis.

STUDYING DINGOES IN THE FIELD

It was one thing to draw up patterns from our colony of dingoes in captivity, another thing to assess patterns in the wild state. With the support of the Pastoralists' Association of Western Australia, a 2000 mile circuit was made in Western Australia extending into the Gibson Desert. Breeding haunts, tracks, evidence of dingo movements about the pastoral properties, on the shores of dry lakes, at the foothills of ranges, or in dry creek beds were explored; 140 skeletons, some with flesh and hide persisting, and the products of trapping or baiting were collected. On that occasion, I found that the CSIRO station near the Kimberley foothills was examining the stomach contents of dingo with surprising results: lizards, insects, eggs, an occasional bird, marsupial mice, and nothing larger were being noted; it appeared that one needed to have second thoughts about this desperate predation of sheep, lambs, and calves. Merrilees (1968) showed that some 30% of the larger marsupial species became extinct during the late Pleistocene when the dingo had apparently not yet arrived. *Thylacinus* and *Sarcophilus* (carnivorous marsupials) apparently became extinct on the mainland (but not in Tasmania) about 3000 B.P., which is when dingo remains begin to appear in archaeological contexts; the corollary, therefore, was that thylacines and sarcophilines could not compete with the dingo. But an extraordinary find of an aboriginal skeleton wearing a necklace of 180 pierced *Sarcophilus* teeth (6820 ± 200 B.P.) (Macintosh 1971) can be interpreted as indicating that sarcophilines were already 7000 years ago on the wane to the extent that the animal was no longer of totemic significance and so the necklace was interred with its last guardian.

In brief, I now doubt the role of thylacines, sarcophilines, and dingoes as predators of anything larger than small kangaroos and wallabies, and while the dingo is responsible for death of lambs, it is due to disturbance of the ewes who in fright desert the lambs, rather than indiscriminate killing and eating of lambs or sheep. This is not to deny that sheep are attacked and bitten, but it is not on the wholesale scale that was formerly and still believed. My views expressed in 1956 have changed appreciably.

The Western Australia survey was followed by similar field surveys from Broken Hill to Yelka and Fort Grey Stations to the South Australian and Queensland border fences. Other regions in Arnhem Land, Central Australia, and Cape York were similarly examined so that I now have some 300 skeletons from about 50 sites across the continent. I have never yet seen a dingo pack hunting and I do not believe it ever occurs; indeed, it is rare to see more than 2 dingoes in company in the daytime. Admittedly, at night one may have the occasional experience of being apparently surrounded by 10 to 20 dingoes about half a mile away arranged in a circle around one's camp fire and holding a howling concert. The dingo disappears with the advance of commercial development. With the whole continent now appearing to be a vast mineral deposit of iron ore, uranium, and bauxite, development proceeds apace. A few months ago on Gove Peninsula (bauxite mining) in northeast Arnhem Land, I didn't see a dingo nor do I think any are left there, whereas 20 years ago there were hundreds.

CHRONOLOGY OF THE DINGO

Firm radio-carbon dates backed up by stratigraphic sequences now indicate 21,000 years B.P. for man's established occupancy in the extreme north of Arnhem Land at Malangangerr near Oenpelli, in the extreme south of South Australia at the Koonalda caverns, in the extreme east coast of New South Wales at Burrill Lake; while at Lake Mungo in far western New South Wales firm dates range from 25,000–30,000 years. Beyond any possible doubt, man arrived in Australia before 30,000 years ago, and he could only get here by sea, which makes the Aboriginal Australian the world's first mariner (Macintosh 1965a, 1965b, 1967a, 1967b).

The last low sea level was 18–20,000 years ago. Distances of necessary sea transport were then shortest. Obviously, the first Aborigines crossed by some form of sea craft when the distances were greater. Gill (1971) points out that the previous low sea level was about 60,000 years B.P., representing a theoretical possibility for the time of first Aboriginal migration.

Jennings (1959), following naval bathymetric surveys of Bass Strait between Australia and Tasmania, identified at 35 fathoms a submarine corridor some 50 miles wide from the Victorian coast to the northeast corner of Tasmania. He later (1971) points out that at glacial low sea level, deep straits persisted between Australia and New Guinea, the Banda Island arc, and the Java-Flores arc. Channels up to 100 km would need to be crossed, but shorter distances and horizon smudges of intervisibility through the Celebes and Moluccas to New Guinea would have existed.

The drowning of Torres Strait land link between New Guinea and Australia occurred probably between 6500 and 8000 B.P., and of the Bass Strait land link probably between 12,000 and 13,500 B.P.

The dingo has never been (to the best of present knowledge) on the Australian offshore islands, hence it could not have been brought to Australia before 13,500 B.P. at the earliest. That would far exceed the age of the earliest archaeologically known dog anywhere.

As the dingo is not reported from New Guinea although *Canis hallstromi* is (Troughton, 1971) as well as at least two other varieties of dogs, one would suppose that the dingo could not have reached Australia earlier than 8000 B.P. Troughton (1971) postulates that the yodelling New Guinea Highland Dog *(C. hallstromi)* became ancestral to the dingo. There are two present difficulties about this postulation: (1) the lack of any archaeological evidence of dog in New Guinea earlier than 1500 years, although man was present there at 26,000 B.P., and (2) *C. hallstromi* is much smaller than the dingo and "appears" to have very different skeletal indices.

These chronological data were not available to Wood Jones. While his thesis that the dingo was brought by sea is not only confirmed but reinforced, it arrived at least some 20,000 years later than the first mariner, named by Jones "the progenitor of Talgai man," with his wife, dog, and dog's wife.

DIFFERENTIATION OF THE DINGO FROM THE DOMESTIC DOG

Differentiation of the Dingo from the Domestic Dog by the Post-cranial Skeleton

This work (Table 7-1) was provoked by my dissatisfaction with the personal age I had assessed for the 3000 years old dingo (Macintosh, 1964) referred to earlier in this paper.

A detailed analysis of 12 metrical and 33 nonmetrical characters of femora in our dingo collection indicated only a narrow range of variation in adult animals. Subsequently, we found the 33 characters could be reduced to 11 to give the same result. Domestic dogs within individual breeds and more strikingly where different breeds are included, give a wider range of variation. While this indicates homogeneity in our dingo collection, it does not help in diagnosis of an individual skeleton, because the domestic dog ranges encompass those of the dingo. Similar results were derived from study of humerus, ulna, and tibia. Indices as used by Mivart (1890) are useful in classifying different types of wild dogs, but are not useful in separating wild dog from domestic dog.

The experience in 1964 alerted us to the need to segregate skeletons into age groups. Our first attempts were based on radiography of corpses from our own colony where the dates of birth and death were known. While radiography was useful for some bones, the degree of fusion could not be determined definitively for the pelvic bones, par-

TABLE 7-1 Age in Months at Near Complete Epiphyseal Fusion

Character	74 Dingo skeletons	Domestic Dog (after Sisson, 1945)
Three bones of os coxae	4	6
Eruption lower M3	4	6–7
Tuber scapulae	4	6–8
Metatarsals and phalanges	5	5–6
Olecranon to ulna	5	15
Distal end humerus	7	6–8
Proximal end radius	8	6–8
Ischiatic crest	9	24
Distal end ulna	10	15
Distal end femur	10½	18
Distal end radius	10½	18
Proximal end tibia	10½	18
Proximal end femur	11	18
Proximal end humerus	11½	18
Iliac crest	14	24
Pelvic symphysis	15	Late

ticularly of the os coxae, nor for the tuber scapulae, metatarsals, and phalanges. Time was far too short to dissect all these corpses in order to verify the radiographic estimates. An attempt, therefore, was made to plot the relative order of epiphyseal fusion in our dingo skeletal collection of which the ages were unknown. Obviously, complete or almost complete skeletons were needed. The epiphyses of each of 74 complete skeletons were recorded for fusion not begun, fusion just begun, fusion almost complete, fusion completed. The 74 sets of records were then shuffled into a progressive order ranging from least to greatest closure and thus were graded in age relative to one another and to an assessed standard of near complete closure.

At this stage, an attempt was made to provide absolute ages for the relative ages so far obtained. The most complete set of x-rays belonged to a 12 month old dingo. Extrapolating from this, ages were then assigned to each of the 74. Less complete x-rays of other dingoes of known age were used as checks to these assigned ages, but very little change was needed.

The series was found to consist of three groups: 22 young (9 months and under); 23 mature (10 to 13 months); 29 postmature (13.5 months and over). Measurements on the main limb bones, i.e., scapula, humerus, radius, ulna, femur, and tibia were made. Range, mean, and standard deviation indicated only moderate variation in the mature and postmature groups, although there was considerable variation in the young group. These skeletons were collected from widely separated geographic areas across the continent. All had been victims of shooting, trapping, or ground or aerial baiting. It is popular opinion that the younger the animal, the more vulnerable it is to any form of destruction, particularly to aerial baiting. This absolutely random series suggests if anything, that the mature and postmature collectively are more vulnerable than the young. Secondly, this diverse geographial collection shows considerable homogeneity. Thirdly, the figures quoted by Mivart in 1890 agree with our results and so appear to confirm my opinion (Macintosh, 1964) that the nearest similarity to dingo (at least in indices) is *C. latrans;* but we are still no closer to differentiating the dingo from the domestic dog.

We have discovered that epiphyseal unions and eruption of third molar teeth are appreciably earlier in dingo than in domestic dog (See Table 7-1), although Sisson's (1945) quoted epiphyseal fusion gives no details about number of specimens and breeds.

As far as postcranial measurements and indices are concerned, the overlapping ranges of domestic dogs allow no differentiation from dingo. It would be more informative to compare the latter with other types of wild dogs where the range of variation is less than in domestic breeds taken collectively.

We therefore decided to undertake a comparative study of the skull based on the methods we had used on Aboriginal human crania (Larnach and Macintosh, 1966, 1970). Our dingo collection contained more skulls than skeletons.

Differentiation of Dingo from Domestic Dog by Skull Characters

Sporadically over a period of 7 years, with the assistance of senior science students, Miss Diane Pratt and Miss Joy Worrall, and of Mr. S. L. Larnach, Curator of the J. T. Wilson Museum in our department, 120 metrical and nonmetrical traits in dingo and domestic dog skulls were assessed. Characters which were obviously not differentiating were progressively eliminated until only 45 remained. More detailed attention was given to these and a variety of statistical experiments finally reduced them to the following 11 (values given in mm):

1. Opisthion — inion height: Under 29 (1), 29–32 (2), over 32 (3).
   Opisthion — highest median point on margin of foramen magnum.
   Inion 8 most posterior point where nuchal and sagittal crests meet.

2. Tympanic bulla length: Under 23 (1), 23–25 (2), over 25 (3).
   From where bulla abuts posteriorly on paroccipital process to eustachian opening anteriorly.

3. Foramen magnum index: Under 132 (1), 132–135 (2), over 135 (3).

   $$\text{Index} = \frac{\text{maximum width} \times 100}{\text{length (basion to opisthion)}}$$

4. Skull length: Under 182 (1), 182–189 (2), over 189 (3).
   Length — inion to prosthion which is most anterior point on inter-premaxillary suture.

5. Bi-poral height index: Over 103 (1), 100–103 (2), under 100 (3).

   $$\text{Index} = \frac{\text{height (basi occip.-sphenoid to bregma)} \times 100}{\text{biporal width}}$$

6. Maxillio-alveolar index: Over 65(1), 62–65 (2), under 62 (3).

   $$\text{Index} = \frac{\text{maximum width} \times 100}{\text{max. length (prosthion to post molar maxillary tubercles).}}$$

7. Maximum post orbital width: Under 50 (1), 50–52 (2), over 52 (3).
   Diameter between most lateral points on post orbital processes.

8. Sagittal crest: 1, 2, or 3 according to prominence judged on its height and on extent forward from the interparietal bone.
9. Medial fossa of mandibular angular process: 1, 2, or 3 according to depth development.
10. Ventral orbital crest: 1, 2, or 3 according to prominence. This sharp crest extends from inferior border of optic foramen towards maxillary foramen.
11. Mandibular diastema between 2nd and 3rd premolars: Distinct 3, ambiguous 1, absent 0.

Values of 1, 2, or 3 were assigned to the grades of development of each character. These were totalled for each skull to determine its score. When the skulls were plotted out on the basis of their scores (Figure 7-6), there was good separation between dingo (with a range of 25 to 33) and domestic dog (with a range of 10 to 22).

Skulls of dingo juveniles, i.e., occipito-sphenoidal suture not synostosed, fell into an intermediate range (19–24) between adult dingo and adult dog; we do not have an adequate sample of juvenile domestic dogs to investigate whether the pattern separates juvenile dingo from juvenile domestic dog.

Recently we have noted also that character No. 10 is only slightly more frequent in dingo than in domestic dog and discrimination might be improved by its removal from the 11 traits used.

There is only one trait which is absolutely differential in our series of crania, i.e., character No. 11, the mandibular diastema, present in every dingo and absent in every domestic dog of our series; but a larger sample of domestic dogs might show exceptions. Identification depends on the total score of the 11 traits. Fewer than that presents an inadequate number for identification.

Troughton (1941) says it is impossible to note different characters in the skull of a well-grown cattle dog and a dingo. Our collection of domestic dog crania included great dane, pomeranian, various terriers and mongrels, cattle dog and kelpie; this presumably accounts for the irregularity of the domestic dog histogram and it is noted that cattle dogs and kelpies more nearly approach the dingo ranges but are still separated. The compactness of the dingo histogram seems to confirm the continent-wide homogeneity, and gives no cause to suspect hybridization.

As there are no comparable figures to be found in the literature, we believe the pattern here described represents an original contribution, which nevertheless may need modification or refinement when larger series are analyzed. It need scarcely be cautioned that this pattern of 11 traits is not of universal application; different characters might have to be used if discrimination is attempted say between dingo and coyote, dingo and wolf, dingo and Tengger dog, and so on.

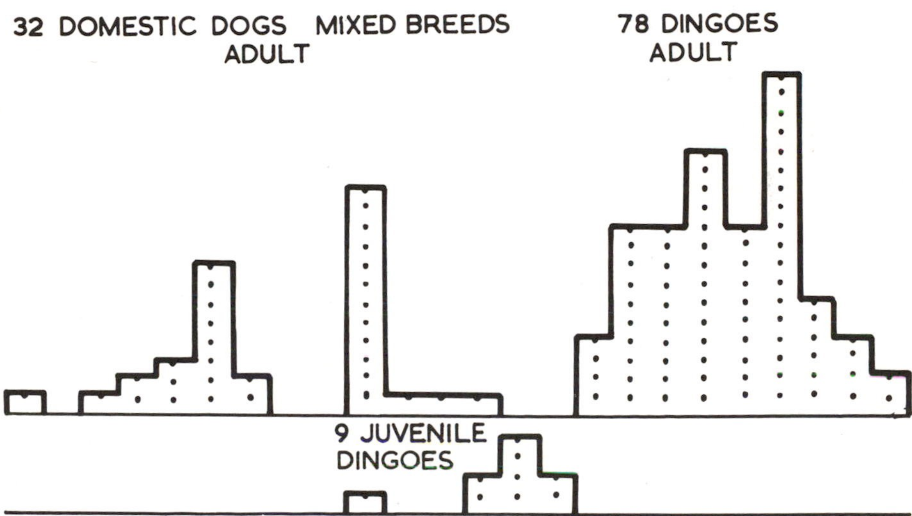

Figure 7-6 Distribution of individual scores of 11 cranial traits.

SUMMARY

A review of the literature shows that the dingo is claimed to resemble closely the pariah dog, Tengger dog, *C. f. poutiatini,* and is linked with the Starr Carr limb bones, the Senckenberg dog (called by Degerbøl (1961) the European Mesolithic dingo) and is therefore also linked with *C. f. matris optimae, C. l. pallipes,* and *C. l. hodophylax* by various authors, and to a *Canis ferus* by Studer (1901) and Zeuner (1963).

Synonym should be *Canis antarticus* Kerr, 1792 followed by *Canis dingo* Meyers, 1793. Dingo may have been the Aboriginal word for white man's dog.

McCoy (1882), Krefft (1865), Gregory (1906), and Etheridge (1916) initiated claims that the dingo is indigenous and of Pliocene antiquity, based on mélanges of extinct marsupial fossil bones and fragments of dingo jaws and teeth. These are invalidated.

The almost complete skeleton of a male dingo of 18 weeks is ^{14}C dated 3000 B.P. from South Australia and proves that dingo morphology is unchanged to the present, so supporting Jones (1921), Tichota (1937), and Tate (1952) that the dingo is homogeneous and not crossed.

At Devil's Lair, W.A., and Mt. Burr, S.A., slight remains accompanied by some dubiety are reported at 8500 and 8000 B.P., thus approaching the 9500 B.P. Starr Carr and Coon's (1951) 11,480 B.P. Iranian canid.

Recordings are derived from a colony of 42 dingoes bred in captivity, and from field surveys right across the continent. Aboriginal rock art and mythology indicate recent introduction of the dingo.

Eustatic and bathymetric studies (Jennings, 1959 and 1971) show that Torres Strait was flooded 6500–8000 B.P. and Basss Strait 12,000–13,500 B.P. As the dingo has not been found in Tasmania, New Guinea, or offshore islands, it must have arrived in Australia after 13,500 B.P. almost certainly not more than 9000 years ago.

Even at glacial low sea level 18,000 years ago, Australia was separated from lands to the northwest by at least two deep straits up to 100 km wide. It is ^{14}C proven that man was established in Australia 30,000 years ago, so the first arrivals had to come by sea. Some 20,000 or more years later the dingo was also brought by sea.

Discrimination of dingo from domestic dog by a pattern of 11 cranial traits is offered as an original contribution and tooth eruption and epiphyseal closure are shown to occur earlier in dingo. Brief, speculative comment is made about the relevance of the advent of man, coyote, and dingo in America and Australia. Aboriginal relations with the dingo reveal only minimal symbiosis. It is concluded that the dingo cannot be domesticated, and never was; it was brought here as a quasi-companion.

The dingo gait, howl, morphology, and some elements of its behavior observed in captivity and reported in this paper are peculiar to it alone. Its ancestry and affinities remain enigmatic.

PART

II

BEHAVIORAL ECOLOGY

Few thorough studies have been undertaken on the behavioral ecology of canids and in this section a number of original studies are detailed.* Most of the canids of South America remain to be studied; others, such as the Indian dhole (*Cuon alpinus*) and the canids of Africa await more in-depth study. "In-depth" implies continuous (seasonal) observations in the wild over several years, and in different habitats, where possible. One paper in this section deals with telemetry, a technique now opening up new possibilities for field biologists. Another paper, more "anthropological," reviews a native hunter's knowledge of applied behavioral ecology.

Many questions will come to mind as one reads these papers; these are questions for further research. Some readers may be frustrated at times by the apparent lack of data and answers in some of the papers. I hope, as editor, that such frustration will provide the stimulus for more intensive research so that in the near future our knowledge of the behavioral ecology of the fascinating and diverse canid family will be more complete.

*For completion of bibliography, references to other species not cited in this text include swift fox (Kilgore, 1969), kit fox (Egoscue, 1956), fennec fox (Gauthier-Pilters, 1962 and Koenig, 1970), Bush dog, *Speothos venaticus* (Linares, 1967).

8
ECOLOGY AND BEHAVIOR OF THE DHOLE OR INDIAN WILD DOG *Cuon alpinus* (Pallas)

E. R. C. DAVIDAR
"Canowie," Coonoor — 1, Nilgiris, India*

Until lately it was said of the tiger that all studies concerning it were made along the sights of rifles. If this is true of the tiger, it is even more so in the case of the dhole, with the difference that the animal was observed even more briefly, and with prejudice amounting to malice. Every reference to this animal concluded with an admonition that it should be shot on sight, and that no effort should be spared to destroy it, whether by fair means or foul. Though there are numerous references to the dhole in sporting literature, if these could be called literature, they are sketchy and for the most part unreliable.

DESCRIPTION

In appearance, the dhole resembles the domestic dog (Figures 8-1 and 8-2) particularly the Indian pariah dog, a nondescript cur of mixed breed, except that its ears, though pricked like the pie-dogs, are more rounded, and its tail is black and bushy, and again, unlike the pariah dog's which is generally curled and carried high, is straight and carried low.

*Mr. Davidar is a naturalist who has spent many years studying the dhole. Perhaps above all canids, this pack hunter warrants concerted efforts for more intensive field research and conservation — Ed.

In coloring, it varies from rusty red to light brownish gray, depending upon its environment. The Trans-Himalayan dogs are paler in color in comparison to the tropical and subtropical animal. The dhole of the peninsular Indian region are uniformly red, shading into dirty white or yellow on the underparts. The points of the hairs along the dorsal ridge and the tips of the ears are often black. Some dogs, especially those from Central India, have a white tip on the tail which in most cases is not visible except on close scrutiny. Because of its red hue, it is often called the "red dog."

Figure 8-1 Juvenile dhole (approximately 12 weeks).

The Indian wild dog (which, of course, is a misnomer, the animal being a cuon) differs from the domestic dog by having 2 true molars on each side of the lower jaw as against the domestic dog's 3, and 12 to 14 teats against the domestic dog's 10. Also, it does not have a true bark. The muzzle is comparatively shorter and the line of the face, when viewed from the side, is slightly convex; whereas, in the case of Canidae, it is straight or concave.

In size, it stands 17 to 22 inches (43 to 55 cm) at the shoulder. Head and body are about 3 feet (90 cm) in length, and the tail 16 to 18 inches (40 to 45 cm). An adult male weighs 40 to 45 pounds (18 to 21 kg) and an adult female 32 to 37 pounds (15 to 17 kg).

Figure 8-2 Adult dholes.

DISTRIBUTION

The dhole is a widely distributed species. From Atlai Mountains and Manchuria in Central and Eastern Asia, it is spread southwards through the forest tracts of India, Burma, and the Malayan Archipelago.

Several races exist (Prater, 1965). In the Indian region alone, 3 races, namely Trans-Himalayan, Himalayan, and Peninsular are recognized. Jungle natives of the Peninsular region believe that 2 separate varieties, a larger and a smaller variety, exists (Burton, 1941).

An inventory of dhole localities has never been prepared. However, from general reports it is possible to state that while the Central Asian wild dogs live in forests and steppes, the dhole of Tibet and Ladakh inhabit open country. In tropical and subtropical countries like India, they live in dense forests and the thick-scrub type of jungle of the plains as well as the hills, but not in the open plains and deserts.

POPULATION

No attempt has so far been made to census the wild dog population, either country-wise or area-wise. Even in places where estimates are available, because of the animal's wandering habits and the chances of the same animal being counted more than once, or not at all, the figures can be misleading.

However, this much can be stated with certainty: due to denudation of forests, the disappearance of the prey species, and persecution by man, the wild dog has totally disappeared from many of its old haunts. Even in the tracts where they do occur, they are greatly reduced in numbers.

The dhole population has been known to fluctuate from a sharp rise to a steep fall (Brander, 1931; Burton, 1941). The decline in population invariably follows when a certain peak is reached. The cause for this has not been established, but from symptoms of sick dogs met with in the jungle during this period, it could be assumed that distemper may be the main cause. Village dogs and jackals are the transmitter of this disease. The dholes are also susceptible to rabies and mange, but distemper takes the heaviest toll.

The dhole is a social animal, associating in packs which number up to 40 individuals, but the usual number is between 5 and 10 or 12, which suggests that the family unit (more likely mother and cubs) forms the basis of the pack. Pack size is not constant, but varies. Larger packs break up and regroup in varying sizes. No information is available on the composition of packs, the ratio between the sexes, and the different age groups.

COMMUNICATION

The dhole has not been observed to scent mark on the trails. For the most part, the dhole is a silent animal, relying more on its nose than on its vocal powers to communicate with each other. However, the dhole is capable of making every kind of noise a domestic dog makes except loud and repeated barking. It howls, snarls, growls, whines, and whimpers. Its bark is rudimentary and can best be described as a "yap." When startled or alarmed or excited, as when confronted suddenly, it yaps so rapidly that the yapping becomes a chatter.* The most common call through which they communicate is the peculiar whistle, which can be imitated by blowing into a medium bore rifle cartridge with a series of three short toots repeated at intervals (Adams, 1949). The animal is often called up by this method, and even vocally by imitating its whistle, as the author has found. The most uncommon cry that is heard occasionally at dusk is a loud hysterical yapping howl made in concert. It sounds weird and can be best described as a cross between the braying of a donkey and the neighing of a horse.

PREY, FOOD HABITS, AND FEEDING BEHAVIOR

The principal prey of the dhole in India is the cheetal or spotted deer. They also hunt the larger sambhur and swamp deer and the big nilgai antelope as well as the smaller deer and antelope of the jungles. Among the larger deer, they prefer to hunt the fawn and hinds, particularly the gravid hinds, which are slow and therefore easy to circumvent. Wild pigs, hares, wild goats and sheep, and even monkeys are hunted when they leave the security of the trees and race across the open (as Nilgiri langur sometimes do; Adams, 1949). Some ground birds like the jungle fowl, when they can get them, form part of their diet. On occasion, they kill domestic cattle, goats, sheep, and young buffalo. Cattle and young buffalo baits tied out for tigers were sometimes taken, when such tying out was common. Occasionally it eats grass and vegetable matter. Though as a rule the dhole kills its own food, it does not refuse carrion, even carrion in a stale and putrid state (Morris, 1937).

They normally demolish their kill at a sitting, though not all at one go if it is large; but do come back if anything is left, provided, of course, they are not unduly disturbed. They are also known to abandon their kill when it gets too late in the evening, and come back in the morning. After the initial scramble following the bringing down of a large prey, they tend to lie around the kill 15 to 20 paces away, and take turns at the kill, making frequent visits. When the prey is small, like a fawn, it is a case of first come, first served and each dog for itself. They are not noisy feeders except when the cubs are around. They tear chunks of meat and bolt it down, along with the skin and hair and soft bones. The hair is not digested and is passed out. The droppings resemble a dog's, except that it is unusually darkish, a tarry substance holding the hairs

*Comparable vocalizations are characteristic of the Cape hunting dog which may well be related to the dhole at some time prior to the continental drift — Ed.

and other parts composing it together. These are invariably found on jungle roads, paths, and game trails.

Hunting

The dhole hunts in packs. Single dogs and pairs also hunt at times. The dogs do not employ a uniform hunting method. They vary their technique to suit the prey and the hunting country. Stayers like sambhur are pursued almost casually at a loping canter which does not tire them, but tires the quarry. Smaller quarry like fawns and hare are rushed and seized with a burst of speed.

In a country capable of lending itself to employment of strategy, that is, a country with good visibility like the undulating downs of the Nilgiri plateau, they seem to work to a definite plan — like beating a jungle for game, complete with "guns," "beaters," and "stops," flushing, tracking, and ambushing the prey. The normal method is to run the prey down in haphazard relays, with no fixed or prearranged order.

The wild dog clearly relies on its sense of smell, since it will cast about and pick up the trail of the prey by scent, and on flushing it, they hunt by sight and sound, using their noses only once in a while. Hunting in numbers as they do, employing their acute sense of smell besides the other senses, places them at an advantage not only over their quarry, but over rival carnivores as well, making them the most effective predators in the jungle.

The dhole runs mute when after a quarry. But they do whimper and yap at times during the final stages of the hunt. When hunting in heavy cover, they keep in touch with each other in a similar fashion.

Most of the prejudice against the wild dog arises from the gruesome manner in which it secures its prey. They literally eat their prey alive. As the quarry flees, the dogs follow, and when they catch up, they take bites off the fleeing animal, which is usually on the flanks and between the hindlegs or the back of the rear thighs. Soon the quarry is disemboweled and big gaping holes appear on the side and back, even before it is brought to a halt. When the quarry stops running or stands at bay, the dogs do not attack from stationary positions, but attack leaping in and out, circling swiftly. The swift, encircling attack and the changing of positions makes 2 dogs appear like 20, and the quarry is clearly confused and disoriented. As the victim turns in one direction, it is attacked from the other. But an odd dog which gets a good and safe hold hangs on.

When the hunted is a stag, the testicles are sometimes seized and removed. This has led to the belief that they emasculate their prey systematically and on purpose. But it does not appear to be so.

There is also another belief that the dhole blinds its prey by whisking its urine into the eyes of its victims after first spraying its tail with it. The fact that the animal, like many others, urinates when excited has probably led to this belief. There is otherwise no basis for this belief, as inspection of several fresh carcasses has revealed.

Sambhur deer have the habit of running into water when attacked and standing at bay in belly deep water, where the dogs follow them, swimming and attacking.

The dogs do almost all their hunting by day, particularly in the early hours of the morning and evening. On rare occasions they have been seen or heard to hunt at night. When they do so, they prefer moonlit nights.

Relationship with prey

As a rule, the dhole does not kill for the sake of killing. But they have been known to kill 2 animals brought to bay together, although 1 is all they may require. Where domestic stock is concerned, there are known instances of the dogs wantonly attacking 3, 5, 6, and once as many as 13 cattle and buffalo calves at a time, killing half of them, and leaving the rest to die. The calves' slow movement and the panic among them may have been a contributory factor for the slaughter.

When their hunger is satisfied, they relax and do not hunt again until they begin to feel hungry. Even the prey species seem to be aware of this, and do not take as much notice of the dogs as they would otherwise do. This, of course, happens only in areas which are well-stocked with game, and the dhole is more or less a permanent resident.

There is a widely-held belief that when wild dogs move into a forest, all the prey species flee. This appears to be so in forests where game is already scarce, as they become scarcer by scattering. But in well-stocked forests the prey species merely move around.

Dholes have been described variously as courageous, rapacious, and intrepid killers. But as Adams (1949) and the author (Davidar, 1965) have observed, sizeable packs of dogs have been known to leave herds of sambhur alone, which are as large as ponies and can use their forefeet with killing effect, prefering to tackle single sambhur or pairs of them. The author has also seen in the Nilgiri forests several sambhur hinds, and sometimes a single hind, rush to the rescue of a fawn attacked by dholes and put a whole pack to flight. This does not mean that sambhur are left alone, or that the dhole is cowardly, but simply that they prefer easier prey where they are available. There are parts of India where the sambhur is the sole prey of the dogs, where these considerations do not arise.

REPRODUCTION

Young (Figure 8-3) are born between November and March. But in South India they are rarely born after December. The period of gestation, according to a person who owned a pair, "was about the same as that of a domestic dog," which is approximately 9 weeks.

Figure 8-3 Approximately 4 week old dhole.

The dhole uses rocky caves, shelving rocks, and earths or burrows of its own digging to litter and bring up its young. A single bitch or pair may bring up their young ones in a cave or earth on their own, but communal nurseries made up of several separate earths, or in a common cave where bitches give birth to and bring up their young, are common. A mother may also shift its earth in a season. The caves are used for 5 to 6 weeks, by which time the young are able to accompany their mothers.* Though most of the members of the pack do not use the nursery grounds, they remain within the vicinity.

The large number of mammae (up to 16) which a bitch is endowed with gives an indication of the dhole's capacity to raise large families. Therefore, it is not surprising that up to 9 well-formed embryos should have been found (Adams, 1949; Burton, 1941). Groups of 12, 11, and 10 cubs have been taken out of single caves or seen at heel with wild dog bitches on different occasions (Burton, 1941; Adams, 1949). These may have been litters of more than one dog. Mother–cub attachment is strong and lasts about a year.

It is believed that the dholes pair off during the breeding season and remain together until the cubs are able to fend for themselves (Brander, 1931). Whether the relationship between the sexes is casual or permanent, as believed, has not been fully established. From observations made by the author, there is strong evidence to discount this theory: mating has been seen by the author to take place within the pack itself and couples do not necessarily pair off.

HABITS, BEHAVIOR, AND GENERAL CHARACTERISTICS

It must be appreciated in this context, and generally, that this creature with such wide distribution and living in such varied conditions is bound to have developed some special local behavior patterns, and therefore what is true of a particular area may not apply elsewhere.

The dholes are hardy and extremely tough and get away with the most frightful wounds. They are either bold or plain stupid, as otherwise they would not stand around and allow hunters to kill two or more of their numbers before bolting. Perhaps it is curiosity that makes them behave in this manner.

They are very fond of water and are often seen lying in water on warm days. Even when drinking they often flop into the water where it is shallow. They are partial to open spaces and are frequently encountered on jungle roads, paths, river beds, and jungle clearings where they rest during the day. They are very seldom seen in the open at night.

They wag their tails when pleased just like domestic dogs. Aggressive contacts between adults in a pack are rare. There also appears to be no bullying. From the manner in which the pack pattern changes, it is reasonable to expect constant quarrels among the dogs, but this is not so. However, there is constant squabbling and rough play among the cubs and subadults. The dogs occasionally growl and snap at each

*Recent observations by the author of an active dens site indicate that whelping bitches are sometimes fed by members of the pack regurgitating food for them.

other when feeding, but no real fighting takes place. They are altogether an amicable lot.

The dhole can be attracted by imitating the frightened squeals of a fawn in distress, which effect can be achieved by blowing on a leaf or grass held between the thumbs.

Dhole packs cover wide territory in the course of their hunts. They migrate locally when prey becomes scarce. Adams (1949) believes that they send one or two scouts in advance to spy the land before moving into new country. There is no other record.

Relationship with the domestic dog

There is no definite pattern of behavior towards the domestic dog. They have been known to give battle to large, aggressive dogs. There are also instances of their having run away from them. Where small dogs are concerned, the tendency is to fraternize (Brander, 1931; Adams, 1949). They have been known to follow small dogs like terriers out of curiosity with no aggressive intention. The dhole has been observed by the author and a few others to hunt with village pariah dogs in the jungles on the lower plateau of the Nilgiris. Such associations are rare and purely temporary, and usually take place when the dhole population is low.

No case of mating between the dhole and the domestic dog in the wild state has been recorded.

Relationship with other predators

The dhole recognizes the tiger and the leopard as its rivals, and emboldened by its strength in numbers, has taken an aggressive role in its dealings with these rival predators. They do not hesitate to drive these larger carnivores out of their way whenever they happen to meet them. These are invariably chance meetings, and the dogs do not deliberately seek a fight, except when it comes to appropriating tiger and leopard kills. When attacked, leopards take to trees, and there are numerous instances of such treed leopards on record. Cases of tigers killed by the dhole are also on record, but these are based on circumstantial rather than direct evidence. Several cases of dholes attacking bears and leopards have been observed. On occasions, leopards have been known to kill stray wild dogs. Hyenas and jackals are simply ignored.

Relationship with man

The dholes are bold and saucy when met, retreating with reluctance. But retreat they do. They do not attack man even when they are in large numbers. In rare cases where a stray dhole has attacked people it was suspected to be rabid. The sportsman considered it a rival and never showed it any mercy. The residents of jungle hamlets, on the other hand, welcome them, as they provide them with a welcome supply of meat. When they see the dhole hunting, they follow the hunt, wait until the dogs have made a kill, chase the dogs away, and appropriate the kill for themselves.

It has been established that the wild dog can be tamed, if taken as a cub. But they never shed their shyness or "wildness." They cringe and cower and do not feel at home and show no real affection.

CONCLUSION

The dhole is a much maligned creature. It has been branded as an outlaw. Not only do no game laws apply to it, but in most places where the dhole operates it carries a bounty on its head.

The dhole is accused of wiping out entire deer populations. This accusation is unjust, as it could not have co-existed with its prey species for thousands of years. However, in a sense it is true in areas where the deer population are low. But the responsibility for this situation squarely rests on modern man who, with his gun, slaughtered deer indiscriminately, and reduced it below the level of survival in those areas. Also where deer populations are not controled and predators are hunted by "sportsmen," the deer populations may explode and then crash, so that the few surviving predators may in turn face starvation. It is hoped that on these pretexts man will not wipe this fascinating creature off the face of the earth.

Editor's Note. Since the completion of this chapter, Mr. Davidar has hand-raised a dhole cub. He notes that it is extremely sociable, outgoing, and aggressive rather than cringing, but is disturbed by novel stimuli such as unfamiliar noises.

9 SOCIAL BEHAVIOR AND ECOLOGY OF THE AFRICAN CANIDAE: A REVIEW

MARC BEKOFF
University of Missouri
Department of Biology
St. Louis, Missouri

Among the African members of the family Canidae there appears to be great diversity in social organization, socio-ecological adaptation, and morphology, and an equally large gap in the literature concerning detailed aspects of their various ways of life. In Eisenberg's (1966) lengthy review of social organization in mammals, little is mentioned about African canids.

As a whole, members of the family Canidae are very adaptable and nonspecialized (Kleiman, 1967) and the diversity that has been phenomenologically observed, both within and between species, is understandable when such variables as habitat, season, and variations in group membership are taken into account. Social behavior and environmental interactions are determined by these and other factors.

Wanton destruction of these little understood canids, particularly the cape hunting dog (*Lycaon pictus*; Estes and Goddard, 1967) and the black-backed jackal (*Canis mesomelas;* van der Merwe, 1953a), has resulted in serious assaults on the balance of nature — e.g., unchecked increases in populations of certain animals. There is increasing evidence from recent field observations that the African canids are very important in the maintenance of the balance of the ecosystem.

The Canidae of Africa are represented by the following species:

A. Foxes:
  1. Sand (*Vulpes pallida*)
  2. Rüppell's *(V. rüppelli)*
  3. Cape (silver) (*V. chama*)
  4. Red (*V. vulpes*)
  5. Fennec (*Fennecus zerda*)
  6. Bat-eared (Delandi's) (*Otocyon megalotis*)

B. Jackals:
  1. Side-striped (*C. adustus*)
  2. Black-backed (*C. mesomelas*)
  3. Golden (common or Asiatic) (*C. aureus*)

C. Simien fox (Abyssinian wolf) (*C. simensis*)

D. Cape hunting dog (wild dog) (*Lycaon pictus*)

In this chapter I should like to concentrate on reviewing what is known about the various African canids and discuss, in the last section, possible research strategies for the future. Much of the information presented concerning very general characteristics of the animals has been extracted from Dorst and Dandelot (1970). These authors also provide useful maps of geographic distribution.

FOXES

Sand Fox (*V. pallida*)

Almost nothing is known about this species (Dorst and Dandelot, 1970). It lives on the savanna of northern Africa, stands about 25.4 cm at the shoulder, is approximately 45.5 cm in length (not including the tail), and weighs about 3.6 kgs. Members of this species are said to be very gregarious (Dorst and Dandelot, 1970) and family parties live in large burrows, feeding on rodents, small reptiles, birds, eggs, and vegetable matter.

Rüppell's Fox *(V. rüppelli)*

As with the sand and the simien fox, not much is known about this species which inhabits stony desert in northern Africa. It stands about 25.4 cm at the shoulder and weighs approximately 3.6 kgs. Without the tail, its length varies between 40–50 cm (Dorst and Dandelot, 1970). There is a conspicuous dark-brown patch on the side of the muzzle, extending towards the eyes, and it has a very soft, thick coat (Dorst and Dandelot, 1970). Rupell's fox appears to be a gregarious species and lives in parties of 3 to 5. It is largely insectivorous.

Cape (Silver) Fox *(V. chama)*

The Cape fox (Figure 9-1) is the only true fox found in southern Africa (Dorst and Dandelot, 1970), inhabiting dry country and open plains. It, like most other African foxes, is quite small, standing approximately 30 cm at the shoulder, roughly 56 cm in length (without the tail), and weighs about 4.1 kgs. The Cape fox is mainly nocturnal and lives singly or in pairs — usually hunting in pairs (van der Merwe, 1953b) — unlike some of its more social congeners. It is characteristic of this species that it leaves little or no scent (van der Merwe, 1953b). The breeding season lasts from September to October, the gestation period is approximately 51–52 days, and litter size varies between 3–5 (Dorst and Dandelot, 1970). Insects constitute the main portion of their diet, and they also prey upon small rodents (Bothma, 1966; cited by Ewer, 1973), fruits, and vegetables. They also eat birds, but do not hunt them. Analyses of silver fox stomachs by Bothma (1971a) revealed no domestic stock. He hopes that these data will help to gain protection for this species from predator control programs.

Figure 9-1 Cape fox, *V. chama*. Photo: courtesy San Diego Zoo.

Red Fox (*V. vulpes*)

The red fox of Africa is smaller and more lightly colored than its European counterpart, standing approximately 30 cm at the shoulder, and being about 65 cm in length (without the tail). It is sandy-red in color, has a white-tipped tail, and lives in stony desert (never in sandy desert) (Dorst and Dandelot, 1970). The most common prey are rodents. The red fox in Africa lives in burrows, supposedly in small family parties (Dorst and Dandelot, 1970) and the average litter numbers 4. In contrast, the red foxes of England and the United States are generally a more solitary species (Type I — see M. W. Fox, Chapter 30) with the family dispersing in the fall (Burrows, 1968; Rue, 1969). It has a vertical pupil, as is generally found in carnivorous predators (Hediger, 1955).

As with most of the other Canidae, the red fox "locks" during the final stage of copulation (Dewsbury, 1972). Neither sex shows any species-typical urination or defecation patterns, as do bat-eared foxes or female bush dogs (*Speothos venaticus*) (Kleiman, 1966). Females in captivity have not been observed to mark (Kleiman, 1966).

Fennec Fox *(Fennecus zerda)*

The fennec (Figure 9-2) is the smallest of all foxes, with very large, triangular ears (approximately 10 cm long; hypertrophy of the bullae), a short muzzle, and a short, bushy tail. It weighs less than 1 kg (Gauthier-Pilters, 1967), stands only approximately 20 cm at the shoulder, and is roughly 40 cm long without the tail. It is very pale (creamy) in color. The fennec lives mainly on the deserts of North Africa and has hairy soles, enabling it to run in loose sand (Burton, 1962). Very little is known about the habits of this animal in the wild, since it is very rarely seen (Gauthier-Pilters, 1967).

The fennec is a nocturnal animal and lives in small groups of up to 10 individuals (Dorst and Dandelot, 1970). Its nocturnal activity combined with its subterranean life (in burrows several yards long) reduces its water needs for thermoregulation, and the animal seems perfectly adapted to desert conditions in that it seems entirely independent of drinking water (and has hairy soles).

The gestation period is between 49–52 days (Gauthier-Pilters, 1967) and approximately 2–5 pups are born per litter. Gangloff (1972) reported gestation periods of 62 and 63 days. Burrows are usually dug in March and young are generally born in February or March. Very little is known about their rutting behavior in the wild. Gauthier-Pilters (1966) caught a 1 year old fennec in a burrow and this led her to believe that the young of the previous year remained with the family. Since they

Figure 9-2 Fennec fox, *Fennecus zerda*. Photo: M. W. Fox.

have been observed to live in small groups, this is possible, but more data are needed.

Prey include small game (rodents, lizards) and insects such as locusts. Food jealousy is common. Gauthier-Pilters (1967) described a back-arched threat (see Fox, 1971a,b; Bekoff, 1972a for description in other canids) when a conspecific approached another fennec already possessing food. The neck-bite is the usual way of killing prey, with feeding beginning at the head, and prey is carried to shelter before it is eaten (Gauthier-Pilters, 1967).

The fennec has been observed in captivity by Gauthier-Pilters (1966) and by Koenig (1970). Koenig (1970) presented protocols of behavioral ontogeny from days 1–135. Social games are composed of fighting, predatory, and running games, with fighting games often ending up in serious fights (Gauthier-Pilters, 1966). Solitary games are mostly predatory in nature and play is subject to seasonal and daily fluctuations, as has been observed in the red fox (Tembrock, 1957).

In captivity, during the rutting season, which lasts from 4–8 weeks (January–February), the male becomes very aggressive and frequently marks his territory with urine (Gauthier-Pilters, 1967). He defends the female before and during the birth, but never enters the whelping box (Koenig, 1970). In captivity, urine and solid wastes are usually covered with sawdust (Gauthier-Pilters, 1967).

The female is in heat for 1–2 days, and during mating, she flags her tail to one side in a horizontal position (Gauthier-Pilters, 1967), similar to that observed in wolves and domestic dogs. The several matings that were observed by Gauthier-Pilters in one pair of fennecs occurred at 2–3 hour intervals, each ending with "typical hanging," which can be interpreted to mean, I believe, the copulatory tie, so common in the Canidae. The female observed by Koenig (1970) was very sensitive to disturbance just after parturition, and actively defended the nest. The aggressive behavior declined as the young got older and disappeared when they were approximately 4 weeks of age. Gangloff (1972) observed an increase in aggressive behavior just prior to parturition, remaining high until weaning. Weaning occurred between 61–70 days and carrying (by grasping the neck — Gauthier-Pilters, 1967), grooming of the young, and suckling by the young was most intense between 28–70 days (Koenig, 1970).

Field observations should nicely supplement the observations on captive subjects that have been cited above. The anatomical and physiological adaptations that the fennec shows towards its environment should also be investigated further.

Bat-eared (Delandi's) Fox (*Otocyon megalotis*)

The bat-eared fox (Figure 9-3) is fairly small, weighing up to 5 kgs, standing approximately 30 cm at the shoulder, and being about 65 cm in length without the tail. Its ears are very large — broad and oval — approximately 10 cm in length, and presumably they have good auditory abilities and can localize sounds quite efficiently. Allen (1939) recognized 4 races of this species.

The bat-eared fox lives in the savanna of eastern and southern Africa and is mainly nocturnal. However, parties (possibly families prior to dispersal of young) of 5–8 have been observed by Smithers (1966) feeding during daylight hours. Adults from East Africa (*O. megalotis virgatus*) are generally more buffy, with their muzzles, ear-tips, legs, and tail-tip tending towards dark brown instead of black, as in their South African counterparts, *O. m. megalotis* (Smithers, 1966). This species of fox was originally called "draakjakkals" or "turning jackals" by early Dutch settlers because of their swiftness, agility, and alertness (Smithers, 1966).

The breeding season lasts from November until April and the gestation period is approximately 60–70 days. Normal litters vary between 3–6 young. Smithers (1966), however, reported that the breeding season in the south was narrow, and that young were born around November at the onset of the rains, when insects were plentiful. Burrows are approximately 60 cm long and have several entrances (Smithers, 1966b).

Figure 9-3 Bat-eared foxes, *Otocyon megalotis*. Photo: Liz Forrestal.

In captivity, Kleiman (1967) observed a social hierarchy (3 males and 1 female), and bat-eared foxes have been observed to rest and sleep in close contact with one another. Mutual grooming (allogrooming) appears to have an important social function in this species, with the face being the area that is groomed most frequently (Kleiman, 1967). The face has conspicuous black markings. Perhaps grooming served to reduce tension (Terry, 1970) in the artificial conditions of captivity in which these animals, who supposedly live in pairs (see below — Kruuk, 1971), were forced to live (larger groups in a small area). In primates, allogrooming also serves an important social function (see Sparks, 1967 for a review of this literature).

Females mark only around the time that they are in heat (Ewer, 1968). During mating, the copulatory tie has been observed (Kleiman, 1967; Dewsbury, 1972). Adults urinate and defecate far from the burrow and single sites are used only 2–3 times (Smithers, 1966b), which is probably important in avoiding unwanted predators. In captivity, they have not been observed to mark vertical objects (Kleiman, 1966), and when they roll on the ground, they roll from side-to-side and do not writhe, as do other canids. The side-to-side movement was not observed by Kleiman (1967) in other canids, but has been by Fox (personal communication).

During expressions of dominance and during attack (which occurs extremely rarely), the tail is held in an inverted "U" position and does not wag, possibly making the black stripe which runs along its tail more conspicuous (Kleiman, 1967). This tail position has not been observed in similar motivational contexts and is unknown in the rest of the Canidae except for the raccoon dog, *Nyctereutes procyonoides* (Kleiman, 1967). During a defensive gape, the lips cover the teeth as they do in all other African canids except for the golden jackal (Seitz, 1959b; van Lawick-Goodall, 1971).

Escape play is typical of the bat-eared fox and von Ketelhodt (1966; cited by Ewer, 1968) has related this to the fact that this species has reduced dentition and escape behavior probably plays a more important role in their lives than does fighting.

Van der Merwe (1953b) noted that packs of 6 or more frequently ran together in search of food, however, Kruuk (1971) wrote that in their natural environment, bat-eared foxes live in pairs. Out of 79 observations, Hendrichs (1972) observed 18 single, 28 pairs, and 24 trios of bat-eared foxes. Perhaps these packs were families, observed prior to dispersal. However, more observations are needed in order to make any definite statement about the social organization of this species. Bat-eared foxes are primarily insectivorous (Bothma, 1959, cited by Ewer, 1973; Burton, 1962; Crandall, 1964; Dorst and Dandelot, 1970), eating termites and beetles, and also feeding upon small rodents, scorpions, eggs of ground-nesting birds, lizards, and fruits. Bat-eared foxes are preyed upon by raptorial birds and very rarely by spotted hyenas (*Crocuta crocuta*) (Kruuk, 1972a). Ewer (1973) reports observations by Leakey (1969, cited by Ewer, 1973), in which *Otocyon* stole prey from falcons.

Many more data are obviously needed on the African foxes, both from observations on captive animals and long-term studies. Developmental studies, best performed in the laboratory for a number of reasons (Bekoff, 1972a), would be of interest, and may aid in understanding the variations in social organization. Predator–prey relationships, which are poorly documented, also require further investigation.

JACKALS

Side-striped Jackal (*C. adustus*)

The side-striped jackal (Figure 9-4) is found throughout a large portion of Africa on open savanna (Dorst and Dandelot, 1970). However, Bueler (1973, *Wild Dogs of The World,* Stein and Day, New York) reports that *C. adustus* prefers thick bush and woods. It is fairly rare in the Serengeti (Kruuk, 1972a), and not very much is known about its

Figure 9-4 *C. adustus,* side-striped jackal. Photo: M. W. Fox.

behavior. The side-striped jackal is greyish-fawn in color, has a white stripe along its flank, and has a bushy tail. It stands about 50 cm at the shoulder, females weigh approximately 11 kgs, and males approximately 14 kgs (van der Merwe, 1953b). In body size, it is slightly smaller than the black-backed jackal, and it has smaller and shorter ears than does the golden jackal (Dorst and Dandelot, 1970).

The female marks all year round and uses a full leg cock (Kleiman, 1966). The breeding season lasts approximately from September to November and the gestation period is roughly 2 months. In the London Zoo, 2 litters had 1 and 3 pups, respectively (Zuckerman, 1953) and Dorst and Dandelot (1970) reported that litters can number up to 6 individuals.

Hunting is usually done singly or in pairs (Kleiman, 1967) and packs have been observed (van der Merwe, 1953a). The side-striped jackal is supposedly a scavenger (Dorst and Dandelot, 1970) and is more nocturnal than the black-backed jackal. They do not hung large animals and feed mostly on carcasses, small mammals, birds, eggs, reptiles, lizards, insects, and vegetable matter. Dentition suggests that *C. adustus* is less predacious than *C. mesomelas* (Ewer, 1956). They do not appear to harm livestock (Dorst and Dandelot, 1970).

Black-backed Jackal (*C. mesomelas*)

The black-backed jackal (Figure 9-5) has been extensively studied by van der Merwe (1953a,b). Recently, Lombaard (1971) has compiled extensive growth curves for *C. mesomelas* Schreber to assist biologists in evaluating populations of these beasts. It is fairly widely distributed in South Africa, Southwest Africa, and in Kenya and the Sudan. Dorst and Dandelot (1970) reported that the black-backed jackal tends to live in open savanna and light woodland and Kruuk (1972a) observed that black-backed jackals occurred more in bush country. Standing approximately 40 cm at the shoulder and weighing approximately 9–14 kgs, it has a dark back which appears silvery at a distance. It is primarily nocturnal (Hendrichs, 1972 observed activity from 5:00–10:00 am) and many become nomadic for a few months a year (van Lawick-Goodall, 1971). Dentition is similar to that of the coyote (van der Merwe, 1953b).

Black-backed jackal pups are born from August to the middle of November with an average litter size of 4 (range: 1–8) and a gestation period of about 60 days. Van der Merwe (1953b) observed a pregnant female with young cubs, possibly indicating a double-estrus period. The average litter size of those born in captivity at the London Zoo was 2 (Zuckerman, 1953). In the wild, black-backed jackals do not dig their own burrows but usually occupy an old aardvark (*Orycteropus afer*) hole (van der Merwe, 1953a). The female has not been observed to mark at any time of the year and does not use the leg cock (Ewer, 1968).

Black-backed jackals demonstrate a very strong pair bond and both the male and the female share in the care of the young (van der Merwe, 1953a; Wyman, 1967). Both animals defend their territory (approximately 7.2 km across, but varying with the habitat) and the male supplies the female with food (van der Merwe, 1953a). If pairs separate, they remain in contact by emitting frequent calls, and the female has also been noted to emit a mating call (van der Merwe, 1953a).

Van der Merwe (1953a) has recorded some data concerning the development of black-backed pups. Eye-opening usually occurs on the 8th day of life. The female has 8 mammae of which only 6 are functional, and the young are weaned at about 8–10 weeks; at this time, they are sometimes already accompanying the adults on hunts. If the pups have to be transported because of possible danger, they are carried by the scruff of the neck. The burrow is abandoned by the pups when they are 2–3 months of age, at which time there is a change in coat color from dull lead to the characteristic brown and white saddle on the back and rufous on the flanks, legs, and lumbar region.

Figure 9-5 Black-backed jackal, *C. mesomelas*. Photo: M. W. Fox.

Before the litter disperses at approximately 4–5 months, mate preferences and pair bonds are established and there appears to be a fair amount of inbreeding. What causes dispersion is not known. At 6 months of age, the pups are practically fully mature.

Hunting is done either singly or in pairs, and prey is generally stalked and then pounced upon (van der Merwe, 1953a). Hendrichs (1972) observed black-backs 250 times and recorded single animals 81 times and pairs 140 times. Small antelope and lambs are killed with a throat-oriented bite. Olfaction appears to be heavily relied upon, since the jackals always hunt against the wind (van der Merwe, 1953a). Shortridge (1934) wrote that black-backed jackals have been observed to hunt in bands of 3–5 individuals, and in packs they are very formidable predators (van der Merwe, 1953a). Black-backed jackals do not feed on the belly — males prefer the hindquarters and females the chest and forequarters (van der Merwe, 1953a). Pack members appear to communicate with one another by means of a fairly elaborate vocal repertoire (Dorst and Dandelot, 1970) and also appear to be able to signal to another party when a kill is located (van der Merwe, 1953a). However, it is possible that their excited vocalizations may serve to attract other congeners anyway.

It is not uncommon to see black-backed jackals scavenge from a previous kill by another animal (e.g., lion) and their relationship with the wild dog (*Lycaon pictus*) appears to be the same as that of the golden jackal (Kruuk, 1972a). Estes and Goddard (1967) had reported that the golden jackals they observed behaved more aggressively and boldly at kills than did black-backed jackals.

The diet of the black-backed jackal consists mainly of rodents, hares, birds, reptiles, insects, young antelope, wild berries, and various fruits (see van der Merwe, 1953a for a detailed list, and also Bothma, 1971b,c and Grafton, 1965). Veld-burning destroyed a good deal of its food, and consequently it took to sheep-killing, which ultimately led to large-scale extermination of the black-backed jackal because of the damage it did to livestock (the animal was in sort of a double-bind situation) (van der Merwe, 1953a). When they were wantonly destroyed, other animals were left without any natural check (van der Merwe, 1953a). As with most of the other African Canidae, more data are sorely needed.

Golden (Common or Asiatic) Jackal *(C. aureus)*

The golden jackal (Figure 9-6) is found in large areas of northern Africa. It is dirty-yellowish in color, stands approximately 30 cm at the shoulder, and weighs in the neighborhood of 11 kgs (Dorst and Dandelot, 1970). It is fairly well-conditioned to the presence of man. Like its congeners, it is a successful scavenger (mostly done at night) (Drimme, 1954; Crandall, 1964), and is often observed surrounding feeding lions (Grzimek, 1970; van Lawick-Goodall, 1971). The golden jackal can scavenge more successfully from lions than can hyenas (Kruuk, 1972a). Its chief asset appears to be its speed (van Lawick-Goodall, 1971) and it is a more persistent runner than a fox (Seitz, 1959b).

Recent field observations by the van Lawick-Goodalls (1971) in the Ngorongoro Crater and Kruuk (1972a) in the Ngorongoro Crater and on the Serengeti have added a considerable amount of data concerning the behavior of this animal in its natural habitat (see also Golani and Keller, Chapter 22 for observations on the Asiatic jackal). The van Lawick-Goodalls (1971) provide interesting observations on various aspects of social behavior (e.g., dominance–subordination relationships, communal howling, courtship, and hunting) and could identify individuals by their distinctive pattern of whisker growth. They observed baring of the teeth during a defensive gape (grin of fear) as has been observed in captivity by Seitz (1959b) and Fox (1971b). In addition to field observations, observations on captive subjects by Seitz (1959b), Golani (1966), Golani and Mendelssohn (1971), and Fox (1971b) have helped to clarify certain aspects of the social ontogeny and social behavior of the golden jackal.

Figure 9-6 Asiatic jackal, *C. aureus*. Photo: M. W. Fox.

Both male and female have been observed to scent mark, usually raising a hindleg (van Lawick-Goodall, 1971). During the rutting season, the male marks more and scratches the ground which serves to spread its mark and produces conspicuous visible marks (Seitz, 1959b). Both male and female have been observed to perform ground-scratching movements in the field by the van Lawick-Goodalls (1971).

Golani and Mendelssohn (1971) have described in great detail the precopulatory behavior of golden jackals, which generally extends over a 4 month period. Males and females reach sexual maturity at approximately the same time (Fox, 1971b), and in the field, the male was observed to remain constantly with the estrous female (van Lawick-Goodall, 1971). During the rutting season, the pair establishes a territory and usually hunts together (Seitz, 1959b; van Lawick-Goodall, 1971), which is a change from their usual solitary hunting behavior (van Lawick-Goodall, 1971). The pair bond may persist over a number of years (Golani and Mendelssohn, 1971). After intromission, the pair "lock." (Golani and Mendelssohn, 1971)

The van Lawick-Goodalls (1971) have made some very interesting observations on the social development of 4 young pups in the wild. They noted that the eyes opened at approximately 10 days and that grooming was important in strengthening affectionate ties. Both parents regurgitated food for the young (also noted by Drimme, 1954). This was elicited by the pups licking the parent's face and lips, similar to that behavior observed during greeting and submission. Seitz (1959b) observed a "greeting ceremony" in the young — low crouch with tail-wagging, snout pushing against the corners of the mouth, whining, and rolling over. Regurgitation near the den as opposed to out in the open reduces the likelihood that birds of prey would be able to swoop down and snatch up young pups.

Furthermore, it was observed that within the litter of 4, there were definite differences in physical appearance and behavior. The most venturesome of the pups and the usual leader after weaning (which occurred just after the pups were 2 months of age), was a male, who was also the first to incorporate body-slamming into his play and was the most likely to transform games into fighting at the least provocation. He could possess and successfully retain food. Similar individual differences have been observed in a litter of coyotes (Bekoff and Jamieson, unpublished) and the complex relationship between within-litter individual differences and later social organization are more fully discussed in Bekoff (unpublished). After weaning, the cubs became more independent and this correlated closely with the first signs of real aggression within the litter. When the cubs were 4 months old, play was rougher, often ending in fighting (van Lawick-Goodall,

1971). Finally, the van Lawick-Goodalls (1971) reported that it is common for female offspring, in particular, to remain with the parents to help care for the next litter.

The diet of the golden jackal varies extensively with its habitat (van Lawick-Goodall, 1971; Kruuk, 1972a). It seems to enjoy a balanced diet throughout the year consisting of insects, small mammals (rodents), some fruit, food scavenged from kills, and the young of small gazelle (both in the Serengeti and in the Ngorongoro Crater). After the wildebeest (*Connochaetes taurinus*) calving season, placenta provide an abundance of food for jackals. Snakes are also a normal part of the jackals' diet (van Lawick-Goodall, 1971). The van Lawick-Goodalls (1971) observed golden jackals chasing hares, but never catching them. One pair of jackals hunted over an area of 1.6 km^2, but some do have larger hunting areas — usually larger on the plains than in the crater.

As mentioned above, golden jackals usually hunt alone. However, they have been observed to hunt in small packs as well as in pairs (Dorst and Dandelot, 1970; van Lawick-Goodall, 1971). Eisenberg and Lockhart (1972) have observed cooperative behavior in a pair of golden jackals hunting deer in Ceylon, and the van Lawick-Goodalls (1971) observed several successful hunts on adult Thompson's gazelle (*Gazella thomsonii*) by small packs of jackals (composed of a full-grown male and female plus a 1 year old offspring) in which the prey were disemboweled (similar to the hunting method of the wild dog). They are more successful in catching gazelle fawn if they hunt in pairs (67% success) than if they hunt alone (16% success) (Kruuk, 1972a). Female gazelles are not usually successful in defending their fawns against a pair (Wyman, 1967).

Kruuk (1972a) observed social relationships between spotted hyenas and both golden and black-backed jackals. Jackal scavenging had no effect on the food supply of the hyena, and since jackals could catch gazelles, hyena were able to scavenge from them. Neither jackals nor hyenas are prey for one another. In the Ngorongoro Crater, jackals appeared to profit the most (Kruuk, 1972a).

As with the foxes of Africa, more data are needed concerning the above three species of jackal. Studies such as Kruuk's (1972a) on the spotted hyena and Schaller's (1972) on the Serengeti lion will hopefully be undertaken.

SIMIEN FOX (ABYSSINIAN WOLF) (*Canis simensis*)

Very little is known about this animal. It stands approximately 60 cm at the shoulder, is rufous in color, and has a long, bushy tail. Hunting is done either singly, in pairs, or in small packs, and the primary prey are rodents. It lives in the high plateaus of East Africa, and is now very rare and is considered to be an endangered species (Guggisberg, 1970), as it has been exterminated as a predator of livestock.

CAPE HUNTING DOG (WILD DOG) (*Lycaon pictus*)

The wild dog, Figure 9-7, perhaps the most gregarious of all canids, inhabits large portions of the African continent, living in wooded and open savanna. It is scarce in the Serengeti (Schaller, 1972). There is geographical variation in morphology (size and coat color — to be discussed below). Allen (1939) recognized 7 different subspecies. Wild dogs have been placed in the subfamily *Simocyonimae* along with the bush dog (*Speothos venaticus*), but this classification has been questioned on behavioral grounds (Kleiman, 1967; Fox, 1971b). Ewer (1973) compares *Lycaon* to *C. lupus* and Mech (Chapter 24), discussing similarities between wolves (*C. lupus*) and wild dogs, suggests that perhaps wild dogs should be classified as *Canis*.

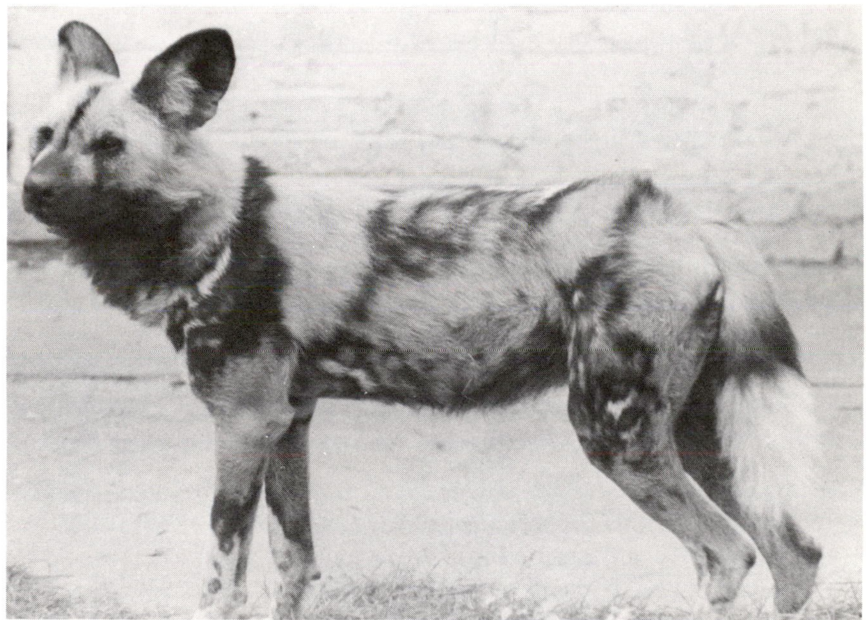

Figure 9-7 Cape hunting dog, *Lycaon pictus*. Photo: M. W. Fox.

The reader should be referred to the writings of Kühme (1965a,b), Estes and Goddard (1967), Kruuk and Turner (1967), Frame (1970), van Lawick-Goodall (1971), Kruuk (1972a,b), Hendrichs (1972), Pfeffer

(1972), and Schaller (1972) for descriptions of many aspects of the social organization of this species. I shall attempt a brief summary of what is known — which is considerably more perhaps than for any of the other African canids.

The wild dog is long-legged, with large, broad, and rounded ears. It is frequently called the "hyena-dog" (German: Hyänenhund) because of its ears (Drimme, 1954). Patches of black, brown, yellow, and white are distributed over the body, and a very consistent body marking appears to be a white-tipped tail, which probably helps maintain visual contact in bush country, high grass, and under crepuscular conditions (Estes and Goddard, 1967). It is very conspicuous in moonlight (Hugo van Lawick, personal communication). White is also common on the lower legs, feet, and chest, and almost all dogs have a patch of yellow on the sides of the neck, the forehead, and the root of the tail (James Malcolm, personal communication). In addition, the wild dog secretes a strong musky odor which may be of aid in olfactory tracking through bush country for dogs who have lost visual contact with the rest of the pack (Estes and Goddard, 1967). This species has a "contact (distress) call" (Kühme, 1965a,b; Estes and Goddard, 1967), as well as an "alarm bark" and a high-pitched "twittering" call, discussed below (Estes and Goddard, 1967).

Body hair is short and coarse and the feet have 4 toes instead of 5 as in true dogs (Dorst and Dandelot, 1970). In East Africa, the wild dog is darker and the average weight is approximately 18 kgs, while Central and South African types weigh between 22–27 kgs, and are several centimeters taller (Estes and Goddard, 1967; Frame, 1970). The latter hunt larger prey.

Wild dogs display diurnal activity rhythms (Kühme, 1965a,b; Estes and Goddard, 1967; Kruuk and Turner, 1967; Hendrichs, 1972) and are nomadic. There is little mixing of separate packs within the same area (Kühme, 1965a,b; Kruuk, 1972a). Only when the pups are young do the adults tend to stay in one place for a while (van Lawick-Goodall, 1971). There is no evidence to suggest that packs are territorial, but it is possible that they defend the area around the den (Schaller, 1972).

There is some information about courtship and social development of the young. The gestation period is approximately 69–72 days and the average litter consists of approximately 7 pups (Frame, 1970), but since more than one female may occupy the same den, exact litter size is often difficult to determine (Crandall, 1964). Large litters of 9–13 have been recorded in conditions where there was only one female present (Crandall, 1964), and the van Lawick-Goodalls (personal communication) observed a litter of 16 pups. The van Lawick-Goodalls (1971) observed a very short copulatory "tie" (approximately 45–50 seconds;

James Malcolm, personal communication) and more information about the reproductive behavior of the male wild dog is presented by Dewsbury (1972).

When the female is ready to bear her young, she seeks a burrow that has been abandoned by another animal (e.g., aardvark, warthog (*Phacochoerus aethiopocus*), or hyena) in which to den (Frame, 1970). Pfeffer (1972) observed that in the packs of wild dogs which he observed north of the Central African Republic, parturition seemed to coincide with the end of the rainy season. Wild dog females have 12–14 teats and young are usually nursed in a standing position, at least at the age when they are out of the den (van Lawick-Goodall, 1971). Weaning takes place at approximately 10–12 weeks (Kühme, 1965a).

Females, according to Kleiman (1966), have not been observed to mark with their urine in captivity, however, the van Lawick-Goodalls (1971) have observed a dominant female marking the area around her own den and also around a subordinate female's den, and also noted an increase in marking by an estrous female just before and during the period of sexual attraction.

Lastly, although Kühme (1965a,b) and Estes and Goddard (1967) failed to observe any dominance hierarchy, but did note that there was a regular leader (male), the van Lawick-Goodalls (1971) observed two separate hierarchies in the pack that they watched. As in wolves, there were separate male and female hierarchies, with the male hierarchy being much more difficult to sort out. The van Lawick-Goodalls' (1971) discussion of neck presentation by the subordinate animal (pp. 93–94) might also be interpreted as the avoidance of eye-contact (cut-off) by the subordinate (see Fox, 1971b). There are other aspects of the social behavior and organization of the wild dog which have been observed by most, if not all, of the investigators noted above, and these appear to be relatively independent of pack size or locale. They are as follows:

1. The wild dog is a gregarious species, with the pack functioning as a highly coordinated unit in rearing of the young, mutual defense, and hunting;
2. The core of the social organization promoting pack harmony is a ritualized greeting ceremony based upon infantile begging behavior (emancipated or socioinfantile behaviors — Fox, 1971c) face-licking, nudging the corners of the mouth, and sometimes audible whining. Accordingly, aggressive behavior is minimal. The greeting ceremony regularly occurs before a hunt ("pep rally");
3. Hunters regurgitate food for the young and their guards, and young have priority at a kill (up to a certain age). Considering the

high death rate of adults, any factor contributing to a high survival rate of pups has selective advantage (Schaller, 1972);

4. Wild dogs display a diurnal activity cycle — most active during early morning and early evening;
5. Wild dogs are visual hunters. Prey are run down and frequently the chase is preceded by a preliminary "open" stalking phase with no attempt at concealment (James Malcolm, personal communication). The wild dog is considered to be a "courser" and not a "stalker" (Schaller, 1972; Ewer, 1973). The dogs do not run in relays (Schaller (1972) observed 3 relay hunts out of 48 observed hunts). Prey are disemboweled where the skin is thinnest, there is no killing-bite, and wild dogs do not scavenge frequently;
6. The wild dog does not possess an elaborate repertoire of facial expressions, and tail postures are rarely used for communication. (More data are needed on this aspect of wild dog behavior. This author recently viewed slides of wild dogs and the possibility remains that very subtle visual signals are used.)

Wild dog packs vary in size. Kruuk and Turner (1967) reported an average of 9.2 individuals in a pack with the most frequent sizes being 6–10 and 1–5 individuals, respectively. Out of 68 observations, Hendrichs (1972) recorded packs of 8 or 9 dogs 17 times, and packs of 11–20, and 21–30 individuals, 22 times each. Estes and Goddard (1967) wrote that a minimum of 4–6 dogs is needed for the pack to function effectively and that large packs seemed to be more efficient in their use of prey. There is no relation between the number of members of a pack and the prey hunted because usually only a few members of the pack do the actual hunting (Kruuk and Turner, 1967).

Many data have been collected concerning the predatory behavior of the wild dog in different locales (Bourliere, 1963; Kühme, 1965a,b; Mitchell, Shenton, and Uys, 1965; Estes and Goddard, 1967; Kruuk and Turner, 1967; van Lawick-Goodall, 1971; Kruuk, 1972a,b; Pfeffer, 1972; Ewer, 1973) and its relationship with other predators (Estes and Goddard, 1967; Kruuk, 1972a). As mentioned above, wild dogs are not frequent scavengers nor do they appear to be very dependent upon water (Kühme, 1965a; van Lawick-Goodall, 1971). Schmidt-Nielsen (1972) has recently offered an anatomical–physiological explanation which accounts for the wild dog's ability to run for long distances (see below) in hot climates. Prey selection varies with geographical location and habitat, and Kruuk and Turner (1967) stressed that habitat is of major importance when trying to explain variations in diet. Wild dogs usually search for prey with the pack trailing behind the leader. If prey double back, then the dogs trailing the leader can intercept them (Kruuk and Turner, 1967; Estes and Goddard, 1967). The wild dog's exploitation of a narrow band of the food spectrum requires remarkable social organization (Kruuk, 1972a). Recent data (James Malcolm, personal communication) indicates that the wild dog has a more varied diet than has previously been thought. As was mentioned above, hunts are regularly preceded by a "pep rally" in which members of the pack greet, and actively submit to one another, frequently emitting a "twittering" call, a high-pitched chatter indicating a high level of excitement (Estes and Goddard, 1967).

East African wild dogs most frequently hunt Thomson's gazelle while Central and South African types (larger than East African congeners) hunt larger prey such as reedbuck and impala (*Aepyceros melamphus*) (Estes and Goddard, 1967). Pfeffer (1972) observed his wild dogs catch a water buffalo in the water. Kruuk and Turner (1967) observed that in the bush on the Serengeti plains medium-sized antelope were the prey of choice, whereas on the plains there were clear preferences for Thomson's gazelle of either sex. Walther (1969) has reported that Thomson's gazelle have a larger fleeing distance from wild dogs than they do from either jackals or lions. On the Serengeti, small Thomson's gazelle constituted 64% of the total prey (3 young, 10 adult males, and 14 females) whereas medium-sized antelope made up only 17% of the diet (Kruuk and Turner, 1967).

In the Ngorongoro Crater, 54% of the prey killed (50 observations) were Thomson's gazelle, 67% of which were territorial males (Estes and Goddard, 1967), caught as they doubled back to their territories (a good example of site attachment in territorial animals; Tinbergen, 1957). This is of selective advantage for the gazelle population in that younger males will be able to breed (Estes and Goddard, 1967). Newborn and juvenile wildebeest (*Connochaetes taurinus*) accounted for 36% of the kills, and Grant's gazelle (*Gazella granti*), 8%, and Kongoni (*Alcelaphus buselaphus*), 2%, comprised the remainder (Estes and Goddard, 1967). Surplus killing has been observed by Kruuk (1972b) but this is a rare phenomenon.

Relations with other predators have been observed. The spotted hyena is a serious competitor of the wild dog, and Asiatic jackals were seen more frequently than black-backed jackals at kills (Estes and Goddard, 1967). However, it appears that wild dogs can usually effectively keep hyenas away (Wright, 1960; Kühme, 1965b; van Lawick-Goodall, 1971; Kruuk, 1972a). Interactions between hyenas and wild dogs are highly variable, probably depending on the number of animals and the extent of their hunger (Schaller, 1972). Various species of vultures scavenge from wild dog kills, but they were not observed to steal food from the dogs as they were eating, as did kites (*Milvus migrans*) (Estes and Goddard, 1967).

The length and success of the hunt is dependent upon habitat and prey choice. Kruuk and Turner (1967) observed that if a pack of wild dogs stayed with one particular gazelle, the chase would last for 6–8 km. The van Lawick-Goodalls (1971) observed chases at speeds up to 48 km per hour for 4.8 km. Generally, if prey were not captured within 3.2–4.8 km, the attempt was abandoned. Estes and Goddard (1967) reported an overall success rate per chase of over 85% (50 kills observed), while the van Lawick-Goodalls (1971) observed 31/91 (34%) successful hunts. Finally, the mean time between starting an activity cycle to capturing the prey was recorded to be 25 minutes by Estes and Goddard (1967); and Schaller (1972) noted than an average of 30 minutes elapsed from the time that a pack left its den to the first kill.

In conclusion, the wild dog appears to be a highly social canid, effectively functioning in packs to provide equally for all members. Their social system provides a good example of the role that ritualized behavior may play in social organization, supporting Ploog's (1970) suggestion that ritualization involves the canalization of expressive behavior. No doubt, more insight will be gained concerning their unique social structure as more data are amassed. Questions that would be of interest include: (1) how does the leader become leader, (2) is their early social ontogeny similar to that of the wolf (*vs* a more solitary species), and (3) how is the social structure conserved from generation to generation (Crook, 1970)?

CONCLUSION

More studies such as those by van der Merwe (1953a) and those cited above for the wild dog are sorely needed. Mech (1970), Kruuk (1972a), and Schaller (1972) provide good examples for the organization of future research. All aspects of behavior must be considered — e.g., the ecological and behavioral specializations of the animals and selective pressures to which they were, and still are subjected (Kleiman and Eisenberg, 1973). Phylogenetic relationships may be compiled based on comparative behavioral data (e.g., Kleiman, 1967), as well as on anatomical similarities and differences (e.g., Atkins and Dillon, 1971).

Developmental and long-term longitudinal studies must be undertaken both in captivity and in the field. Studies conducted on captive subjects may yield data that is highly relevant to social behavior observed in the wild. Also developmental studies are frequently more successfully undertaken in captivity (see Bekoff, 1972b), and furthermore, behavior observed in captivity may be of great import in the normal environment, even though field workers have not yet described it (Chance, 1962).

Geographical differences, seasonal and daily fluctuations, spatial relationships (within and between species — e.g., McBride, 1964; Brown and Orians, 1970; Kummer, 1971a,b), interrelations with other species, individual and sex differences, intralitter, adult–infant, and adult–adult relationships are among those variables which must be accounted for when studying the social behavior and ecology of these canids. Normal ranges of variability should be determined and species differences understood. The compilation of ethograms and sociograms (e.g., "who" does "what" to "whom," "how many times" and "when" and "where") is essential.

Andrewartha (1971; p. 21) noted five components of the environment which must be considered in any study (resources; mates; predators, pathogens, and aggressors; weather; and malentities), and Eisenberg (1966) listed eight relevant factors that must be recorded in the description of social organization. These include: are the individuals dispersed or aggregated into groups; what are the sex and age classes within a group; is there a gradient of dominance within the social grouping; is there a leadership status; is there a group territory; what is the permeability; how cohesive is the group; and how permanent are the relations among group members? It is essential to understand the social group within which an individual lives because presumably it is adapted to the environment of the species and represents the social organization within which the individual survives and procreates the best (Crook, 1970). The great variability that is already evident within the family Canidae is warning enough that "procrustean" lumping is misleading.

In addition, as stressed by Leyhausen (1965), more studies are needed on solitary animals — how do they refrain from contact with one another and how do they change from solitary to group life and vice versa. Further investigations of nocturnal species are also necessary.

Finally, in order to be able to maintain animals in captivity, knowledge of their behavior in the field is essential (Hediger, 1950). To avoid further wanton destruction of important predators and of the ecosystems within which they are an important link (e.g., Guggisberg, 1970; p. 128), in-depth field and laboratory studies are urgently needed. Anderson and Herlocker (1973, Soil factors affecting the distribution of the vegetation types and their utilization by wild animals in Ngorongoro Crater, Tanzania. *Journal of Ecology,* 61, p. 627–651) have recently published a paper relating soil types, vegetation, and utilization by lightly, he may alter the natural order and indirectly threaten his own place in the unnatural order thus created (McCabe and Kozicky, 1972).

Acknowledgments I would like to thank Hugo van Lawick (Gombe Stream Research Centre) and R. D. Estes for reading an earlier draft of this chapter. Also, thanks to Wanda Meek for typing the manuscript. I recently had the pleasure of speaking with James Malcolm (currently at Harvard University, Biological Laboratories), who spent 14 months studying wild dogs. He helped to clarify many aspects of *Lycaon* behavior. The author's present address is: University of Colorado, Department of Environmental, Population and Organic Biology, Boulder, Colorado 80302.

10 ECOLOGY OF THE ARCTIC FOX *(Alopex lagopus)* IN NORTH AMERICA — A REVIEW

DAVID L. CHESEMORE*
Oklahoma Cooperative Wildlife Research Unit
Oklahoma State University
Stillwater, Oklahoma

ECONOMIC IMPORTANCE

Macpherson (1969) notes that the pelt of the Arctic fox attracted many of the first European explorers and traders to the lands bordering the Arctic Ocean. Each year Canadians sell 10,000 to 68,000 Arctic fox pelts that have an annual value of 50,000 to 2 million dollars (Anon., 1962). In Canada, Arctic fox populations vitally affect the settlement and economy of large, inhospitable areas.

The Arctic fox is the primary and often only source of earned income for many Alaskan Eskimos living in northern and western Alaska (Chesemore, 1967). Although encroachment of modern civilization continues to reduce the dependence of Alaskan natives on existing wildlife resources, fish, game, and furbearers still play a major economic role in their lives. Trade in Arctic fox pelts developed in northern and western Alaska in conjunction with whaling, but became

*Present address: Department of Biology, California State University, Fresno, California 93710.

important only after 1900 when whaling was no longer profitable. Intensive fox trapping occurred in Alaska during the "fur boom" of the 1920s until the 1929 economic depression destroyed the market for wild furs. Since 1929, a steady decline in pelt value and total harvest of white foxes has occurred, and today, with a few local exceptions, the trapping of this furbearer has little impact on the economy of northern Alaska.

Siberian traders were shipping pelts of the blue phase of the Arctic fox from Alaska as early as 1745 but not until 1798 are there records of the white phase of the Arctic fox in the Russian fur trade (Petroff, 1898). From 1912 to 1963, Alaska produced an average annual harvest of approximately 4000 white fox pelts. Ranch raising of Arctic foxes, especially the blue phase, began about 1865 on the coast of Alaska and it soon spread to various points in eastern Canada (Peterson, 1966).

PHYSICAL CHARACTERISTICS

The Arctic fox is small, ranging in weight from 1.4 to 3.2 kg (Figure 10-1). Morphologically, this animal is well-adapted to arctic conditions with its reduced limb size, short snout, short, rounded ears, thickly-haired feet, and dense winter coat which may turn white in winter. Scholander, *et al.* (1950) discusses physiological adaptations of the Arctic fox to its cold environment. Physical features and body measurements of the *Alopex* group have been analyzed by several authors (Hilderbrand, 1952; 1952a; 1954; Tsalkin, 1944; Vibe, 1967). Aging techniques for these foxes are discussed by Grigor'ev and Popov (1952), Dolgov (1966), Macpherson (1969), and others. Closure of skull sutures and eruption of canine teeth are accurate guides to fox age.

The Arctic fox occurs in two color phases, white and blue, which differ in both their summer and winter pelages. This dichromatism is controlled by recessive genetic inheritance (Johansson, 1960). The blue phase is rare in the central Arctic region of North America but becomes common in the eastern Arctic, Greenland, and western Alaska (Anderson, 1937; Fetherston, 1947; Chesemore, 1970). In Alaska, the blue phase, introduced by man with the possible exception of those found on Attu Island (Murie, 1959), predominates on the Aleutian and Pribilof Islands; northward it is replaced by the indigenous white phase (Nelson, 1887). Detailed description of the pelage patterns and molt sequence in *Alopex* may be found in several reports (Freuchen, 1935; Lavrov, 1932; Chesemore, 1970).

Considerable variation exists in the timing of the spring molt of *Alopex* but generally the molt period extends from March or April until June or July (McEwen, 1951). Pocock (1912), Lavrov (1932), and

Figure 10-1 Arctic fox in summer pelage on St. Lawrence Island, Alaska. Photo: Bob Stephenson.

Barabash-Nikiforov (1938) discuss hair replacement and pelage change during the sprint molt. The autumn molt, changing the brown-grey summer pelage to the pure white coat of the white forms, begins in September and may extend until mid-October on Baffin Island (Soper, 1928), November on the southwest coast of the Taymyr Peninsula (Kirpichnikov, 1937) or December on Novaya Zemlya (Dubrovskii, 1937). Pocock (1912) reported on the autumn molt of captive white foxes in London's Zoological Garden. The back and tail were the last parts of the fox pelage to turn white during the autumn molt.

Studies on the factors influencing changes in fox pelage are lacking, but probably climatic and geographic factors, particularly light, as well as intrinsic physical and physiological conditions all interact to govern the periodicity of priming in this species. Lavrov (1932) found differences in the time of priming between the foxes of Novaya Zemlya and northwestern Siberia, indicating latitudinal influences. Although Faester (1943) correlated the warmer annual temperature in Greenland with the annual increase of white foxes taken in "summer skins," Vibe (1967) believed a lower death rate among these animals, attributed to favorable weather conditions, was the reason for the increase in the occurrence of "summer skins." Eskimo trappers from the Barrow, Alaska area think that a warm autumn produces poorer fur during the following winter trapping season, but no data exist to support or refute this statement.

Preble (1915 in Osgood, *et al.*) found that about 6% of the harvest of Arctic foxes consisted of white pelts. This was prior to the initiation of intensive measures to extirpate the white phase from the Pribilof Islands in the 1800s. Dr. F. H. Fay, biologist at the Arctic Health Research Center, College, Alaska, (personal communication) noted that 80% (44 of 55) of the Arctic foxes seen on St. Matthew Island in July and August, 1963 and 99% (2125 of 2137) of those noted on St. Lawrence Island from 1956 to 1961 were white. Stephenson (1970) found 2 litters of fox pups on St. Lawrence containing both white and blue forms. Few blue Arctic foxes are taken at Barrow, Alaska, although occasionally they are seen in the area.

EVOLUTION AND TAXONOMY

Arctic fox remains have been found in European Pleistocene deposits, being especially common in deposits containing abundant reindeer and lemming remains (Trouvessart, 1899; Zittel, 1925). Beddard (1902) states that the remains of Arctic foxes have been found as far south as Germany and England.

Lavrov (1932) sums up the geologic history of this group in the following statement:

> In early times, . . . the distribution of the Arctic fox covered a much greater area than it does at present. In excavations and examinations of caves, bones have been discovered belonging to the prehistoric Arctic fox (in France, Switzerland, Austria-Hungary, Germany, Poland, etc.). Bones have also been found in Siberia, near Krasnoyarsk, Nizhne-Udinsk, and other places. These remains show that . . . the Arctic fox inhabited the whole of Europe and apparently a large part of the territory of Siberia.

Arctic foxes have a probable Old World or Nearctic origin in the Pleistocene but a definite fossil record is lacking (Savage, 1958).

Simpson (1945) divides the Canidae into two subfamilies, placing the Arctic fox, *Alopex,* into the subfamily Caninae. He also includes the red and kit fox, *Vulpes*, the fennec fox, *Fennecus,* and the gray fox, *Urocyon* in the Caninae. Miller (1912) and Lavrov (1932) describe the anatomical differences between *Alopex* and other members of the Caninae.

Taxonomists divide the Arctic fox group *(Alopex)* into four different species (Miller and Kellogg, 1955), into two species (Ognev, 1962) or consider all Arctic foxes as one circumpolar species while recognizing several subspecific forms (Braestrup, 1941; Rausch, 1953a; Walker, 1964; Vibe, 1967). A thorough revision of the nomenclature of this group is needed to resolve present taxonomic inconsistencies. *Alopex lagopus* is used in this review to denote all Arctic foxes. Some of the common names used for the *Alopex* group include: Arctic fox, blue fox, white fox, renard arctique, polarfuchse, podrossen, and polar fox.

DISTRIBUTION

The Arctic fox inhabits the arctic or tundra zone of North America (Figure 10-2), Eurasia, and portions of the alpine zone in the mountains of Norway and Sweden. Also it breeds on the barren islands of the Arctic, North Atlantic, and North Pacific Oceans (Ognev, 1962; Vibe, 1967; Macpherson, 1969).

In Alaska, the white Arctic fox is found along the coast from the Kuskokwim River north to Point Barrow and eastward along the arctic coast to the Alaskan–Canadian border (Petroff, 1898; Dufresne, 1946). White foxes predominate on St. Lawrence, St. Matthew, Hall, and Diomede Islands as well as on most of the other islands in the Bering Sea. They also occur occasionally among the predominantly blue phase of the populations on the Pribilof and Aleutian Islands.

The erratic movements of white foxes between the islands and coasts of Alaska and Siberia are apparently dependent upon ice conditions existing in those respective regions. Vibe (1967) discusses fox distribution and its relationship to sea ice conditions (Figure 10-3) in the eastern Arctic. This species is capable of travel for long distances over the sea ice. Foxes have been observed by Naval Arctic Research Laboratory (Barrow) personnel far out on the polar ice pack, approximately 640 km north of the arctic coast of northern Alaska. During the 1962–63 trapping season, an Eskimo from Wainwright, Alaska trapped a white fox bearing a Russian neck tag. Assuming the fox was tagged in Russian territory, perhaps on Wrangel Island, it had completed a considerable journey before finally being trapped near Wainwright.

Figure 10-2 Winter habitat of tundra foxes, northern Alaska, December, 1962. Photo: Dave Chesemore. Reproduced by permission of the National Research Council of Canada from the *Candian Journal of Zoology,* **47,** 121–129 (1969).

Records exist of white foxes being found as far south as the Brooks Range and in Interior Alaska, but most Arctic foxes are found on the tundra and coastal plain areas of northern and western Alaska. Dependence on arctic tundra conditions and perhaps interspecific competition with the red fox (*Vulpes vulpes*) are the most probable reasons for the existing distribution of Arctic foxes in Alaska. Formozov (1946) discussed the effect of snow on the life of foxes and related it to their present tundra distribution. Sdobnikov (1967) analyzes *Alopex* distribution on the Asiatic tundra areas.

In Canada, the Arctic fox ranges from the northern Yukon, Mackenzie, and Keewatin Districts of the Northwest Territories to the western shores of Hudson Bay and Baffin Island, the northern parts of Alberta, Saskatchewan, and Manitoba, and that part of northwestern Ontario bordering on Hudson Bay (Anderson, 1946). Arctic foxes are also found on most of the islands of the Canadian Arctic and range far out onto the sea ice in winter. In eastern Canada, it is primarily a coastal species occurring around the Hudson and James Bay coasts, along Hudson

Strait, and on the Labrador coast (Peterson, 1966). Sporadic movements place foxes on Newfoundland where specimens have been taken as far south as St. Shotts on the Avalon Peninsula (Peterson, 1966).

Figure 10-3 Winter habitat of sea ice foxes in sea environment, December, 1962. Photo: Dave Chesemore.

MOVEMENTS AND MIGRATIONS

Arctic fox movements may be classified into four general categories: local movements, the general, daily travels of individual foxes; sporadic movements, those involving the unpredictable occurrence of individual foxes many miles from their normal range; seasonal movements, those that are correlated with seasonal changes in the environment; and migrations, periodic movements involving many foxes traveling long distances in one sustained direction (McEwen, 1951).

On the Alaskan Arctic Slope, two distinct seasonal movements occur, with the first in the fall and early winter months when foxes move seaward towards the coast and sea ice, and the second in late winter and early spring months when they return inland to mate and occupy summer den sites. Similar seasonal movements have been noted in the Hudson and James Bay regions, on Southern Baffin Island, and on the Taymyr Peninsula (Richardson, 1839; Preble, 1902; Seton, 1929; Soper, 1944; Chitty and Chitty, 1945; Alekseev, 1957).

The fall and winter movement is probably triggered by food scarcity (Chitty and Elton, 1937; Barabash-Nikiforov, 1938). Rausch (1958) found foxes often abundant along the Arctic Coast of Alaska during the fall and early winter after an abrupt lemming (*Lemmus* and *Dicrostonyx*) decline. White foxes were abundant in the Barrow coastal region in 1949, 1953, 1954, and 1956, the year of the greatest concentration of foxes on the coast (Rausch, 1958). Instances of sporadic movements of Alaskan white foxes have been summarized by Chesemore (1968). McEwen (1951) attributes his 40 records of this species occurring far south of their normal Canadian range to sporadic movements.

Arctic foxes are capable of traveling long distances over sea ice and being carried by ice floes to many areas that they do not normally occupy (Bailey and Hendee, 1926; Lewis, 1942; Anderson, 1946; Elton, 1949; Cameron, 1950; Banfield, 1954; Vibe, 1967). Foxes commonly immigrate to the Bering Sea islands via the pack ice from the north (Preble and McAtee, 1923; Murie, 1936). Nansen saw fox tracks on polar ice 240 km north of Franz Josef Land and McClintock saw a blue fox in March, 1958 on the winter ice of Baffin Bay about 208 km from Greenland (Elton, 1949).

Migrations of white foxes, as described in other areas of the Arctic, have not been recorded from Alaska. Reports of fox migrations note that foxes follow the coastline southward, and it is expected if such a marked, mass movement involving many foxes occurred in Alaska, it would either be recorded in the literature or at least observed by local trappers. Neither has occurred. Definite fox migrations have been reported from Novaya Zemlya, the Yamal and Taymyr Peninsulas, Yenisey Tundra, and in northern Europe (Lavrov, 1932; Kalashnikov, 1936, Vasil'ev, 1938; Chirkova, 1951; Yakushkin, 1963).

Chirkova (1955) stated that the shortage of food is the trigger of white fox migrations and this view is supported by other Russian workers (Lavrov, 1932; Dubrovskii, 1937; Kirpichnikov, 1937). Seton (1929) believes that the migration of white foxes consists of only the young of the year, the surplus population that the range cannot support, and that this movement is truly emigration rather than migration since few of the young animals survive to return to their home areas. Recent data by Pulliainen (1965) indicates that it is mainly male foxes that are involved in these movements. Vibe (1967) also notes that roaming foxes are more likely to be males. Emigrating foxes are reported to follow caribou (*Rangifer tarandus*) herds and to feed on caribou killed and lef by wolves (*Canis lupus*) (Manning, 1943). Other biologists have also recorded foxes following caribou herds (Critchell-Bullock, 1930; Dubrovskii, 1937; Clarke, 1940).

FOOD HABITS

The food habits of the Arctic fox necessarily reflect the seasonal distribution and abundance of available prey species within the animal's habitat. Distinct regional differences in the summer diet of inland and island foxes and in the winter diet of inland and sea ice foxes exist. The fluctuations of primary prey species density must periodically alter the eating habits of the foxes.

Small mammals, particularly lemmings, are the major fox prey during the summer months with many studies noting the importance of these animals in fox diets (Braestrup, 1941; Elton, 1942; Pitelka, *et al.*, 1955; Chesemore, 1968a; Macpherson, 1969; Stephenson, 1970). Birds and eggs also form a large part of the foxes' summer diet. Foxes prey heavily on nesting birds and are extremely effective predators, often governing the nest locations of sea bird colonies (Figure 10-4) (Turner, 1886; Nelson, 1887; Manniche, 1912; Osgood, *et al.*, 1915; Lavrov, 1932; Dubrovskii, 1937; Kirpichnikov, 1937; Bertram and Lack, 1938; Braestrup, 1941; McEwen, 1951; Fay and Cade, 1959). Ptarmigan (*Lagopus* spp.) are also eaten by foxes during the summer (Gross, 1931; Barabash-Nikiforov, 1938; Gunderson, *et al.*, 1955). Other summer food items in the fox diet include fish, caribou, sea mammals, insects, berries, grasses, and various herbaceous plants (Osgood, *et al.*, 1915; Freuchen, 1935; Barabash-Nikiforov, 1938; Braestrup, 1941). Macpherson (1969) discusses summer food habits in detail in his work on Arctic foxes in the Keewatin District, Canada. Lavrov (1932) found that marine invertebrates were an important part of the summer diet of island foxes. Stephenson (1970) compared summer food habits of foxes in inland and bird-cliff habitats on St. Lawrence Island and found that the summer diet of Arctic foxes clearly reflected the regional differences in the composition of the prey population. Inland foxes depended mainly on small mammals for food while those near the cliffs were dependent primarily on alcids for food.

Many reports of foxes caching food for winter use exist in the literature (Feilden, 1877; Beddard, 1902; Osgood, *et al.*, 1915; Seton, 1929; Freuchen, 1935; Dubrovskii, 1937; Braestrup, 1941; Soper, 1944; Tuck, 1960; Barry, 1967). Chesemore (1968a) found no evidence of food caching during field work in northern Alaska but Stephenson (1970) found several such caches on St. Lawrence Island and felt that the foxes on

Figure 10-4 Arctic fox in bird cliff habitat, St. Lawrence Island, Alaska. Photo: Bob Stephenson.

the island depend on prey cached in the summer months for part of their winter diet. The caches found during his study indicate that the foxes inhabiting lowland areas may take advantage of a temporary increase in the vulnerability of microtine rodents due to loss of snow cover and the concurrent flooding of burrows during the spring thaw. These caches were located well away from the more extensive dens where pups could not interfere with them. Although no caches were found in the bird-cliff habitats, local residents reported that in previous years they had found caches of "little auklets and other birds" in these areas (Stephenson, 1970). Additional study of this caching behavior is needed.

The winter diet of foxes depends primarily upon the habitat in which it spends the winter season. Food secured by foxes on sea ice is restricted to marine mammals and invertebrates, resident sea birds, seaweeds, algae, and marine fishes (Manniche, 1912; Osgood, *et al.*, 1915; Lavrov, 1932; Barabash-Nikiforov, 1938; Dubrovskii, 1937; Braestrup, 1941). Foxes may follow polar bears (*Ursus maritimus*) on the sea ice to feed on the remains of bear kills (Soper, 1928; Freuchen, 1935; Dufresne, 1946; Cahalane, 1947; Bee and Hall, 1956). Foxes certainly could benefit from such behavior, and their ability to locate meager food items under harsh conditions is essential for survival on the sea ice. Forty white foxes were observed feeding on a dead walrus (*Odobenus rosmarus*) frozen into the sea ice off of Tangent Point, Alaska, on December 5, 1961. The foxes had dug through 0.6 m of ice to reach the carcass, chewed a 30 cm diameter hole through the thick walrus hide, and were feeding inside the carcass. Schiller (1954) also observed foxes near St. Lawrence Island feeding on dead walruses frozen into the ice.

Winter diets of inland or tundra fox consist mainly of lemmings. This lemming diet may be supplemented by other microtines, caribou, arctic hares (*Lepus arcticus*) and ptarmigan (Soper, 1928; Freuchen, 1935; Pedersen, 1926). Foxes have followed reindeer herds and attacked weak fawns (Blanchet, 1925; Porsild, 1945).

REPRODUCTIVE BIOLOGY

Arctic foxes are monestrous, monogamous, and may mate for life (Seton, 1929; Dement'yev, 1955). Males mature at 10 months, while having a minimal, functioning testes weight of 2.6 gms (Sokolov, 1957). Spermatogenesis begins in January and mature spermatozoa are found in the epididymis from March until the end of July (Sokolov, 1957; Asdell, 1964). Large follicles occur in the ovaries in February and

March; corpora lutea are well-developed until the beginning of August (Asdell, 1964).

Fox mating, indicated by sets of paired tracks in the snow, begins in March and early April in northern Alaska, at the end of April or early May in Siberia (Sokolov, 1957), and in March or April on Southhampton Island, Canada (Sutton and Hamilton, 1932). Sokolov (1957) found no indication of estrus from January until April 10.

After a gestation period of 52 days (Barabash-Nikiforov, 1938; Johnson, 1946), the annual litter is born in May, June, or early July. A high reproductive potential for this species is indicated by the reports summarized by Chesemore (1967). Alaska trappers usually noted 4 to 8 pups per Arctic fox litter but they had seen as many as 12 pups in a single family. Placental scar counts closely resembled actual litter sizes in a group of Arctic foxes examined by Macpherson (1969). The morphology of Arctic fox placental scars has been described by McEwen and Scott (1956). In his study, Macpherson (1969) found that 33% of the 1 and 2 year old vixens bore placental scars while 85% of those 3 years and older had produced young. He felt that the 1 and 2 year old vixens breed in their first year if from a large cohort or in their second year if from a smaller one. Based on a sample of 118 foxes, he found an average of 10.5 placental scars per female with no statistical evidence of variation in reproductive potential between years, although the mean size of weaned litters was varied greatly.

Braestrup (1941) indicates that a real difference in pup production between the blue and white phases of the Arctic fox exists in Greenland, with the white phase being more prolific. Reduced production of fox pups during periods of food scarcity has been noted (Braestrup, 1941; Chirkova, 1951; Dement'yev, 1955). Macpherson (1969) found that at Resolute Bay and in the Central District of Keewatin, Canada, weaned litter counts varied directly with lemming numbers. The average litter size was 9.7 pups in 1960, 4.6 in 1961, and 0 in 1962, while associated midsummer lemming indices were 85, 57, and 13, respectively. It is evident that the litter suffers considerable mortality after implantation and before trapping season begins. Macpherson's view is that food scarcity may increase pup mortality indirectly by increasing intrasibling strife. During this conflict resulting from competition for limited food supplies brought to the den by the vixen, the stronger siblings kill the weaker ones. Similar reproductive losses are well-documented in other animals, especially in ungulate populations (Severinghaus and Cheatum, 1956), and it appears reasonable that similar mechanisms, in this case increased sibling rivalry, operate to control population levels of Arctic foxes during times of food scarcity.

POPULATION CYCLES

Analysis of Arctic fox harvest data from Alaska indicate that periodic highs and lows in the fox population occur approximately every 4 years. An average 4 year population cycle is well-established for Arctic foxes in Ungava and Labrador, Greenland, Novaya Zemlya, northern Quebec and in the Northwest Territories, Canada (Elton, 1942; Braestrup, 1941; Dubrovskii, 1937; Chitty and Chitty, 1945; Butler, 1951). While essentially the same duration, the Alaskan Arctic fox cycle is not simultaneously in-phase within these regions.

These periodic increases in foxes are usually closely correlated with a similar population increase in the primary prey of the foxes, small rodents. In the Barrow, Alaska area, the fox population highs in the past 30 years have usually coincided with or preceded brown lemming (*Lemmus trimucronatus*) population peaks (Pitelka, *et al.,* 1955; Thompson, 1955; Rausch, 1958). The southward migration of snowy owls (*Nyctea nyctea*) also coincides with the abrupt lemming decline, and usually indicates a high, if not peak, Arctic fox population in the north (Gross, 1931, 1947; Kirpichnikov, 1937).

It is clear that the fox and lemming cycles are interrelated, but exact data on the mechanisms involved in these interactions is lacking. Diseases, parasites, predation, competition, migrations, and seasonal movements all may play important roles in these population fluctuations. Litter size reduction and den abandonment are the overriding determinants of cohort size and it is believed that their scale is determined by the relative scarcity of lemmings, with sibling aggression the proximate cause of at least some of this reduction (Macpherson, 1969).

DEN ECOLOGY

Dens of Arctic foxes have been studied in the U.S.S.R. (Tsetsevinski, 1940; Danilov, 1958; Skrobov, 1960a, 1961, 1961a), in Canada (Macpherson, 1969), and in Alaska (Chesemore, 1969; Stephenson, 1970). Fox dens differ considerably in size, complexity, and development with these differences related to usage and age. Sdobnikov (1967), Skrobov (1961), and Macpherson (1969) present classification systems for Arctic fox den sites.

Skrobov (1961) reports that on the Yamal Tundra the percentage of occupied dens is a reliable indicator of the coming harvest. He notes a 3% occupancy indicates a low trapping harvest while 30% occupancy a moderate harvest during the next season.

Figure 10-5 Group of young foxes at den site, St. Lawrence Island, Alaska. Note protective coloration of summer pelage (4 foxes in picture). Photo: Bob Stephenson.

In the early spring after pairing, Arctic foxes occupy underground burrows (Figure 10-5) where the female gives birth to her pups. These dens are used in northern Alaska from late spring until mid-August, when pup dispersal occurs, and then are abandoned until the next breeding season. Macpherson (1969) reports that in Canada foxes often utilize dens throughout the winter; various trappers on St. Lawrence Island also reported similar observations (Stephenson, 1970). Chesemore (1969) found no usage of dens during the winter along the northern coast of Alaska but fox use of temporary burrows dug in snow drifts was noted.

Terrain determines the type of den site which the Arctic fox will occupy. Foxes may establish dens in rocky areas near cliffs with the arrangement of the den being dependent upon the structure of the rock and talus. Stephenson (1970) describes this type of den on St. Lawrence Island as does Barabash-Nikiforov (1938) on the Commander Islands. Foxes in tundra habitat of northern Alaska established their dens in low mounds 1 to 4 m high. The depth to the permafrost's surface below the ground's surface determined if an Arctic fox could successfully den in northern Alaska; many unfinished burrows 14 to 18 cm deep were found terminated at this permafrost layer. A minimum mound height of 1 m appeared necessary for an Arctic fox to successfully dig a den. The shallow, ice-free layer and a wetter, heavier soil in the mounds less than 1 m high prevented the establishment of suitable fox dens. Generally, habitats with dens had warmer soil temperatures than did those habitats lacking fox dens. The location of the den site in relation to that of lakes and rivers on the tundra apparently varies with the locality. Danilov (1958) noted a relationship between den sites and rivers on the Bol'shezemel'skaya Tundra but no such relationship was noted in northern Alaska. Examination of unstabilized dune areas along rivers in northern Alaska revealed no fox dens; the sand at these sites appeared too loose to support the construction of a tunnel system. However, dens were common in the sandy, stabilized soils along the river systems.

The presence of a fox den radically alters the floristic community near the burrow (Chesemore, 1969). The typical dry tundra community is replaced by a lush growth of Poaceae 30 to 50 cm high. This is probably due to the addition of organic materials to the soil plus the physical disturbance, aeration, and mixing of the soil that occurs when the fox digs and uses a den site. During late summer, the lush vegetation surrounding fox dens facilitated their location by ground observers and they were also readily visible from the air.

An average of 4 den entrances was found at each site, but the largest den found during the study in northern Alaska had 26 entrances and occupied an area of more than 100 m^2. Den sites, essentially circular in area, usually occupied less than 30 m^2 per den site. From the large size of several dens (Figure 10-6), it appears that they are used repeatedly year after year, either by the same or other foxes. A tendency to position den entrances on the warmer exposures was indicated.

The influence that territorialism has in determining the density of occupied dens and the home range of Arctic foxes is not known but it may be important in determining population densities of certain areas. Arctic fox dens were not very abundant in the Northwest Territories (Macpherson, 1969) with the average density of dens at the Aberdeen Lake study area being 1 den per 36 km^2. Den density appears greater in the U.S.S.R. with densities of 1 to 6 dens per 10 km^2 having been reported in the Bol'shezemel'skaya region (Danilov, 1958; Skrobov, 1961). Dement'yev (1955) found that the upland tundra favorable for denning had an average density of 1 den per 12 km^2. Sdobnikov (1958) found high densities of dens in the maritime tundra belt of Taymyr, where up to 2 dens per km^2 were recorded along the major river valleys that formed the favored denning habitat of foxes. Boitzov (1937) esti-

Figure 10-6 Large fox den at edge of tundra stream, Northern Alaska. Fox usage and wind erosion create bare areas at den site (July 5, 1962). Photo: Dave Chesemore. Reproduced by permission of The National Research Council of Canada from the *Canadian Journal of Zoology*, **47**, 121–129 (1969).

mated an average density of 1 den per 32 km^2 for the tundra zone of the U.S.S.R., which is similar to the estimated den density found at Aberdeen Lake.

Conspicuous groupings of Arctic fox dens were found in several areas near Barrow, Alaska, indicating that foxes will locate dens here with only a few meters separating them from neighboring den sites. A group of 9 dens was found within a 30 m section of low mound habitat, and another concentration of 13 dens all within 65 m of one another was located in a series of 2 to 4 m high mounds along the shore of Tusikvoak Lake.

Very high densities of breeding foxes, though coupled with small litters, are obtained on the open, island "ranches" of the North Pacific Ocean. Ashbrook and Walker (1925) indicate that up to 200 foxes were maintained on island "fox farms" of 16 hectares.

Macpherson (1969) studied the spatial dispersion of fox dens in the Aberdeen Lake area. He found that in 2 years out of 3, fox dens were more widely spaced than they would probably have been if they were randomly dispersed. This suggests territoriality in denning foxes. However, there was no tendency toward a more uniform dispersion with an increase in the breeding population, nor was spacing maximal. The minimum distance between occupied dens, irrespective of the density of occupied dens, was about 2 km. He concluded that the number of occupied dens in the Aberdeen Lake area is limited neither by habitat nor territorial behavior.

COMPETITORS AND PREDATORS

Alaskan trappers consider red foxes, wolves, wolverines (*Gulo gulo*), and polar bear to be the primary predators of Arctic fox; dogs (*Canis familiaris*) also may kill some foxes (Seton, 1929; Lavrov, 1932). Manniche (1912) found a fox killed by a polar bear in Greenland. Soper (1928) considers the red fox the most aggressive Arctic fox predator, but the human trapper may deserve this title.

Skrobov (1960) notes the general northward advance of the red fox and its replacement of the Arctic fox wherever their ranges overlap. Marsh (1938) reports a similar occurrence at Eskimo Point, 368 km north of Churchill, Manitoba. There, Eskimos found the red fox occupying Arctic fox dens on the tundra. Dubrovskii (1937) also noted red fox occupancy of Arctic fox den sites. Chirkova (1967) discusses the relationships between the Arctic fox and red fox in the far north.

Avian predators probably include snowy owls, large hawks, and jaegers. Sutton (1932) saw a white fox killed by an owl and Bee and Hall (1956) recovered an Arctic fox radius and ulna from a snowy owl pellet at a den site near Teshekpuk Lake, Alaska. Both parasitic and long-tailed jaegers (*Stercorius parasiticus* and *S. longicaudus*) have attacked Arctic fox pups (Collett and Nansen, 1900; Birulya, 1907). Foster (1955) considered rough-legged hawks (*Buteo lagopus*) to be a main enemy of foxes. Avian predation on the Arctic fox is probably prevalent only on fox pups at den sites but recorded instances of these interactions are lacking.

Intraspecific strife between Arctic foxes may occur and during periods of food scarcity or stress, Arctic foxes may be cannibalistic (Lavrov, 1932; Dubrovskii, 1937). Adult foxes may kill and feed on other adults (Osgood, *et al.*, 1915) or pups (Lembley and Lucas, 1902; Pedersen, 1934). Devold (1940) has found young foxes feeding on their littermates, while Macpherson (1969) ascribes this strife a major role in *Alopex* population regulation.*

*Which would seem to be a logical deduction in view of the apparent lack of prenatal regulation, since Arctic foxes characteristically have large litters — Ed.

An Arctic fox was observed on St. Matthew Island driving away others from a reindeer carcass on which they were feeding. Never more than one fox fed at the carcass at any one time. Arctic foxes also exhibited antagonistic behavior towards one another while at a garbage pit near the campsite on the same island (Chesemore, 1967). Freuchen (1935) reports he saw a white fox actively defending a territory during the summer, driving away all strange foxes. Stephenson (1970) observed territorial behavior among Arctic foxes on St. Lawrence Island in mid-June. This involved one fox sporadically chasing another for a distance of about 200 m along a plateau with both foxes vocalizing frequently during the 10 minute chase. Bédard (1967) noted that the periphery of the Kongkok basin, a large cirque located on St. Lawrence Island, was also divided among 5 or 6 foxes and he observed territorial clashes among them. Territoriality among Arctic foxes needs more study to determine its effect on fox distribution and population levels.

Interspecific competition between Arctic foxes and other animals is probably most common in competition for food. Avian predators, primarily snowy owls and jaegers, and the least weasel (*Mustela rixosa*) all compete with the foxes in northern Alaska for lemmings (Pitelka, *et al.*, 1955). Also, wolves, red foxes, wolverines, and dogs actively compete for food, especially during the winter when food supplies are limited. Instead of competing with Arctic foxes, polar bears may augment the meager fox food on the sea ice by unintentionally leaving remains of their kills which the foxes find and utilize for food.

DISEASES

Rabies was recorded in Arctic foxes from northern Alaska during the high fox populations of 1949–50, 1954, and 1956–57, but also was found in foxes during years of population lows (Rausch, 1958). Most rabies cases in foxes occur during the cold months of the year but their source of infection is unknown. Elton (1931) has speculated on a relationship between a rabies-like disease in lemmings, which if assumed to exist during lemming die-offs, could be transmitted to foxes. However, no laboratory proof of the existence of such a disease in lemmings has been found.

Rabies could be an important factor in fox mortality. During 1953–54, a rabies outbreak occurred among Arctic foxes on St. Lawrence Island, Alaska. It was more extensive than the Barrow rabies outbreak of 1956–57, and possibly caused a significant reduction in the numbers of foxes on the island. Freuchen (1935) regarded rabies as an important factor in the population decline in eastern Canada. Rabies in this species has also been reported from the western Canadian Arctic, Greenland, Novay Zemlya, and Siberia (Braestrup, 1941; Elton, 1931, 1942; Lavrov, 1932; Plummer, 1947a, 1947). Syuzyumova (1967) discusses the epizootiology of rabies among Arctic foxes on the Yamal Peninsula. In addition to rabies, Lavrov (1932) found Arctic foxes infected with encephalitis, paratyphoid fever, and coccidiosis. Chirkova (1953) reported finding tularemia among Arctic foxes in the Voronezh Province.

Considerable literature is devoted to "sledge-dog disease" in the Arctic. Elton (1931) felt that this disease, affecting the central nervous system in dogs and Arctic foxes, was neither ordinary rabies nor dog distemper. Research by Kantorovich (1956) revealed similarities between an etiologic agent isolated from the Arctic fox and the rabies virus. After study, he decided this agent was a variety of the rabies virus (Kantorovich, 1957). Between 1958 and 1960, Strogov (1961) found 6 out of 94 Arctic foxes infected with this rabies variant. This variant may be identical to the sledge-dog disease of Elton and relatively widespread in the Arctic. Cowan (1949) suggests that the 1949 decline of Arctic foxes in the Mackenzie District, Canada, may have been caused by an epidemic resembling rabies. Autopsies of wolves and dogs attacked by infected animals showed positive evidence of rabies.

Chirkova (1951) noted the relationship between the lack of food, fox migrations, and the start of fox epizootics in her studies. In North America no such relationships have been found.

PARASITES

Abundant literature on the parasite fauna of Arctic foxes has been published. Chesemore (1967) recovered. *Taenia crassiceps, T. polyacantha, Echinococcus multilocularis, Toxascaris leonina,* and *Trichinella spiralis* from foxes collected near Barrow, Alaska. Other cestodes found in Arctic foxes include *Taenia pisiformis* (Dubrovskii, 1937; Dubnitskii, 1953; Lavrov, 1932; Thomas and Babero, 1956), *Dipylidium caninum* (Dubrovskii, 1937; Lavrov, 1932); *Diphylobothrium latum* (Lavrov, 1932; Rausch, 1953), and *D. erinacei-europaei* (Dubinina, 1951). Nematodes found in foxes include *Uncinaria stenocephala* (Dubnitskii, 1956; Lavrov, 1932; Parnell, 1934; Dubrovskii, 1937; Thomas and Babero, 1956; Olsen, 1958), *Eucoleus aerophilus* (Lavrov, 1932), *Toxascaris canis* (Lavrov, 1932, Kirpichnikov, 1937), *Spirocera lupi* (Luzhkov, 1960), and *Strongyloides vulpis* (Petrov and Dubnitskii, 1946). Only two trematodes, *Opistohorchis felineus* (Lavrov, 1932; Luzhkov, 1961) and *Microphallus pirium* (Schiller, 1959) have been reported from Arctic foxes. *Listeria monocytogenes* was found in foxes by Cromwell, Sweebe, and Camp (1939) and by Nordland (1955). Few ecotoparasites have been recovered from Arctic foxes, with *Linognathus setosus* being found by Ferris and Nuttall (1918), Lavrov (1932), Critchell-Bullock (1930), and Sutton and Hamilton (1932).

The infection of humans with *Echinococcus multilocularis* results in development of alveolar hydatid disease in man (Rausch, 1967) and creates a public health problem in many northern areas. The Arctic fox is the most important primary host of *E. multilocularis* in North America (Rausch, 1958). Distribution of this cestode from Asia via Arctic foxes may have been accomplished by fox movements over Arctic pack ice. Rausch (1956) believes this cestode was transported to St. Lawrence Island from Asia in this manner. Similar movements from St. Lawrence to the Alaskan mainland followed, increasing the distribution of this cestode. *E. granulosus* has never been recorded from the Arctic fox (Rausch, 1956).

SUMMARY

The trapping of Arctic foxes is often the only source of earned income for many Eskimos and Indians living in northern areas. Circumpolar in distribution, *Alopex* is a relatively small fox physiologically adapted to its cold environment. Dichromatism in this species is probably controlled by recessive genetic inheritance. A thorough revision of the *Alopex* group is needed to resolve present taxonomic inconsistencies.

Foxes utilize tundra habitat as well as the sea ice environment with fox movements being apparently governed by sea ice conditions. Preference for arctic conditions and interspecific competition with red foxes may determine the present distribution of Arctic foxes in northern areas. Mass emigrations of this species have been observed in some arctic areas but no such movements have been recorded in Alaska. A shortage of food may be the trigger of these fox emigrations, with males more likely to roam than females.

Food habits of Artic foxes reflect their seasonal distribution and abundance of available prey species within the animal's habitat. Small mammals, particularly lemmings, are the major fox prey during summer months. In certain areas, birds and eggs also form a large part of the summer diet. Foxes cache food for winter use in some areas. The winter diet of foxes on sea ice is restricted to marine mammals and invertebrates, resident sea birds, and other marine life. Utilization of the remains of polar bear kills may be an important source of food for foxes on sea ice. Inland, the winter diet consists mainly of small mammals, caribou, Arctic hares, and ptarmigan.

Artic foxes are monestrous, monogamous, and mate in the early spring, producing a litter of pups in May, June, or July. A relationship between litter size and small mammal abundance has been noted in northern Canada. Periodic highs and lows in the fox population occur approximately every 4 years. Litter size reduction and den abandonment determine final cohort size with increased sibling aggression occurring with the decrease in availability of food. Increased sibling aggression may trigger the periodic population decline noted in *Alopex*.

In early spring, foxes occupy burrows where the female gives birth to her pups. Dens may be used all year in some areas or abandoned in late summer when pup dispersal occurs. Terrain determines the type of den site Arctic foxes will occupy. The depth of the permafrost's surface below the ground's surface determines if an Arctic fox could successfully den in northern Alaska. The presence of a fox den radically alters the floristic community near the burrow. The typical dry tundra community is replaced by a lush growth of grasses which makes the den area readily visible from the air. Territorialism may determine the density of occupied dens in certain areas but its exact impact is unknown. Territoriality among Arctic foxes needs more study to determine its affect on fox distribution and population levels.

The occurrence of rabies and *Echinococcus multilocularis* in foxes, the cause of alveolar hydatid disease in man, creates a public health problem in many northern areas.

11
ECOLOGY OF THE GRAY FOX *(Urocyon cinereoargenteus)*: A REVIEW

GENE R. TRAPP
and
DONALD L. HALLBERG
Department of Biological Sciences
California State University
Sacramento, California

The range of the gray fox (*Urocyon cinereoargenteus*) extends from northern Oregon and southeastern and southcentral Canada to northern South America; however, it excludes the central and northern Great Plains and the northern Rocky Mountains, Great Basin, Cascades, and Coast Range (Hall and Kelson, 1959). Although common in many parts of its range, the gray fox is secretive and therefore inconspicuous. Knowledge of gray foxes has come mostly from studies conducted in the eastern, southeastern, and northcentral United States. There have been few formal ecological studies of this species in the western United States or Latin America, and almost no behavioral studies of it anywhere in its range. The purpose of the present paper is to survey the literature on the gray fox for ecology, life history, and social behavior. The closely related island foxes (*Urocyon littoralis*) that inhabit islands off the coast of southern California are not discussed in this survey except to say the following: (1) little is known of this species other than taxonomic descriptions (Laughrin, 1971), (2) Lyndal Laughrin, University of California, Santa Barbara, is currently studying its ecology, (3) for a few natural history observations on this species see Grinnell, Dixon, and Linsdale (1937), and Laughrin (1971), (4) see Von Bloeker (1967) for comments on origin, and Leach and Fisk (1972) for brief comments on its status as an endangered species.

NICHE: SPATIAL AND TEMPORAL ASPECTS

Habitat

Throughout most of the gray fox's range its habitat could be described as shrublands and brushy woodland, on hilly, rough, rocky, or broken terrain. The gray fox probably functions best in this type of habitat, possibly because of competition (but this is not documented) from the other foxes that live in different, but adjacent, habitats depending on geographical location, *viz.,* red fox (*Vulpes fulva*), swift fox (*V. velox*), and kit fox (*V. macrotis*). Where such possible competitors are absent, one may find exceptions to the general statement above.

For the eastern, northeastern and northcentral United States, the general statement holds. See Layne and McKeon (1956b) for New York, Failor (1969) for Pennsylvania, Leopold (1931), Errington (1935), and Richards and Hine (1953) for the Wisconsin–Iowa region, and Schwartz and Schwartz (1959) for Missouri. In southern Georgia, Wood, Davis, and Komarek (1958) found that gray foxes are most abundant in mixed woods and cultivated areas, less abundant in savanna pine, and least abundant in mixed woods with dense underbrush, paralleling the relative prevalence of foods. This seems to be an exception to the rule and might be explained by the fact that red foxes do not occur in the extreme southeast, so in the absence of competitive exclusion pressures, the gray fox has changed its use of the available habitats.

Palmer (1956) says that gray foxes occurred in portions of the northeastern United States in precolonial and colonial times, then became scarce for many years, but appear to have been reinvading that area since the 1930s. It is curious that the same apparent disappearance of the gray fox in southern Ontario for almost 300 years appears to coincide roughly with the introduction (in about 1650) and establishment of the European red fox (*Vulpes vulpes*) in the northeast, the native red fox (*V. fulva*) apparently having been absent from the eastern coastal states prior to this introduction (Peterson, Stanfield, McEwen, and Brooks, 1953, and citations therein). Whether the introduction of *V. vulpes* and apparent disappearance of the gray fox are related is not clear. Palmer (1956) suggests that the reappearance of the gray fox may be due to the range extension and increase in abundance of the prey species *Sylvilagus floridanus* and *S. transitionalis,* following increases in agricultural land at the expense of forest. Palmer did not specify whether the habits or habitat of gray foxes had changed since colonial times. For more comments on the early records of gray and red foxes in Ontario, see Peterson, *et al.* (1953).

Urocyon apparently benefits from "edge effects" created by man. Unlike the kit fox and swift fox, the gray fox does nearly as well on the outskirts of cities as it does in less disturbed habitats (Leopold, 1959). Wood and Davis (1959) believe that in Georgia the trend to decrease cultivated land, and increase timber and pulp forests or pasture land, will reduce that state's gray fox population.

Gray fox habitat in Texas is similar to that of the eastern United States, but also included are wooded sections of the short-grass plains and pinyon–juniper woodlands above the low-lying desert (Davis, 1960). The open sections of the southern plains were occupied by the swift fox before trapping and poisoning reduced its numbers. Habitats of gray fox in the western United States and Latin America include chaparral, woodlands of pinyon–juniper or oak, and rocky hillsides, mountainsides, and washes (Nelson, 1930; Grinnell, *et al.*, 1937; Cabrera and Yepes, 1940; Hardy, 1945; Johnson, Bryant, and Miller, 1948; Leopold, 1959). These habitats usually coincide with the "Upper Sonoran Life Zone," but gray foxes may inhabit areas somewhat below or above this zone, depending on the region (Grinnell, *et al.*, 1937). In Zion National Park, gray foxes are abundant in the blackbrush (*Coleogyne ramosissima*), brushy meadows (*Atriplex canescens, Bromus* spp.), open meadows (*Bromus* spp.) and pinyon–juniper (*Pinus monophylla–Juniperus osteosperma*) communities to the lower portion of the ponderosa pine (*Pinus ponderosa*) zone at about 1554 m. Comparative use of these communities by *Urocyon* is described by Trapp (1973). Lack of red fox in the park may have led to greater use by gray fox of the open meadows and relatively open brush meadows.

Denning

Places used for dens include slab, scrap, or brush piles, space under old houses, holes in rocky outcroppings, cavities in hollow trees, and less frequently, underground burrows dug by other animals or by foxes themselves (Seton, 1929; Grinnell, *et al.*, 1937; Sullivan, 1956, and references therein; Layne and McKeon, 1956b; Davis, 1960). Dens dug by foxes generally are prepared where little difficult digging is required, e.g., in loose soil, or in burrows (dug by other animals) that need only slight enlargement (Sullivan, 1956). Dens in hollow trees may be as high as 7.6 m above ground (Grinnell, *et al.*, 1937) or 9.1 m (Davis, 1960). A whelping den may contain grass, leaves, or shredded bark as nest material (Seton, 1929; Grinnell, *et al.*, 1937). In the East, den sites are commonly located in dense cover, less than .40 km (¼ mile) from water (Sullivan, 1956; Layne and McKeon, 1956b). Biotelemetry data from Zion National Park (Trapp, 1973) revealed that gray foxes often rest in different parts of their home range each day for several days in a row.

Foxes have often been seen resting in the sun on top of a large rock, or up on a limb or in the crotch of a tree (Grinnell, *et al.*, 1937; Yeager, 1938). One was seen resting in the top of a tree in the abandoned nest of a hawk (Seton, 1929).

Home Range

Richards and Hine (1953) report that *Urocyon's* home range in Wisconsin varied from .40 to 2 km in diameter (0.13–3.10 km^2). From recapture distances in northern Florida, Lord (1961a) estimated a home range diameter of 3.2 km (7.7 km^2). From biotelemetry data on tagged foxes in semiarid Zion National Park, Trapp (1973) obtained a mean home range (lumping all seasons) for 4 adult males and 4 adult females of 1.0 km^2. The home range for a juvenile male was .52 km^2 in July and August, 1967. Home ranges in the canyons of Zion National Park tend to be elongate, because they are restricted laterally by the steep canyon sides and clifflike outcroppings.

If home range sizes were compared over the geographic range of the gray fox, one would reasonably expect that they would become relatively smaller as the food resources increased. But more measurements of home range will be needed before this can be clarified since the three above-mentioned studies, taken alone, only serve to confuse the picture. The population estimates discussed later in this paper do seem to support this hypothesis however. Why should home ranges in arid Zion National Park be smaller than in northern Florida, unless Lord's (1961a) study area was indeed less productive, even with its greater rainfall? One explanation might be that fox home range sizes may vary with population density so that local population fluctuations would serve to confuse the picture when comparing them with different parts of the continent.

Territoriality

No mention was found in the literature of territoriality in *Urocyon,* nor of fighting other than in the mating season (Grinnell, *et al.*, 1937). From evidence collected by trapping and recaptures in Florida, Lord (1961a) found that at least 21 "family aggregations" of gray foxes appeared to be spatially separate from other family aggregations. An aggregation appeared to be a family unit consisting of an adult male and female, plus a number of juveniles. Evidence collected via biotelemetry studies in Zion National Park (Trapp, 1973) supports Lord's descriptions. The apparent separateness of 4 adjacent ranges occupied by adult males suggested territoriality. The mapped home ranges of 3 adult females, 1 presumed subadult female, and 1 juvenile male coincided closely in their essential features with one or the other of two of the adult male ranges (Trapp, 1973). Two apparent agonistic encounters between gray foxes were observed by Trapp.

Scent posts may aid in marking extent of territories. Urine and scats are left on or next to prominent objects within the natural habitat, often along travel routes (Richards and Hine, 1953; Grinnell, *et al.*, 1937). In Zion National Park, Trapp (1973) collected 1196 *Urocyon* scats over a 16 month period in 1967–68. Scats were commonly seen along roadsides, in or beside trails, often in small concentrations. The scats were often placed directly on top of a small rock, a log, a stump, or some other elevation beside the trail, but also could be found on top of large boulders. Scats were found most easily by following trails or road edges. In the annual sample of 240 scats, 72% occurred as singles, 16% with another scat, 5% in groups of 3, 3% in groups of 4, 2% in groups of 5, and 1% in groups of 8 or more. Aside from concentrations of scats near dens, fox "latrines" sometimes develop. On September 17, 1967, at the onset of the *Opuntia* fruit harvest, 220 relatively fresh fox scats were found in a grassy area about 35 × 35 feet (10 × 10 m). Most contained red *Opuntia* fruit remains.

Movements and Use of Habitat

Sheldon (1953) reported dispersal movements of 2 gray foxes in New York. One juvenile female first caught in October was recaptured 3 years later 83.7 km away, and another juvenile female caught in October was recaptured 17.7 km away just over a month later. In Alabama, a banding study showed a juvenile female moved 7.2 km during its first winter (Sullivan, 1956).

Nightly movements were studied in Zion National Park by Trapp (1973) using biotelemetry. Using the method of Storm (1965) (who studied *Vulpes*), each distance between daytime rest areas (or early-evening starting points) and each of the individual nighttime activity "fixes" was measured. A mean of such distances per tracking night for each individual was calculated, as was a mean distance for the entire tracking period. Based on a total of 362 measurements, the mean distance for each of 4 adult females (2 studied in winter, 2 in spring–summer) was 600.5 m. For 369 measurements on 5 adult males (4 studied in spring–summer, 1 in winter), the mean distance was 475.5 m. For 68 measurements taken on a juvenile male in July and August, the mean distance was 259.1 m.

Mobility seems to be greatest in the fall (Richards and Hine, 1953) or winter (Wood, 1954b). Pooling adult fox data, Trapp (1973) found a mean winter distance for 2 females and 1 male of 673.6 m and a mean spring–summer distance for 2 females and 4 males of 460.2 m.

While traveling from one location to another, the gray fox often follows an old road or some open trail (Richards and Hine, 1953; Wood, 1954b). However, Grinnell, *et al.* (1937) point out that when foraging the gray fox winds and twists its way through thickets and through crevices beneath boulders. Many of the 172 visual sightings of foxes in Zion National Park between July, 1966 and September, 1969 were of foxes working their way among bushes and boulders, apparently foraging (Trapp, 1973). The direction of the path followed by a given fox during its nightly travels was difficult to predict: foxes' foraging paths are erratic with many abrupt turns and reversals of direction.

Unusual for canids, the gray fox is adept at tree climbing. Taylor (1943) raised 5 young foxes, and found that within a month after they began leaving the nest box, the pups could climb a vertical trunk unaided. Although foxes usually climb or run up sloping trunks, climbing of vertical trunks has also been reported by Seton (1929), Bailey (1941), and Grinnell, *et al.* (1937). The gray fox climbs by grasping the trunk with its forepaws and boosting itself up with the hind feet in a scrambling motion (Terres, 1939). Apparently the claws are well-suited for climbing (Bailey, 1941). Goldman (1938) noted that the tropical gray fox is more arboreal than the larger northern races and that in the southern latitudes the claws are sharper and more recurved.

After reaching the branches of a tree, the gray fox may climb or jump from limb to limb to heights of 6 m or more (Seton, 1929; Nelson, 1930; Grinnell, *et al.*, 1937; Yeager, 1938). In addition to occasionally using trees for den sites and resting, this dimension of its habitat may be used as a lookout site, an escape avenue from danger, or as a place to seek food (Nelson, 1930; Grinnell, *et al.*, 1937; Yeager, 1938; Terres, 1939; Bailey, 1941; Davis, 1960; Gunderson, 1961). When leaving the tree, the fox may run down head first (Nelson, 1930; Yeager, 1938) or back down to the ground (Seton, 1929). When shaken from a tree, the fox is agile enough to land on its feet by using its laterally flattened tail as a balancing instrument or rudder (Bailey, 1941). The laterally flattened effect is due partly to the tail's black dorsal "mane," overlying the supracaudal gland (see Seton, 1923, and Fox, 1971c).

Activity Periods

The gray fox is most active at night, and somewhat less active during the early morning and late afternoon (Seton, 1929; Grinnell, *et al.*,

1937; Taylor, 1943; Trapp, 1973). However, Gander (1966) observed foxes coming to his feeding station at all times of the day and night. In Zion National Park, *Urocyon*'s activity in the twilight and daylight hours adds an important parameter to its trophic niche, allowing it to exploit diurnally-active prey species, such as lizards and certain small mammals (Trapp, 1973).

NICHE: TROPHIC ASPECTS

Food Habits

Determinations of gray fox food habits have been based primarily upon examination of stomach contents. Data are most often stated in % volume or occasionally as % frequency of occurrence. In a comparative study, Wood (1954a) determined that contents of the stomach, small and large intestine, as well as droppings, all give reliable indication of the foods eaten. An examination of the major studies on food habits shows that the gray fox is an opportunistic consumer. The principal components of the diet vary with the season, with the relative abundance of different foods, and to a certain extent with geography. It appears that in the eastern and northcentral regions of the U.S., where snow falls, the autumn and winter diet consists of the following, listed in order of decreasing % volume: mammals, plants, and birds or carrion. Winter studies were conducted in Virginia (Nelson, 1933), Pennsylvania (Bennett and English, 1942). Wisconsin (Errington, 1935; Besadny, 1966), Minnesota (Hatfield, 1939), and in Iowa (Scott, 1955b). The volumes varied from 42.8 to 70.1% for mammals, 14.3 to 34.8% for plants, 2.1 to 26.2% for birds, and 0 to 11.2% for carrion. Although Besadny's (1966) frequency figures appear to approximate this order of importance, Errington's (1935) frequency data show the mammalian portion conspicuously high in each of 3 winters. During winter months, the difference from north to south is the appearance of arthropods (mostly insects) in the diet (1.1–2.3%), with mammals still making up the greatest volume (see Wood, *et al.*, 1958, for Georgia; Davis, 1960, Taylor, 1953, for westcentral Texas).

Few food habit studies have been made in spring and summer. In the spring in Texas, arthropods, plants, and reptiles become more important, although mammals still remain most important (Davis, 1960; Taylor, 1953). Insects and plants increase in importance during summer and fall months. The relative order of importance of food types in Georgia and westcentral Texas is arthropods, plants or mammals, and birds (Wood, *et al.*, 1958; Davis, 1960). The same order was observed in some of the seasonal analyses in westcentral Texas made by Taylor

(1953, including Sperry's data therein), but his results for different years show that the order can also be mammals, arthropods or plants, birds.

Studies reporting food habits lumped for the entire year are less informative than seasonal analyses. In these reports mammals are the dominant food item, volumes ranging from 61.4 to 77.7% (see Grinnell, *et al.*, 1937, for California; Chaddock, 1939, for Wisconsin; Taylor, 1953, for westcentral Texas; Wood, 1954a, for eastern Texas; and Schwartz and Schwartz, 1959, for Missouri).

Lagomorphs, principally *Sylvilagus,* dominate the mammalian portion of the gray fox's diet in the central and eastern parts of its range where studies have been conducted. Rodents are next in importance (*Peromyscus, Microtus, Sigmodon, Geomys, Neotoma, Thomomys, Citellus, Marmota,* and *Sciurus*), and appear to equal or exceed lagomorphs in importance in the drier western regions of the U.S. (Grinnell, *et al.*, 1937; Trapp, 1973). The plant portion of the diet includes fruits, nuts, and grains, e.g., apples, grapes, persimmons, acorns, peanuts, hickory nuts, beechnuts; mesquite, manzanita juniper, and prickly pear fruits; corn, and miscellaneous grasses. Arthropods include primarily Orthoptera, Coleoptera, and Lepidoptera. During peak periods, insects comprise as much as 41% of the volume of food consumed (Wood, *et al.*, 1958). Avian foods are dominated by Passeriformes, with other birds, such as Galliformes, in lesser amounts. In Wisconsin, Errington (1933) conducted a winter study of bobwhite quail (*Colinus virginianus*) populations on ranges occupied and unoccupied by foxes and found that quail mortality was primarily associated with the carrying capacity of the environment, rather than with the density of the fox population. A few other *Urocyon*–avian coactions have been described. Nelson and Handley (1938) reported on fox behavior in raiding bobwhite quail nests. Wetmore (1952) noted an active interest by gray foxes in artificial crow calls and crow decoys. At a watering station, Marsh (1962) observed relative tolerance by Gambel's quail (*Lophortyx gambelii*) of a nearby gray fox, but intolerance to a domestic cat, which perhaps suggests that the gray fox is less dangerous from the quail's point of view.

Domestic animals and deer (*Odocoileus*) are often fed upon as carrion by gray foxes (Errington, 1935; Leopold, 1959; Trapp, 1973). Predation on domestic stock seems to be the exception (Nelson, 1933; Taylor, 1953; Davis, 1960).

In Zion National Park, Trapp (1973) found that the gray fox appears to be an herbivore, insectivore, and scavenger, more than a carnivore (see Table 11-1). The fox diet in that area showed marked seasonal changes. Some foods that dominated the diet seasonally were:

TABLE 11-1 Percent frequency of occurrence of food items for 1967–68 in 240 *Urocyon* scats (20/month) for 3-month periods and on an annual basis in Zion National Park (Trapp, 1973).

Food Type	D–J–F	M–A–M	J–J–A	S–O–N	Annual
Fruits	83	37	48	98	67
Arthropoda	22	68	88	52	57
Mammalia[a]	65	62	30	25	45
Miscellaneous[b]	45	55	37	45	45
Grasses and Sedges	7	37	32	27	25
Poikilotherms, etc.[c]	3	17	8	5	8
Aves	5	2	12	2	5

[a]Includes carrion (primarily from December through March).
[b]Unidentified matter, foil, paper, sand, unidentified vegetable matter, *Juniperus* leaves, and Pisces.
[c]Including unidentified tetrapods.

Juniperus fruits in winter, arthropods, rodents, and *Odocoileus* carrion in late winter and spring, arthropods in summer, and fleshy *Opuntia* fruits in autumn.

Gray Foxes as Prey

Gray foxes may serve as food for other carnivores more often than is realized. Grinnell, *et al.* (1937) believed that the larger hawks may take pups occasionally and they noted that adults have been attacked or eaten by golden eagles, coyotes, and bobcats. There is some indication that, where numerous, coyotes may limit gray fox populations (Davis, 1960). Gander (1966) observed 2 attacks by bobcats on gray foxes.

POPULATION BIOLOGY

Breeding Season and Natality

There is only a small amount of variation from north to south in the peak of the breeding season. There is some indication that at the latitude of New York or Wisconsin the peak of breeding occurs in late February to mid-March, (Sheldon, 1949; Richards and Hine, 1953; Layne and McKeon, 1956a). This is about ½ to 1 month later than at more southerly locations where the peak occurs in mid-February in Illinois (Layne, 1958), February in California (Grinnell, *et al.*, 1937), and late January–early February on the Florida–Georgia boundary (Wood, 1958). From less precise reports, February to March is suggested to be the peak for breeding in Missouri (Schwartz and Schwartz, 1959), Texas (Davis, 1960), and Mexico (Leopold, 1959). The breeding period may extend over several months, specifically, from late December to March on the Florida–Georgia boundary (Wood, 1958) and from mid-January to the end of May in New York (Sheldon, 1949; Layne and McKeon, 1956a).

It is surprising that of all the papers reporting on aspects of reproduction, it appears that the gestation period has not been clearly determined for the gray fox. Although some authors of the more popular publications say that it is 63 days (Leopold, 1959; Davis, 1960), they may be basing this statement on Grinnell's, *et al.* (1937) assumption that it was the same as the domestic dog. Sheldon (1949) disagreed with this and assumed that the gestation period was 53 days, as in the red fox, since these species have similar growth rates. Schwartz and Schwartz (1959) report a gestation period for the gray fox of 53 days (51–63), but in the text of their semipopular publication they do not cite their sources of information, so whether this is from their Missouri data or an assumption based on the red fox is not clear.

A survey of the many reports on litter size reveals no perceptible geographic trends; however, all the most precise statements on litter size come only from the eastern and northcentral U.S. Based primarily on placental scar counts or number of embryos, these are 3.7 for New York (Sheldon, 1949), 4.4 for New York (Layne and McKeon, 1956a), 3.9 for Wisconsin (Richards and Hine, 1953), 3.8 for Illinois (Layne, 1958), 3.8 for Alabama (Sullivan, 1956), and 4.6 for the Florida–Georgia boundary (Wood, 1958). There is only 1 litter per year and the range in litter size usually is about 2–6, often 3–5 (Seton, 1929; Grinnell, *et al.*, 1937; Cabrera and Yepes, 1940; Sheldon, 1949; Richards and Hine, 1953; Layne and McKeon, 1956a; Sullivan, 1956; Layne, 1958; Wood, 1958; Leopold, 1959; and Davis, 1960). Schwartz and Schwartz (1959) suggest extremes of 1–10.

Reproductive Potential

Most female gray fox breed in their first season (Wood, 1958). The percentage of barren females found in samples taken during the breeding season was 3.3% of 90 (Sheldon, 1949), 6.5% of 31 (Sullivan, 1956), 3.8% of 53 (Layne and McKeon, 1956a), 2.0% of 49 (Layne, 1958), and 6.4% of 151 (Wood, 1958). Wood (1958) found no difference in fecundity between age groups, but due to numerical dominance, the younger cohorts had a greater overall productivity; 54.5% of young were produced by 1 year old females.

Sheldon (1949) calculated the annual reproductive potential to be 1.5 young per adult in New York. Also in New York, Layne and McKeon (1956a) calculated an average annual productivity over a 3 year period of 1.8 per adult. Wood's (1958) mean of 4.5 young per litter, based on embryo and placental scar counts, indicates a potential productivity of 2.3 young per adult per year on the Florida–Georgia boundary.

Mortality and Age Structure

Corresponding to high fertility, mortality is also high. Wood (1958) calculated the probability that a pup would die during its first summer at about 0.5; during its first winter, about 0.9; and for succeeding years, about 0.5. Davis and Wood (1959) concluded that fox populations are essentially an "annual crop." This conclusion was supported by data from 5 sites that showed 48–61% of the population to be less than a year old. Richards and Hine (1953) reported 72, 62, and 72% immature females in all-female samples collected in 3 different years in Wisconsin. In Alabama, Sullivan and Haugen (1956) used x-ray aging techniques to examine epiphyseal closure of long bones, and found that 73% of a sample of 60 foxes were juveniles. For more on determining age and sex see Petrides (1950); Richards and Hine (1953); Wood (1958); and Lord (1961a, 1961b).

Sex Ratio

After a 4 year study in Georgia and Florida, Wood (1958) concluded that the sex ratio was about 50:50. In New York, Layne and McKeon (1956a) found a consistent and significant preponderance of males (116.7–187.2 males per 100 females). However, this may have been due to greater mobility of males, resulting in a greater probability of being caught in traps. (For sex ratios, see also Richards and Hine, 1953; Layne, 1958; Linhart, 1959; Schwartz and Schwartz, 1959.)

Population Indices and Density

Various indices have been used to determine fox population trends. Richmond (1952) studied bounty statistics, Richards and Hine (1953) studied trapping–hunting data and made den counts, and Lemke and Thompson (1960) analyzed questionnaires that had been sent to farmers. Wood (1958) used standardized trap-lines and scent post lines (see also Richards and Hine, 1953), recording the number of captures or visits, respectively. An increased proportion of stations capturing foxes in the fall and winter reflected increased fox mobility and dispersal of the year's young.

Reported densities range from 1 to 27 per 2.6 km^2 (1 $mile^2$) and are probably dependent on the productivity of the habitat. An unusually high density of foxes was noted by Errington (1933) in southern Wisconsin in December, 1932. By tracking, he estimated 29 foxes (27 grays, 2 reds) per 2.6 km^2. Grinnell, *et al.* (1937) estimated a gray fox population of about 1 per 2.6 km^2 on typical California chaparral lands at the beginning of a trapping season. Gier (1948) mentioned that in an area of Ohio, foxes had an average occurrence of at least 3 red and 5 gray foxes per 2.6 km^2 with concentrations of 12 or more foxes per 2.6 km^2 in a favorable habitat. To determine fox population density in Florida, Lord (1961a) used the age-ratio-reduction technique and calculated density to be 3–4 foxes per 2.6 km^2.

Population Fluctuations and Regulation

Fox population fluctuations have been studied by Richmond (1952), Richards and Hine (1953), Wood (1954b, 1959), and Wood and Odum (1964). Richmond (1952) examined bounty data from a 35 year period in Pennsylvania and found a positive, but inexplicable correlation between apparent marked increases in the gray fox population and wetter and warmer-than-average late winters (January, February, and March). Wood and Odum (1964) described a 9 year history of furbearer populations on a 20,000 acre (49,400 hectare) area in South Carolina. The relative proportions of *Urocyon* (54%), *Procyon lotor* (21%), *Lynx rufus* (11%), *Vulpes fulva* (7%), *Mephitis mephitis* (3%), and *Didelphis marsupialis* (3%) did not vary significantly, although there was a threefold variation in numbers of animals caught.

Rabies appears to be a density-dependent control mechanism in many fox populations; it is closely associated with fox population fluctuations (Grinnell, *et al.*, 1937; Richards and Hine, 1953; Wood, 1954b; Davis and Wood, 1959; Wood and Davis, 1959). Rabies seems to show a localized distribution in *Urocyon* populations (Davis and Wood, 1959), and also species-specificity in some areas (Wood, 1954b; Parker, 1962; Verts and Storm, 1966).

Gier (1948) reviews rabies outbreaks in the U.S.: the occurrence of the disease, symptoms in infected animals, transmission, and control. Rabies outbreaks in the Southeast seem to occur most frequently during the winter when increased movements of foxes result in greater frequency of intraspecific contact (Wood, 1954b). Wood also described a density-contact index (foxes trapped per trap-mile (1.6 km)) to measure this phenomenon. His data suggested that an index of 4.1 is sufficiently high to allow an epizootic, whereas populations represented by indices of less than 1.6 have a contact rate too low to support an epizootic even though they contain infected foxes. (For more information on fox rabies, see Bryant, 1924; Johnson, 1945; Burr, 1947; Rietz, 1947; Jennings, *et al.*, 1960.)

That foxes are essentially an annual crop probably explains why rabies control programs involving animal removal may reduce fox populations in local areas of the Southeast only for a limited period (Wood, 1954a; Davis and Wood, 1959). Layne and McKeon (1956a) suggest that under certain conditions, control programs might actually help to increase fox productivity.

ONTOGENY AND SOCIAL BEHAVIOR

Information on physical and behavioral ontogeny is not well-documented, but some data can be found in Grinnell, *et al.* (1937); Taylor (1943); Sheldon (1949); Wood (1958); and Layne (1958). Taylor (1943) presents an interesting account of behavior, personality, and activities of 5 young foxes raised in captivity. (See Grinnell, *et al.*, 1937;

Figure 11-1 Threat display of the gray fox. Photo: M. W. Fox.

Carr, 1945; and Richards and Hine, 1953 for more information on behavior in captivity.)

Weaning takes place at the end of June in Illinois' gray foxes (Layne, 1958). Adult weight is reached at 5–6 months (Sheldon, 1949; Wood, 1958). Seton (1929) and Grinnell, *et al.* (1937) reported that monogamy prevails in *Urocyon*, although Sheldon (1949) suggested that polygamy may occur at times, with more than 1 litter in a den. Wood's (1958) data suggested that the young spend 3 months in the den. However, observations by Grinnell, *et al.* (1937) indicated that the young spend 4–6 weeks in the den, then forage with the parents until at least July 1, family ties weakening in July, and pups are foraging independently of each other by September. Pups appeared to remain in the parents' range until January or February. Whether subadults are forced to leave the parents' range or leave of their own accord is not clear.

Fox (1969a, 1969b, 1971c) has recently published a series of papers dealing with social behavior in captive canids, among which he includes *Urocyon* (Figure 11-1). However, comparable detailed information on the gray fox's social behavior under natural conditions is essentially nonexistent.

CONCLUSIONS

Concerning basic biology, we have a picture of the gray fox sketched for us from various sources of information, primarily from the eastern half of the United States. Ecologically, the gray fox appears to be an omnivorous (often insectivorous–herbivorous) consumer at the primary, secondary, and tertiary trophic levels. It obtains food by canid-like searching at ground level in rough brushlands during the night, and to a lesser, but important, extent in the twilight and day. Though food habits and certain aspects of reproduction are better understood than other areas of its biology, more studies are still needed in the western and Latin American portions of the gray fox range. Still poorly known, even in the eastern half of the United States, are many other features of gray fox biology, such as its home range size in areas of differing productivity, the nature of its territoriality, its activity periods, its numerical and trophic relationships to other sympatric Carnivora, its physical and behavioral ontogeny, and its ethology. Still needed are basic studies in diverse parts of the gray fox range; in southeastern Canada, the North Dakota–Minnesota area, southern Florida, the Southwest, Oregon, the Great Basin, Baja California, Mexico, Central America, and northern South America. Little is known of the physiological adaptations of this widespread canid to arid conditions, or to other stressful environments. Finally, ethological studies under natural or seminatural conditions would contribute to our understanding of gray fox social behavior and canid behavior in general.

Acknowledgments We are indebted to Dr. Dan W. Anderson, U.S. Fish and Wildlife Service, Dr. Miklos D. F. Udvardy, California State University, and Carolyn Trapp, Center for Primate Biology (Davis), for critically reading the manuscript. Thanks also are due to Sharon Hallberg for proofreading assistance.

12
ECOLOGY OF THE PAMPAS GRAY FOX AND THE LARGE FOX (CULPEO)

JORGE A. CRESPO
Argentine Museum of Natural Sciences
Buenos Aires, Argentina

THE PAMPAS FOX *(Dusicyon gymnocercus)*

This paper reviews original field studies of two species of South American canids, the gray fox and the colored or large fox (Crespo, 1971; Crespo and de Carlo, 1963). The gray fox or pampas fox *(Dusicyon gymnocerus)* is one of the more common canids in the central and eastern regions of Argentina, and is possibly the principle wild canid that was always more in contact with human populations since they are extremely adaptable.

Natural Environment — Habitats

In Argentina, the pampas gray fox which averages 4.4 kg weight, lives primarily in prairie environments from sea level up to approximately 1000 m, being a typical representative of the zoogeographic region known as the pampas. These same prairies in the east are covered by dense cultivated soil, but towards the west the environment becomes gradually dryer, surging xerophyle forests and dense, spiny thickets (briar and thorn bushes) that offer excellent natural refuge to this fox, particularly in open country where trees have been cut down and the country abandoned.

In the province of Buenos Aires, where great pressure from man occurs conjunctly with a decreased availabilty of refuges and natural food supply, the foxes' numbers have been considerably diminished.

Sexual and Reproductive Cycles

Studies performed in the central province of the Pampa show that this species has 1 litter a year (monestrous) with an annual reproductive cycle circumscribed by the month of August in which they begin the manifestations of pairing and copulation, and end in October and November, when birth takes place. Such has been deduced by the analysis of development or degree of maturity of ovaries, follicles, corpses, swelling of uterine horns, embryos, scarred placentas and lactation, in addition to development of testicles and presence of free spermatozoa.

Pregnancy occurs from the end of August up until the middle of November, having been exhibited in a sample of 30 females in October with the following distribution: 15 with swollen uterine horns up to 12 mm in diameter but without embryos visible to the naked eye (of these, 3 with scarred placentas and 2 lactating); 15 with visible embryos of 10 or more mm (of these, 2 with embryonic readsorption).

Of 72 females in September and October, 20 presented a total of 67 visible embryos and viables up to that moment, which shows a median value of 3.35 ± 0.40 embryos per female, with a standard deviation of 1.74, a coefficient of variability of 54.3 and a range of 1–8; the number of embryos calculated by turgid uterine or postpartum uterine scars would be somewhat larger, 4.5 and 4.2, respectively, which would give a final mean to 4.01 embryos per female; as the arithmetic mean of dead gestates was 5.95 per female, those values already would indicate a rough prenatal mortality of 34%.

The population of females of this species presents for the reproductive months a median value of prevalence of pregnancy of roughly 0.61 (number of gestating females above the total number of females in the sample). The incidence of pregnancy had a value of 0.84, taking the reproductive time or ecological opportunity of reproduction for the species as 80 days (between mid-August and the beginning of November) and a gestation period estimated at 58 days. The chronology of reproduction in the female is summarized in Figure 12-1.

In the males, the seasonal increase in the weight of the testicles is very notable in mid-June (winter in South America) and is maintained up till August–September, when it declines, a phenomenon which occurs through all ages. The epididymis tissue shows spermatozoids in all the months of the year for the semiadult males (13 to 24 months) and the adults (more than 24 months); the young have exhibited premature

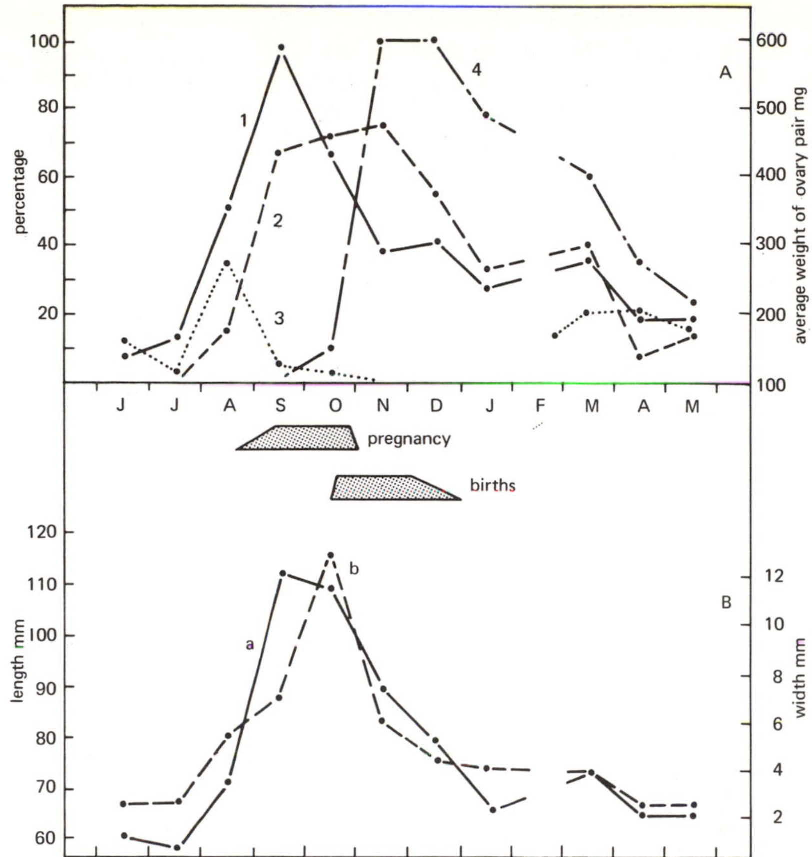

Figure 12-1 Chronology of reproduction in the female Pampas gray fox.
A. (1) Average monthly weight of the ovary pair.
(2) Percentage of females with corpora lutea or recent atresic follicles.
(3) Percentage of females with vesiculous follicles in the ovary.
(4) Percentage of females with recent postpartum scars in the ovary.
B. Monthly development of the uterine horns: (a) length, (b) width.
In both cases the sample size varies between 162 and 172 individuals.
Reprinted with permission from J. A. Crespo (1971). Ecologia del zorro gris en La Pampa. *Rev: Mus. Arg. Cs. Nat. Ecol.*, **1**(5), 147–205.

cases, since some 120 days old (approximate) samples have possessed free spermatozoids (March).

Productivity

The first interpretation of the biological productivity of this species (obtained by computing all the evident cases and presumptions of fertility based on the aforementioned signs) in a sample of 92 females showed that 86 were in one reproductive stage or another. This brings the rough fertility total for the population to 93.5%.

A detailed analysis for classes of ages (presented later) indicates that although the partial value of prevalence of pregnancy (Pp) for classes increasing in age (class A 0.55; B 0.91; C 1.00), or being that fertility does not seem to diminish, the youngest classes A and B in the final production dominate the major population density. The mean of embryos per litter increases from the young females to the semiadults, from 2.91 to 4.84 and diminishes to 1.50 for the adults and the old females.

Continuing, they (females) show some values for age classes connecting the prevalence of net pregnancy (Ppn) with the index of embryo production per class (pre), which offers us another index that we can call "growth of the net reproduction per class," (Prn) (see Table 12-1).

TABLE 12-1 Percentage Productivity in Different Age Classes of a Female Gray Fox Population

Class	Ppn		Pre	Prn	Percentage
A	0.47	×	0.52	0.24	61.0
B	0.35	×	0.43	0.15	38.0
C	0.16	×	0.04	0.006	1.0

These values demonstrate that the major part of the reproductively positive females in the natural population is made up of the younger, 8 to 12 month old subjects. Moreover, this cross section of the population will constitute the principal purveyor of the new offspring that will join the population; following the previous class are the semiadults with values somewhat less, while the adults per se would increase the population in a very insignificant manner.

Structure of the Population; Dynamics and Mortality

Because this species has a well-defined time of reproduction, it is possible to follow the development of the young. The rapid development of these foxes, from 10 to 12 months in age makes it more difficult to determine their absolute ages or chronological age. The use of knowledge of the process of ossification or the closing of cranial sutures in 1 year olds (like the basioccipital–basisphenoid, basisphenoid–presphenoid and the maxillary–premaxillary united by the degree of attrition of the protoconulos and meta-conulos of the M^1) permit classification of the population into 3 age classes that we call:

Class A: young; specimens of 2–12 months in absolute age; November is considered month 0, being the month of birth.
Class B: semiadults; specimens of 13–24 months in age.
Class C: adults; those of 25 months and over (this class is heterogenous by including all the ages above 25 months).

Through the period of August 1966–September 1967, a sample of 324 specimens of all months, collected from the center of the province of the Pampa, showed the following distribution (Figure 12-2): Class A — 171 specimens =52.8%, Class B — 111 specimens =34.3%, Class C — 42 specimens =12.9%.

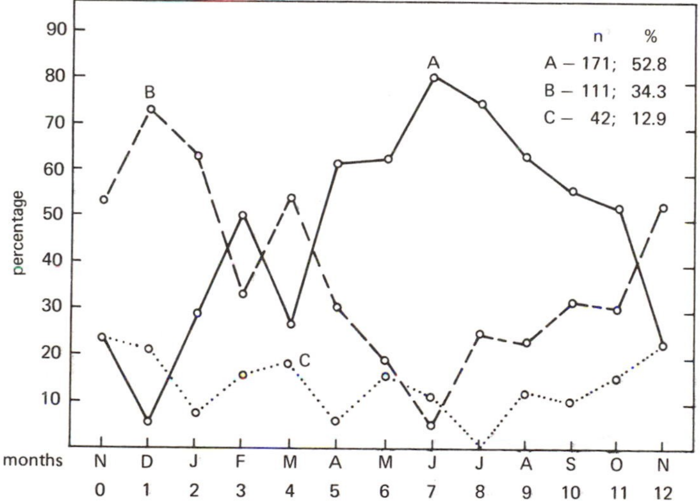

Figure 12-2 Variation of the monthly proportions of each age class of the pampas gray fox during one year. The sample includes the 324 individuals available. Reprinted with permission from J. A. Crespo (1971). Ecologia del zorro gris en La Pampa. *Rev. Mus. Arg. Cs. Nat. Ecol.*, **1**(5), 147–205.

On studying the monthly variations of the proportions of these 3 classes, it is possible to reconstruct the pyramids of age and calculate the number of specimens that pass from one age class to the following class, or the number of surviving members of each class. Since 85% of the individuals of class A graduate to class B, there is clearly very little mortality in this period; on the other hand, the graduation of those in class B to class C is less (32%), and we conclude that the proportion of individuals that exceed 36 months of life is very low, no more than 7% of the total population. This then indicates, in general, a very brief ecological longevity for this canid.

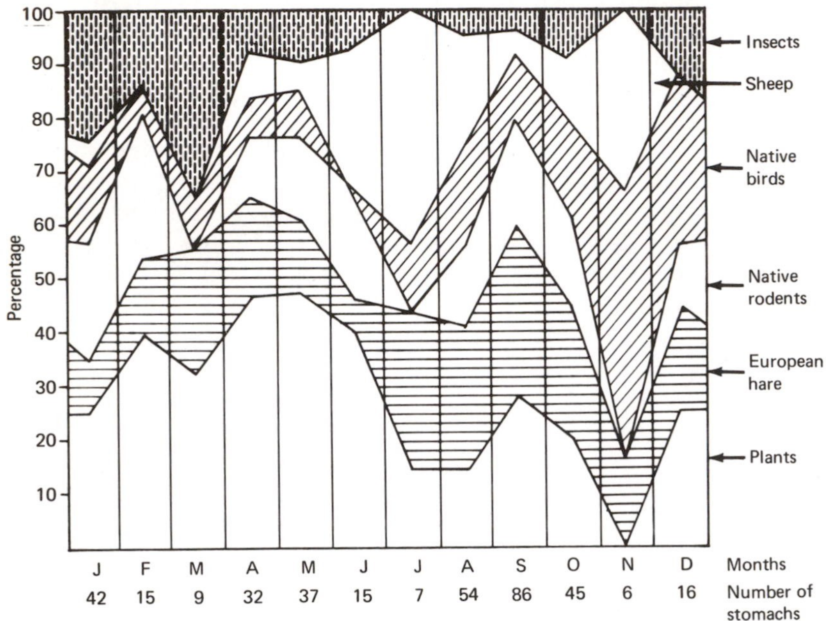

Figure 12-3 Percentages for each month of the absolute frequencies of the principal kinds of food found in the 230 pampas gray fox stomachs containing food. Reprinted with permission from J. A. Crespo (1971). Ecologia del zorro gris en La Pampa. *Rev. Mus. Arg. Cs. Nat. Ecol.* **1**(5), 147–205.

Food Habits of the Gray Fox

The gray fox, at least in the province of the Pampa, proves to be highly omnivorous in its eating habits, utilizing a large variety of items equally balanced during the year between vegetables and animals. A total of 37 different classes of foods reveals how the gray fox utilizes the ecosystem with which it integrates; an ecosystem of multiple ecological niches of primary, secondary, and tertiary consumers.

Taking as ground work the analysis of stomach contents of 230 young foxes (Figure 12-3) gathered between August, 1966 and August, 1967 in similar localities and expressing the results in frequency of appearance and percentages of the diverse nutritious items, one can see that the elements of vegetable origin, especially fruits from trees and wild shrubs of the region, constituted ¼ of the total diet and are important throughout the year, though womewhat more in autumn. The other ¾ is of animal origin. Of this, 1/7 is from domestic animals (principally sheep), while the remaining 6/7 is from wild animals; mammals — 54%, birds — 31%, insects — 10%. Among the mammals, the European hare *(Lepus europaeus)* stands out with a general frequency of 33%: next comes the small rodents of the country *(Cricetidae)*. Of the birds, the most affected are the partridges (Tinamidae), and of the insects, the Acridiidae. In general, it is deduced that the gray fox diet is some 9/10 derived from natural (non-domestic) sources, which indicates a truly exhaustive hunting and foraging for over 35 different food items that vary from small insects to wild mammals of weight approximately that of its own, as in the case of the large hare of the prairie *(Lagostomus maximus)*.

THE COLORED FOX OR LARGE FOX *Dusicyon culpaeus culpaeus* (Molina 1782)

The colored fox (Figure 12-4) is a typical species of the Andes-Patagonial region, with a distribution extending from Equador through Patagonia up to Tierra del Fuego the southernmost tip of South America). After the large fox *(Chrysocyon brachyurus)*, the colored fox is the second largest canid in South America, averaging 7.35 kg. It is also the one that reaches the highest habitats in South America, easily reaching 4500 m above sea level, and has in general always been less in contact with man than the pampas fox.

Natural Environments: Habitats

The totality of its large area of dispersion offers the colored fox abundant and varied habitats and natural refuges; this can be well-appreciated in Argentina, from the province of Neuguen and towards the south, where the varied topography consists of irregular mountains and plateaus (Figure 12-5); large forests and dense thicket areas occur in the west.

The extensiveness of its geographic dispersion presents numerous and various habitats, but all of them remain within a general physiography of arid or semiarid regions, as in the mountains, with less rough reliefs and plateaus as in Patagonia. Nevertheless, in the Patagonic Andes (mountains — Ed.) it frequently penetrates the dense subantarctic forests without being qualified as a representative species there;

Figure 12-4 Adult example of the colored fox caught in a steel trap. Catan-Lil River, Neuguen, Argentina. April, 1959. Reprinted with permission from J. A. Crespo and J. M. DeCarlo (1963). Estudio ecologico de una población de zorros colorados. *Rev. Mus. Arg. Cs. Nat. Ecol.*, **1**(1), 1–53.

this type of habitat is interposed as a variant between the semiarid Andes and elevations of the north and the Patagonic steppes to the south.

The ecological factors that interfere with the fox's advance eastward in the arid plains of western Argentina are not known; in Neuguen, after rainy and very favorable years, it tended to advance toward the east during the last 40 years. It is pertinent to note that related forms of *Dusicyon culpaeus,* during the Pleistocene in Argentina, had a much greater dispersion, embracing all that we call in zoogeography, pampas terrain.

Sexual and Reproductive

The colored fox is a monestrous species (1 litter per year) with a preestrus period from October to July, a postestrus period from the end of July to the middle of October, and an estrus from August to October.

Figure 12-5 General appearance of an environment of transition near the Catan-Lil River, Nueguen, Argentina. Altitude: 100 m. February, 1960. Reprinted with permission from J. A. Crespo and J. M. DeCarlo (1963). Estudio ecologico de una población de zorros colorados. *Rev. Mus. Arg. Cs. Nat. Ecol.*, **1**(1), 1–53.

The various states of pregnancy (estimated from 55 to 60 days in duration) occur from August up to the beginning of December, including states of pseudopregnancy. The principal time of birth is from October to December, and lactation is extended up until February. In 6 pregnant females, a median value of 5.16 embryos was recorded (range 3–8).

Since July is a period of sexual arousal and violent male rivalry fights, males show a loss in body weight at this time. Also, a seasonal cycle is seen in the increase of testicular weight that lasts from July to October, a unique time in which they show free spermatozoids, contrary to what was observed in the gray fox, a species that has free spermatozoids all year.

Structure of the Population

According to a study of 119 specimens gathered between April, 1959 and March, 1960 (originating from a single locality in the province of Neuguen), and taking, to establish the age classes, the states of fusion of cranial articulations and level of attrition of the M^1, we arrive at the following age class distribution:

Class A: young; 4–12 months; 92 specimens — 77.5%.
Class B: semiadults; 13–24 months; 21 specimens — 17.5%.
Class C: adults; more than 24 months; 6 specimens — 5.0%.

In consequence, it is deduced that a high proportion of the population, principally between February and June, consists of specimens in their first year, and that it is this nonreproducing segment of the population that constitutes an important reproductive potential in the maintenance of the population.

Food Habits

The examination of stomach contents from 96 specimens from the Neuguen province and representing all the months of the year reveals that this canid is strictly carnivorous, as shown by the nature of the 15 items identified, the majority being remains of mammals (Figure 12–6).

They can be grouped in two categories: a) rodents in general, with a frequency of 61.4%, including the European hare *(Lepus europaeus)*, and b) domestic mammals, 27.4% (sheep, cattle, and horses); there are other smaller items like the wild birds, 6.0%. It is interesting to note how the colored fox nowadays favors the nonaboriginal mammals (63%) like the European hare and livestock, affecting only 37% of the indigenous species (rodents like *Ctenomys* sp. and *Lagidium* sp.). The food spectrum is ample inside the limitations that the local fauna of vertebrates impose, but it is much less varied or abundant than that of the gray fox *(Dusicyon gymnocercus)* in the Pampas with its 37 classes of animal and vegetable foods.

It is pertinent to note that the colored fox did not always have domestic livestock and European hares at its disposal and in the abundance as they now occur. Up to a time period between the years of 1910–1915, the Neuguen province had intense exploitations of equine livestock, very little bovine, and nothing of sheep. Moreover, the European hare had not arrived in the region. The colored fox was then maintained at a low population density, without interfering in man's interests, eating basically species of aboriginal fauna. After 1915 began the great development of sheep ranching and the appearance of the European hare. Almost simultaneously and before this single variation of the conditions, the colored fox increased its numbers rapidly and was converted to a destructive species in relation to man.

This great change in its diet also permitted it to elude the competition of other wild carnivorous species that utilize the same sources of natural foods such as the small gray fox of Patagonia *(Dusicyon griseus griseus)* of a similar ecological niche although more omnivorous, the ferret *(Galictis cuja cuja),* the cat of the grass *(Felis colocolo pajeros),* and the mountain cat *(Felis geoffroyi* subsp.) as well as other smaller carnivores and birds of prey.

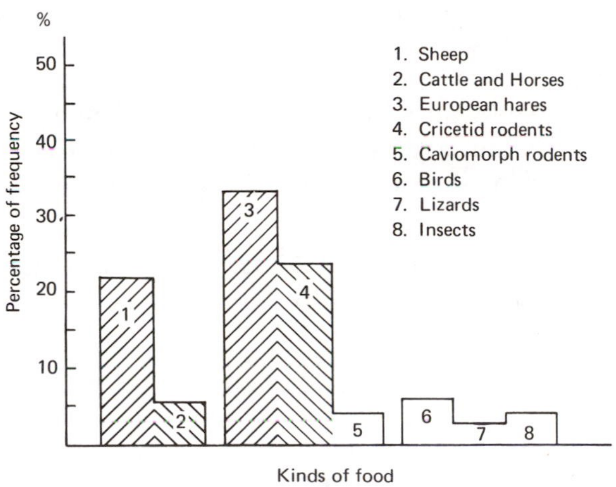

Figure 12-6 Kinds of food and their percentage of frequency in 96 colored fox stomachs. Reprinted with permission from J. A. Crespo and J. M. DeCarlo (1963). Estudio ecologico de una población de zorros colorados. *Rev. Mus. Arg. Cs. Nat. Ecol.,* **1**(1), 1–53.

Community

Nowadays the colored fox plays an important role as the dominant, wild predator in the balance of this seminatural and semiartificial community that man tries hard to maintain and which presents the apparent but not real aspect of a closed and balanced community. In reality, the major part of the energy flow of the ecosystem is displaced across the trophic plot that is formed by a few dominants: the producer, vegetables–grasses, the primary consumers, sheep and hares, and secondary consumers, colored fox and man, removing the remainder of the consumers to a secondary role. The colored fox serves as a regulator species in this community by maintaining predation pressure upon the population of European hares; the latter may be regarded as being an important "buffer" species, protecting indirectly the ovine livestock from over-predation.

Activity

We are able to observe two types of activity and movement in this canid, the first being that it responds to a daily rhythm of 24 hours related to the search for food and that it is active inside a home range, which we have calculated to be 4 km in diameter; activity is almost totally nocturnal.

The other type of movement is a displacement that occurs seasonally and is related to the altitudinal movement of sheep and hare; in the spring these animals are displaced, climbing up to the high country summer pastures, and in autumn descending to the lower winter pastures. A considerable number of foxes demonstrate this same migration of 15 to 20 km in pursuit of sheep and hares, although many foxes instead remain all year in the high country or low country.

Ecological Comparisons between *Dusicyon gymnocercus* and *Dusicyon culpaeus* and Actual Status of Both Species

At present in Argentina, these two species are definitely native of one area only, the colored fox the species of the mountains and plateaus of Patagonia, and the gray fox the species of the prairie with sclerophyle forests. Only the subspecies *Dusicyon c. smithers,* which occurs in the high Sierras of Cordoba (around 2000m) establishes areas of contact with *Dusicyon gymnocercus* subsp. In the west of Argentina, the small gray fox of Patagonia (*D. griseus griseus*) is a subspecies that occupies a narrow and remote region ecologically comparable to desert conditions.

Those distinct, geographic distributions seem to have influenced the adaptations to the kinds of food available. *D. gymnocercus,* extending from Paraguay, north and central Argentina, southwest Brazil, and Uraguay, always lived in forest areas or savannas (treeless plains) that were very rich in vegetation and fauna; the omnivore temperament evolved as reflected in its diet of 37 different food items. In contrast, *Dusicyon culpaeus,* at least in the province of Neuguen, is exclusively carnivorous, with only 17 identified food items in its diet.

The introduction of exotic mammals like the sheep and the European hare by man into the natural scenario must be considered as a crucial event in the ecology of those canids. As a consequence of the intense land exploitation in the east and central parts of the country, the gray fox drew back from more favorable areas as in nearly all of the province of Buenos Aires, south of Sante Fe, Cordoba, Entre Rios, and Uruguay.

Also, it must be noted that *D. gymnocercus* has to compete in a great part of its distribution with the mountain fox (*Cerdocyon thous entrerionus*) of similar eating habits, while *D. culpaeus,* due to its great size, dominates physically and gets a better profit of the foods of greater bulk with respect to the small gray fox, *D. griseus* subsp. In other, more biological aspects, both species here are very similar, as is deduced by comparing their reproductive modalities, ecological longevity, and structure and dynamics of their natural populations.

At present, in spite of continuous control exercised by man over both species by considering them noxious, particularly in zones of livestock development, these canids have still tenaciously maintained dense populations year after year without it appearing so. They would recuperate strongly within a few years if such unsuccessful attempts to totally exterminate them were relaxed.

13
ECOLOGY AND EVOLUTION IN THE SOUTH AMERICAN CANIDS

ALFREDO LANGGUTH
Department of Mammalogy
American Museum of Natural History
New York
and
Departamento de Zoologia Vertebrados
Facultad de H. y Ciencias
Montevideo, Uruguay

This paper is a summary of the information in the literature on the ecology of South American Canidae and an interpretation of these data in terms of canid evolution. Except for the culpeo and the pampas gray fox (Chapter 12), little is known of the ecology and behavior of most of these dogs. Therefore, some of the hypotheses and interpretations presented here may have little factual support and are subject to change when new evidence becomes available. Hopefully new ideas and further research in this field will be stimulated by this paper.

The classification used here is similar to that I proposed earlier (Langguth, 1969). The main difference is the use of *Canis* instead of *Dusicyon* as the proper generic name for the Patagonian dogs (i.e., the dogs living in the Patagonian zoogeographic subregion). The reasons for this will be explained in a forthcoming paper by the author.*

THE ARRIVAL OF DOGS TO SOUTH AMERICA

Representatives of the dog family probably arrived in South America toward the end of the Pliocene or the beginning of the Pleistocene

*See footnote on opposite page.

period. The earliest remains have been found in sediments of Uquian age, early Pleistocene, on the Atlantic coast of the Province of Buenos Aires, Argentina (Kraglievich, 1952; Pascual, *et al.*,1965). Since this record is separated by three-fourths of the continent from the Panama

---

*Classifications change because they reflect our changing knowledge about the interrelationships of animals. It seems, and this has been confirmed by the study of fossils, that there are two main groups of recent canids, 1) the *Vulpes*-like foxes including *Vulpes, Otocyon, Urocyon, Fennecus,* and *Alopex,* 2) the *Canis*-like canids including *Canis,* most South American dogs, *Cuon, Lycaon,* etc. Both groups derive from a common ancestor that is more similar to *Vulpes.*

Part of the dogs in the *Canis* group such as the forms commonly referred to as *Canis, Thos,* and *Dusicyon* have maintained a basic canid pattern in their structure that I call the generalized pattern. Some of the species, although living in different continents, are very similar to each other. For example, it is difficult to distinguish a skull of the African *Canis adustus* from a skull of an average-sized South American *C. (Pseudalopex) culpaeus.* They both share the same basic canid features and differ only in the relative size of the M^1 and M^2. The smallest South American foxes, *C. (Pseudalopex) griseus,* differ to a greater extent from the large *Canis* (lupus of the north; they represent, however, extremes in the range of the genus *Canis.* The smallest South American species are to some extent similar to the foxes of the genus *Vulpes* as you may suspect since they have a common origin and since the smaller species may be the more primitive.

Other dogs of group 2 departed from the generalized stock, probably independently (we do not know) and I call them differentiated dogs. They show (among others) modifications in skull morphology and in color pattern of the skin. The angular region of the mandible, the number of teeth, and the proportions of the facial part of the skull are some of the characters that became modified. The distribution of colors in the fur is originally complex. A pattern of distinct body markings as seen in *Vulpes* and *Canis* represent the primitive condition since they are present in the more primitive living forms. In the differentiated dogs this pattern tends to be modified, usually simplified, as you see in *Cuon, Speothos, Cerdocyon, Lycaon,* etc. It would be interesting to see if some differentiated dogs with a simple color pattern still show behavior patterns corresponding to a complex color pattern.

In my classification I put all the generalized dogs in the same genus *(Canis,* for reasons of nomenclature), granting the smaller differences among the generalized dogs a subgeneric rank. The necessity and advantages of using subgenera have been well-explained by Osgood (1934) and Simpson (1945). Each differentiated dog is also granted full generic rank, for instance *Lycaon, Cuon, Cerdocyon, Speothos, Lycalopex,* and *Atelocynus.*

Regarding the relationships at a subgeneric level in South American canids, the first point is to use the name *Dusicyon* exclusively for the extinct Falkland Island wolf. Neither Osgood (1934) nor Simpson (1945) and the other mammalogists that used so enthusiastically the generic name *Dusicyon* for almost all the South American canids gave any indication of having examined carefully specimens of the type species of the genus, *D. australis* (10 of the 11 preserved specimens of the Falkland wolf are in European museums). I have seen them all and found that they differ from *C. (Pseudalopex) culpaeus* in the dentition and in features of the frontal and palatal regions of the skull so much as the latter differs from *Canis latrans. Canis (Dusicyon) australis* cannot be put in the same subgenus as *Canis (Pseudalopex) culpaeus.* Regarding the small *Lycalopex vetulus,* it shows particular differentiations in the dentition and the skeleton that merit generic rank according to my criteria. When new information on this little-known fox becomes available, his status in the classification may, of course, change.

I am preparing now a paper where I discuss this classification of the Canidae and related problems together with the description of a new fossil dog from Southern Patagonia that shows a mosaic of characters from *Canis* and *Dusicyon.*

landbridge, the moment of arrival might be placed at an earlier date. As the Panama landbridge was permeable to land mammals during most of the Pliocene (Nygren, 1950), it would not be surprising to find earlier dog remains in South America. The factors determining the dispersal of dogs to South America probably were changes in the environment in the areas adjacent to the Panama bridge in Middle America and northwestern South America. The availability of grassland may have been the most important factor in canid dispersal.

GENERALIZED VERSUS DIFFERENTIATED CANIDS: THE ANCESTORS OF THE SOUTH AMERICAN DOGS

Typical modern dogs (the term "dog" is used in this paper as a synonym of "Caninae") are grassland dwellers adapted for cursorial habits in open country. A generalized dog, i.e., a dog possessing features common to the majority of modern canid species, shows the following characters: a) limbs of average length, not the differentiated short legs of *Speothos* or the long legs of *Chrysocyon,* b) mandible and skull showing a smooth ventral mandibular profile (without a subangular lobe as seen in *Urocyon* and *Otocyon* and less marked than in *Cerdocyon*, 2 upper and 3 lower molars with the typical cusp pattern seen in *Vulpes* (not with reduced number of cusps as found in *Lycaon* or *Speothos* nor with the small carnassials and relatively large molars of *Lycalopex (Duscicyon ventulus)* and *Otocyon),* temporal lines limiting a lyriform area or building a saggital crest depending upon size of the skull (but not beaded temporal ridges as seen in *Urocyon* and *Otocyon*, and c) a pattern of color distribution in the pelage that enhances particular regions of the body by means of distinctively colored areas or patches (not the uniform color seen in *Cuon* or the peculiar pattern seen in *Speothos, Atelocynus,* or *Lycaon,* considered here to be differentiated forms).

South America has both generalized and differentiated dogs. The first are species currently included in the subgenera *Dusicyon* and *Pseudalopex* and the second are species included in the genera *Chrysocyon, Cerdocyon, Atelocynus,* and *Speothos.* The fossil record reveals that the generalized type of dog is also the more primitive.

The first canid invaders of South America, ancestors of the living species, were very probably generalized dogs. It is not likely that living generalized forms like *Pseudalopex (D. gymnocercus)* evolved from differentiated ones. On the other hand, if differentiated dogs like *Speothos, Atelocynus, Cerdocyon,* or *Chrysocyon* had evolved in Middle America, and later dispersed over South America, they should still live in Middle America today since suitable habitats have been available in Middle America throughout the Pleistocene. No fossil remains of these genera have been found outside Brazil.

THE EVOLUTION OF DOGS IN SOUTH AMERICA

The present pattern of diversity and geographic distribution of the South American dogs may be explained assuming that a species or group of species of generalized grassland dogs came to South America and spread over suitable grasslands along the Andes, reaching the southern pampas and Patagonian grasslands as well as spreading over the Brazilian highlands. Those living in the Brazilian highlands underwent a broad adaptive radiation, some lineages of which were unparalleled by canids of other parts of the world.

That the radiation of the dogs now living in the Brazilian zoogeographic subregion of South America ("Brazilian dogs") occurred in connection with the Brazilian highlands is suggested by the following facts: a) the Brazilian highlands (Sauer, 1950) is the geographic area where the greatest number of subgenera and species of South American dogs are found, b) in spite of the fact that suitable habitats are available for them in other regions (i.e., in the Guianan highlands) two genera, *Chrysocyon* and *Lycalopex* are endemic in this area, c) fossil forms intermediate in characters between *Atelocynus* and *Speothos* as well as a generalized form close to *Pseudalopex* have been recovered from caves in this area (Winge, 1895), suggesting that it was the actual center of evolution of at least some of the species, d) of 10 South American orders of living mammals, 9 are present in the Brazilian highlands — of 43 land mammal families only 11, comprising fewer than 30 well-defined species, are absent (the number of endemic genera in this area is high, despite the fact that the highlands constantly contribute samples of their fauna to surrounding districts (Hershkovitz, 1969), and e) climatic changes in this area have been marked during the Quaternary, producing significant changes in the vegetation (Simpson Vuilleumier, 1971).

According to Bigarella and Andrade (1965), most of eastern Brazil was semiarid during the Pleistocene glaciations. The Amazonian forest had different extensions at different times during the Pleistocene, forcing mammals including dogs of the adjacent grasslands whether to move into higher and less humid habitats, adapt to the new forest environment, or become extinct. Such cyclic changes in habitat are conditions that speed the evolution of animals and are very probably responsible for peculiar patterns of speciation seen in neotropical birds and reptiles (Haffer, 1969, Vanzolini and Williams, 1970).

These facts suggest that the Brazilian highlands and its surrounding lowlands are a major evolutionary center not only for dogs but also for other mammals.

RESULT OF THE BRAZILIAN CANID RADIATION: THE SPECIES AND THEIR ECOLOGICAL NICHES

Cerdocyon thous

This species (Figure 13-1) is distributed all over the Brazilian subregion of Neotropica (Middle American province of Hershkovitz (1958) not included) except the lowlands of the Amazonian basin. Specimens have been collected in the gallery forest and tree savannas of Venezuela and Guiana (Tate, 1939; Schomburgk, 1848). Carriker (in Allen, 1911) mentions specimens from "the tangled woodland scattered about in the savannas," the animal being abundant "in the whole savanna country of Venezuela." Collector's notes recorded on labels of museum specimens from Venezuela usually say "shot in savanna," but reference is also made to "thorn forest," "gallery forest of savanna river," "in pampas," "caught with steel trap at forest edge," or "beside main wood." In Colombia, Smith (in Allen, 1904) maintains that *Cerdocyon* "seems to belong properly in the dry forest region" of

Figure 13-1 *Cerdocyon thous* (crab-eating fox). Photo: M. W. Fox.

Santa Marta, but has also been reported from the high savanna of Bogota (Allen, 1923; Hershkovitz, 1957).

South of the Amazon River in Brazil, according to Coimbra Filho (1966), *Cerdocyon* inhabits "the southeastern Brazilian forest." Specimen labels of *Cerdocyon* from Paraguay, Bolivia, and southeastern and northeastern Brazil bear the following remarks: "caught in the forest," "lives in the forest," "captured at the forest edge on a banana plantage," "caught in thick forest at the wheat fields edge," "captured in thick and dry forest," "dry gallery forest," etc. In this area the information taken from collector's labels always refers to the forest as the habitat of the species, and the Brazilian vernacular name "cachorro do mato" (forest dog) points out its habitat preferences.

In Argentina and Uruguay, at the southern border of its distribution, *Cerdocyon* is sympatric with the pampas gray fox, *Canis (Pseudalopex) gymnocercus,* and the habitat preferences are clear: *Pseudalopex* is found in the open country while *Cerdocyon* lives in the gallery forest that accompanies the water courses across the grasslands. Trapping results and sight records suggest that the home range of *Cerdocyon* in this area includes the open country at the very edge of the forest.

The only analysis of stomach contents of *Cerdocyon* reported are those by Coimbra Filho (1966) and Mondolfi (in Walker, 1968). Two specimens caught at Guanabara and studied by the first author revealed large amounts of *Artocarpus integrifolia* and *Sysygyum jabolana* seeds respectively, in addition to grasshopper fragments. The 19 specimens from Venezuela examined by Mondolfi included, in order of abundance, small rodents (field mice and rats), insects (mostly grasshoppers), fruits (figs, small berries, bananas, guasimo, and mangos), lizards (2 different genera), frogs and crabs, and in one case, a few bird feathers were present. According to Walker (1968), these foxes "dig for turtle eggs of a species of *Podocnemys* called 'galapago,' who bury their eggs in small holes on the savanna." Since *Cerdocyon* is usually trapped by hanging a spoiled fish, bird, or mammal over a steel trap, one must assume that it includes carrion in its diet. According to Smith (in Allen, 1904), in Colombia the animal preys on small rodents, lizards, perhaps also crustacea, and eats many fruits. Hershkovitz (1957) says that its principal prey is the cottontail rabbit *(Sylvilagus floridanus),* and is also a voracious fruit and maize eater. In eastern Brazil, according to Wied (1826), *Cerdocyon* visits the river shores close to the ocean and during low tide looks for crabs and similar animals among the mangroves. Dead animals are also eaten.

All this information suggests that the ecological niche of *Cerdocyon* is that of an omnivorous ground-dwelling carnivore, active both in the forest as well as the forest edge, open woodlands or wooded savannas. It is a predator of small mammals and an eater of fruits and invertebrates as well as a scavenger.

Atelocynus microtis

Nothing is known of the habitat of *Atelocynus* (Figure 13-2). The localities where specimens have been collected are all within lowland tropical rainforest. This species seems to be the only member of the living Caninae that has been able to colonize this habitat. The food habits of the short-eared dog are unknown.

Figure 13-2 *Atelocynus microtis* (small-eared dog). Photo: M. W. Fox.

Speothos venaticus

The bush dog (Figure 13-3) is distributed throughout the Brazilian subregion. Its habitat in Guiana has been described by Sanderson (1949) as the open wet savannas and the meandering tongues of grass that extend from these areas into the surrounding forest. He found his specimen hidden underneath dry bushes. Goldman (1920) reported a *Speothos* burrow on a steep hillside covered with tall forest on Mount Pirre, southern Panama. P. O. Simons collected specimens that were

Figure 13-3 *Speothos venaticus* (bush dog). Photo: M. W. Fox.

found running alongside a creek in the Department of Beni, Bolivia. Linares (1967) comments on specimens caught close to a river and to thick forest in Venezuela. Specimens collected by Lund (1950) had been found at the edge of the forest, and according to Roth (1941), the species in British Guiana is apparently confined to the forest. It may be concluded that *Speothos* is primarily a forest dweller that also visits open country at the forest edge and generally is found close to a water course. It is semiaquatic and Tate (1931) reported that they are able to follow "pacas" into the water and kill them there. The captive animal studied by Bates (1944) spent a good deal of time in the pool and could dive and swim under water with great facility. A specimen collected in British Guiana was swimming in a river together with two other bush dogs (Bridges, 1954). Santos (1945) also comments upon the bush dogs preference for water and its swimming abilities, and Walker (1968) mentions a female with two juvenile offspring that were observed swimming across the Rio Negro.*

No specific analysis of the stomach contents in *Speothos* has been recorded in the literature. According to Santos (1945), *Speothos* feeds on any animal of small size but hunts also the "nandu" (rhea) and the "capibara" *(Hydrochaeris)*. At the Lincoln Park Zoo in Chicago, Illinois,

*Bush dogs have webbed feet —Ed.

they delight in attacking and eating live pigeons, chicks, rats, and mice (Kitchner, 1971). Hershkovitz (1957) was told that it pursues game as large as deer and an informant of Tate (1931) said that its principal prey is the "paca." The animal observed by Bates (1944) in captivity regarded monkeys and any animal, except dogs and men, as potential food. This has been confirmed by Kitchner (1971).

In summary, *Speothos* seems to play the role of a ground-dwelling carnivore predator in the forest and forest edge community, preying on small and, relative to its own size, large mammals and birds and being able to pursue its prey into water.

Lycalopex (Dusicyon)

The habitat of *Lycalopex* is the "campo," grassy savannas on smooth uplands, or the "campos cerrados," scattered tree savannas in the States of Mato Grosso, Goias, and Minas Gerais in central Brazil. The three references available agree with this point (Lund, 1950; Santos, 1945; Coimbra Filho, 1966). The vernacular name of *Lycalopex* in Brazil is "raposa do campo" (campos fox) in contrast with the name "cachorro do mato" used for *Cerdocyon*.

No analysis of stomach contents of *Lycalopex* has been recorded in the literature. According to Santos (1945), it feeds on small rodents and birds, as well as insects, especially grasshoppers. Lund (1950), who seems to be the source of all printed information on food habits of *Lycalopex,* observed his captive animal eating any kind of animal food, cooked or raw, and easily capturing rats, mice, and insects such as grasshoppers, cicadas, and ants. It never accepted vegetable food.

Coimbra Filho (1966) stresses that in the wild this animal eats a larger proportion of animal food than does the omnivorous *Cerdocyon*. They are persecuted, because many presumably prey on domestic fowl. However, the small carnassial and wide crushing molars of *Lycalopex* suggest a more insectivorous diet than one of the meat of large vertebrates. Additional evidence on food habits will be needed to clarify this point and to understand *Lycalopex's* ecological relationships to the other sympatric dogs, *Chrysocyon* and *Cerdocyon*. This small and little known fox may represent the typical grassland canid in the open country of central Brazil.

Chrysocyon brachyurus

Chrysocyon (Figure 13-4) is distributed over central and southeastern Brazil, reaching to the south as far as Paraguay, eastern Bolivia and northern Argentina. The typical habitat in Brazil is, according to Lund

Figure 13-4 *Chrysocyon brachyurus* (maned wolf). Photo: M. W. Fox.

(1950), Wied (1826), and Santos (1945), the open grassy savannas or the scattered tree and scrubs savannas, or the "campos" and "campos cerrados." In Paraguay, according to Rengger (1830), it lives at the edge of the forests especially those close to swamps or rivers, but also in fields covered with tall grass. The description of the habitat given by Krieg (1948) agrees with that of Rengger, confirming the presence of swamps. The very long limbs of *Chrysocyon* have been explained as an adaptation for locomotion in the mud of swampy areas. This is probably not the case. The Brazilian habitat of this species does not always include swamps. It is very likely that the unusual height of the maned wolf is useful in hunting and other activities in the extensive tall grass savannas that existed prior to the introduction of cattle in South America.

Krieg (1948) examined the stomach contents of several specimens and found remains of various small rodents, especially wild guinea pigs (*Cavia*), and on one occasion, a bird of the genus *Nothura* that lives in the savanna. He also found great quantities of chewed nuts of the palm *Copernicia australis* in one maned wolf. He believes that the maned wolf also eats lizards, frogs, and grasshoppers, besides several fruits. Silveira (1968) gives a list of items on which *Chrysocyon* feeds in the wild. He does not indicate, however, if it is based on analysis of stomach contents or simply an estimation of diet on the basis of other kinds of evidence. His list includes the above-mentioned items plus terrestrial gastropods (*Strophocheilus*), bird eggs, turtle eggs (*Podocnemys*), honey, sugar cane stems and various fruits (bananas, guavas, *Solanum lycocarpum, S. auriculatum S. crinitum, S. grandiflorm, Myrciaria cauliflora, M. jaboticaba, Anona paludosa, A. pisonis, A. reticulata, Mauritia vinifera)* and roots *(Manihor palmata, M. aipo, M. utilissima*). Lund (1950) states that *Chrysocyon* clearly shows a preference for the "fruta do lobo" (*Solanum lycocarpum*) and according to Silveira (1969) these solanacea have therapeutic properties against the maned wolf's kidney parasite, *Dioctophyma renale*. In captivity, *Chrysocyon* accepts all the above mentioned food items in addition to fish (Encke, 1964,1970).

The ecological niche of the maned wolf is that of a terrestrial omnivore, a predator of small mammals and birds, and is especially adapted to live in the tall grass of some savannas and swampy areas.

THE CANIDS OF THE PATAGONIAN SUBREGION

The living wild dogs of the Patagonian zoogeographic subregion fall within the generalized canid pattern. Their ecology is better known, because they came into closer contact with man. As a result they have a greater economic importance than their Brazilian relatives. The biggest species is the culpeo fox, *Canis* (*Pseudalopex*) *culpaeus,* which inhabits mainly steppes but is found also in parts of the deciduous forest from southern Patagonia along the Andes to Ecuador. Its food habits are markedly carnivorous (Crespo and De Carlo, 1963; see also Chapter 12). The second species, the pampas gray fox, *Canis (Pseudalopex) gymnocercus,* inhabits the more humid pampas grasslands in southern Brazil, northern Argentina, part of Paraguay, and Uruguay. These habitats become dryer to the west and south, and in the almost desert-like steppe in Patagonia this species is replaced by its close relative, the Patagonian gray fox, *Canis* (*Pseudalopex*) *griseus*. The small foxes occurring in the low open grasslands and forest edge of Chile belong to the latter species. *Canis* (*Pseudalopex*) *gymnocercus* has typically omnivorous food habits (Crespo, 1971). The same is probably true for C. (*Pseudalopex) griseus* (Housse,1949) but little information is available. The fourth species, C. (*Pseudalopex*) *sechurae*, is the fox of the deserts on the Pacific coast of Peru and Ecuador. Its habitat, sandy soil with a very low density of plant cover, has been described by Huey (1969) who also gives information on this animal's diet on the basis of scat analysis. The Sechura fox shows also a typical omnivorous diet but with marked emphasis on seeds and insects. To a lesser extent, they also prey and scavenge on live and dead vertebrates at the seashore (Koepcke and Koepcke, 1952).

The Falkland Islands wolf, *Canis (Dusicyon) australis,* exterminated by man in the last century, lived in the treeless Falkland Islands. It had a specialized dentition and preyed on marine birds (penguins) and seals.

DISCUSSION AND SUMMARY

The canids colonizing the Brazilian subregion of Neotropica evolved along two different lines. One line invaded the available forest habitats and produced several adaptive types unique among the family Canidae such as the short-eared dog (*Atelocynus*). This animal seems to be the only canid occupying an ecological niche in the rain forest and is the least known South American wild dog from an ecological point of view. Another unusual type is the bush dog (*Speothos*). Its food habits are poorly known but it seems to be able to prey on relatively large animals and probably possesses the unique ability of carrying its hunting activities into the water.

Movement in dense forest requires certain adaptative features, like strength and weight of body and shortness of limbs (Hesse, *et al.*, 1947). It is precisely these kinds of differentiated characters that are found in the Brazilian forest dogs; their limb bones present an increasing degree of shortness and robustness from *Cerdocyon* through *Atelocynus* to *Speothos* (Langguth, 1969). Other characteristics of this group are short ears, short tail, short rostrum, and a modified pattern of pelage coloration.

Adaptations in the anatomy of the masticatory and locomotory apparatus of *Speothos* (Langguth, 1969) agree with its assumed ecological niche. The molars are reduced in number and size, the precarnassial dentition is relatively heavy, the rostrum is short, and the temporal crests meet at the sagittal plane of the skull. These features parallel those observed in other big prey hunters such as the Cape hunting dog, *Lycaon pictus.* The small molars would be less useful in crushing and chewing the omnivorous, varied diet typical of *Cerdocyon.* On the other hand, the short and broad legs also resemble the adaptations seen in aquatic mammals like the capibara *(Hydrochaeris)* and the otters.

Cerdocyon is apparently less advanced in adaptation to the forest because it is the least differentiated of the three forms. In fact, according to the available data, it seems to frequent the grassland habitat north of the Amazon river but still is restricted mainly to the forest or the forest edge south of that river. The reason for this may be that *Cerdocyon* does not have a canid competitor in the grasslands north of the Amazon (*Urocyon* has a restricted distribution and a low population density). South of the Amazon, however, *Cerdocyon* would have to compete with the maned wolf and/or the hoary fox (*Lycalopex*) in order to occupy the Brazilian grassland habitat. Further south, *Cerdocyon* will find *Pseudalopex* occupying similar ecological niches in the grasslands of southern Brazil, Paraguay, Uruguay, and Argentina.

The dogs of the open country represent the other evolutionary line. Among them, the specialized and long-legged maned wolf represents an extreme in the typically canid adaptation to the grasslands. It is able to pursue its role as an omnivore, preying on small mammals and birds even in the difficult conditions of the tall grass habitat.

The actual ecological significance of the smaller *Lycalopex* is less understood. If little is known of the habitat and relationships to the environment of the rare maned wolf, even less is known of the common "campos" or hoary fox. Their lightly-built extremities, elongated only in the metapodia, and the peculiar dentition, having large molars and very small carnassials (Langguth, 1969), suggest special adaptations quite different from those of the typical small dogs of the Patagonian subregion. This species and the maned wolf are sympatric in parts of their distribution ranges and it would be very interesting to investigate what the mutual ecological relationships are.

As different as the maned wolf and the three species of Brazilian forest dogs are, they still share one distinctive character: a short and simple caecum, in contrast to the long and coiled caecum of the other North and South American canids (Langguth, 1969). This apparently insignificant feature may be a clue to an understanding of the phylogenetic relationships of the Brazilian canids. It suggests a common ancestry rather than an independent origin for each individual genus from different generalized ancestors. However, more supporting evidence is needed. Although several specimens of rare Brazilian dogs have died in zoological gardens in the last few decades, only 1 caecum of each species has been reported thus far. The morphology of this organ in *Lycalopex* is still undescribed. Special attention should be paid to save the caecum by those workers who have access to animals that die in the zoos.

Among the living Patagonian dogs, the largest, *C. (Pseudalopex) culpaeus,* shifted from presumably an omnivorous ancestral condition to an almost exclusively carnivorous diet, preying more on larger mammals than the others. The three smaller species are similar and do not represent a significant departure from the generalized, and probably more primitive, type. Their habitat ranges from the humid pampas, which supports a richer fauna and flora and where the bigger subspecies of *C. (Pseudalopex) gymnocercus* live, to the dryer and less productive Patagonian steppes inhabited by *C. (Pseudalopex) griseus*, to the true desert of Sechura in Peru and Ecuador, the habitat of *C. (Pseudalopex) sechurae*. Probably the easiest evolutionary change for small, generalized dogs like these is the adaptation to desert conditions. It may signify little more than a shift in the omnivorous diet, becoming primary consumers, when this is the only way to ensure a reliable supply of food and water.

We have seen that most of the South American canids, with the exception of *C. (Pseudalopex) culpaeus*, *Speothos*, and the Falkland wolf, are omnivorous. This generalized condition is of great advantage because it ensures flexibility of food habits and enables the species to adapt more easily to changing conditions in the environment.

In addition to canids very similar to recent living forms, the South American fossil record shows other coyotelike and some large dhole- and wolflike animals (Kraglievich, 1928; 1930). Some of them, for instance, *Canis dirus*, are of North American origin. Other may have evolved in South America separately. In any event, the evolution of the large fossil canids must be seen in relation to the dispersal and evolution of their prey, namely, the large Pleistocene herbivores.

Some insight into the present and future evolution of the South American wild dogs may be gained by studying the impact produced on their ecology by man through his changing of the environment. Agriculture and cattle raising have modified and will in the future further modify the typical canid habitat. Man has already reduced considerably the extent of the tall grass savannas, sometimes turning them in steppe because of overgrazing by livestock. Generalized omnivore species can more easily adapt to the new environment. Agriculture increases rodent and bird populations, thus providing additional food resources for the carnivores. In this connection, foxes may be tolerated or even favored by man, since they share common interests. Sheep farming supplies the culpeo foxes, mainly at lambing time, with an additional and important food resource, as did the introduction of the European hare in Patagonia by man. Cattle and sheep raising also have negative consequences since they promote competition with man. Predation on sheep and cattle was the reason for the extermination of the Falkland wolf and other species like the red wolf and the wolf are now endangered.

Destruction of the habitat may become a cause of extinction for specialized forms like the Amazonian short-eared dog *Atelocynus microtis* and the maned wolf, both of which, unlike the smaller Patagonian foxes, seem to avoid contact with man. On the other hand, *C. (Pseudalopex) sechurae* originally adapted to desert conditions, is now invading cultivated areas on the edge of its range (Birdseye, 1956). This is another example of adaptability to changing environment, one of the features of generalized dogs that makes the family Canidae so successful.

Acknowledgments I am indebted to R. G. Van Gelder, R. H. Tedford, and A. Rosenberger for their helpful comments and criticisms on the manuscript, and I am grateful to the John Simon Guggenheim Memorial Foundation for granting me the Fellowship under which this paper was prepared.

Editor's Note. The reader will note the discrepancies in the nomenclature and classification of some of these South American canids in the Stains and Langguth chapters. Although an editor likes consistency in a book, these different classifications reflecting different conclusions based on actual research emphasize how complex the South American canids are and also how little we know about them.

14
THE RED FOX IN BRITAIN

H. G. LLOYD
Ministry of Agriculture Fisheries & Food
Pest Infestation Control Laboratory
Guildford, Surrey, Great Britain

In Britain the fox (*Vulpes vulpes*)enjoys ambivalent status. According to locality, it is fostered for sport, pursued as a pest, harassed as a nuisance, and tolerated or ignored where it is none of these. It is the only mammal in Britain subject to a Government approved and aided bounty in some sheep rearing areas. It owes its success in almost all situations in Britain to its unspecialized requirements and flexible adaptability.

HOME RANGE AND HABITAT

The home ranges of the fox — variable in size according to habitat, but of which little is yet known for British foxes — are sufficiently large to encompass a variety of biotopes in much of England and Wales. Foxes are most numerous where the diversity and fragmentation of the habitat within small areas ensures variety of resources; food, harborage, denning, and breeding places. Conversely, large tracts of hill land unrelieved by deep valleys, afforestation or cultivated areas, as in parts of Scotland, extensive artificial conifer forest as in northern England, or the predominantly arable areas of the eastern English counties do not provide this variety and carry fewer foxes. Foxes are exceptionally numerous in the county of Pembrokeshire, where fields are small and each 100 acres (40.4 hectares) of agricultural land had in 1962 a mean length of 4.6 miles (7.3 km) of hedgebank. These earth banks, frequently eroded by livestock, are usually topped by hedges sometimes overgrown 2 or 3 m wide; the county abounds with noneconomic woodlots, scrubby overgrown areas of gorse (*Ulex europaeus*), bramble (*Rubus fructicosus*), blackthorn (*Prunus spinosus*), hawthorn

(*Crataegus oxycantha*) and bracken (*Pteris aquilina*), and steep-sided dingles; on the coastline are boulder-strewn, often densely blackthorn-covered sea-cliff slopes. Small sized farms, approximately 50% being under 40 acres (16.2 hectares), are predominantly pasture. The agricultural acreage is 46% permanent pasture, 19% temporary leys, 17% rough grazings, and 17% under arable crops or fallow. Contrast this with Cambridgeshire where farms and fields are large, field boundaries thin, affording little cover, with corresponding crop acreage proportions of 10%, 3%, 1% and 86%, respectively. In addition to the high proportion of land not under the plough, Pembrokeshire has an abundance of derelict and overgrown land, which provides suitable habitat for prey and secure cover for foxes. The numerous small farms and villages may also be of significance in providing food resources for foxes both normally and during hard times. Each cow in the county produces annually a mean of about 5.44 kg of potential fox food in their foetal membranes, or the equivalent of 1.32 kg per acre throughout the county; in addition, foxes have access to 0.20 kg per acre of (seasonal) sheep afterbirths (data based on Ministry of Agriculture, Annual Agricultural Census). Absolute population densities are not known with accuracy for any part of Britain, but as an example of the numbers encountered in the better habitats, a study area of 800 acres in Pembrokeshire yielded 76 foxes (53 males and 23 females in January–February, 1971. Dispersal movements, predominantly of males, occur in late autumn–winter, and doubtless many of those killed were itinerant at the time of trapping.

NICHE OF THE RED FOX

Being a nonspecific predator of prey species weighing up to about 3 kg (brown hares *Lepus europaeus* or neonatal deer), a scavenger on carrion as diverse as the afterbirths of cattle, dead livestock, animal road casualties, and human waste, frugivorous and insectivorous seasonally, vegetarian in emergency (silage and stored potatoes), and being able to exploit prey in many habitats regularly, such as voles and mice, or occasionally, such as spawning salmon (*Salmo salar*), the fox would best be described as a facultative predator occupying an opportunist niche. Unlike predators which are able to exploit prey in particular habitats only, for example, vole predation by tawny owls (*Strix aluco*) in deciduous woodland, and by kestrels (*Falco tinnunculus*) on open grassland, the fox is able to exploit prey widely in many habitats, albeit more efficiently in some than in others. Furthermore, its hunting activities are not restricted by daylight or darkness.

POPULATION DENSITY

Judging by the values of fox bounties formerly administered by the Church from the 15th to 19th century and by the number of bounties claimed, foxes then must have been considerably fewer than today. Changes in habitat created by man, notably the enclosure of land, the erection of hedges and banks, and the adoption of crop rotation culminating in the Norfolk four-course rotation favored prey species and provided conditions which led to dramatic increases in numbers of wild rabbits, but habitat diversity providing a wide variety of food and other resources were mainly responsible for promoting the status of foxes from this time of agricultural expansion. At present, however, high density fox populations are not related to high rabbit numbers and since the myxomatosis epizootic of 1953–54 the number of rabbits has remained comparatively low, nationally being probably not more than 5% of former numbers, though rabbits occurred on 59% (± 6%) of all farm holdings in England and Wales in 1970 and can be abundant locally (Ministry of Agriculture unpublished report). Foxes now appear to be more numerous than in 1953 and their current success is clearly not dependent on large numbers of rabbits. Foxes derive benefit from past and present rabbit communities since many fox dens originated as rabbit burrows, and disused warrens still provide foxes with an abundance of ready-made exploitable denning places.

TABLE 14-1 Population age-structure of foxes, mid-Wales. (Data based on autopsies of 600 foxes killed December–April)

Age	Frequency or % Occurrence of Age Groups	% Mortality
9 to 12 months	59	
		59
1 to 2 years	24	
		58
2 to 3 years	10	
		60
3 to 4 years	4	
		50
4 to 6 years	2	
Total	99	

The sex ratio of 15,679 adult foxes killed by a variety of methods in Wales was 56% males. The age-structure of a late winter–early spring sample of mid-Wales foxes, based upon radiological and microscopic tooth examination (pulp cavity size and annulation of cementum, respectively) has been determined and the proportions of foxes in given age-groups are shown in Table 14-1.

FOOD HABITS

Investigations reporting the diets of foxes are many but most studies do not relate food availability to exploitation. Exceptions are the investigations of Lever (1959), following those of Southern and Watson (1941), which showed changes in diet associated with a drastic change in the abundance of rabbits before and after the myxomatosis epizootic of 1953–54. A synopsis of available data clearly indicates small mammals, notably *Microtus agrestis, Clethrionomys glareolus,* and *Apodemus sylvaticus* as the staple diet of foxes, but other foods identified provide a large and varied catalogue of species eaten, indicating little more than the opportunist nature of fox feeding behavior. It has proved difficult to collect sufficient supplies of scats and stomach contents throughout the year from defined habitats to permit a comparison of intake with prey and other food availability. (Bait acceptability trials in the field in winter have shown that up to 88% of bait stations were marked by fox scats, a behavioral trait which can perhaps be exploited for study purposes.) It is not possible by fecal examination to distinguish between carrion and animals taken alive. Also, the origin of sheep wool fiber found in stomachs or scats cannot be assigned to adult or lamb, to carrion, or to an animal killed by the fox, not even more misleading, to a ball of wool fibers eaten by the fox. Feces of unweaned and newly weaned lambs are eaten by foxes and may provide a significant item of diet for a short period of the year, but these, as with afterbirths of sheep or cattle, would not be revealed by scat analysis.

Some potential prey species are only occasionally eaten by foxes; notably, moles *(Talpa europaea)* and shrews *(Sorex spp)*. Though little information is available on food preferences and exploitability of food in the wild, seasonal feeding patterns are well-documented. Lever (1959), investigating fox diets in areas of low rabbit numbers, showed that most rodents were eaten in autumn and winter, birds in spring and summer, insects in summer, and fruit in summer and autumn.

RELATIONSHIP AND IMPACT ON PREY

Being a facultative predator living in nonuniform fragmentary habitats and able to exploit a wide variety of prey and other foods, the fox in Britain is unlikely to be adversely affected by changes in abundance of a single prey species. Where few prey species occur and where the supply of other foods may be seasonal but liable to yearly fluctuations in abundance (sheep or deer carrion, for example), other prey species and other foods might not occur in sufficient abundance to provide a buffer against food shortage, but habitat diversity tends to ensure against this over most of Britain, with the possible exception of the extensive hill areas of Scotland. Abundance of rabbits followed by a sudden decline locally is characteristic at the present time when outbreaks of now enzootic myxomatosis occur: the biomass of potential food resources can be drastically changed within a short space of time at such outbreaks but the decrease in availability of rabbits can be redressed by increased exploitation of other species. Little, however, is known of food preferences and the relative ease of exploitation of the different prey species in different habitats, but only exceptionally are foxes of poor body condition found at autopsy which suggests that subadult or adult foxes are rarely exposed to severe food shortage in Britain. Cubs and juveniles exhibit poor body condition more frequently than older foxes. Whilst the fox can and does thrive in the absence or scarcity of rabbits, there is evidence to suggest that predation by foxes may have a considerable influence opposing rabbit population increase. In 1970, rabbits occurred on 59% of a 2.5% sample of agricultural holdings in England and Wales, and only a negligible proportion of these holdings had high rabbit population densities (unpublished Ministry of Agriculture report). Immediately before the myxomatosis epizootic, 94% of these same holdings had rabbits, 45% at high densities. Throughout Britain at the present time, rabbits are probably not more than 5% of former pre-myxomatosis numbers and enzootic myxomatosis is the major factor opposing rabbit increase, but predation also has its effect. During an outbreak of the disease, infected rabbits, including those that would have recovered, become vulnerable to predation and to indiscriminate or surplus killing which is characteristic of foxes presented with highly vulnerable prey. Where rabbit numbers are insufficient to support myxomatosis, predation by a facultative predator such as the fox may severely retard the growth of the rabbit community, and areas where rabbits do not regularly occur or where they are scarce may be only marginally suitable for rabbits while they remain at low numbers. Mortality rates of juvenile rabbits, even in the absence of myxomatosis, is as high as 80% in many areas studied. Whilst the fox competes with stoats (*Mustela erminea*), weasels (*Mustela nivalis*), domestic and feral cats for the same prey species, the fox does not, along with the domestic and feral cat, occupy a specialized and narrow niche and is not dependent upon any single prey complex for its survival, but its numbers are lower in species-poor habitats.

MOVEMENT PATTERNS

Dispersal of juvenile foxes occurs during the period September to December; the distances moved by males and females differ but it is not possible to generalize on the extent of movements in all areas. Current knowledge based on the tagging of 272 cubs suggests that the dispersal and the extent for movements seem to be related to fox population density or to some parameter related to it. In areas of moderate to low density, on predominantly sheep-rearing hill areas of mid-Wales, the mean linear distance moved from the rearing den at the time of tagging as a cub, to the place where the fox was killed by hunters was 6.5 miles (10.4 km) for males and 1.4 miles (2.2 km) for females. In high population density areas of west Wales (Pembrokeshire and Cardiganshire), the corresponding distances were 2.9 (4.6 km) and 1.2 miles (1.9 km) respectively. In these two areas, respectively 69% and 66% of the female cubs had moved less than 1 mile (1.6 km), and among males 11% had traveled more than 6 miles (9.6 km) in west Wales and 39% in mid-Wales. Thus, vixens tend to be sedentary in both areas but the dispersal movements of emigrant males are greater in mid-Wales. Movements of the few tagged adults recaptured (16) are more restricted, with a mean of 1.2 miles (1.9 km) for males and 1.6 miles (2.6 km) for females.

The mean ages of tagged cubs at death (at the hands of man) also varied in the two areas, being 8.5 months in mid-Wales and 18.6 months in west Wales. These data have some bearing upon the observed dispersal movements since it is clear that many of the foxes recaptured in mid-Wales may not have completed or even begun to emigrate when they were killed, thus the figures for mid-Wales probably underestimate the extent of dispersal. The stimulus to dispersal is not known, nor what resource or environmental feature promotes the settlement of an emigrant fox in any particular place. Tagging and recapture evidence suggests that the dispersal movement of individual foxes in mid-Wales occurs suddenly and is not a gradual movement away from home over a long period of time. Emigration seems to be completed by January with most activity in November and December.

From February (after the mating season) to June, foxes tend to be found singly. From July to October, groups of foxes are encountered and constitute family groups not yet dispersed. In hill areas, there is also some movement of foxes to higher ground from June to September. In November–December, more noticeable clumping of foxes occurs, and from evidence obtained by shooting, a high proportion of these are males. In January, clumping is less pronounced and in the breeding season foxes are distributed fairly evenly again throughout available or suitable cover. By the time when most vixens are in mid-term stages of gestation, foxes appear to be solitary and males then appear to have least contact with females.

During the post-weaning period vixens and cubs live in family groups and contact with male foxes becomes less frequent. After dispersal of juveniles, the adults tend to become solitary, but if the juveniles have not moved far, contact with them seems to be maintained or at least cannot be avoided.

REPRODUCTION

During the 6 weeks before estrus, until the first vixens become pregnant, congregations of foxes are encountered, and within these groups some form of competition, probably for vixens, ensues as they move from place to place. Vixens are frequently reported to be attended by several foxes, usually identified as males (an observation possibly not without error). These congregations are not uncommonly seen in daylight. Ritualized fighting, with the contenders adopting an upright stance, face to face, open-mouthed, uttering short, sharp, high-pitched grunts, snarls, or yaps are a common feature.

Autopsies of vixens killed from mid-December to late January have revealed spermatozoa in about 8% of vaginal or uterine smears of anestrous vixens. The densities of spermatozoa in these smears are never as high as in smears taken from recently-mated estrous vixens. Clearly, intromission has occurred but it is not known whether coupling had taken place as in estrous matings, and possibly the spermatozoa identified were the product of the mating attempts reported to occur at the congregations described. If these anestrous matings are common to all vixens, they may serve to provide a physiological or-psychological stimulus hastening proestrus and estrus in vixens which have male partners, thus modifying the proximate abiotic stimulus to estrus. Such behavior might be of advantage to a solitary monestrous species, having an estrus period of short duration, in ensuring the maximum probability of fertile matings than might otherwise be the case if the onset of estrus was determined strictly by external physical, as opposed to biotic, factors. If the agonistic displays of males, directly or indirectly provide this stimulus, and one of the contenders remains with the vixen (there is no evidence to confirm this) the probability of maximum fertility of the populations might be obtained. Autopsy evidence indicates a high pregnancy rate of 92%.

Mating, with coupling, at estrus (mid-January to mid-February) occurs in daylight and darkness and is often accompanied by much vocal clamor. Sometimes more than 2 foxes may be seen on these occasions, the supernumeraries possibly being unattached males attracted by olfactory and vocal stimuli.

Birth litter sizes range from 1–11, the mode is 5 and the mean 4.3 per litter. The proportion of nonproductive barren vixens varies according to location and year and has been observed over a period of 4 years to range from 14% to 29%. In a season of moderate fertility, the mean number of cubs per adult vixen was 3.2. In the field, the mean litter sizes of cubs up to about 6 weeks of age found by excavating dens was 5.2, but this decreased quite sharply in litters of cubs over 6 weeks old. Thereafter, until litters took to surface at about 10 weeks, litter sizes were a mean of 2.7.

The significance of this sudden decrease in litter size is not understood — it might represent partial mortality of litters, the partitioning of litters by the parents, or the accidental wandering away from the maternal den by some of the cubs, subsequently to be deposited in the nearest den by the parents. From the time of weaning when cubs are partially fed on regurgitated food until they are about 10 weeks old when the cubs can seek some of their food themselves, the parents may often be hard-pressed to provide food for themselves and their young. Judging by the sometimes large quantities of unconsumed food at cubbing earths, it seems that the parent or parents hunt indiscriminately and take all that they can. Surplus food is more conspicuous at earths containing small litters of 2 or 3 cubs than in those containing 8 or 9 cubs. A litter of 8 cubs of 8 weeks of age, weighing jointly about 32 lb (14 kg) would require about 6 lb (2.7 kg) of food daily.

It is alleged by many that have experience with foxes that the male of a breeding pair will forage for the vixen for the first 10 to 14 days after parturition, but there is little valid evidence for or against this observation. During this postnatal period, however, the vixen lies with the cubs in daytime, and from autopsy and trapping evidence, emerges at some time during the night. From about the time when the eyes and ears of cubs open at 12–14 days, vixens seek daytime cover outside the nursing den. In some areas as in west Wales, cubs tend to stay in the birth den until they are weaned and later begin to seek surface cover at about 10 weeks of age. In other areas such as mid-Wales, cubs often are moved from one earth to another even before being weaned at 5 to 6 weeks, but it is not known whether these translocations are spontaneous or precipitated by interference from man, his dogs, or by the presence of some natural, potential or real, hazard such as the presence of or a visitation to the earth by a polecat (*Putorius putorius*). Certainly foxes will move cubs with little provocation in mid-Wales, and this is characteristic behavior throughout much of the range of the fox.

The density of occurrence of tenable fox earths varies considerably from place to place, so also does the proportion of used to unused earths during the 8 week postnatal period. In parts of Pembrokeshire, a mean of 1 examined earth per 5.2 is occupied by cubs while in more sparsely populated (1 earth per 12.6 hectares) hill areas of mid-Wales, where much clumping of the distribution of earths occurs as revealed by the use of terriers, the ratio was 1 occupied den to 120 unoccupied dens in March, April, and May, 1970 for an area covering 26,304 hectares.

Occupation of dens in soil is seasonal but it can be modified by topographical and climatic features. The period when dens are mainly used by adults is from November to January with peak vixen occupation after parturition in late March–early April. At other times, foxes seem to prefer to lie above ground depending upon availability of cover, but they will occupy earths at any period of the year following prolonged heavy rainfall. At present, little is known about the availability of earths and their use within the ranges of individual foxes. From January to April, many earths show signs of fresh digging by foxes, but only a small proportion of those seem to be occupied subsequently by cubs during the following few months.

CONCLUSION

At present little is known of the factors regulating reproductive productivity, mortality, longevity, and population density of foxes in Britain. Throughout much of its range the fox is pursued for one reason or another, and in areas where its destruction is sought with unflagging persistence as in predominantly sheep-rearing areas, it is not too colorful to say that few adult foxes survive long enough to die of natural causes. In spite of this destruction, the fox is able to hold its numbers at a moderate level of abundance in these areas and will probably continue to do so, here and elsewhere, providing that nothing more drastic than traditional means of control are employed.

15
ECOLOGY OF THE RED FOX IN NORTH AMERICA

E. D. ABLES
College of Forestry
Wildlife and Range Sciences
University of Idaho
Moscow, Idaho

PHYSICAL CHARACTERISTICS

The red fox (*Vulpes fulva* Desmarest) is one of the most handsome of the wild canids. Typically the color is pale yellowish-red to deep reddish-brown on the back and sides, blending to lighter shades and even white on the undersides. Ears and lower legs are black and the tail is always tipped in white. Red foxes are smaller than most people realize with females averaging 4.1–4.5 kg and males 4.5–5.4 kg. Those in Canada and Alaska are somewhat heavier.

In addition to the common red, there are several other color phases, the most common ones being the silver which has black hair with white bands near the tip, and the cross fox which has a dark cross formed by a stripe down the back and across the shoulders. These two color variants are rare to nonexistent in the southern range of the red fox in North America, but become increasingly common northward. All three have been observed in the same litter in Alaska (Murie, 1944). Genetics of the various color phases are described by Butler (1945, 1947). The silver fox fur industry evolved through selectively breeding the most desirable recessive silvers until the offspring bred true (Samet, 1950).

Considerable variation exists in texture of guard hairs and fur fibers, in luster of the pelt, and in body size among foxes from different geographic locations in the species range (Bachrach, 1947). The largest pelts are from Alaska, and decrease in size eastward across Canada. In the United States the largest pelts are from the northwestern states, the next largest from the eastern states, and the smallest from the central region of the country. Pelts from temperate regions in the United States are smaller than those from subarctic Canada and Alaska. This suggests an increase in size toward the northwest in North America. Standard measurements from foxes in the Pacific states (Ingles, 1965), Michigan (Burt, 1946), eastern Canada (Peterson, 1966), and Alaska (Osgood, 1904) support this hypothesis.

GEOGRAPHIC RANGE AND TAXONOMIC STATUS

Twelve subspecies of red fox range over a major portion of the North American continent north of Mexico and south of the ice sheet (Figure 15-1). They are rare or absent from pine forests of southeastern United States, moist conifer forests along the Pacific coast, semiarid grasslands, and deserts. Range extensions have been noted northward in Canada (Macpherson, 1964) and westward in the United States (Hall, 1955).

Red foxes were apparently absent in the eastern deciduous forests of New England at the time of settlement (Seton, 1929). No evidence of skeletal remains were found in archeological excavations or in caves in Pennsylvania (Gilmore, 1946, 1949). However, recent evidence from New England archeological sites suggests that red foxes may have been present during glacial periods as late as 1000 years ago (Waters, 1964, 1967). Numerous accounts are available which substantiate the introduction of English red foxes (*Vulpes vulpes*) into New England between 1650 and 1750 (Mansueti, 1955). Thus, the evidence suggests that the red fox in the eastern United States is a direct descendant of the English fox (Peterson, *et al.*, 1953) or is a hybrid between the English fox and the red fox of southeastern Canada (*Vulpes fulva fulva*)and/or the northern plains red fox (*Vulpes fulva regalis*). Churcher (1959) discussed at length the specific status of the red fox in North America and concluded that *Vulpes fulva* Desmarest should be considered synonymous with *Vulpes vulpes* Linn. The foxes which have been studied most intensively and which have contributed the bulk of data in this chapter are the eastern red fox and the northern plains red fox.

HABITAT PREFERENCES

Red foxes are adaptable and occur in a variety of terrain and vegetation types. In British Columbia they range from deep forests to the most

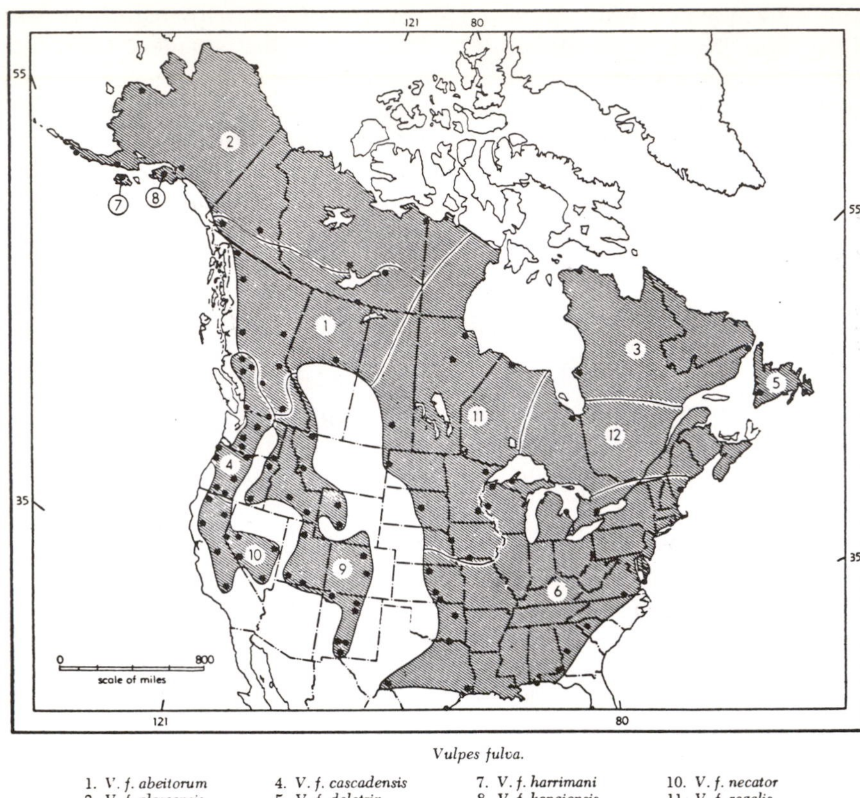

Figure 15-1 Geographic distribution of subspecies of *Vulpes fulva* in North America (reprinted by permission from E. Raymond Hall, Ph.D., and Keith R. Kelson, Ph.D. *The Mammals of North America,* Vol. II, Copyright © 1959. The Ronald Press Company, New York).

exposed country with highest populations in a region of mixed forest, meadow, and plains, but are scarce in lowlands (Soper, 1942). The Alaska red fox prefers broken country such as hillsides, sides of valleys, and canyons (Bee and Hall, 1956). Iowa foxes were most numerous in hilly, wooded regions but were also common in the flatter prairie corn belt (Errington, 1937). One of the densest populations of foxes in North America is in southwestern Wisconsin, described by Richards and Hine (1953) as inhabiting a patchwork of woodlots, cropland, pasture, and stream bottoms.

Intensive studies of habitat use by individual foxes have revealed some of the intricacies involved. Schofield (1960) followed fox trails in snow in Michigan and counted the number of paces in each of several vegetative types. There was a preference for lowland brush and oak woods and an avoidance of swamps. Storm (1965) used radio-tracking to monitor the activities of foxes in Illinois but could detect no clear-cut preference for croplands versus woodlands. In southern Wisconsin, foxes were radio-tracked in an area of ecological diversity and in less diverse farmland. In the diverse area, vegetative types containing brush and a ground cover of grasses and sedges were used most heavily at all times. Deciduous woods, mowed or grazed lands, and residential areas contained fewer radio fixes than expected.

In the rural farming district, croplands were used least, while deciduous woods, marsh, and brush were used heavily. Significantly more time was spent near an edge or transition zone between vegetative types than would be expected by chance (Figure 15-2). The overall implication, based on all available data, is that red foxes prefer a physiographic type composed of a mixture of vegetative components, but avoid large homogeneous tracts of any single type. Within this general habitat foxes select areas of greatest diversity. The determining factors are probably food availability and cover requirements. Woodlands are a desirable habitat component but are not necessary, especially in large blocks.

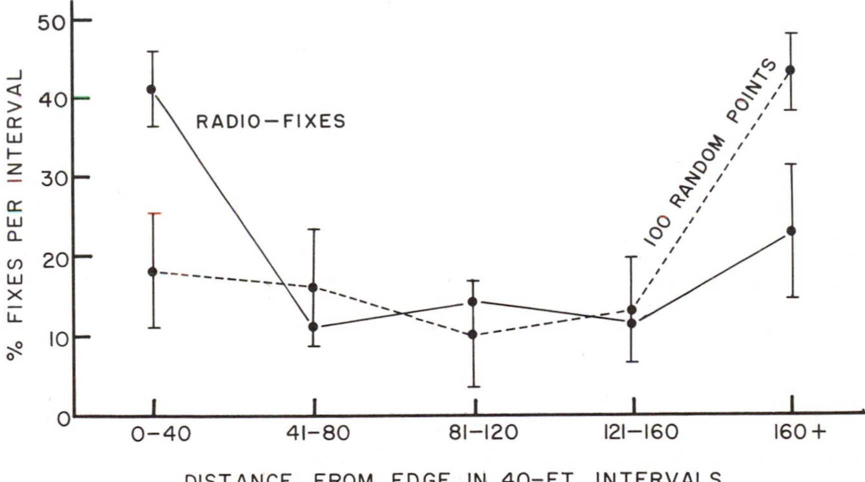

Figure 15-2 Relationship of radio-fixes to distance from a vegetative boundary; for an adult male red fox in southern Wisconsin during October–November, 1964. C.I. = 0.05.

REPRODUCTION IN THE FOX

Most of the early information on reproduction in red foxes was derived from captive herds of silver foxes (Pearson and Enders, 1943; Asdell, 1946; Pearson and Bassett, 1946; Venge, 1959). These data were accepted as applying equally to wild foxes. Subsequent studies (Sheldon 1949, 1950; Layne and McKeon 1956, 1956a) suggest that the above conclusion was justified.

Spermatogenesis begins in late fall or early winter and males are fertile for several weeks before the breeding season commences. Development of mature sperm in young males lags behind that of mature males (Asdell, 1946), but most males achieve full reproductive capability before the end of the breeding season. Testes size is indicative of reproductive condition with individual testes weights of approximately 4 grams during the peak of breeding.

Females are monestrous with estrus periods of 1–6 days and periods of sexual receptivity of 2–4 days (Pearson and Enders, 1943; Asdell, 1946; Pearson and Bassett, 1946). Young are capable of breeding at 10 months of age and usually breed later than adults. Success of fertilization and litter size are greatest if mating occurs on the second day of receptivity (Asdell, 1946). Among silver foxes, 16% of the females failed to breed during their first year, a condition that seems to apply among wild foxes (Layne and McKeon, 1956a; Ables, 1968). Fetal development is described by Layne and McKeon (1956a), and measurements at birth by Hoffman and Kirkpatrick (1954) and Storm and Ables (1966).

A latitudinal variation in mean breeding dates seems to exist as suggested by Sheldon (1949). Breeding peaks vary from late December to early January in Iowa (Scott, 1943), late January in Wisconsin (Ables, 1968), to late January and early February in New York (Sheldon, 1949; Layne and McKeon 1956a). Earliest recorded breeding dates for foxes in the United States (exclusive of Alaska) are in December and the latest are in April.

Litter sizes have been estimated from examination of reproductive tracts and litters removed from dens. Variations in mean litter sizes between geographic regions and between years are to be expected, but on the whole the average number of young per litter is remarkably constant. In New York, mean litter sizes were 5.37 and 5.4 (Sheldon, 1949; Layne and McKeon 1956a), in Wisconsin, 5.1 and 5.4 (Richards and Hine, 1953; Ables, 1968), and in Michigan 4.92 (Switzenberg, 1950) and 4.6 and 5.5 (Schofield, 1958). The data of Schofield were obtained from 1809 litters over a 7 year period, the largest sample of any study on record. Hoffman and Kirkpatrick (1954) reported an average of 6.8 fetuses in Indiana foxes, slightly greater than the mean from other states. Estimates and calculations of mean litter sizes are not always comparable since some are based on placental scars, some on fetuses, and others on pups of various ages. Extremes in observed litter sizes and numbers of fetuses range from 1–11 (Switzenberg, 1950), 4–13 (Hoffman and Kirkpatrick, 1954), and 1–10 (Ables, 1968). One den in

Michigan contained 17 pups, all of which were believed to belong to the same litter (Holcomb, 1965). Other investigators have noted large numbers of pups in one den but size differences indicated two different ages (Sheldon, 1949; Switzenberg, 1950).

Dens are prepared in late winter at which time the female restricts her activities to the vicinity of the den site. There is a preference for loose soils on well-drained sites near or within vegetative cover. Most fox dens were located on slopes in Iowa (Scott and Selko, 1939), on southerly facing slopes in woods in Wisconsin (Richards and Hine, 1953), in sandy soils near the edges of woods in New York (Sheldon, 1950), and on islands in Maryland marshes (Heit, 1944).

Dens of other burrowing mammals may be enlarged by foxes or they may dig their own. Most dens have at least 2 openings which are interconnected. Murie (1944) described dens with up to 19 entrances used in successive years by Alaska foxes. I have examined many fox dens with 12 or more openings in the vicinity of Horicon Marsh in Wisconsin. These dens have been occupied more or less continually for periods of at least 35 years. Such dens are traditional and at a premium because of a scarcity of suitable sites.

Red fox pups first open their eyes at 9 days of age, appear outside of the den at about 1 month, and are weaned at 8–10 weeks. Both parents may bring solid food to the dens before the pups are capable of consuming it. Most investigators report that the pups are moved from the natal den to an alternate site at least once. As the pups become more mobile, their ties to the den sites weaken and the center of activity changes to some location within the home range referred to by Scott (1943) as a "rallying station." The family remains together as a unit until early fall when the young are full-sized and dispersal occurs.

Our knowledge of the role of the male in the family is less than exact. Seton (1929) states that during the latter stages of pregnancy and while nursing, females are hostile toward males, but that males bring food to the nursing female and later actively participate in care and rearing of the young. Sheldon (1949) reported many dens with vixens only, but observed males playing with the pups. Scott (1943) does not specifically describe the role of the male. Radio-tracking studies have detected males carrying food to the den and associating with the pups until autumn (Storm, 1965). However, other males, radio-tracked in the ranges of a female with young, have made little contact with the family group (Sargeant, 1965). I have observed families at play and around dens but have never seen a male in attendance. These apparent discrepancies may be due to incomplete data or to real differences in behavior. In regions where foxes are hunted or trapped heavily during winter months, females may lose their mates.

DISPERSAL MOVEMENTS

During fall and winter months, family groups break up and the young foxes disperse from their home areas. Trappers and hunters concentrate their efforts at this time when foxes are most vulnerable. Consequently, many investigators have taken advantage of this opportunity to tag the young during summer and recover the marked animals during fall and winter, thereby gathering information on distance movements.

Dispersal apparently takes place in easy stages with the young wandering progressively further afield (Errington and Berry, 1937). Male pups disperse before females and move greater distances. Average distances for recoveries of males and females, respectively are as follows: 43 and 8 (Arnold, 1956), 26 and 13.3 (Ables, 1968), and 26.3 and 7.1 (Phillips, 1970). Sheldon (1950, 1953) recorded the longest distance for a male as 40 miles (64.0 km) and only 2 miles (3.2 km) for a female. The majority of recoveries in most studies were within 10 miles (16 km) of the release site. However, some very long movements have been reported; 160 miles (257 km) (Errington and Berry, 1937), 126 miles (203 km) (Longley, 1962), and 245 miles (394 km) (Ables, 1965). There is a possibility of bias in studies where recoveries are made in the same year of tagging. Young foxes killed during the fall and winter are usually wandering and have not selected a home range. Their final destination may have been further from their tagging site than their recovery.

HOME RANGE

There is strong evidence that adult foxes remain in the same home area for life (Seton, 1929; Murie, 1944; Sargeant, 1965). The size of this home range varies with terrain, complexity of the habitat, and food supply. In addition, the area of use changes daily and seasonally.

Seton (1929) states that the home range is usually not more than 5 miles (8 km) in diameter. Murie (1936) found that most fox activity was contained within an area of 2 mile^2 (6 km^2). Scott (1943, 1947) believed that a l mile (1.6 km) arc would encompass the normal activities of individual foxes or family groups. Results of subsequent studies agree with this conclusion (Richards and Hine, 1953; Schofield, 1960; Storm, 1965; Ables, 1969a).

During the period of parturition and for a few weeks thereafter, adult foxes usually remain within 0.5 mile (.8 km) of the den (Scott, 1943). Ranges are largest during the winter season (Sheldon, 1950) probably in response to the added difficulty of securing adequate food. On the other hand, concentrations of food and deep snows which make movement difficult sometimes result in a fox remaining in one small area for several days.

Internal patterns of home range use were studied with radio-tracking

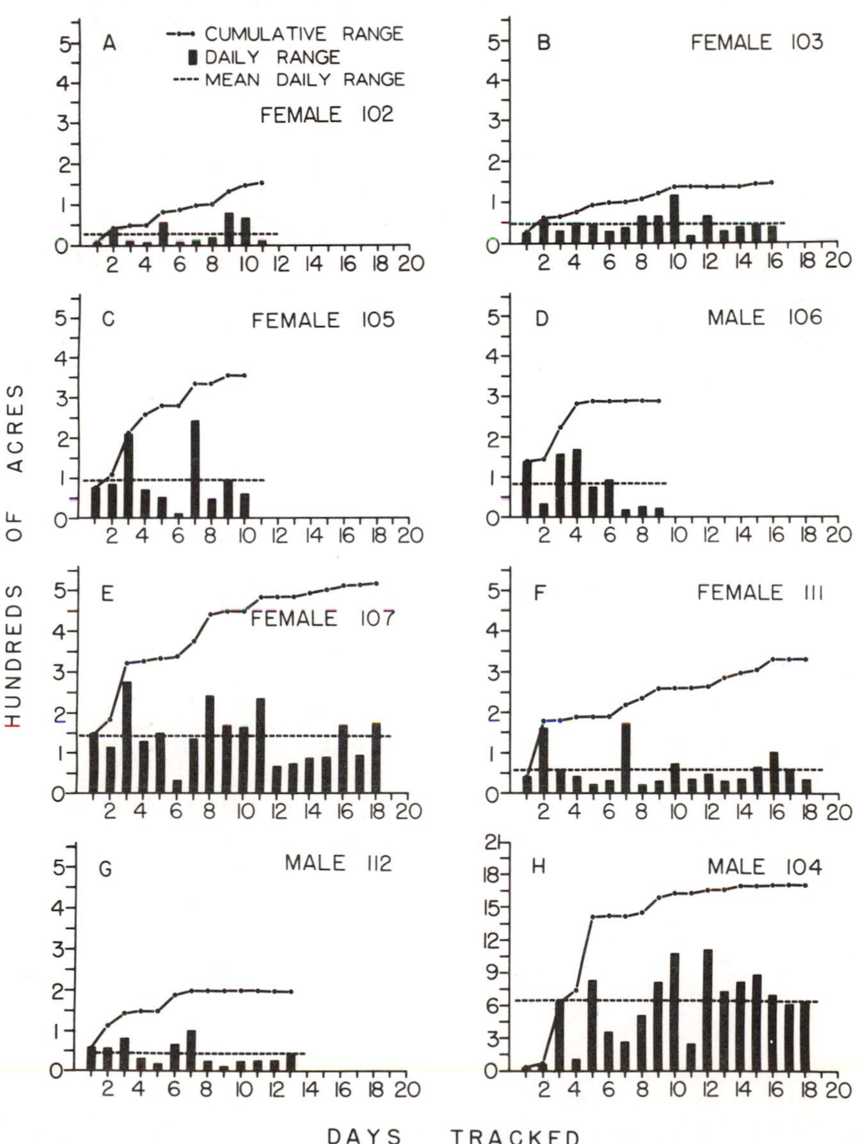

Figure 15-3 Daily, cumulative, and mean home ranges for 8 red foxes radio-tracked in southern Wisconsin during 1964–65.

techniques by Storm (1965) in Illinois and Ables (1969a) in southern Wisconsin. Storm found that the home range consists of a series of pathways which connect areas of intensive use. Adult foxes regularly returned to familiar rest areas during the day but seldom reused a particular bed. Intraspecific activity and availability of food importantly influenced daily movements.

In the southern Wisconsin studies, home range use patterns were similar to those in Illinois. There were defined centers of activity associated with dens, favorite hunting areas, abundant food supplies, and daytime resting sites. Mean areas of daily ranges were less than ½ the total home range sizes (Figure 15-3). Home range shapes were linear, with linearity not always associated with obvious features of the habitat. Some foxes clearly used roads and other physical features, both natural and man-made, as limits to their movements. Most activity was confined to areas rather easily recognized by the investigator with occasional unexplained trips outside the usual home range.

TERRITORIALITY

Territorial behavior is difficult to document in the red fox, if it indeed exists. Early references mention the territory of a fox, but in most instances the authors are referring to the home range rather than to a social phenomenon. Because of our lack of sufficient detailed observations, territorial behavior must be inferred from indirect evidence. Thus, Scott (1943) states, "If. . . positive reaction to a particular place and familiarity with the environment are manifestations of territory, then territorialism is characteristic of the red fox."

The best evidence we have of territorial behavior is the existence of nonoverlapping home ranges. Scott (1947) observed that theaters of activity of neighboring families of foxes did not overlap during the denning and rallying station period. Sargeant (1965) observed very little overlap in home ranges among 3 females in the same area for 2 successive years. I radio-tracked 2 adult females, 1 of which had young, whose home ranges adjoined but did not overlap an oak woods. Thus, the evidence is inconclusive but suggests at least seasonal territorial behavior among females.

DAILY ACTIVITY PATTERNS

Diel activity patterns of red foxes and the environmental parameters which influence them have been reported by Storm (1965) and Ables (1969). In Illinois, Storm found that foxes were mainly nocturnal and

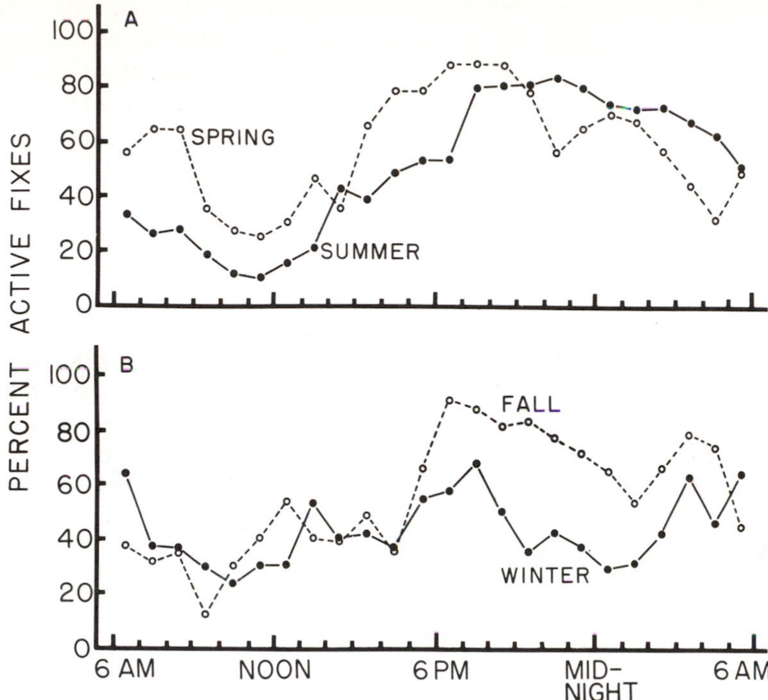

Figure 15-4 Diel activity patterns of 8 red foxes radio-tracked in southern Wisconsin during 1964–65.

that journeys began as early as 2 hours before dark, continued throughout most of the night up to as late as 4 hours after dawn, with rest periods during midday. Temperature, cloud cover, and precipitation had no detectable effect on nocturnal activity. Availability and distribution of foods had the greatest influence on daily movements.

In Wisconsin the diel activity patterns (Figure 15-4) were monitored and related to diel patterns in availability of prey. Foxes were most active at night with a tendency toward crepuscular peaks. Activity peaks of foxes corresponded to peaks in prey activity. During winter, the level of daytime activity was higher than during other seasons and may have been due to greater difficulty in obtaining food.

Fox activities were correlated both positively and negatively with various climatological factors, but the mode of operation of these factors was not always clear. Foxes were more active on colder, windier days and nights during summer. They sought shelter in dense pine plantations during summer thunderstorms. During winter, movement was restricted by deep snows and by strong winds in combination with subzero temperatures. All of these reactions can be explained as comfort-seeking behaviors, or even survival mechanisms. However, a negative correlation between fox activity and barometric pressure during winter is not easily explained. Perhaps low pressures are associated with storm systems and no true cause–effect relationship with fox activity exists.

FOOD HABITS

No other aspect of red fox ecology has been the subject of such numerous and intensive research efforts as have food habits and predatory relationships. Controversies have raged between the public and wildlife biologists for many years over the effects of foxes on populations of game birds and mammals. At least 100 food habits investigations have been published. Space permits listing only a few of these. A fairly complete picture of what foxes eat can be obtained by consulting Errington (1935, 1937, 1937a), Murie (1936), Scott (1943, 1947), Seagers (1944), Latham (1950), Richards and Hine (1953), Scott and Klimstra (1955), Sullivan (1955),Korschgen (1959), and Besadny (1966).

The majority of studies have shown that the major food items of foxes are small rodents, rabbits (*Sylvilagus* spp.), wild fruits and berries, and insects. The upper size limit of prey is usually an animal no larger than a hare (*Lepus* spp.), while abundance and ease of capture seem to govern whether or not small food items will be eaten. Foxes eat large quantities of insects when they are available.

One of the most widely applicable statements that can be made is that foxes are opportunistic feeders and take any acceptable food in proportion to its availability. This is supported by seasonal changes in foods eaten and by studies in which availability of foods were compared to items consumed. Within this broad generalization, there are apparent preferences for certain food items and rejections of others. Microtine rodents are preferred over mice of the genus *Peromyscus,* though part of this apparent selection may be due to ease of capture. Rabbits are high priority items in the diets of foxes and have been taken in higher numbers than their populations would suggest (Richards and Hine, 1953). At other times, preferred prey such as voles (*Microtus* spp.) may attain saturation levels and are not eaten in proportion to their high numbers.

Foxes have clearly demonstrated strong preferences for certain wild berries and fruits. During seasons of abundance, blueberries, raspberries, and wild black cherries constitute almost 100% of the diet (Hamilton, *et al.,* 1937). In the University of Wisconsin Arboretum at Madison, an abundant supply of frozen apples supported a family of foxes until late winter, much past the time of normal dispersal of the young.

Certain prey species, especially the insectivores and mustelids, are disliked as dietary items. Such animals are frequently killed but left uneaten unless other preferred foods are not easily obtained. Karpuleon (1957) states that even snowshoe hares (*Lepus americanus*) were not eaten though the preferred cottontail rabbits were low in numbers. In other studies, however, snowshoe hares were readily eaten.

Effects of fox predation on prey populations are difficult to assess. Often the fact of predation is confused with the effect of predation. In those studies dealing with the extent of predation by foxes on game birds, it has been generally concluded that foxes have little effect on population levels of the ring-necked pheasant (*Phasianus colchicus*) (Robeson, 1950; Arnold, 1952; New York Cons. Dept., 1951; Wilcomb, 1956). In New York, ruffed grouse (*Bonaza umbellus*) studies (Bump, *et al.*, 1947) state that 39% of nests were destroyed by predators, and a major portion of these losses were attributed to the red fox. Predator control, however, failed to bring about an increase in fall populations of grouse. Birds as a group are not staple items in the diets of foxes and are taken in significant numbers when made vulnerable by severe weather conditions or when alternate foods are scarce (Errington, 1937; Seagers, 1944; Besadny, 1966). Large ground nesting birds such as pheasants and ducks (Anatinae) are sometimes found in considerable numbers around fox dens (N. Dakota State Game and Fish, 1949), but the effects on these prey populations are not known. Foxes are frequent pests around poultry farms, but many of the domestic fowl eaten by foxes are carrion (Latham, 1950).

Mammalian prey populations are sometimes subject to heavy pressure from foxes. Muskrats (*Ondatra zibethica*) exposed by droughts were preyed upon almost exclusively (Scott, 1947). Latham (1952) demonstrated that the numbers of foxes and weasels (*Mustela* spp.) were inversely correlated and implied that foxes were responsible for reducing weasel populations. Richards and Hine (1953) and Korschgen (1959) have calculated the probable numbers of rabbits eaten by foxes but were unable to show the actual effects on the rabbit populations. Most of the usual mammalian prey species have high reproductive capabilities and it is doubtful if foxes can control these populations under normal circumstances. For further appraisal of the red fox as a predator, see Seagers (1944), Latham (1951), and Scott (1955).

Some feeding activities are of limited overall importance but demonstrate the opportunistic nature of red foxes. Schofield (1960) was able to locate many dead white-tailed deer (*Odocoileus virginianus*) not previously discovered until foxes were tracked in snow. These deer carcasses constituted the bulk of winter food for foxes in one part of Michigan. Scott (1947) found that foxes ate corn (*Zea mays*) in the soft stage, acorns (*Quercus* spp.), and oats (*Avena sativa* L.). Rausch (1958) observed foxes feeding along beaches in Alaska, apparently scavaging dead fish and other marine organisms. I counted several dozen winter-killed carp (*Cyprinus carpio*) around fox dens at Horicon Marsh, Wisconsin.

HUNTING AND FEEDING BEHAVIOR

Few animals are as fascinating to observe as a red fox hunting for mice. I have watched them standing motionless with tail arched stiffly to the rear, ears erect, listening and watching intently. The capture or attempted capture is executed by a leap in which the stiffened forelegs are brought down sharply. Often there is much frantic striking with the feet and searching with the nose as the intended victim attempts to escape. The more mobile insects are captured by pinning them down with the paws. These activities are described in detail by Seton (1929), Scott (1943), and Fisher (1951).

Murie (1936) observed that foxes captured rabbits by outsprinting them. Scott (1947) states that foxes use stealth; they stalk and capture the prey with a final rush, and that sprints for cottontails are generally unsuccessful. Methods obviously vary and probably depend on conditions such as snow depth, vegetative cover, and on past training or experience. It seems evident that the sense of hearing is used to a great extent especially for locating mice. Foxes hunting in grasslands frequently stop and remain motionless. I have followed fox trails in snow and observed where leaps of several feet have been made at right angles to the direction of travel. The sense of smell is also very keen as experienced fox trappers will attest. Foxes hunting at night commonly approach brushpiles, fence corners, etc. from the downwind side. Such behavior would favor hearing but is more likely for olfactory reasons.

Prey killed by foxes can often be identified by tooth marks through the pectoral region. Small items such as mice are swallowed whole while larger prey such as rabbits and muskrats are sometimes skinned partially or completely before being eaten. Viscera, feet, and tail of larger prey and wing tips of birds are usually discarded. Caching of uneaten food items is common among foxes. Excavations are made with the feet, the food deposited with the mouth, and covered with the nose. Cached items may be retrieved at a later time, but are commonly detected and taken by other species.

POPULATION ECOLOGY

Aging Techniques

Aging techniques are a prerequisite for any study of population structure. No method that is entirely suitable for rapid field aging of red foxes has been developed. Criteria for distinguishing foxes less than 1 year from older animals include ossification of proximal epiphysis of the humerus (Reilly and Curren, 1961), x-rays of the forefeet (Sullivan and Haugen, 1956), and use of eye-lens weights (Friend and Linhart, 1964). Churcher (1960) used a combination of cranial characters to separate year classes among adults and Linhart (1968) described criteria for aging pups.

Age ratios

Average litter sizes of 4.5 to 5.5 pups would result in autumn age ratios of approximately 2.2 to 2.8 young per adult, or 70 to 73% young in the population, if no mortality occurred. Observed fall and winter age ratios correspond closely with litter sizes and suggest little spring–fall mortality among young foxes. Phillips (1970) found 2.8 and 2.4 young per adult, respectively in two areas in Iowa and Storm (personal communication cited by Phillips, 1970) found 2.1 young per adult in northwest Illinois. Richards and Hine (1953) recorded 73.3% immatures among females in Wisconsin. I found 77% young in Wisconsin foxes trapped in October–November and 66% among foxes shot in January–March. Schofield (1958) reported 3.12 young per adult for 76,027 foxes over a 7 year period in Michigan. This ratio is higher than could be obtained by known litter sizes and supports the presence of vulnerability among the young.

Mortality

The average life span of a pup in the wild is less than 1 year (Arnold, 1956) and few foxes live beyond the age of 3–4 years, at least where they are hunted and trapped heavily. Causes of mortality other than hunting and trapping are poorly known. Based on band recoveries of foxes in Wisconsin, man-caused mortality is 30% for the first year of life and 53% during the life span of a fox. The estimated total annual mortality rate of 64% is slightly less than would be expected if the population were stationary. However, fox populations are dynamic and are seldom if ever stationary. Schofield (1958) observed that they are resilient and increase their productivity when mortality increases. In Wisconsin, fox populations have increased in spite of heavy harvests (Lemke, *et al.*, 1967). In Canada, red fox numbers fluctuate according to the dictates of the 10 year cycle (Cross, 1940; Calhoun, 1950; Butler, 1951; Keith, 1963).

Sex ratios

Observed sex ratios show much variability between and within studies (Table 15-1). Reasons for these discrepancies are not clear. There is an indication that the age ratios at birth favor females. With few exceptions, apparent sex ratios among subadult foxes in the fall are biased in favor of males because of differential vulnerability. Among adults in shot or trapped samples there are more females than males. In samples of foxes of unknown age there is a shift toward a higher proportion of females from fall to spring, but exceptions occur (Layne and McKeon, 1956a). Thus, one may tentatively conclude that a true sex imbalance favoring females exists among adult foxes in populations that have been subjected to hunting or trapping. Caution must be used when interpreting any set of sex or age ratio data for red foxes. Possible biases can occur because of method of collecting, differential vulnerability in sex and age classes, intent of trappers or hunters, and season of the year.

TABLE 15-1 Some sex ratios of red foxes in North America

Location	Fetuses (N)	Pups	Subadults	Adults	Age Unknown	Season
Wisconsin[a]	—	96:100 (54)	—	—	117:100 (811)	Sept–Dec
	—	—	—	—	82:100 (226)	Jan–Mar
[b]	100:100 (30)	156:100 (79)	213:100 (120)	100:100 (48)	156:100 (712)	Dec–Mar
Central Iowa[c]	—	—	92:100 (123)	144:100 (22)	—	Oct–Dec
Northeast Iowa[c]	—	—	152:100 (146)	97:100(127)	—	Oct–Dec
New York[d]	95:100 (117)	Fetuses + pups	—	—	—	Spring
[e]	154:100 (142)	94:100 (4551)	—	—	125:100 (4890)	Sept–Dec
[f]	—	—	—	—	152:100 (904)	Jan–Apr
	—	—	—	—	121:100 (3988)	May–Aug
[f]	—	—	142:100 (242)	60:100 (125)	—	Sept–Oct
	—	—	—	66:100 (91)	—	Apr–May

[a]Richards and Hine (1953), [b]Ables (1968), [c]Phillips (1970), [d]Sheldon (1949), [e]Layne and McKeon (1956a), [f]Friend and Linhart (1964).

A sex ratio favoring females provides a possible explanation for the inconsistencies in the observed participation of males in rearing the family. There are undoubtedly many females with young that have lost their mates. Also, I am inclined to question strict monogamy in red foxes. Heavily hunted or trapped fox populations become more productive, not less so as a sex imbalance among monogamous animals would suggest.

Numbers

Densities of red foxes have been estimated from bounty records, other hunting and trapping data, and from den surveys (Seagers, 1944; Sheldon, 1950; Fisher, 1951; Colson and McKeon, 1952; Scott, 1955; Korschgen, 1959; Phillips, 1970). Numbers of foxes harvested from large geographic areas have ranged from 0.75 per mile^2 (2.56 km^2) to 2.5 per mile^2 in Iowa, approximately 2 per mile^2 in New York, and 0.91 per mile^2 for the entire 54,705 mile^2 (~142,200 km^2) in Wisconsin. The percentages removed are unknown but the original fall densities must have been at least 2–3 times the harvest, or 2–8 foxes per mile^2 during the fall. Scott (1955) estimated spring densities of 1.6 per mile^2 over 576 mile^2 (~1500 km^2) of Iowa.

In local areas of good fox range, up to 12–20 foxes per mile^2 have been estimated. In a 10 county area of southwestern Wisconsin 2.1 foxes per mile^2 were bountied during the 1961–62 fiscal year (Wisconsin Conservation Department, 1962). Localized very high densities of 5 litters in 200 acres (81 hectares) (Sheldon, 1950) and 6 families 0.25 mile (.4 km) apart (Fisher, 1951) have been reported. All of the above estimates and counts were made in the best fox range in North America and are much higher than those found in Canada even during cyclic highs.

DISEASES

Of all diseases deadly to foxes, rabies is the most widely known. It has been responsible for extensive mortality in foxes throughout North America (Gier, 1948; Cowan, 1949). Epizootics of rabies occur most frequently when fox populations are high, but the epidemiology of this disease is not well-understood. Fredrickson and Thomas (1965) found a positive relationship between the incidence of fox rabies and the number of caves in Tennessee, which suggests that bats transmit rabies to foxes. Rabies can apparently be latent in skunks, (*Mephitis mephitis*), some species of bats, and possibly foxes until some unknown mechanism triggers an outbreak (Davis and Wood, 1959; Colson and McKeon, 1952).

Canine distemper and infectious canine hepatitis are known to kill foxes. Both are more prevalent when foxes are abundant (Parker, *et al.*,

1957) and serve as controls on fox populations. Foxes infected with either distemper or hepatitis or any other highly virulent disease generally die. Therefore, the prevalence and effects of such diseases are very difficult to assess.

TABLE 15-2 Results of bacteriological and serological tests for selected diseases in red foxes in southern Wisconsin during 1963–64. The number of positive reactors (numerator) are listed over the number tested (denominator)

Diseases	1963		1964	
	Males	Females	Males	Females
CEV [a]	9/32	9/16	1/30	0/20
Staphyloccocus	8/34	8/16	Not tested	
Tularemia	3/29	3/22	1/34	1/26
Leptospirosis [b]	8/16	5/7	8/29	6/19
Distemper	0/7	1/4	Not tested	
Hepatitis	0/7	1/4	Not tested	

[a]No positive reactors for other encephalitis viruses.
[b]Largest number of positive reactors were for *L. autumnalis.*

There are many disease organisms, especially endo- and ectoparasites, which affect the welfare of foxes but generally do not kill directly. Olive and Riley (1948) described the deaths of an unknown number of foxes infected with the mange mite (*Sarcoptes scabei*). They believed that the feeding habits of the foxes were affected and contributed to malnutrition. Haberman, *et al.* (1958) trapped a female fox in poor condition and parasitized heavily with intestinal worms. Erickson (1944), Smith (1943), and Babero and Rausch (1952) reported on the incidence of nematodes and other internal parasites.

In southern Wisconsin during 1963–64 I conducted a survey to determine the presence and prevalence of diseases primarily of concern to the welfare of man and domestic animals. Diseases tested for included 5 encephalitis viruses (EEF, WEV, CEV, SLEV, EMC), 6 serotypes of leptospirosis (*Leptospira autumnalis, grippotyphosa, canicola, icterohaemorrhagiae, pomona, Hardjo,* and *hebdomalis*), Staphylococcus aureus, tularemia, infectious canine hepatitis, and canine distemper. Positive reactors are summarized in Table 15-2. There is a tendency for females to have a higher incidence of all diseases. This may be related to denning by females during the period of caring for the young. Behavioral traits (scent posts, sexual behavior) would tend to favor the spread of leptospires in proportion to population densities. Antibiotic resistance patterns of *Staphylococcus* isolates suggests that foxes had become infected with organisms of human and domestic animal origin. Foxes are most numerous in populated areas of North America where they interact with man and domestic animals and may play important roles in the epidemiology of domestic animal and human diseases. The importance of most of these diseases to foxes is not known and illustrates the need for further research in disease relationships.

INTELLIGENCE AND ADAPTABILITY

Accounts of cunning and extraordinary reasoning ability of red foxes have been so firmly entrenched in our folklore that one has difficulty conducting an objective study of the animal. We can only speculate at the feats related by Seton (1929). Regardless of the true or alleged mental powers of the fox, this animal is one of the most intelligent and adaptable of the wild canids. It has increased in numbers under severe harassment, has resisted all efforts at extermination, and lives in close proximity, though sometimes not in complete harmony, with man. Some of the highest fox populations in North America are in heavily populated farm and dairy communities. Foxes can live and thrive in suitable areas within the edges of large cities. Stanley (1963) found a den within 50 yards (46 m) of a residential area of Lawrence, Kansas. As many as 12 foxes live surrounded on three sides by residential developments in the University of Wisconsin Arboretum at Madison. It was obvious that these foxes were well-acquainted with pedestrian and vehicular traffic patterns and showed no alarm unless a deviation from the usual routine occurred. These animals regularly used a pedestrian underpass beneath a four-lane highway, and once a radio-tagged fox traveled the length of a city block in a storm sewer. Foxes in rural farming areas quickly become habituated to farming practices. Fisher (1951) describes foxes hunting in plowed fields close to a tractor in operation.

SOCIAL BEHAVIOR

Red foxes have been described as only slightly social (Seton, 1929). The family which includes the vixen, young, and sometimes the male, is the social unit. After dispersal in the fall, foxes remain more or less solitary until they pair and travel together for several days during the mating season. As stated previously, the exact role of the male in the family is not clear. Males seem to be more solitary than females. Sibling females may remain together into adult life. No intensive studies of social behavior among wild red foxes are known to the author and the interested reader should consult Seton (1929), Murie (1936, 1944), Scott (1943), and Sheldon (1950).

Voice was listed as the principal method of communication by Seton (1929), and several characteristic and distinctive calls were described. However, vocalizations by foxes have been rarely mentioned in subsequent writings. I have heard sharp alarm or warning yaps while approaching dens, but the shrill calls described by Seton seem rare among foxes living in close proximity to man. The principal mechanism for communication between individuals and between the sexes is probably the scent post.

EGOCENTRIC BEHAVIOR

Behaviors directed toward maintenance and preservation of the individual have been studied intensively and have been discussed above — food habits, hunting behavior, habitat selection and use, movements, and daily activity cycles. Other behavioral patterns directed toward individual comfort have received only incidental study. Grooming must occupy a considerable part of a fox's time. Balls of fur removed during grooming with the teeth serve to identify the presence of foxes (Scott, 1943). Foxes avoid getting wet; they seek shelter during thunderstorms and move less when dews are heavy. They also seek warmth during cold weather and are frequently observed sunning on south-facing slopes in winter.

INTERSPECIFIC RELATIONS

Relationships with species other than prey are poorly known. Murie (1944) described encounters between foxes and wolves (*Canis lupus*), and Mech (1970) stated that wolves sometimes killed foxes. Trappers believe that red foxes and gray foxes (*Urocyon cinereoargenteus*) are antagonistic toward one another and do not occupy the same habitat. However, red foxes and gray foxes are sometimes trapped in the same spots on successive nights. There is strong evidence that red foxes interbreed with kit foxes (*V. velox*) (Creel and Thornton, 1971; Thornton, *et al.*, 1971). Foxes, raccoons, skunks, opossums, badgers (*Taxidea taxus),* bobcats (*Lynx rufus*), lynxes (*Lynx canadensis*), and other carnivores with similar food habits but slightly different ecological roles occur in the same area presumably with a minimum of conflict.

Seton (1929) states that foxes were seen traveling in herds of wild sheep (*Ovis canadensis*), and caribou (*Rangifer arcticus*), and at other times seemed to deliberately tease a bull caribou as well as domestic livestock and farm dogs. Red foxes and nesting black vultures (*Coragyps atratus*) were observed sharing a common den entrance (Barkalow, 1940). Hock (1952) described an encounter between a red fox and a golden eagle (*Aquila chrysaetos*) that seemed to be something other than attempted predation by the eagle.

ECONOMIC STATUS

No discussion of red fox ecology in North America can be complete without considering the economic values, both positive and negative, of foxes. Predatory habits which result in depredations on small game birds and mammals, domestic poultry, pigs and lambs, and the role of the fox in rabies epizootics regularly bring about retaliatory actions at considerable monetary costs to man. On the positive side, foxes prey on rodents and provide much recreation to sportsmen. Fox chasing with hounds and traditional mounted hunts are common in the eastern United States. In the northcentral United States, fox hunting in which the fox is usually killed is a popular winter sport. Commercial fox fur farms are no longer the lucrative enterprises they once were because of trends away from long furs in fashion. At present, a balance sheet would probably show the fox to be neutral.

Bounties require special mention. Several states in the Midwest and New England have paid out millions of dollars in fox bounties in the past 30 years. After examining several evaluations of bounty systems, the general conclusion is that bounties have been ineffective in reducing fox populations. Long-term increases have actually occurred in spite of bounties. Because of the administrative structure of bounty system, they tend to be self-defeating. For detailed discussions of bounties and the economic status of foxes, see Wright (1949), Haller (1951), Switzenberg (1951), Latham (1951), Richards and Hine (1953), Arnold (1956), Scott (1955), Bednarik (1959), and Lemke and Thompson (1960). Perhaps the greatest benefit from fox control programs has been the large number of specimens from which have been obtained substantial portions of the existing knowledge on food habits, reproduction, population structure, and movements.

RESEARCH NEEDS

One of the largest existing voids in the ecology of North American red foxes is the lack of behavioral data. Social behavior and the role of the male are not clear. There is doubt of the existence of monogamy or at least which one of its many forms exists among foxes. Territorial behavior between females during the reproductive season may exist but remains unproven. Thus, there is much yet to be learned about one of the most thoroughly studied canids in North America.

16
DISPERSAL AND SOCIAL CONTACT AMONG RED FOXES: RESULTS FROM TELEMETRY AND COMPUTER SIMULATION

G. L. STORM*
Dept. of Ecology and Behavioral Biology
University of Minnesota
Minneapolis, Minnesota

and

G. G. MONTGOMERY
National Zoological Park
Smithsonian Institution
Washington, D.C.

The occurrence of dispersal in the red fox (*Vulpes vulpes*) is well-documented (Errington and Berry, 1937; Sheldon, 1950; Arnold and Schofield, 1956). However, published reports provide few details of dispersal such as seasonal timing, rates and routes of travel, proportions of the sex and age segments of the population which disperse, and amounts of social contact between transient and resident foxes. These details are essential for understanding gene flow among fox

*Present address: Pennsylvania Cooperative Wildlife Research Unit, The Pennsylvania State University, University Park, Pa.

populations, range extensions, epidemiology of fox-borne diseases, and social control of dispersal. This paper reports on details of dispersal of red fox through use of radiotelemetry and other field techniques, and of social contact through computer simulation of red fox movements.

FIELD RESEARCH

Field work was conducted in the northcentral United States (Illinois, Iowa, Wisconsin, and Minnesota) during the period of 1965 through 1971 using two general methods. In the first method, foxes (mainly young of the year) were captured from underground dens (Storm and Dauphin, 1965) ear-tagged, and released at the dens. When the tags were recovered, mainly from hunters and trappers during the October–March period of each year, the location and date of recovery were recorded. The recovery information from 786 foxes, all tagged as juveniles, provided most of the data to estimate timing of dispersal, proportions of the sex and age classes which dispersed, dispersal direction, and distances traveled during dispersal. The second method involved capturing foxes in steel traps or from dens in late summer, fitting each with a radio-transmitter, and releasing the foxes where captured. Signals from the transmitters were monitored daily prior to dispersal, and dispersing animals were followed with a combination of aircraft-mounted and truck-mounted receiving equipment (Storm, 1972).

The red fox in North America tends to be restricted to discrete spatial areas (home ranges),at least during spring, summer, and early fall. Each home range is typically occupied by a family group consisting of an adult male, 1 or 2 adult females, and seasonally by young born to the females. Home ranges are about 259 to 777 hectares in size (Scott, 1943; Storm, 1965; Ables, 1969a). Sargeant (1972) indicated that neighboring family groups occupy well-defined, nonoverlapping home ranges; mutual avoidance between members of neighboring groups is apparent.

DISPERSAL PATTERNS

Beginning in late September and early October, and continuing through fall and winter, some foxes, primarily young born during the preceding spring (see below), leave the home ranges which they occupied during spring and summer, and disperse through unfamiliar terrain. By mid-October, some juveniles have moved more than 32 km from their natal home range (Phillips, *et al.*, 1972).

The onset of dispersal coincides with a marked increase in the size of testes during September and October. Venge (1959) reported that the male fox reached sexual maturity by late November and December of their first year. Since the cyclic pattern of gonadal activity is similar for adults and juveniles, the onset of dispersal corresponds not only to puberty, but also to increased gonadal activity in adult males (Storm, 1972).

Most juvenile males and a substantial proportion of juvenile females dispersed from their natal home ranges. Of males which were tagged as juveniles and were recovered during their first year, 80% had moved more than 8 km from their natal home range. The corresponding figure for juvenile females was 37%. Foxes tagged as juveniles whose tags were recovered before they were 2 years old showed that 96% of the males and 58% of the females went more than 8 km from their natal home range. Thus, less than 5 and 45% of the males and females, respectively, remained near the locality where they were born.

Dispersal occurred, although less commonly, among adult red fox. For those ear-tagged as adults, 30% of 22 adult males and 21% of 49 adult females whose ear-tags were recovered had moved more than 8 km from where they were tagged, within the year following tagging. Only 2 of the 26 adults which dispersed were recovered more than 24 km from where they were tagged, indicating that adults tended to disperse shorter distances than juveniles. Some of the movements made by adults outside their home ranges in fall and winter may have been related to mating behavior rather than to dispersal *per se*.

The onset of dispersal by radio-marked animals was sudden; a disperser left its home range by traveling in a nearby straight-line course which appeared to be in a random direction. During dispersal, some individuals moved along this course for 3 to 6 consecutive nights. Others dispersed during 1 to 7 nights, remained in an area for 1 to 6 days and nights, then resumed dispersal. The longest recorded period between the onset and apparent cessation of dispersal was 18 days.

The pattern of movement preceding the cessation of dispersal varied among the radio-marked animals. Some simply stopped and began to occupy a new home range after dispersing for several nights. Others traveled in a circular manner for 1 or 2 nights, returned to an area through which they had passed previously, and then apparently settled on a home range.

Travel paths during dispersal were generally straight, although the path of travel was often temporarily changed by cities and lakes. For example, the travel path of one dispersing juvenile was influenced by at least four lakes (Figure 16-1). After passing around such a barrier, the animals usually resumed the course along which they had previously

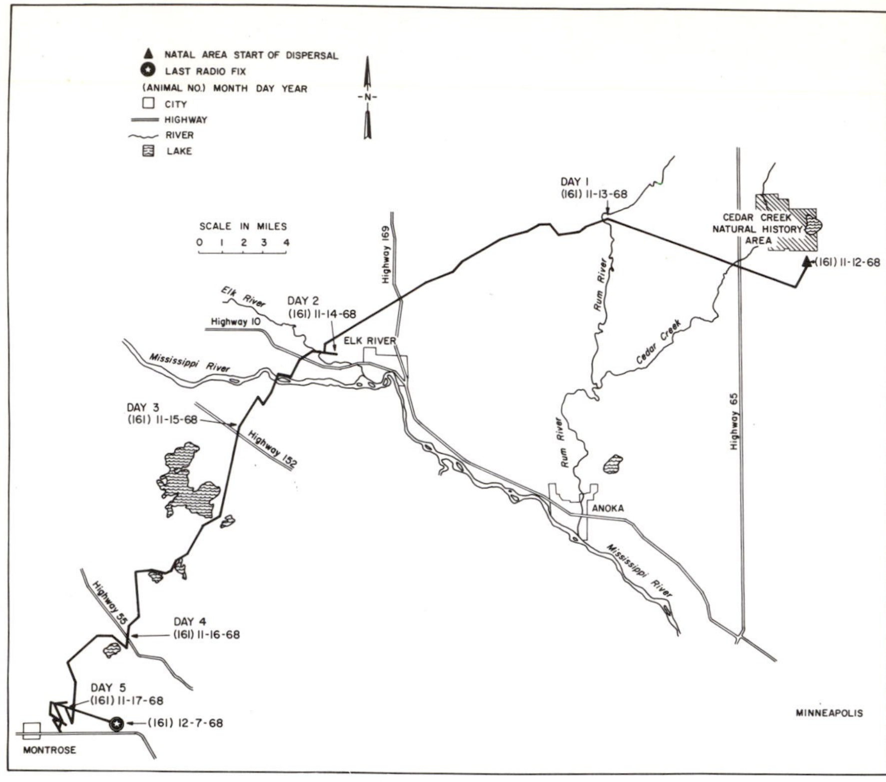

Figure 16-1 Dispersal route of a juvenile red-for radio-tracked in Minnesota.

traveled. Large rivers apparently formed a barrier to dispersal, while smaller rivers and creeks did not. None of the marked foxes, including both radio-marked and ear-tagged individuals, was known to have crossed the Mississippi River where it forms the boundary between Illinois and Iowa. Dispersing foxes in Minnesota, however, crossed the Mississippi River above Minneapolis. Other smaller creeks and rivers encountered by dispersing foxes in Minnesota, Illinois, and Iowa were readily crossed; movements of the animals sometimes became more erratic, and the average rate of travel decreased before the river was crossed.

Dispersing foxes traveled 10 to 16 km during a typical night of dispersal, with males moving further than females. During 33 nights of dispersal, 5 radio-marked males traveled an average of 15 km per night; 3 females traveled an average of 9 km per night during 14 nights of dispersal. When it is considered that a very high proportion of the foxes from a local population disperse, and that an individual often travels 81 km in less than 1 week, it is apparent that red fox can readily invade formerly occupied home ranges or invade new habitats. It is also apparent that dispersing foxes have a great potential for social contact with resident foxes. For example, if one assumes a high fox population density and uniform spacing of the animals, there would be about 1 male–female pair of residents per 4 km of dispersal path. A transient moving 81 km during 5 nights could thus interact with 40 or more individual resident animals.

COMPUTER SIMULATION

A field study of social contact between dispersers and residents would be extremely difficult since a very large number of foxes would have to be observed (perhaps by radio-tracking) to insure that an observed disperser would pass through the home range of an observed resident. The field problems may well be insurmountable. Thus, computer simulation was used to give preliminary estimates of the magnitude of such contact.

A computer-simulation program was developed to allow us to control the movements of dispersing and resident foxes, and to measure the potential for social contact between the simulated animals. The program (Montgomery, 1973) was written in Fortran IV, and simulations were run on a CDC 6600 computer at the University of Minnesota. The program used the same means for controlling movements of the animals as those described by Siniff and Jessen (1969).

In each simulation, a disperser attempted to pass through the home range of a resident without being close enough to the resident so that communication between them could occur. To simplify interpretation of the results, some of the factors which could influence such social contact were standardized. These included the size and shape of the home range (Figure 16-2) and the disperser's initial location. All home ranges corresponded to an ellipse with axes of 1.6 and 2.4 km. Dispersers began their movements 0.3 km from the right home range boundary on the line of the major axis. The initial location of the resident was fixed by random processes, and could be anywhere within the elliptical home range boundary.

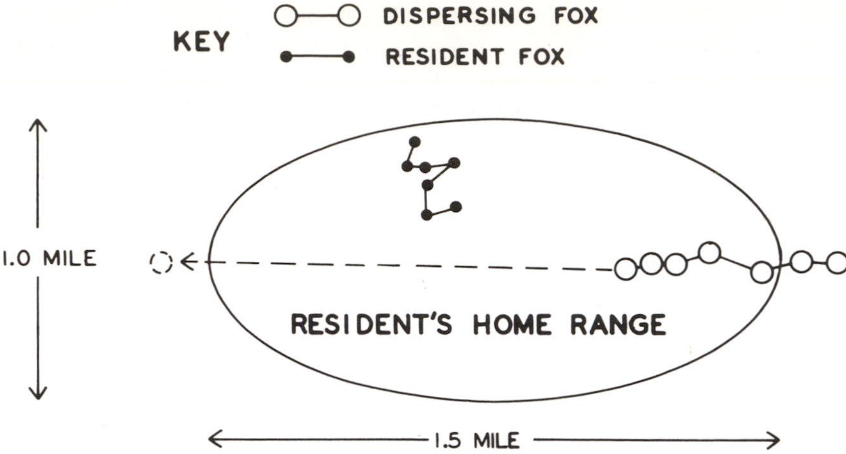

Figure 16-2 The beginning of each simulation. The disperser begins his movements to the right of the home range; his movements are biased so that he tends toward a point to the left of the home range (dashed circle). The resident, beginning from any point on the home range, moves about in a manner similar to the movements of real resident foxes on their home ranges. As the simulation proceeds, both the resident and the disperser move once each 5 minutes. The method for controlling movements of the simulated animals is given in the text.

Movements of the animals during the simulation were controlled by a modified random walk procedure (Siniff and Jessen, 1969; Montgomery, 1973). Both disperser and resident moved from location to location simultaneously, once per 5 minutes. At each move, directions and distances which the animals moved were determined by random choices from appropriate frequency distributions. These distributions of direction of travel and distance moved per 5 minutes were derived by radio-tracking real red fox (Siniff and Jessen, 1969).

Direction of travel of dispersers was biased by causing them to seek a point beyond the left home range boundary (Figure 16-2), so that they tended to move in a straight line across the home range at its maximum dimension. Residents moved about the home range in a manner similar to the movements of real resident foxes. A disperser could traverse the maximum dimension of the home range with about 25 moves, thus remaining on a home range for about 2 hours while traveling at about 1.2 km per hour.

Social contact between the animals depended on them being close enough to each other so that visual, vocal, and direct olfactory communication could occur. Social contact was assumed to occur each time the animals were within some distance (contact range) of each other. Use of the contact range allowed us to measure potential for visual, vocal, and direct olfactory communication; scent trails or marks could not affect the disperser. Different values were used for the "contact range" distance (see Figure 16-4); these distances are reasonable guesses to obtain some notion of the probability of social contact between transient and the resident. In a more complete study (Montgomery, 1973), the role of various means of communication, including scent trails and marks between a resident and a transient fox was studied by computer simulation; such scent communication was not included in this report.

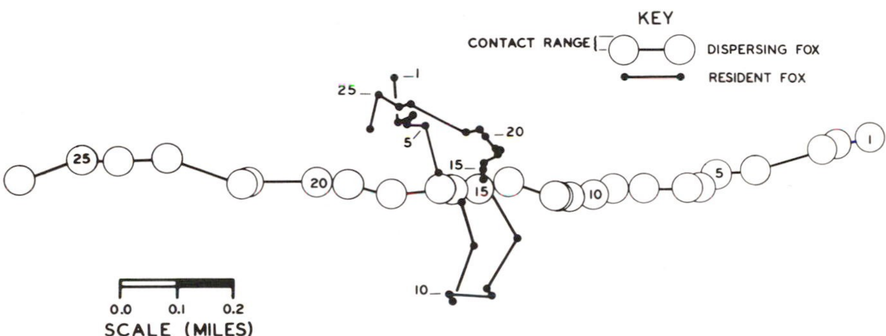

Figure 16-3 A typical sequence of movements of resident and invader taken from one of the simulations. Note that social contact would have occurred when the animals reached their respective locations marked 15, if the contact range had been approximately ⅓ greater.

Social contact between the animals was measured in the following way. At the end of each simulated 5 minutes (after each move) the program measured the distance between the animals and compared this distance with the contact range programmed for the simulation. When the distance between the animals was greater than the contact range, the animals continued to move. A simulation was terminated when the distance was less than the contact range, or when the disperser had passed through the home range.

The sequence of movements of a resident and a disperser during a typical simulation is shown in Figure 16-3. Twenty-six moves were required for the disperser to traverse the home range. As the disperser moved across the home range from right to left, the resident moved about near the center of the home range. None of the distances between disperser and resident was less than the contact range, thus the simulation terminated when the disperser passed the left home range boundary. Had the animals been slightly closer together at the 15th location used by each, the distance between them would have been less than the contact range, and the simulation would have terminated.

We tested 8 different contact ranges within the limits 0.05 to 0.50 miles (81 to 805 m). For each contact range, 8 different movement patterns were programmed for both resident and disperser. Because of the biased travel directions, however, the movement patterns of all dispersers were very similar. In the results of the simulations, we present data for 8 contact ranges; for each contact range, 1 disperser attempted to pass through the home ranges of 8 different residents.

No social contact between a disperser and a resident occurred when the contact range was 0.01 miles (16 m) or less, and social contact was unlikely with contact ranges of about 35 m or less (Figure 16-4). This result suggests that transfer of disease organisms such as those causing rabies, which require direct contact between the animals, is unlikely unless one or both of the foxes alters its movements to seek out the other.

Likewise, direct confrontation and communication by short-range means (reviewed by Tembrock, 1968) is unlikely to occur. These means of communication (changes in facial expression, position of the tail or other body parts, sniffing of the anal region, etc.) require that the individuals be close enough to each other to perceive fine details by sight or smell. Red fox disperse entirely at night, and often travel through heavy ground cover. The disperser would be invisible to the resident unless movements of one or both of the animals changed so as to bring them closer together.

Some means of communication which regularly reveals the location(s) of the animal(s) is necessary before altered movement pattern(s) can bring the animals closer together. Communication between disperser and resident became increasingly more likely as the contact range increased(Figure 16-4), but *only* if one or both of the animals chose to emit signals or to respond to signals emitted by the other. Only vocal signals would be effective; poor visibility would prevent visual communication, time required for scent dispersal would prevent direct olfactory communication, and distance between the animals would prevent communication by mechanical auditory signals produced by movement of the animals on the substrate.

If we assume that a dispersing fox attempts to travel without being detected by resident fox, it becomes clear that the disperser could avoid detection by not vocalizing as he travels. The resident might be aware that a disperser had traversed his home range by detecting the scent of a strange animal. However, by the time detection of his scent occurred, the disperser would usually have passed out of the home range.

If behavioral changes which occur when a fox begins dispersal include a reduction in response to signals which would normally lead to

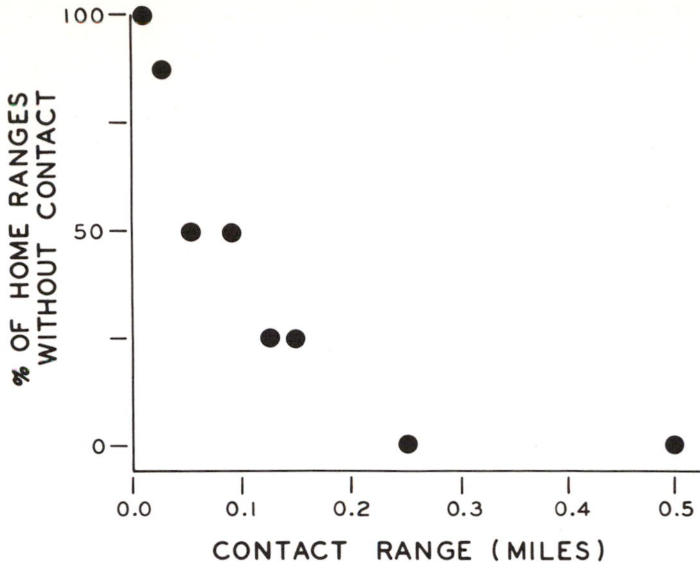

Figure 16-4 The effect of increasing the contact range on the probability that a disperser would pass completely through a home range without social contact with the resident fox.

avoidance of a resident fox, we can see how a disperser may move rapidly cross-country with few deviations in course resulting from social contact with residents. If this reduction occurs, a disperser can travel a relatively straight path through areas occupied by fox by ignoring vocal signals from residents, regardless of how often residents vocalize or the distance over which their vocalizations can be heard.

The straight travel paths taken by radio-marked dispersers support the view that little social control over the dispersers was exerted by residents. Had control been exerted, we would expect many changes in the direction which a disperser took, each change corresponding to an attempt to avoid further social contact with a resident. On the contrary, radio-tracking data show that foxes dispersed along relatively true courses, until they met geographical barriers such as lakes, rivers, or cities.

A better understanding of the social interactions between resident- and transient foxes will require more research on the ability of foxes to detect each other and respond to one another under various environmental conditions. This kind of information will provide further input for detailed computer simulation and ultimately provide further insight into the relationships between social behavior and the ecology of red foxes.

Acknowledgments Financial support for this study was provided by NIH Training Grant No. 5T01GMO1779, and the United States Atomic Energy Commission (C00-1332-83) directed by Dr. J. R. Tester, by the Iowa Conservation Commission, and by the Smithsonian Research Foundation Grant No. 435090, directed by Dr. J. Eisenberg. We thank Dr. D. B. Siniff for advice and suggestions on all phases of this study, Drs. V. Dirks and C. Jessen for help with computer programs. Mr. M. Sunquist reviewed the manuscript.

17
ECOLOGY AND BEHAVIOR OF THE COYOTE *(Canis latrans)*

H. T. GIER
Kansas State University
Manhattan, Kansas

The plains coyote (prairie wolf, brush wolf, little wolf) has met the challenges of the modern world and is winning the battle for survival. It has the size, strength, reproductive capacity, and general adaptability to meet all the challenges yet presented except government hunters and extreme human concentrations. Modern coyotes vary from an average of 11.5 kg in the Mexican deserts to 18 kg in Alaska.

Canis latrans was present during the Pleistocene as a species distinct from *Canis lupus* (Lane, 1948). Evidence of distribution and numbers of Pleistocene coyotes is so fragmentary as to be of little interpretative value except to limit the species to North America.

The coyote (or coyotyl) was well-represented in Aztec art and folklore (Dobie, 1950) indicating an understanding of its biology but attaching many mystical properties to it. Indians of the Great Plains had "dogs" that served as beasts of burden or as food for special feasts when the Spaniards visited the area in the 17th century. Some "Indian dogs" appeared to be only partially domesticated coyotes, somewhat modified by selection. Whether or not Indians moved or bartered their coyote-type "dogs" beyond the range of coyotes is not known.

Coyotes were first recognized as a distinct species of canids by Say in 1832 (Young and Jackson, 1951). Consequently, all reports before then are vague and inseparable from comparably hazy reports of lobo, timber wolves, and swift foxes. By 1850, the range of the "prairie wolf" was quite clearly established as extending from the Mississippi River, west into the Sierra Nevada Mountains, north into Alberta, and south to southern Mexico. They were apparently present throughout California and Mexico, but were excluded from what is now the Gulf

States. Numbers of coyotes present throughout the range during the 1840s must have been astronomical, as Hastings (Dobie, 1950) reported that several hundred coyotes could be seen during one day's travel and others reported coyotes alone or in small bands as being common sights. Decimation of such coyote populations within half a century resulted from the general reduction of large mammals coupled with specific efforts to destroy all "wolves."

About 1850, coyotes extended their range into Illinois and Michigan, Yucatan, northern California and the Pacific Northwest. The 1880s saw the great northern expansion during the Alaskan gold rush as the coyotes followed the trail of dead horses headed for the Yukon (Dobie, 1950). By 1950, coyotes had occupied much of Yukon Territory and Alaska, all the way to Point Barrow. Since 1925, the "plains coyote" has expanded to the Atlantic coast, to Hudson Bay (Young and Jackson, 1951), and into Florida and the Gulf coast (Gipson, 1972).

Such extensions of range are unique in modern mammals and demonstrate extreme versatility. West and northwest extensions followed reduction of gray wolves,coupled with provision of a special, although temporary, food supply of horse flesh, and the opening of forested areas, by logging and burning. East and northeast extension of the range followed removal of gray wolves and return of millions of acres to pastures and brush land. The range extension southeastward occurred after near elimination of the red wolf (Gipson, 1972). Such range extensions label the former "plains coyote" as the ultimate opportunist.

The only contractions of coyote range within historical times are local and temporary. "Government trappers" have, since 1915, progressively reduced coyote populations on both federal and adjoining private lands to near local extinction. Not only coyotes, but foxes and remnants of cougar, wolf, and bear populations have been relentlessly persecuted, under the guise of "predator control" for the "protection of sheep and cattle." Coyotes have been practically eliminated from areas of Colorado, Wyoming, Montana, Utah, and Nevada in the past 45 years, with many of the Rocky Mountain areas kept clear of all predators by continuous trapping and use of cyanide guns, strychnine, and compound 1080 (sodium fluoroacetate). In North Dakota, intensive "coyote control" (1948–1960) was followed by a 10 to 20-fold increase of red fox with, as yet, no great rebound of coyotes. It will be interesting to watch the picture of predator population recovery in those areas now that cyanide guns and poison baits have been outlawed. Can the coyote recover before his niche is taken over by other predators? These observations point up some basic relationships: (1) coyotes are limited in range by absence of open areas, (2) coyote range has not greatly overlapped the ranges of either the gray or the red wolf except in the Great Plains, (3) when coyotes are selectively eliminated from an area, other predators take their place, and (4) coyotes are no match for modern man with specialized techniques of destruction.

ECOLOGY OF COYOTES

Coyotes developed as denizens of the open prairies and deserts, restricted from forest areas by the larger wolves and puma, with whom they cannot compete in either a close fight or open chase. Consequently, coyotes are generally considered to be "adapted" for plains living. Recent developments contradict such limitations.

Habitat

The versatility of coyotes is well-illustrated by consideration of the original habitat. Apparently, coyotes were most numerous in grasslands where bison, antelope, elk, and deer were numerous. They did well in short grass areas in which prairie dogs dominated or they were just as much at home in semiarid lands with sagebrush and jackrabbits or in deserts dominated by cactus, kangaroo rats, and rattlesnakes. They ranged indiscriminately from the Sonoran Deserts to the Alpine regions of adjoining mountains or the plains and mountains of Alberta, so long as they were not in direct competition with *Canis lupus*. Recent extensions of range indicate that they can be successful in broken forests, from the tropics of Guatemala to the north slope of Alaska. Thus, neither altitude nor latitude restricts their survival any more than does the vegetation. Expansion of coyote range was made possible by elimination of effective populations of gray wolves in the mountain forests and encouraged by increasing numbers of men in the valleys. The range of no other species of wild mammal extends over such latitude as does the modern coyote (72°, southern Guatemala to northern Alaska). Coyotes of the far north or the high mountains are somewhat larger and more heavily furred than are those of the Arizona deserts and Mexico, but otherwise they are alike and must be considered as one *species* throughout their range. How the reproductive response of coyotes is affected by such varying light conditions over the wide range is yet a puzzle, but the critical factor seems to be increasing day length, not light intensity or hours of light.

FOOD OF COYOTES

Coyotes are as versatile in their eating habits as are rats, pigs, and man. They eat almost anything that any other mammal will eat, although they show some distinct preferences.

Coyotes are carnivores, and their diet is basically meat. The type of meat they eat depends on what is available: bison, deer, elk, sheep, rabbits, rodents, birds, amphibians other than toads, lizards, most snakes, fish, crustaceans, and insects (Sperry, 1941; Gier, 1968). Approximately 90% of all stomach contents examined was animal matter with as much as 75% consisting of one species, jackrabbit (Clark, 1972). A few mammals such as moles, shrews, and brown rats are not found in stomachs in proportion to their occurrence in the habitat. Although some coyotes are known to kill rattlesnakes, rarely have these been found in coyote stomachs (Dobie, 1950), definitely indicating a taste selectivity. Coyotes seem to prefer fresh meat, but they consume large quantities of carrion, even when it is badly decomposed or filled with maggots.

Fruits comprise a significant part of the food of most coyotes. Blackberries, blueberries, peach, pear, apple, prickly pear apples, chapotes, persimmons, and peanuts have all been reported as main repasts for coyotes at times (Sperry, 1941; Dobie, 1950; Gier, 1968). Watermelon, cantaloupe, and even carrots come in for their share of selection. Coyotes are adept at selecting ripe watermelons and opening them to get the red "heart meat." Grasses, particularly green wheat blades, are consumed in considerable quantities during winter and early spring, probably as a source of vitamins and added bulk.

Some unusual food items include cotton cake and soybean meal that had been distributed for cattle feed; droppings from domestic animals, particularly those of young animals which contain a high proportion of undigested food; corn, wheat, sorghum, beans, and other cultivated grain; and afterbirth of cows and sheep.

Other "miscellaneous materials" as listed in most food study reports consist of almost anything that is chewable. Harness straps, rubber or leather shoe material, scraps of auto tires, paper wrappings, and other such non-nourishing items must have been swallowed during play.

A coyote may frequently investigate such odoriferous materials as crude oil or rotting skunk and rattlesnake carcasses, and then he may roll in the mess, sliding his shoulders in it. Frequently coyotes have skunk odor on them but seldom eat any part of the skunk carcass.

Most of the conflicts between coyotes and man result from the coyote's food habits. Sportsmen accuse coyotes of excessive destruction of nests of game birds and of the birds themselves, even though neither accusation has been substantiated. Farmers repeatedly report losses of calves, lambs, chickens, turkeys, and ducks to coyotes. Undoubtedly, coyotes kill some newborn lambs, calves, fawns, and antelope in their daily foraging, as well as some older animals that are sick, crippled, or weakened by malnutrition or old age. Some coyotes even prefer the more easily killed domestic animals. There are records of coyotes killing full grown sheep, cows, deer, and antelope, for which 2 or 3 coyotes may team up to make the kill.

The versatility of coyotes in changing to the most readily available food resource has created so much antagonism from the humans who own or consider themselves the protectors of the food materials, that all sorts of restrictive measures have been placed on coyotes: from bounties, hunting parties, dog packs, flood lights, and "coyote proof" fences, to Congressional action providing manpower with cyanide guns, strychnine, and 1080 to annihilate the "varmints." Only the most intensive hunting and poisoning have made serious inroads on local coyote populations, but in many cases the removal of one or a few coyotes has resulted in cessation of the depredations.

The quantity of food required by coyotes has been grossly exaggerated by various authors (Dobie, 1950; Clark, 1972) on the basis that a coyote can eat several pounds of food at one meal. Captive coyotes have been maintained for long periods in my pens on an average of less than 400 grams of meat per day. Mature animals that ate over 1 kg of meat one day would not eat throughout the day following the big meal. My report of food passing through the stomach (Gier, 1968) within 20 hours was based on a normal meal of 300 to 400 grams. There are no published reports of how long a "super-meal" remains in the digestive tract. On the basis that field activities require more food than cage activities, possibly 600 grams of meat is required per day, or 250 kg per year, plus the insects, fruit, and grass that the animal would consume. Such a quantity would indicate an annual requirement of about 75 jackrabbits, or 160 cottontails, or 5000 *Microtus*, or 3000 *Dipydomys,* or various combinations of these. A pregnant or lactating bitch requires more food by a factor of approximately 1.5. I consider these figures as a maximal requirement, which is a long way from the 1.5 rabbits (30% of the coyote's body weight) per day, estimated by Wagner and Stoddard (1972). This would indicate 550 rabbits or approximately 650 kg if cottontails were implicated, or if jackrabbits were intended, over 1350 kg of "rabbit" annually per coyote. That quantity of food would maintain an entire coyote family!

SPACE REQUIREMENTS

Territoriality of coyotes is poorly understood. Apparently there is no defense of territory except during the denning season, so the general concentration of coyotes is naturally limited by (1) adequacy of the food supply, (2) denning territory, (3) intraspecific strife dependent on frequency of contact and competition for food, and (4) adequate space devoid of interspecific rivalry with other carnivores.

If food is inadequate, as is the case following rabbit–rodent crashes, multiple occupation of the same hunting areas results in many conflicts between hungry coyotes. If such food shortages and intraspecific strife do not result in death, they may be debilitating enough to severely reduce reproduction the next year (Gier, 1968). Therefore, the hunting area must be large enough to provide each coyote with about 250 kg of meat each year, with a minimum of 400 grams per day, without which that animal will be hurt or will migrate into unknown territory in which the local population may not tolerate invaders. Gipson (1972) showed, by use of radio tracking, that individual coyotes' feeding territories "vary from 0.1 to 24 mile^2 (0.38–62 km^2)" and there is much overlap.

"Denning territory" is not well-defined although numerous workers (Dobie, 1950; Fichter, 1950; Gier, 1968) have commented on the wide space between dens. My observations indicate that naturally limited areas, such as a ravine, a segment of prairie or desert separated by a sharp "wash" from other areas, or one delimited hillside may be claimed and defended by 1 pair of coyotes. Occupied dens may be as close as 0.4 km apart in separate valleys, or 0.8 km in the same valley. In the plains of Kansas and Colorado, 1 denning pair per km^2 seems to be about maximum, but in the geologically cut section of the Flint Hills in Kansas there may be as many as 3 pairs denning in any 1 km^2.

Except during the denning season, food supply being adequate, the limit to possible coyote concentrations is high. In Yellowstone National Park, 6 to 10 coyotes may feed simultaneously on a single elk or deer carcass. I have recorded as many as 26 animals in a 12 km^2 area in central Kansas in midwinter, with no indication of any special concentration in that area. One animal per km^2 over 50 to 75 km^2 areas have been killed in "circle hunts," with at least that many more known to escape. The report of Bynum of 876 coyotes (Dobie, 1950) from a 1000 km^2 tract in Texas is invalid as a population index because of (1) the 3 month trapping period, (2) no isolation from the surrounding "Texas," and (3) the trapping was done at a time in which the population was highly mobile. Consequently, "drift" was unrestricted and indeterminable and the number trapped was probably derived from a much larger area.

SOCIAL STRUCTURE

There is no known "social structure" above the family within any coyote population. The occurrence of alpha and beta males is hypothetical at best. Observation of coyotes in zoos has given no clue to any overall social structure. It appears that the social unit among coyotes is the family, revolving around a reproductive female.

Nonfamily coyotes are loners that occasionally join forces for companionship or for killing prey too large or too fast for one coyote. Bachelor males, nonreproductive females and near-mature young constitute the "nonfamily" category. There may be 2 to 6 such animals banded together loosely, with obviously 1 dominant animal in the group, thus forming coyote "packs," but since none of the animals seem to be restricted to that group, the "pack" is temporary.

A coyote family has its inception in midwinter when a female comes into estrus and attracts one or more sexually active males. Normally over 90% of the females 20 months or older become sexually active and emit the sex attractant by the end of January; a smaller proportion (0 to 60%) of the 9 to 10 month old females by mid-February. It appears that all the reactive, unattached males within an attractive female's territory join the parade and follow that female for days. I have seen as many as 7 males following 1 estrous female. Copulation between old males and old females may occur (in central Kansas) as early as Januray 12, but the earliest record of ovulation that I have from 20 years of study is January 31. The males follow the estrous female for as much as 4 weeks. Copulation is frequent and involves more than a single male although observations from distances sufficient not to disturb the activities do not allow definite identification of individual males. In January, 1954, I observed a female copulating 3 times in a 2 hour period in midafternoon. There was some threatening, snarling, and jostling for position among the 6 males. When the female chose to move along, the males took up either a single file line behind her, or a bunched arrangement, with one close behind, one on either flank, and the others behind the lead group. Copulation occurred when the female stopped, nuzzled a male, then positioned herself to that male and lifted her tail. The selection was either the male directly behind her or the lead male at her side. When a male mounted, the others stood attentively around, not interfering with the process in any way. From observations in the field and on penned animals, I am forced to the conclusion that the female makes the choice as to which male will be allowed to mate. The strength, persistence, and cunning of the individual male to get himself into the proper position undoubtedly plays a part in the selection.

As the breeding season progresses, other bitches in the area become attractive, and one by one, the male entourage of the earlier female drift to new females until a single male is left with that particular female and the "wedding" is consummated, with the female and her selected mate having the last few days of her estrous period alone. Ovulation occurs 2 or 3 days *before the end* of receptivity, so in most cases, elimination of the suitors, either by discouragement by stronger males, dissipation of stamina, or rejection by the bitch, has been effective in limiting the sire of the pups to the strongest, most cunning male available. It is possible that 2 or even 3 males will persist with a given female to the end of her estrous period. Then how does she select *one* mate? More observations are needed.

Old females in Kansas ovulate between January 31 and February 20. Old males mate with old females, and most of the old males will find a mate. Estrus of the earlier maturing young females (those born in April) overlaps that of the late-breeding old females so the last of the unmated old males may proceed on down the line to follow and mate with the young females, if their stamina holds. Males of the year enter the breeding pool in late January or early February, so there is little reduction in the number of available males when the last females select their mates in mid-March. Old males are reproductively active from late December to early April; males of the year from late January into May.

At the end of the breeding season, there will be unmated males to the extent of noncompetitive old males plus young males equivalent in numbers to the nonbreeding females. The bachelor males disperse, possibly maintaining a loose association with others of their kind, wandering and trying to survive until the next breeding season. Unmated females seem to have other possibilities, such as serving as nursemaid and "baby-sitter" for mother or sister, or teaming up with an unmated sister or brother until "next year."

The pair, consisting of a pregnant female and compatible male, select a territory, clean out old badger, marmot, or skunk dens within their territory or dig a new den, and otherwise prepare for their family responsibilities. They hunt together, sleep near each other, and in late pregnancy, the male frequently hunts alone and brings food to his heavy mate. Possibly as a means of strengthening the tie between mates, there is another round of copulating in the last 10 days of pregnancy. How frequently preparturition copulation occurs, I do not know, but I have seen it 2 times in the wild and with 3 out of 4 pregnant coyote bitches in the laboratory.

We have been told (Young and Jackson, 1951; Dobie, 1950) that the male coyote is excluded from the den before, during, and after parturition. I consider this idea to be more folklore than fact, as I have observed both male and female entering the den in which there were young pups, and on some occasions, both male and female were taken with the pups when their den was dug out. Again, more direct observations are needed.

Normal litter size is 2 to 12. Dens containing over 12 young always have 2 litters. Of 27 extralarge litters I have personally examined, 15 had 2 sets of pups 5 to 15 days difference in age; the others were more nearly alike. In central Kansas, approximately 10% of all litters dug from dens consist of 2 age groups. Two sets of pups in a single den may be explained by 2 bitches using the same den, either from a mother–daughter or sister–sister relationship, or by 2 bitches having been won by a single male, or it is possible that the mother has adopted the pups of another bitch that had been killed or was away from home when visited. I have no records of 2 bitches and a male sharing a den, although I do have records of 2 bitches being killed in a den with a double litter and of a male and a female in a den with a double litter. Double litters cannot be explained by consecutive births from 1 female. There are no records of any canid giving birth to young over a period exceeding a few hours. Spaced ovulation may result in different aged pups within the uterus by as much as 2 days, but they are always born as 1 litter.

Lost litters cannot be compensated for during that year: all canid females have an ovarian cycle that cannot be repeated until the full cycle is completed. One set of follicles grows to near maturity, produces the estrogen that brings the bitch into estrus, then in due time (1 to 4 weeks in coyotes) most of those follicles ovulate and form corpora lutea, and all remaining vesicular (tertiary) follicles degenerate. The corpora lutea persist for 60 days whether the ova were fertilized or not, then slowly degenerate, allowing a new set of follicles to develop, which takes 10 months in coyotes, wolves, and foxes. I have been able to modify the ovarian cycle to some extent by hormonal destruction of growing follicles, or by slightly hastening ovulation, but not in producing a new set of ovulatable follicles.

DEVELOPMENT OF PUPS

Newborn coyote pups weigh 200 to 250 grams, depending on the number in the litter. They are helpless, toothless, blind, essentially hairless, and are, therefore, totally dependent for food and protection. Most coyote pups are born in a den, a hollow tree, or under a ledge or other protection. Some females prepare a grass or leaf-lined nest, others deposit the young on the bare floor. Some females pull the fur off their bellies for bedding much as rabbits do, and in the process completely bare the nipples. For the first 10 days, the pups are nourished strictly by milk, offered at intervals of 2 to 12 hours (in the laboratory, pups nurse at 1 or 2 hour intervals).

The pup's incisors appear at about 12 days, canines at 16, and second premolars at 21. Their teeth are of little use for tearing or cutting for at least 6 weeks, so milk continues to be the staple food for nearly 2 months. The eyes open on the 10th day, and at that time the pups can move around to some extent, becoming progressively more motile until they can walk quite well by 20 days and can run before they are 6 weeks old.

At 12 to 15 days after the pups are born, the adults begin supplementary feeding by regurgitation. At first, the female regurgitates only a cupful or so, then helps the pups eat it. Progressively, she regurgitates more, and at about 4 weeks, the male joins in feeding by regurgitation. I have not determined the stimulus for the first regurgitations, but as soon as the pups learn what it means, they stand on their hind legs and claw at the sides of the parent's mouth, thus stimulating regurgitation. Male coyotes are similarly stimulated to regurgitate by the pups clawing, scratching, and caressing the lips. In the laboratory, every coyote, male or female, and coyote–dog hybrid that I have raised responded to lip stimulation by regurgitation. I consider feeding by regurgitation to be a regular and natural means of supplementing the milk supply, and not a biological oddity. Also, feeding by regurgitation provides a convenient means of carrying and delivering food to the pups. Predigestion is not essential to regurgitation, as I have seen the adult coyote on many occasions (in the laboratory) eat her (or his) fill at the feed tray, walk a few steps to the pup nest, and on stimulation, regurgitate the entire mass into the nest, then enthusiastically share the meal with the pups. The male can successfully feed the brood after feeding by regurgitation is established.

When the pups are 4 to 6 weeks old and their "milk teeth" are functional, the parents bring in small food items such as mice, then rabbits, and later even "legs of lamb" so by the time the pups are 3 to 3.5 months old, they have been effectively transferred from milk and regurgitate, to regurgitate and fresh items, to fresh items only. Lactation is progressively reduced after 2 months, but may be continued to 4 months.

The male is an integral part of the family throughout the denning period, helping with the grooming and feeding, and moving the pups to a new den if necessary (Fichter, 1950). He stands guard usually within sight of the den. If the female does not return from a hunting trip, the male attempts to maintain the "home," but if the pups are not old enough to eat regurgitate, he soon gives up and leaves them to their fate. If both parents survive, they probably remain together until the next breeding season, during which the old male has the advantage.

If the den or nest area is disturbed by man, the parent coyotes move the pups to another shelter 50 m to 2 km away, carrying them one by one if they are too small to walk, or coaxing them along in a group after they can walk (3 to 4 weeks). Sometimes the pups are moved without apparent reason, maybe to get away from their fleas (Dobie, 1950).

The months of May, June, and July constitute the basic training period for the pups. They learn progressively, whether from "instinctive development" or from purposeful training by the parents, to stimulate the parents to regurgitate, climb in and out of the den, respond to specific vocal signals from the parents, catch insects, and finally hunt and catch larger prey. The sound signals are specific for (1) "food, come and get it," (2) "lie low and keep quiet," and (3) "follow [illegible] that I have not been able to discern. [illegible] obedience to signals is more strictly [illegible] human families. By late June or early [illegible] l the family sets out to explore their [illegible] 'survival" training. The details of this [illegible] ertainly deserving of the long, tedious [illegible] comprehend the procedures. Such [illegible] r natural conditions because wild ani- [illegible] an-fed, quickly develop the "welfare [illegible]

[illegible] away from the family in August but it [illegible] y units remain intact for much longer [illegible] ies seem to be strong for months. [illegible] ne months for wandering, and some of [illegible] s, as indicated by tagging experiments [illegible]; Robinson and Grand, 1958). The pups [illegible] heir 8th month and attain adult weights [illegible] er, January). The weak, the dull, and the poorly trained usually succumb to the hazards that they encounter, thus selecting for those animals that can meet the current challenges.

POPULATION DYNAMICS

Mortality of coyotes, and thus reduction in numbers, begins before ovulation and continues on a regressing logarithmic scale until the last howl, possibly at age 10. The number of ova produced is determined by subtle factors of nutrition (Gier, 1968). Corpora lutea in the ovaries vary from an average of 6.5 in over 80% of the females when conditions are optimal (correlated with high rodent populations in Kansas) to less than 4.5 with only 50% of the females ovulating when conditions are poor (low rodent populations). Approximately 10% of the ova are not fertilized or the embryos do not live long enough to claim a spot in the uterus (14 days postovulation), with a higher proportion lost if nutrition, weather, disease, and so on are adverse. Another 10 to 15% is lost by death and resorption before birth, so the average number born is about 80% of the ovulated ova; less under adverse environmental conditions. Another 10 to 15% is lost at or within a few days after whelping, as the number of young in the dens rarely equals the number of placental scars in the uteri of the bitches killed in those dens. All the pups that survive the first 5 days after birth are subjected to normal attrition resulting from (1) death of parents, (2) parasitic infections (hookworms, intestinal roundworms, etc.), (3) accidents, predation, and neighboring coyotes, (4) inherent weaknesses, and (5) man's activities. Coyote pups are particularly vulnerable to hawks, owls, eagles, and neighboring coyotes during their first few days out of the den (their 3rd to 10th weeks of life); then to dogs, automobiles, and other coyotes when they stray from the family groups. In Kansas, from 1945 to 1965, with a bounty on pups in effect, no more than 50% of the pups survived to July 1 of the year in which they were born, and 10% of the adults were killed April 1 to June 30. Now, with no bounty, the 50% loss may take until the end of August. Hunters take relatively large numbers of coyotes from early November to late March. If a balanced population is to be maintained, the annual increase (pups born in April and May) must equal the annual loss (pups of the year plus older animals). Such a balance was maintained in Kansas for nearly two decades with the aid of pup bounties in addition to normal hunting, plus the kills by predators, disease, and intraspecific strife (Gier, 1968). Now, with cyanide guns outlawed and no bounties, coyote populations are bound for a new balance at a higher level. Likewise, with the recent elimination of all poison for predator control on public lands, the coyote populations throughout the Rocky Mountain States will undoubtedly reach a much higher level than has been maintained for the past half century.

I have shown (Gier, 1968) that adverse food and climatic conditions may affect coyote populations almost as quickly and more effectively by reduction of birth rate than can be attained by an intensive poisoning program. The animals that survive adverse environmental conditions are weakened and may take a year or more to recover, while those that do not get the poison (cyanide, strychnine, 1080, etc.) are completely virile and may even profit by removal of competition, *if* the populations were previously great enough to limit reproduction and survival.

COYOTE COMMUNICATION

Coyotes are the most vocal of all North American wild mammals but unfortunately, coyote communications have not been analyzed. Some of the older pseudonaturalists interpreted coyote calls as supplementing their arthritic pains in forecasting storms, death in the family, good luck, and so forth (Dobie, 1950) with no real basis for their interpretations.

There is no doubt that coyotes have voice communications among themselves. Parents communicate effectively with their pups, with visual results, by specific sounds for freeze, hide, run away, come, dinner is served, and possibly others. Young coyotes emit specific sounds indicating hunger, pain, fear, and pleasure. The wavering calls of coyotes on clear nights are undoubtedly primarily for communication with other coyotes, and clearly announce the position, hunting success, and probably some deeper feelings and needs of the caller. I am convinced that study of coyote calls by modern recorder-analysis methods would be as rewarding as is that of bat and porpoise calls, but at present few people can distinguish the difference between coyote songs of joy and calls for help. Although coyote "songs" are more frequent during the breeding season than at other times of the year, they are not limited to mating calls or declaration of territory. Conversation between coyotes on separate hills are interesting to humans: we are sure they carry some meaning, but an understanding necessitates concentrated effort.

Coyotes have another, more lasting means of communication in the form of "scent posts" on or near which a visiting coyote urinates or defecates. The "scent post" may be a post, stump, bush, rock, dried cow dung, or a bare spot. The visiting animal may go many meters up-wind to investigate a scent post, and leave his mark. Whether these actions indicate a claim to territory, a warning to keep off, or only an indication that "I was here" has not been determined but the trait of every passerby to investigate "scent posts" has been used in preparation of lures and setting traps for coyotes for at least a century. Males are attracted more frequently by female urine and females by male urine. I have collected large quantities of coyote urine and distributed it to trappers with consistent results: traps scented with female urine catch about 75% males, while those with male urine catch 75% females. Urine from females in estrus is most attractive, and results in a catch as high as 85% males.

INTERSPECIES RELATIONSHIPS

Coyotes have many interactions with mammalian species other than their prey. Dogs and coyotes are mortal enemies throughout 10 months, but there seems to be a near-complete truce between sexes during the breeding season. Male coyotes may approach the farm building to mate with a domestic bitch, or domestic bitches may be lured into the coyote domain during their estrous periods and become "wild dogs." Likewise, a male dog may desert domestication and become a full mate to a coyote bitch, or more often stay with her only while she is in estrus. Coyote–dog hybrids have been reported throughout the range of the coyote (Dobie, 1950; Young, 1951; Gier, 1968; Gipson, 1972), many of which have coyote behavioral patterns, and remain in competition with coyotes (Gipson, 1972) or extend the coyote range, as the New Hampshire "wild canids" (Silver and Silver, 1969). The breeding season of hybrids is random, so many of the hybrids are unable to breed back into the coyote population (Mengel, 1971). Thus there may develop hybrid populations, as in New Hampshire. Others have breeding seasons that coincide with those of coyotes and integrate with the parent populations (Gipson, 1972). A high percentage of coyote–dog hybrids are fertile (Gier, 1968; Mengel, 1971; Gipson, 1972) giving proof that domestic dogs and coyotes are genetically closely related.

Although coyotes have been historically precluded from wolf ranges, demise of the *Canis rufus* population in the southeastern United States permitted a complete invasion of coyotes into what had been exclusively wolf range, with complete interbreeding of the remaining red wolves with the invading coyotes (Gipson, 1972). There were probably many incidences of interbreeding along the margin of their ranges before man became cognizant of the possibility, and the presence of this buffer zone of hybrids possibly played a major role in the swamping of the red wolf remnant population. In Arkansas, red wolves and coyotes now are thoroughly integrated, with only three areas left with "strong red wolf influence" (Gipson, 1972).

Hybridization of coyotes with *Canis lupus* has not been so extensive nor so noticeable as with *Canis rufus,* but it must have occurred to some extent in areas of overlapping range with coincidence of breeding seasons. Since dogs readily interbreed with both wolves and coyotes, and both dogs and coyotes interbreed with *Canis rufus,* it is inconceivable that coyotes cannot consummate fertile matings with *Canis lupus* in some of its forms. Even though the coyote and the wolf were present in America during the Pleistocene period, their separation must have been geographic or ecological rather than genetic and reproductive, otherwise interbreeding could not be so readily accomplished and the offspring would not be so fertile.

Interactions of coyotes and foxes needs some clarification. In North Dakota, coyote populations were drastically reduced by "predator control" activities, and red foxes reproduced and expanded their range excessively, filling the predator's niche within 3 years (Robinson, 1961). There are no records of more than an occasional killing of foxes by coyotes and no records of coyote–fox hybrids. In the Flint Hills of

Kansas, coyotes, red foxes, and gray foxes are all present within a single ravine with no evidence of agonistic reactions between them. In western Kansas and neighboring Colorado, New Mexico, and Oklahoma, swift foxes and coyotes occur in the same areas again with no indication of conflict.

Badgers have notoriously been friends with coyotes, with numerous reports of companionship and cooperation between the two species, particularly for hunting rats (Dobie, 1950). There seems to be little conflict between them, as they are too equally matched for fighting and there is relatively little competition between them. Coyotes regularly den in abandoned badger dens or in holes dug by badgers in search of food.

Likewise, marmots provide dens that need only slight modification. In contrast to their reactions to badgers, however, coyotes readily kill any marmot that can be caught away from its den.

Coyotes have only occasional fatal contacts with animals, except with cougars, wolves, and large dogs. Coyotes have learned to cope with domestic dogs, automobiles, and farm equipment. Trained hunting dogs and men concentrating on coyote destruction have not been met successfully.

Coyotes have benefited greatly by man's efforts in removing timber and establishing meadows and grain fields with subsequent increase of rabbits, rodents, and game birds. Coyotes profit by the presence of poultry and sheep when predation on such animals is moderate, but excessive predation on any domestic animals ultimately raises the ire of the owner and leads to death of the raiders and their young. Coyotes have likewise benefited by man's assistance in removal of their common enemies: wolves, cougars, and bears.

Coyotes have adapted to the presence of man in moderate concentrations. They may successfully raise a litter of young within 100 m of a farm house as well as in the back pasture or on the undisturbed range. They have learned to pick up animals killed on our highways with little danger to themselves. They have adapted to scavenging from humans, even within a few meters of the farmhouse, as they formerly scavenged from the gray wolves.

Man has been directly responsible for much of what the coyote is today. The slow and crippled fall readily to hunting dogs. The dull of wit are caught in traps, hit by cars, or run down by antelope, deer, or jeeps. The unwary are shot by the rancher or sportsman. We, with our persecution of the coyote, have added another parameter to natural selection, with the result that coyotes are now larger, smarter, more adaptable, faster, and more cunning than when white men first entered the coyote's territory.

With such selection and adaptability, we can expect the maintenance of a reasonable population of coyotes throughout North America wherever the concentration of humans is not so great as to totally usurp their territory. The only foreseeable serious threat to the coyote is a concentrated effort by man armed with more than the weapons so far employed.

18
A SURVEY OF THE RED WOLF (*Canis rufus*)*

G. A. RILEY
R. T. McBRIDE
U.S. Department of the Interior
Fish and Wildlife Service
Bureau of Sport Fisheries and Wildlife
Washington, D.C.

This paper discusses the red wolf's (*Canis rufus*) status, distribution, and ecology and describes and differentiates the red wolf from other closely related canids. Difficulties in distinguishing red wolves from coyotes (*Canis latrans*) and red wolf–coyote hybrids have resulted in much confusion over the range and status of the red wolf.

The paper is based on information gathered as part of the Bureau of Sport Fisheries and Wildlife's red wolf program which began in 1968. The purposes of the program are: (1) to determine the red wolf's range, population size, food habits, and ecology, (2) to determine the actual extent of red wolf predation on livestock, and (3) to gain the understanding and cooperation of local people throughout the red wolf's range in the effort to preserve the species.

DISTRIBUTION

The red wolf formerly occurred from central Texas eastward to the coasts of Florida and Georgia, and along the Mississippi River Valley north to central Illinois and Indiana (Hall and Kelson, 1959). Presently, the red wolf occurs in Liberty, Chambers, Jefferson, Brazoria, Galves-

*Reprinted with permission from G. A. Riley and R. T. McBride, A Survey of the Red Wolf *(Canis rufus)*, Special Scientific Report No. 162 (1972), Fish and Wildlife Service, Bureau of Sport Fisheries and Wildlife, Washington, D.C.

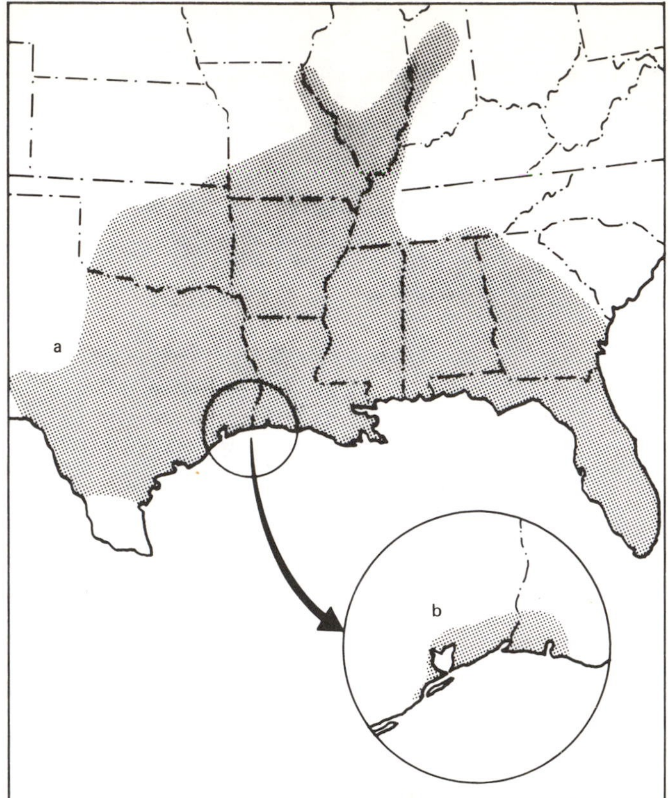

Figure 18-1 Historic range (a) and present known range (b) of the red wolf (*Canis rufus*).

ton, and Harris Counties in southeastern Texas, and in Cameron Parish in southwestern Louisiana (Figure 18-1).

Two of three described subspecies of the red wolf, *Canis rufus rufus* and *Canis rufus gregoryi,* apparently occur over the present range (Paradiso and Nowak, 1971). Animals matching the description of the former subspecies occur in portions of Brazoria, Harris, and Galveston Counties. Animals matching the description of the latter subspecies occur in portions of Liberty, Chambers, and Jefferson Counties, and in portions of Cameron Parish. Galveston Bay and the Houston metropolitan area restrict contact between the two populations.

The primary portion of the remaining red wolf range is found in Chambers, Jefferson, and southern Liberty Counties — an area of approximately 1,260,000 acres (509,903 hectares). Virtually all of this land is in private ownership. Much of the land is used for livestock grazing (521,874 acres), and for producing rice (248,657 acres) and other crops

(39,370 acres). Forested areas encompass approximately 90,000 acres. We estimate that there are 300 red wolves in this area.

The present known range of the red wolf lies within the coastal prairie and coastal marsh areas — habitat markedly different from the forest habitat found over the majority of its historic range. Although the present range extends to the very edge of the heavily forested "Big Thicket" area of southeast Texas, we have not yet located any red wolves within that area.

Vegetation on the prairie consists of tall bunchgrasses such as big bluestem (*Andropogon gerardi*), little bluestem (*Andropogon scoparius*), Indiangrass (*Sorghastrum nutans*), eastern gamagrass (*Tripsacum dactyloides*), switchgrass (*Panicum virgatum*), and gulf cordgrass (*Spartina spartinae*). Marsh vegetation is composed of various species of *Carex, Scirpus, Rhynchospora, Juncus,* and marshhay cordgrass (*Spartina patens*).

There are also "islands" of loblolly pine *(Pinus taeda)* mixed with hardwoods. The predominant hardwood species are oaks *(Quercus),* magnolia *(Magnolia grandiflora),* and sweet gum *(Liquidambar styraciflua)* (Figure 18-2).

Figure 18-2 Photograph of coastal prairie red wolf habitat. An "island" of brush in the background. Photo: R. W. Clapper, BSFW.

DESCRIPTION

Even the most complete accounts of the external characteristics of the red wolf (Young and Goldman, 1944) are sketchy, and may give the impression that this animal closely resembles the coyote except for size and color variations. However, there are certain definite characteristics which we believe are so pronounced that the two species can be readily distinguished in the field.

This proposition is supported by: (1) data gathered in the field and through examinations of skeletal material and skins in the National Museum of Natural History and in private collections, (2) evidence gathered from interviews with persons who were familiar with red wolves when they were more common in southeast Texas and Louisiana, and (3) reviews of photographs of known red wolves from Texas, Louisiana, and Arkansas.

Coloration

The term "red wolf" is misleading, and suggests an animal with a distinct red color. Residents of areas that have been inhabited by red wolves have seldom referred to the animals as "red," but rather as "gray," "yellow," or "black" wolves. Goldman (Young and Goldman, 1944) describes the color of the red wolf as a varying mixture of "cinnamon-buff," "cinnamon," or "tawny" with gray and black, the dorsal area more or less overlaid with black. Recent observations of red wolves in the coastal prairies of southeast Texas and southwest Louisiana reveal that present populations vary from tawny to grayish. The black phase is probably nonexistent at the present time even though it was reported occasionally in the historic range of the animal.

John Knight, a retired trapper from Segno, Texas, who trapped in the southeast Texas area from the late 1920s to the early 1940s, and T. E. (Doc) Harris, supervisor of predator control operations for the State of Louisiana since 1952, both report that the black phase was never common on the coastal prairies (personal communication). We have been able to locate records of only 3 black wolves from that area. One was recorded in Chambers County in 1935, and 2 in Cameron Parish in 1963. Evidently, the red wolf of the coastal prairie and marshland is a lighter-colored animal than its forest-oriented counterparts.

Facial color patterns of red wolves often resemble those of gray wolves (*Canis lupus*) and are distinctive:

(a) Muzzle coloration: like the gray wolf, the muzzle of the red wolf often tends to be very light. The area of white around the lips may extend well up on the sides of the muzzle, leaving only the

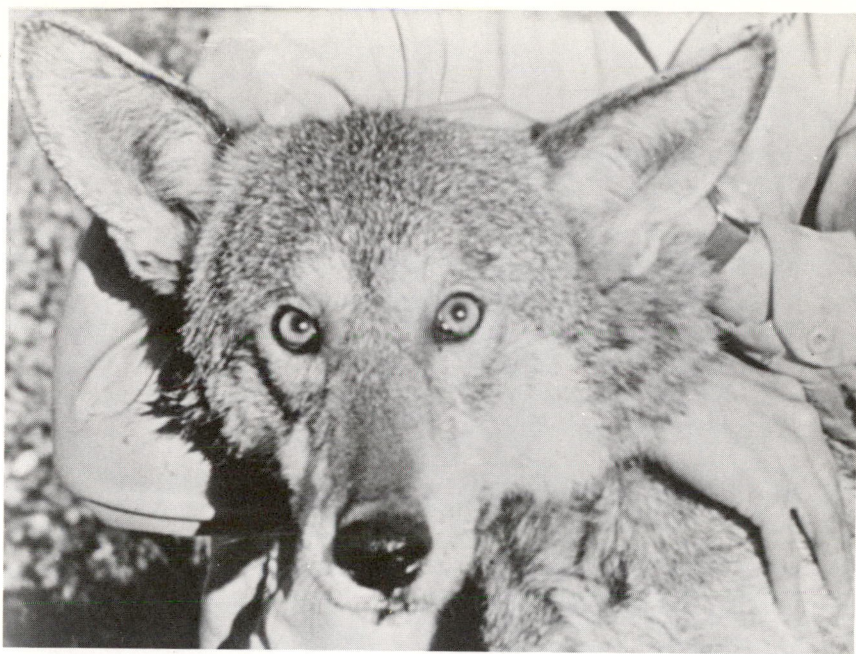

Figure 18-3 Photograph of a red wolf (a) and a coyote (b). Notice the facial markings, the length of the ears, and the width of the nose pad and muzzle. Photos: C. J. Carley, BSFW.

bridge of the nose with a tawny to cinnamon coloration. In contrast, the area of white around the lips of coyotes is thin and sharply demarcated.

(b) Coloration around eyes: on many red wolves, light areas occur around the eyes. A light tan spot may be present over each eye adding to the almond or slanted eye effect.

Because of a deeper profile and broader head, the facial appearance of the red wolf is more massive than the more foxlike coyote. While the red wolf has a less prominent ruff than the gray wolf, the almond-shaped eyes, broad muzzle, and wide nose pad contribute to its wolf-like appearance (Figure 18-3).

Measurements and Weight

Young (Young and Goldman, 1944) quotes Andy Ray of the U.S. Fish and Wildlife Service, who reported that red wolves from Arkansas averaged between 45 and 65 pounds (20.41–29.48 kg). T. E. (Doc) Harris reports that red wolves from Louisiana have averaged between 40 and 65 pounds (18.14–29.48 kg); the largest individual male red wolf he recalls weighed 74 pounds (33.56 kg) (personal communication, Oct. 5, 1971). John Knight reports that red wolves from the southeastern Texas area averaged between 40 and 60 pounds (18.14–27.21 kg); a few very large males weighed as much as 80 pounds (36.28 kg) (personal communication, July 7, 1970).

Weights of recently collected red wolves from the Texas Gulf Coast prairies fall within the 18.14 to 27.21 kg range. Individuals weighing over 27.21 kg are rare, the heaviest recorded being 34.47 kg. Of 14 adult red wolves captured in Chambers County, Texas between 1968 and 1970, weights of males ranged from 20.86 to 28.12 kg (average 52.25 pounds (23.70 kg) and those of females ranged from 20.41 to 24.49 kg (average 46.71 pounds (21.18 kg)).

Our observations agree with Young's (Young and Goldman, 1944) description of the red wolf as a long-legged, rangy animal. Of all the distinguishing external characteristics, the long legs are one of the most striking. Persons familiar with North American wild canids invariably notice and remark on the "legginess" of the animal. Ranchers differentiate between coyotes and red wolves in the same manner, referring to the latter as "long-legged" or "tall" wolves. This feature appears to be due to the increased length of the tibia,fibula, and radius and ulna of the red wolf.

Observations of gray wolves and red wolves suggest that the "legginess" of the red wolf may stem from differences in body conformation as well as the length of the legs. Measurements from the tip of the toes to the top of the shoulders indicate red wolves may be as tall as the

TICKET WEB

www.ticketweb.com

HOLDER VOLUNTARILY ASSUMES ALL RISKS AND DANGER INCIDENTAL TO THE EVENT FOR WHICH THE TICKET IS ISSUED, WHETHER OCCURRING PRIOR TO, DURING OR AFTER THE EVENT. HOLDER VOLUNTARILY AGREES THAT THE MANAGEMENT, FACILITY, LEAGUE, PARTICIPANTS, PARTICIPATING
CLUBS, TICKETWEB, AND ALL OF THEIR RESPECTIVE AGENTS, OFFICERS, DIRECTORS, OWNERS AND EMPLOYEES ARE EXPRESSLY RELEASED BY HOLDER FROM ANY CLAIMS ARISING FROM SUCH CAUSES. Management reserves the right, without the refund of any portion of the ticket purchase
price, to refuse admission to or eject any person whose conduct is deemed by management to be disorderly or who fails to comply with these or any other management rules. Breach of any of the foregoing will automatically terminate this license. NO REFUNDS. NO EXCHANGES EXCEPT AS
AUTHORIZED BY THE SELLER OF THE TICKET OR AS REQUIRED BY APPLICABLE LAW. EVENT DATE & TIME SUBJECT TO CHANGE. ALL RIGHTS RESERVED. This ticket is a revocable license and admission may be refused at any time for any reason, subject to applicable law. Unlawful resale or
attempted resale is grounds for seizure and cancellation without compensation. This ticket is not valid unless purchased from an authorized ticket seller. Tickets obtained from unauthorized sources may be lost, stolen or counterfeit, and if so are void. DO NOT EXPOSE TO EXTREME HEAT.
Subject to applicable law, if this ticket is for an event that fails to occur for any reason whatsoever, including a cancellation of such event, then this ticket can only be exchanged or refunded, if at all, by and from the place, and the company, from where the ticket was originally purchased,
subject to availability and the policies and procedures of the seller of the ticket. TicketWeb is not the seller of this ticket and, subject to applicable law, shall have no responsibility for any refund or exchange of this ticket. Certain maximum resale premiums and restrictions may apply such
as: PA - $5 or 25% of the ticket price, whichever is greater, plus lawful taxes; MA - $2; CT - a reasonable charge, but not more than $3; NJ - $3 or 20% of the ticket price (or 50% of acquisition price if registered broker or season ticket holder), whichever is greater, plus lawful taxes. In NY: if
the venue to which this ticket grants admission seats 6000 or fewer persons, this ticket may not be resold for more than 20% above the price printed on the face of this ticket, whereas if the venue to which this ticket grants admission seats more than 6000 persons, this ticket may not be
resold for more than 45% above the price printed on the face of this ticket; this ticket may not be resold within one thousand five hundred feet from the physical structure of this place of entertainment under penalty of law. Holder agrees by use of this ticket, not to transmit or aid in transmitting
any description, account, picture, or reproduction of the game, performance, exhibition or event for which this ticket is issued. Holder acknowledges that the event may be broadcast or otherwise publicized,
and hereby grants permission to utilize holder's image or likeness in connection with any live or recorded transmission or reproduction of such event in conformance with some local requirements or certain
facility rules, certain items, may not be brought into the premises, including without limitation, alcoholic beverages, illegal drugs, controlled substances, cameras, recording devices, bundles and
containers of any kind. This ticket cannot be replaced if lost, stolen or destroyed, and is valid only for the event and seat for which it is issued. This ticket is not redeemable for cash and bears no cash
value. It is unlawful to reproduce this ticket in any form. Unless indicated otherwise, price includes all applicable taxes and/or cash discounts (if available).

C2226680

GENERAL A
SECTION RO
16 NOV06

THE ORANGE PEEL

101 Biltmore Ave., Asheville, NC

16 NOV06 9:00PM DOORS@8:00PM

PRICE:18.00 FEES:1.00 ORDER:MWK4LLLQQ

* Fee or tax is average per ticket for the entire order.

Table 18-1 Ear Length of *Canis lupus, Canis rufus,* and *Canis latrans* in centimeters.

Canis lupus		Canis rufus[a]		Canis latrans	
Length	Sample Size	Length	Sample Size	Length	Sample Size
13.97	1	13.97	2	11.43	6
12.70	1	12.70	10	10.795	10
12.54125	1	12.065	1	10.4775	1
12.065	1	11.43	1	10.16	4
10.95375	1			9.525	2
				9.8425	1
				8.89	1
		Average lengths			
12.446		12.7254		10.6172	

[a]Measurements for this species are mainly from animals from Liberty and Chambers Counties, and some from Galveston County.

gray wolf, but because they are not as massive through the thoracic region, they have a rangy, long-legged appearance. The weights of 5 gray wolves from Minnesota, eastern timber wolf subspecies (*Canis lupus lycaon*), varied from 25.85–38.10 kg (average 65.4 pounds (29.66 kg), and ranged from 68.58 to 76.20 cm (average 28.4 inches (72.03 cm)) from the tip of the toes to the top of the shoulders. Ten red wolves from Liberty and Chambers Counties weighed from 18.14 to 34.47 kg (average 53.9 pounds (24.44 kg)), and ranged from 62.23 to 74.93 cm (average 27.6 inches (70.10 cm)) from the tip of the toes to the top of the shoulders.

Another distinctive characteristic of the red wolf is its proportionately large ears and the angle at which they are normally carried (see Figure 18-3). Although the ears of a gray wolf may equal those of a red wolf, they are less prominent because the head of the gray wolf is more massive. Also, the angle at which the red wolf carries its ears creates an accentuated triangular facial appearance markedly different from that of the gray wolf or coyote, whose ears are more erect. Table 18-1 compares the ear lengths of these three species.

Voice

The voice of the red wolf has never been adequately described. The only early description of the vocalizations known to us is an unpublished field note written by Vernon Bailey (Biological Survey) while at Sour Lake, Hardin County, Texas: "They howled repeatedly not far from our camp. Their voice is a compromise between that of the coyote and lobo, or rather a deep voiced yap-yap and howl of the coyote. It suggests the coyote much more than the lobo." Interviews with people familiar with the red wolf during the first half of the century seem to support Bailey's description. John Knight stated: "They howl long and mournful but break off into yapping. I kept one for seven years, raised it from a pup (1928–1935); he would howl a deep coarse howl and break off into a yap-yap" (personal communication, July 7, 1970).

Using sirens, Dennis Russell and Jim Shaw of the Texas Parks and Wildlife Department elicited a prolonged coarse howl from red wolves that was similar to, but of a higher pitch, than that of the gray wolf (Russell and Shaw, 1971). Using sirens, we have elicited a bark, a howl ending with 1 or 2 short barks, and a prolonged howl.

A recent recording of spontaneous vocalizations of a group of red wolves in southwestern Louisiana indicates that elicited and spontaneous vocalizations may differ from one another. The group vocalizations consisted of a series of blended coarse yaps and prolonged howls of different tone and pitch. The harmony seemed more controlled and deliberate than that of coyotes, yet was not blended as well as the vocalizations of gray wolves.

Tracks and Signs

Since red wolves are larger and heavier than coyotes, it is not difficult to distinguish between the tracks and scats of the two; experienced field workers familiar with the tracks and signs of coyotes immediately notice the difference.

A red wolf track is larger, the stride longer, and the pattern different from that of a coyote. In fact, they closely resemble those of the gray wolf. Red wolf tracks measured in Liberty and Chambers Counties, Texas and Cameron Parish, Louisiana ranged from 8.89 to 12.70 cm from the back of the heel pad to the end of the longest claw (average 4.0 inches (10.16 cm)); the stride ranged from 55.28 to 76.20 cm (average 25.9 inches (65.78 cm)). Coyote tracks measured in west Texas varied from 5.71 to 7.24 cm (average 2.6 inches (6.60 cm)); the stride ranged from 32.38 to 48.26 cm (average 16.3 inches (41.40 cm)). The tracks of red wolves are more compact than those of the average domestic dog, being proportionately narrower with a more elongated heel pad. Differences in the track pattern further differentiate the red wolves and domestic dogs.

Thompson (1952) notes that coyote scats rarely exceed 2.5 cm in diameter, while most eastern timber wolf scats range from 2.5 to 3.8 cm. Our data indicate that red wolf and eastern timber wolf scats are of comparable size.

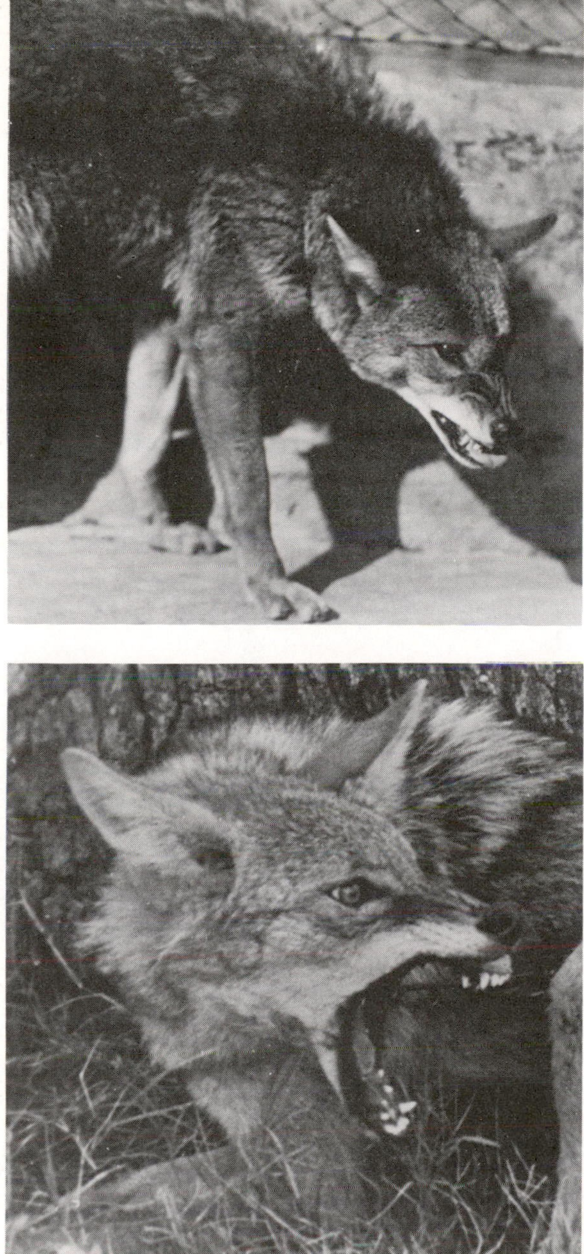

Figure 18-4 Red wolf (a) and coyote (b) threat behavior. A red wolf's tongue does not normally protrude as shown here.

BEHAVIOR

Trapped red wolves are more aggressive than trapped coyotes or gray wolves. With tail in an upright position and canine teeth bared, they often attempt to attack. They frequently bark or howl as they are approached. This may occur when the trapper is 90–100 meters or more distant. Coyotes bark under similar circumstances, but they rarely howl. When in similar situations, gray wolves seldom bark or howl (L. D. Mech, personal communication, Oct., 1971; Robert Himes, personal communication, Oct., 1971).

The threat postures of red and gray wolves differ from those of coyotes. Wolves bare their canine teeth and raise the fur along their necks and backs (Figure 18-4a). Coyotes, on the other hand, assume a wide-mouthed posture with teeth showing, back arched, and tail held between the legs. The fur along their necks and backs may or may not be raised (Figure 18-4b).

Red wolves readily swim. Nilo Esquivel, a wolf hunter from Alvin, Texas, relates that they frequently take refuge in water when pursued by hounds (personal communication, August, 1970).

Sociability

Red wolves are more sociable than coyotes, but probably less so than gray wolves. It is not unusual to find three or more traveling throughout their range as a group. Numerous observations of groups and group signs convince us that red wolves maintain a group structure throughout the year. There is little reason, however, for them to hunt in packs since their food is composed of small prey species.

Home Range

Roads, canals, flooded rice fields, and bayous constitute travel barriers of varying degrees. Large canals and bayous seem to isolate certain family groups to some extent. In some cases, the wolves do not complete a circle in their nightly hunting forays, simply returning over the same route taken earlier in the night, which explains why their tracks are often found coming and going on the same trail.

Based on the systematic tracking of 3 adult red wolves (1 for 1 year, the others for 2 years), we estimated the home range of a red wolf to be approximately 64.75 to 129.50 km^2. James H. Shaw has found through radio telemetry that the home range of an adult red wolf averages 35 mile^2 (90.65 km^2) (personal communication, April, 1972).

Reproduction

Breeding occurs in January and February, and the pups are born in March and April. Based on litters born in captivity, the average litter size seems to be 3 or 4.

Red wolves normally rear their young in dens dug in the slopes or crests of the low natural sand mounds common in the coastal prairie. Dens have also been found in drain pipes and culverts, and in the banks of irrigation and drainage ditches. The dens average about 2.43 m in length and are normally no deeper than .91 m. The den entrance varies from .60 to .76 m in diameter, and is normally fairly well-concealed.

Both male and female take part in rearing the young. Frequently, young of the previous year are found in the vicinity of dens but they do not appear to participate in the guarding, feeding, or training of the pups. Evidence suggests that the pups actually spend more time in "beds" located in areas of good cover than in the den, especially after they are 6 weeks of age.

Food Habits

From scat analyses and observations, we find that the predominant prey species are nutria (*Myocastor coypus*), swamp rabbit (*Sylvilagus aquaticus*), and cottontail rabbit (*Sylvilagus floridanus*). Other common prey species are rice rat (*Oryzomys palustris*), cotton rat (*Sigmodon hispidus*), and muskrat (*Ondatra zibethicus*). Scat analyses by Russell and Shaw (1971) and by Stutzenbacker (1968) also indicate that rabbits and nutria are major prey species. We believe that the nutria is a buffer species between red wolves and domestic livestock.

With constant exposure to large herds of cattle, it is to be expected that cattle will sometimes be killed and eaten by red wolves. Predation upon newborn calves occurs occasionally. Actually, there is much disagreement among local ranchers regarding the seriousness of red wolves as killers of cattle, a disagreement that never existed with the gray wolf. This disagreement is, in itself, an indication that red wolves are not as serious a predator on cattle as were gray wolves.

It is interesting to note that it was not predation on cattle that precipitated large-scale reduction efforts against the red wolf, but predation upon smaller, more easily obtained domestic prey such as hogs, which in earlier times were allowed to run free.

LIMITING FACTORS

Human activities appear to be a major limiting factor of the red wolf. Each year, agricultural and commercial use of the land intensifies, and little by little more wolf habitat is lost. Prime denning areas are plowed and converted to rice fields, and industrialization creeps across the land. A favored red wolf rendezvous site in a given year may be an industrial development the next year. The habitat changes have been most drastic in Brazoria, Harris, and Galveston Counties.

Commercial hunting and shooting preserves flourish on the coastal prairies. On the opening day of waterfowl season, hundreds of hunters may be afield on portions of prime red wolf habitat. Each year some wolves are killed by these hunters. On the northern edge of the range, deer hunters also kill a few wolves each year. With these pressures and the high natural mortality, the red wolf faces a dim future unless it receives help from its greatest enemy — man.

Since man and his activities are constantly within sight or hearing, the red wolf is conditioned to man's presence and may be very tolerant of him. Red wolves are unquestionably easier to capture than gray wolves or coyotes. On several occasions we have seen where wolves traveled within easy rifle shot of residences and farm buildings. They exhibit little fear of farm tractors, at times following them in search of rodents plowed out of the ground, and exhibit little fear of men on horseback.

Parasites appear to be another major limiting factor of red wolves. Heartworms (*Dirofilaria immitis*) have been present in all 27 wolves examined for internal parasites during our study. Infestation probably increases with the age of the host due to the constant exposure to the mosquito vectors. Red wolves 3 years of age and older usually were heavily parasitized by heartworms, sometimes to the point that the heart valves could not close. Such animals cannot tolerate stress situations and may die from incidents that would be of little or no consequence to animals with a lesser degree of parasitism. Other internal parasites commonly found include hookworms (*Ancylostoma*), tapeworms (*Taenia*), and occasionally spiney headed worms of the class *Archiancathocephala*.

The sarcoptic mange mite (*Sarcoptes scabei*) infests a considerable portion of the red wolf population. Several red wolves observed in the field were 90% devoid of fur. These animals were in very poor physical condition and probably did not survive very long. Dr. U. S. Seal of the

Veterans Administration reports that all of the 27 red wolves he has received from us for blood studies have been affected by mange (personal communication, Oct. 19, 1971).

HYBRIDIZATION

Genetic barriers which separated *Canis rufus* from *Canis latrans* undoubtedly have eroded. This has resulted in the red wolf being genetically swamped by coyotes. McCarley (1962) reports that hybridization had taken place over most of the original range of the red wolf.

Hybridization with coyotes poses one of the greatest threats to the remaining red wolf population. While our understanding of the dynamics involved is not clear, we do know, however, that coyotes and hybrids are found along the periphery of the remaining range of *Canis rufus* and apparently are progressively invading the remaining range.

The controversy over identification of red wolves is due in part to the former lack of knowledge as to the identity of hybrids. Recent efforts to correct this matter have dealt with brain morphology, chromosome analysis, and electrophoretic tests of blood samples. To date, none of these methods have provided a means of positively confirming identification of live specimens. We state again that the confusion can be resolved because coyotes, hybrids, and red wolves are distinguishable by certain external characteristics which, in the final analysis, can be supported by skull examination.

In dealing with hybrids *(Canis rufus* × *Canis latrans)* one must expect an occasional phenotype that approaches that of *Canis rufus*. However, only where populations of canids are uniform in phenotype may the genotype be considered pure. When an occasional phenotype approaching that of *Canis rufus* appears in a group of apparent hybrids, it must be assumed that it is a hybrid.

Hybrids can be distinguished from the red wolf by the following characteristics:

(1) Smaller feet and legs, both in length and breadth;
(2) Shorter ears;
(3) Less massive muzzle;
(4) Overall, generally smaller in size;
(5) Coyotelike threat posture.

The sizes of the feet, legs, and ears appear to be the first external characteristics to change with hybridization. Table 18-2 gives comparative weights and external body measurements of red wolves, coyotes, and hybrids.

TABLE 18-2 Comparative Weights and Measurements of Red Wolves, Hybrids, and Coyotes from East Texas, 1969–1972.

	Red Wolf[a]		Hybrid		Coyote	
Sample size	26 ♂	26 ♀	21 ♂	18 ♀	15 ♂	17 ♀
Weight (kg)						
$\bar{x}$ ± s.d.	22.67 ± 3.90	19.95 ± 2.67	18.14 ± 1.90	15.87 ± 1.95	14.96 ± 1.99	13.15 ± 1.76
Range	17.23–34.47	16.32–24.49	13.60–20.41	12.70–20.41	9.97–17.23	9.52–15.87
Total length (cm)						
$\bar{x}$ ± s.d.	141.98 ± 6.60	134.11 ± 5.84	133.85 ± 4.82	126.74 ± 3.55	127.25 ± 4.31	120.39 ± 5.84
Range	132.08–160.02	121.92–142.24	128.27–139.70	119.38–132.08	120.65–134.62	111.76–129.54
Hind foot length (cm)						
$\bar{x}$ ± s.d.	23.11 ± .73	22.09 ± .38	21.59 ± .93	20.82 ± 2.10	20.57 ± .71	19.81 ± .91
Range	21.08–24.89	20.32–24.13	19.05–22.86	20.06–22.60	19.05–21.33	17.78–21.59
Ear length (cm)						
$\bar{x}$ ± s.d.	12.70 ± .60	12.19 ± 1.54	11.43 ± .81	11.17 ± .66	11.68 ± 1.11	10.92 ± .96
Range	11.43–13.97	11.43–12.70	9.65–13.46	10.16–12.44	10.66–12.19	8.63–12.19

[a]These specimens were from Liberty and Chambers Counties, except 2 from Galveston County.

SUMMARY

In this paper we have described the red wolf as it occurs today. The animal's relationship to its environment is discussed briefly. We stress that the red wolf is a strikingly different animal from the coyote. We feel the difficulties described in the literature in distinguishing this animal from coyotes or hybrids are exaggerated from the true situation in the field.

Canis rufus still exists as a separate entity, but its numbers have been reduced. Evidence from a recently completed computer-based multivariate analysis of skulls indicates that *Canis rufus* still survives as a pure species in Chambers County and probably in southern Liberty County (R. M. Nowak, personal communication, June, 1972). We believe that this is the *gregoryi* subspecies and that it also survives in portions of Jefferson County and Cameron Parish. Also, there are the populations of canids in portions of Brazoria, Harris, and Galveston Counties matching the description of *Canis rufus rufus.* While the identity of *Canis rufus* is still being pondered by some, the future of the animal is in a delicate balance which could be tipped by chance. The Bureau is presently expanding its red wolf recovery program in cooperation with others to restore the species to secure population levels. The State of Louisiana has granted the species a protected status.

Acknowledgments We wish to thank the following people for their encouragement and assistance in preparing this paper: Milton Caroline, Dr. Frederick F. Knowlton, Dr. L. David Mech, Curtis J. Carley, and Stephen P. Atzert, all of the Bureau of Sport Fisheries and Wildlife, and Dr. Howard McCarley of Austin College.

ADDENDUM

Some Contributions to the Ecology and Systematic Position of the Red Wolf: An Interim Report

J. H. SHAW
School of Forestry and Environmental Studies
Yale University
New Haven, Connecticut

The red wolf (*Canis rufus*), America's most endangered mammal today, had never been systematically studied in the wild. From June, 1971 through September, 1972, as part of my doctoral dissertation, I measured home range, social structure, population density, activity patterns, and food habits in the red wolves of Chambers County, Texas. In the narrow coastal strip of Chambers County eastward through Jefferson County and into Cameron Parish, Louisiana, there remain the last populations of *C. r. gregoryi*. The research included an analysis of taxonomic relationships using new criteria of ecology, behavior, and morphology.

Home range, measured by telemetry, varies from 2300 to 9000 hectares. Animals are active primarily from dusk through early night and again at dawn. Group size can be as high as 7 but is usually observed between 1 and 3. Food of the Chambers County wolves, based on scats collected year round, consists primarily of the nutria — an introduced rodent (*Myocastor coypus*), rabbits (*Sylvilagus* spp.) and the cotton rat (*Sigmodon hispidus*). From Cameron Parish, scats collected in September showed similar food habits despite the presence of deer, the presumed primary prey of red wolves within most of their original distribution. Thus, in terms of home range size, activity patterns, group size, and prey selection, these wolves appear more similar to coyotes than to gray wolves.

While there is overlap in all morphological characteristics measured, samples showed the following significant differences. Body size differs among *C. r. gregoryi, C. r. rufus,* and *C. latrans*. The glucose-phosphate isomerase locus, according to analysis by R. Storez of Yale, shows different allelic combinations among *C. lupus, C. r. gregoryi,* and either *C. r. r.* or *C. latrans*, but not between these last two. A difference in threat posture, discovered in this study, was consistent without exception: *C. r. gregoryi* snarls as in *lupus* and *familiaris,* while *C. r. r.* shows the gaping mouth posture of *C. latrans*. These results suggest that *C. r. gregoryi* is distinct from both *C. lupus and C. latrans* as well as from *C. r. rufus. Canis r. rufus,* though perhaps once distinct, appears by these criteria so close to *C. latrans* now as to suggest widespread hybridization.

Evidence from this and other studies demonstrates that the red wolf is a distinct form which deserves full protection as a "species." It is feasible to preserve free-living populations within at least one small corner of the species' original range. As proposed by the Bureau of Sport Fisheries and Wildlife's Office of Endangered Species, conservation efforts should be focused on the restricted coastal strip where *Canis rufus gregoryi* still exists. These wolves need protection against further persecution by man, loss of their habitat, and against possible introgression with the coyote.

Support for this research was provided by the U.S. Bureau of Sport Fisheries and Wildlife, World Wildlife Fund, Boone and Crockett Club, Explorer's Club, and the National Science Foundation.

19

THE ECOLOGY OF THE WOLF IN NORTH AMERICA

DOUGLAS H. PIMLOTT
University of Toronto
Toronto, Ontario, Canada

The wolf (*Canis lupus*) in North America has been shown to be a wide-ranging, highly-adaptable animal which occurs in relatively low densities throughout its range. The range of the species originally extended from Central Mexico to Ellesmere Island in the high arctic and to the coastal areas adjacent to the ice cap in Greenland. Deserts and rain forests were the only environments that the species could not inhabit.

FOOD AND PREY INTERRELATIONSHIPS

Food

The wolf is primarily a predator of large mammals. The species has exhibited an exclusive dependence on large ungulates for its food during the winter in all areas where it has been intensively studied in North America. During the summer they kill a much wider variety of prey, though in the majority of areas 75% or more of the food consumed comes from the ungulates (Murie, 1944; Cowan, 1947; Thompson, 1952; Mech, 1966; Pimlott, *et al.*, 1969; Kuyt, 1971; Clark, 1971).

The greatest variation from the use of ungulates has been reported for an area of eastcentral Ontario where beaver (*Castor canadensis*) occurred in excess of 75% of the wolf scats that were examined during a period in the mid and late 1960s (Kolenosky, 1972). It was considered that the dependence of wolves on beaver had developed in response to a marked change which had occurred in the relative abundance of beaver and white-tailed deer *(Odocoileus virginianus)* during the decade (Hall, 1971). Arctic hare predominated in scats from Ellesmere Island (Tener, 1954). They were the only cases reported where a single species was more important than the ungulates of the area in the diet of wolves. However, significant use is made of other species in Mt. McKinley National Park in Alaska (Murie, 1944) and in the Mackenzie District of the Northwest Territories of Canada (Kuyt, 1971). In both these areas, caribou hair occurred in 80% or more of the wolf scats which were examined. On Baffin Island in the eastern Canadian Arctic, the remains of caribou (*Rangifer tarandus*) occurred in 98.5% of scats collected in summer and comprised 92% of all animal food items which were identified (Clark, 1971).

Food Consumption

Quantitative data on the food consumption rates of wolves in the wild have been obtained for Isle Royale and for an area in eastern Ontario. The daily rate of consumption for the wolves of Isle Royale was 0.18, 0.13 and 0.19 kg of moose per kg of wolf for the winters of 1959, 1960, and 1961 (Mech, 1966). The daily rate of consumption for the Ontario wolves was calculated for a pack of 8 wolves for a period of 63 days. The average consumption was estimated at 0.10 kg/kg per wolf per day (Kolenosky, 1972). Similar rates of consumption have been recorded for some of the large predators in Africa (e.g., Estes and Goddard, 1967). Some factors which may influence the accuracy of the estimates are discussed by Pimlott, *et al.* (1969), Mech (1970), and Kolenosky (1972).

The rate of kill on Isle Royale varied from 1 moose per 3.0 to 1 moose per 3.7 days. The maximum period between kills was from 118 to 137 hours. The Ontario pack had a kill rate of 1 deer every 2 days in January and early February. The rate accelerated to 1 deer per day for 2 periods during the study. The longest period between kills was 102 hours.

Selection of Prey

It has become evident that in terms of a particular species, the selection of prey by wolves can be described as primarily dependent on the relative ease with which animals of different sexes and ages can be caught. The most common pattern that has emerged is that young animals of the year are highly vulnerable during the spring and summer; a change in relative vulnerability occurs during the fall and by the time lakes and streams are frozen over, animals in the older ages classes are most commonly killed. However, juvenile animals usually continue to be vulnerable to predation throughout their first year of life. Animals between 1 and 3 years of age usually suffer a relatively low rate of predation, particularly when very large animals, such as moose (*Alces alces*) and bison (*Bison bison*) are involved (Pimlott, 1967; Mech, 1970).

The most significant variation to the general pattern was reported for an area in eastcentral Ontario where many of the animals killed in winter were in the prime age class (Kolenosky, 1972). The area was open to human hunters and Hall (1971) believed ". . . that the different patterns of mortality that arise from hunting and wolf predation resulted in a higher occurrence of old deer in the Algonquin population."

The question of whether wolves prey more heavily on males or females has not been clear (Mech, 1970). However, the work in Ontario showed a sex ratio of 250 males:100 females for 42 adult deer killed by wolves from 1964–69. The ratio for 290 deer killed by hunters during the same period was 93 males:100 females. The difference was significant $(P > 0.01)$ (Kolenosky, 1972).

Efficiency of Hunting

Isle Royale provided the first detailed data on the efficiency of wolves as predators. Of 87 moose that were directly encountered by wolves in winter, 8 were attacked and 7 were killed for a predation efficiency of 7.8% (Mech, 1966; Shelton, 1966). In eastern Ontario, wolves were successful in killing 16 of 35 deer which were chased during two winters. The rate of success was significantly different (25% and 63%) between years. The difference was attributed to variation which existed in snow conditions (Kolenosky, 1972). The data from Isle Royale were based on direct observation, while those from Ontario were based on interpretation of tracks in snow. Wolves on Baffin Island were observed to have an efficiency rate of 5% for summer hunting; 55% of the encounters observed were made by single wolves (Clark, 1971).

The Influence of Predation on Prey Species

The rates of food consumption and the selection of animals of different sexes and age classes in the population are variables which influence the effect of predation by wolves on their prey. Additional factors which are also of importance are densities of wolf and prey populations and the density and quality of alternate foods which are available to the wolves (Pimlott, 1967).

Wolf–prey interrelationships have been studied in three areas in eastern North America where simple predator–prey systems exist (Isle

Royale, Algonquin Park, and Baffin Island) and where humans were exerting relatively little influence on either the predator or prey populations. The results indicate that wolves constitute a major mortality factor in such situations. They also indicate that a high percentage of the biomass of the primary prey contributes to the food of the wolf population, since the wolves are effective both as predators and as scavengers (Mech, 1966; Jordan, *et al.*, 1967; Pimlott, *et al.*, 1969; Clark, 1971).

The effect of wolf predation on more complex predator–prey systems such as those that were studied in Mt. McKinley National Park in Alaska (Murie, 1944) and the Rocky Mountain National Parks of western Canada (Cowan, 1947) is not clear. In the latter study it was reported that the wolves were not even removing the diseased and injured animals from the population. Intensive studies are being conducted in both areas at the present time. Preliminary results indicate that they will provide a better understanding of wolf–prey interrelationships in complex systems. However, in both simple and complex situations where human interference is minimal, it is evident that the effect of predation as a depressant on the population is minimized as a result of being concentrated on old animals and on juveniles which have not entered the breeding segment of the population (Pimlott, 1967; Jordan, *et al.*, 1971).

Where wolves exist in areas where large mammals are hunted by humans, the influence of wolf predation on the availability of deer to hunters is much debated. The work in Ontario suggests that human hunting is less selective for calves, males, and old adults than wolf predation (Hall, 1971; Kolenosky, 1972). Heavy hunting appears to result in big game populations which contain relatively few old animals. In such situations, it appears that predation by wolves in winter may be concentrated more on juveniles and on adult males in prime age classes.

It seems likely that in some situations where ungulates exist on marginal range and are heavily hunted (e.g., in Quebec and Ontario) that wolves are in competition and influence the availability of deer to hunters. However, in eastcentral Ontario where beaver were the primary food of wolves in summer, Kolenosky (1972) concluded that competition between sport hunters and wolves was not great.

POPULATION DYNAMICS

The density of wolves has been shown to vary a great deal throughout its range in North America (Pimlott, 1967). However, except for special stituation when their prey, for instance, caribou, are concentrated on winter range, it appears that the density of 1 wolf per 8–10 mile^2 (20.7–25.6 km^2) achieved on Isle Royale and Algonquin Park is as high as is likely to be tolerated by wolf society in the wild. The Isle Royale study is particularly informative since it represents a situation where both wolves and their prey were completely protected from exploitation by humans and where the moose population was very high. In that situation, the population averaged about 24 animals and increased to 28 in only 1 year. The area of the island is 212 mile^2 (544 km^2).

The capability of exercising control over their own numbers is one of the significant adaptations that wolves evolved over the centuries. The intrinsic mechanisms which appear to operate most effectively in populations which are not subjected to control by humans are a lower rate of reproduction and high mortality of pups and yearlings. Mech (1970) reviewed the existing data in some detail. The greatest body of data on the subject was provided from Alaska where over 4100 wolves that had been poisoned, trapped, and shot were examined (Rausch, 1967). The causes for the high mortality rates among pups and yearlings has not been demonstrated for wild populations, however, it is rather widely believed to result from social stress within the population. The data from Alaska (Rausch, 1967) indicate that a wolf population has a strong capability to adjust to mortality factors that result from human attempts to control or exterminate it.

SOCIAL ORGANIZATION, RANGE, AND TERRITORY

It was originally believed that the social organization of the wolf pack broke down in summer with members of the pack living in pairs or as singles during the period when the pups were being raised (Young, 1944). It is now evident from the studies in Mt. McKinley National Park (Murie, 1944), Algonquin Provincial Park (Pimlott, *et al.*, 1969), and Baffin Island (Clark, 1971) that the pack persists throughout the year. During the spring and summer, the home area, den, or rendezvous site, becomes the focal point around which the activity of members of the pack is centered.

The principal change that occurs is in the activity pattern of the members of the pack when away from the home site. Much of the hunting activity is conducted by subunits of the pack. These vary in size from a single animal to most or all members of the pack.

At the home site, no more than half the members of the pack may be present for days at a time; this appears to result from a fairly frequent change as members of the pack come and go.

The travel of the full pack is established by early fall when the pups begin to travel with the group. However, variation exists even during the winter and some packs appear to maintain a much tighter organization than others.

The difference between pack organization in summer and winter can probably be explained by the presence of pups that are too small to travel with the pack and by the fact that wolves prey on young animals and smaller prey during the summer. These smaller animals do not require the combined effort of all members of the pack to the same extent as does the killing of large, older prey during the winter.

Details of the social organization within packs have to a large extent been inferred from studies of wolves in captive situations (Schenkel, 1947; Rabb, *et al.*, 1967). It is widely believed that they are directly relevant to wild situations and some elements of organizational patterns have been reported for wolves on Isle Royale (Jordan, 1969; Wolfe and Allen, unpublished).

A wolf pack has a strong affinity for a specific area or range. Studies in Algonquin Park (Pimlott, *et al.*, 1969) and Isle Royale (Mech, 1966; Jordan, *et al.*, 1967; Wolfe and Allen, unpublished) show that packs are regularly spaced over the existing area. This fact and limited observations of agonistic behavior between groups has led to the assumption that territorial relationships, in the sense of an area which is defended as an exclusive preserve, exist between packs. There are, however, some records of wolf packs having met and parted without apparent conflict (Pimlott, *et al.*, 1969; Wolfe and Allen, unpublished). The best presumption at this time would appear to be that territorial behavior exists but it is subject to considerable variation depending on past relationships between individuals of the different packs and on the geographical area where interactions occur. It may be, for example, that there is defense of a central area of range but that packs of equal status make joint use of peripheral or boundary areas without serious conflict (Pimlott, *et al.*, 1969).*

There seems to be little substance to the belief that wolves travel over their ranges following a cyclical, predictable pattern (Kolenosky, 1972). Relatively little is known for certain about the size of home ranges of packs which live in areas where no physical barriers exist to their movements (Mech, 1970).

*Mech (personal communication) has observed interpack conflict during winter when food was scarce — Ed.

20 THE ESKIMO HUNTER'S VIEW OF WOLF ECOLOGY AND BEHAVIOR*

ROBERT O. STEPHENSON
Alaska Department of Fish and Game
Fairbanks, Alaska
and
ROBERT T. AHGOOK
Anaktuvuk Pass, Alaska

The present paper discusses some aspects of the Eskimo hunter's view of wolf ecology that augment the view held by modern science.* In particular, we will discuss the view of wolf ecology developed by the Nunamiut ("people of the land") Eskimo, now residing in Anaktuvuk Pass, Alaska. We will give only an elementary treatment of a subject that, to be fully understood, requires a thorough knowledge of Nunamiut ethnohistory and culture. Though the Nunamiut's view of wolf ecology is in most respects similar to that so far developed by western science, there are a few differences in general theory and many in details of natural history. We will deal with the former.

Accounts of the Eskimo's extraordinary practical knowledge and astute perception are numerous in the writings of early explorers, archaeologists, and anthropologists (Chance, 1966; Freuchen, 1915; Jenness, 1957; Spencer, 1959; Stefansson, 1919, 1922). The nature of his relationship with his natural environment has been effectively illumi-

*Information presented in this paper was obtained during studies supported by Federal Aid in Wildlife Restoration Funds, Project W-17-3.

nated by Nelson (1969) in his study of sea ice lore and zoological knowledge among Alaskan coastal Eskimos. As Nelson shows, ". . . there is no 'mystical inherited germ' in the Eskimo's mind that allows him to sense the mood of an animal, to anticipate the fickle movements of the ocean ice, or to sense a change in the weather." Through daily, functional interaction with the environment on which he has been vitally dependent, the Eskimo has acquired an impressive store of empirically derived knowledge. The Nunamiut's veracity with regard to matters of animal ecology has been relied upon and warmly praised by zoologists (Irving, 1953, 1958, 1960; Rausch, 1951, 1953).

Prior to 1949, the Nunamiut mode of life was seminomadic. Since then this group has become almost sedentary and in the last decade the economy of the remaining mountain Nunamiut, now numbering about 115, has been converted from one of mobile hunting and trapping to an economy of localized sporadic hunting, trapping, and odd jobs though still within the framework of traditional Nunamiut culture (Gubser, 1965). Formerly, hunters roamed the country in search of game almost daily even in summer; the winter months were and still are characterized by hunting and trapping over an even wider area. Prior to bounty hunting, which began in earnest in the late 1930s and continued until 1967, wolves were sought almost exclusively during winter. The bounty prompted an additional annual effort to obtain wolf pups during summer. Each year in late May, about four groups of from 2 to 5 hunters set out during the period of 24 hour daylight, traveling and observing at night. They searched for dens in an area of about 20,700 km^2 in the northcentral Brooks Range until early July.

The experience of the Nunamiut, in terms of time spent studying the wolf and its environment, is probably unparalleled in other aboriginal peoples or in western science. Although the northcentral Brooks Range supports a population of wolves comparable in density to the timbered areas further south, the terrain is open, offering excellent opportunities for observation. This and other circumstances mentioned above have enabled the Nunamiut to witness the full gamut of wolf natural history events under virtually all types of environmental conditions. Many hunters have demonstrated a remarkable ability to distinguish, at a distance, wolves of different sexes and ages, thus facilitating their interpretation. The validity of the resulting view of wolf ecology (i.e., the ability to interpret aspects of wolf behavior) is well-substantiated by the outstanding success of the Nunamiut as wolf hunters and the sophistication of their hunting methods.

Nunamiut show an impressive ability to predict the behavior of wolves. For example, by imitating certain wolf howls, hunters routinely call wolves to within rifle range from distances of 1.5–6.5 km or more.

They also stalk sleeping wolves. On several occasions hunters have crept within rifle range and killed as many as 5 wolves before they could escape. Using these and other hunting techniques (including steel traps), some hunters have taken as many as 40 wolves in a single winter during periods of high wolf population density. Several older Nunamiut, now in their sixties and seventies, have taken roughly 500 adult wolves in their 50 or more years of hunting; some younger hunters in their thirties and forties have taken 200–300 wolves. The ability of the Nunamiut to locate wolf dens has been amply demonstrated by the location of well over 100 dens in the last 30 years. Essentially, the den locations were deduced from observations of movements and behavior of adult wolves on hunting forays miles from the den, based on knowledge of hunting activity patterns.

The Nunamiut view the wolf as a highly intelligent predator that, with the sole exception of the occasional rabid animal, poses no threat to man. They are, in fact, quite amused by the tales of bloodthirsty wolves lurking beyond the glow of a trapper's campfire. There are, however, a few accounts of wolves attacking Nunamiut traveling alone or in small groups prior to the introduction of firearms in the late 1800s. The intelligent and social nature of the wolf is cause for real admiration among the Nunamiut. One of the first things they point out to outsiders inquiring about wolves is that they are a very smart animal. It is instructive to note that one of the very few remaining societies that has been, and to a considerable degree still is, in direct competition with the wolf for essential food items harbors no animosity toward the animal. The Nunamiut do not begrudge the wolf its prey. Though they hunt the wolf and value his fur, they take his life without the hate, rancor, or guilt that has so often accompanied the killing of wolves by other human societies.

The Nunamiut recognize broad variation in the physical and behavioral attributes of wolves. Some variation is correlated with sex and age and environmental conditions, but a good deal is attributed to inherent individual differences in morphology and temperament. The Nunamiut learned long ago what has been recently stated by Ayala (1971): "Species are not monolithic entities composed of identical copies of the same model; rather, in species that experience sexual outbreeding, no two individuals, with the trivial exception of monozygotic twins, are likely to be genetically identical." They often refer to *that* wolf or wolves under *those* particular environmental conditions, using the collective "wolf" much less often than does the modern wolf ecologist. Though individual variation in temperament is known to scientists from studies of captive wolves (Rutter and Pimlott, 1968) it has been extended only rarely to interpretations of behavioral phenomenon among free-roaming wolves. Inherent temperamental variability is a subtle phenomenon to the unpracticed eye but is employed by the Nunamiut to partially explain the diversity observed in many categories of wolf behavior including response to man, alertness, hunting ability, sociality, aggressiveness, and so on.

For example, from literally hundreds of encounters with wolves at natal dens, Nunamiut know there are real differences in the individual responses of adult wolves to human intruders. Mothers, for instance, usually evidence highly anxious behavior, first barking and then howling at intruders, staying within a few hundred meters of the den. In some cases, however, females have simply left the den area without vocalizing or otherwise behaving anxiously. Male wolves, on the other hand, characteristically show less anxiety than females, moving further from the den, vocalizing less, and sometimes leaving the area entirely. However, some have responded to intrusion in a manner similar to the "typical" female parent.

Inherent physical and temperamental differences, modified by experience, probably also explain many of the differences the Nunamiut have observed in the roles wolves play in killing ungulate prey. From the countless wolf–prey encounters they have observed, including repeated observations of certain packs, the Nunamiut know that not all wolves play equal roles in killing prey. They say, as a rule of thumb, that usually 2 wolves in a pack of 5 do the great majority of killing, the remaining wolves playing peripheral roles in the hunt. The very young and very old, of course, rarely take an active part in killing large prey.

The Nunamiut also stress the importance of recognizing the wolf's capability for high order behavior, showing both insight and purposiveness (Fox, 1971e) and his great learning capacity. One unpleasant experience with a given phenomenon, such as an encounter with a hunter, is said to be sufficient for a wolf to avoid any similar situation in the future. Adults having one experience with traps are known to be very difficult to trap and will not permit their pups to visit, for instance, a carcass that man has tampered with. Most Nunamiut agree that it is nearly impossible to capture a really experienced wolf. Examples of high order behavior in wolves are many in the Nunamiut experience and have been observed in varied behavioral contexts. The use of fairly elaborate "ambushes" in hunting is a notable example. Another is the response of yearling or adult wolves to the presence of humans in the vicinity of a rendezvous site. On one occasion, two Nunamiut watched a yearling wolf rush to a rendezvous site and lead a litter of pups away upon discovering the hunters about 1.5 km from the pups. On the same occasion, just prior to the wolf's discovery of them, the hunters watched two adults, including the parent female, leaving the rendezvous ste on a hunt. When one pup attempted to follow the adult wolves, the female led it back to the area where where the other pups waited, then again left in the direction the other adult had taken. After going only 90–100 meters, however, she lay down for about a half hour, then rose abruptly and left, following the trail of the other adult. The Nunamiut surmised that the female wolf had waited in case the pup should again decide to follow.

In assessing the ability of wolves to capture ungulate prey of various sexes, ages, and in various states of health, the Nunamiut point out that one of the more important elements bearing on the outcome of a wolf–prey encounter is the wolf's determination to capture the animal. In the case of caribou, for example, Nunamiut hunters agree with the assertion that wolves kill a good number of "inferior" animals and are beneficial in maintaining vigor in caribou populations (as well as in sheep and moose populations). They also adamantly maintain that an experienced adult wolf in reasonably good condition can capture any caribou it decides to, the sole exception being an animal that takes refuge in deep water. It is during summer and winter, when caribou are present in moderate or low numbers in the northcentral Brooks Range, that the wolf most often displays its ability to catch even the healthiest caribou. In recent months we have documented successful chases of 7 and 13 km. Each involved one wolf and one caribou. In both instances, caribou were present in very low numbers. The 7 km chase, involving an adult bull caribou, occurred on July 30, 1971. The latter event will be recounted in some detail since it provides another interesting example of high order behavior.

On March 21, 1970, Ahgook and another Nunamiut hunter followed the fresh trail of a chase involving one wolf and one cow caribou across the foothills 64 km northeast of Anaktuvuk Pass. Both wolf and caribou had run for 10 km across an area of windswept tundra and packed snow. Then, upon entering a 3 km expanse of soft snow about 75 cm deep, both animals had slowed to a walk. While traversing this deeper snow they alternately ran and walked, changing pace about every 90–100 meters. There was no indication that the wolf attempted to kill the caribou as they crossed this area of deeper snow but, instead, simply kept pace with the animal. When the caribou left the deep snow and ran down a windswept slope, the wolf overtook and killed the animal in a short, 70 meter chase. The entire chase covered at least 13 km (from the point the hunters intersected the tracks to the kill). This example illustrates the longer distances that wolves may chase caribou when only a few, healthy animals are available and also shows how they may purposefully adapt their pursuit to topographic conditions. The Nunamiut have witnessed even longer chases.

These observations differ from the short chases of a few hundred meters or less (Crisler, 1956) which are often cited as evidence of the wolf's inability to catch other than "inferior" animals. Most of Crisler's observations were made at times when caribou were relatively abundant, during migrations in spring and fall. In addition, some of her observations involved young human-reared wolves. In this connection it should be mentioned that the Nunamiut are quite certain that wolves, particularly the younger and more energetic individuals, do not always pursue caribou with an intent to kill. Many chases have been witnessed, primarily during times of caribou abundance, in which it appeared to the Nunamiut that the wolves were "playing." They have often watched a wolf cease pursuing a caribou, just when it appeared to have a good chance of capturing it, only to then begin chasing another caribou.

The Nunamiut have occasionally witnessed wolves killing wolves; this seems to occur mainly during years of high population density. Also trapped wolves may be killed and sometimes eaten by others, even by members of their own pack.

It would seem to have been an uncommon practice for Eskimos to keep wolf cubs as pets, nor were any concerted efforts made to crossbreed their own dogs with wolves.

The above discussion shows that in the Nunamiut view the substance of wolf behavior is comprised of the interaction of many individual behavior patterns, usually in the context of a social order, with environmental conditions and learning continually modifying behavior. As a result of the vast array of behavioral events they have witnessed, the Nunamiut interpret wolf behavior in a broader, yet more intricate theoretical framework than that heretofore used by modern science; their in-depth knowledge gained from patient, on-the-ground observation has taught them that the adaptability and elasticity inherent in wolf behavior rivals that in human behavior.

21
WOLF ECOLOGY IN NORTHERN EUROPE

ERKKI PULLIAINEN
Univ. Helsinki
Helsinki, Finland

HISTORICAL BACKGROUND

During the deglaciation phase of the Würm glaciation wild reindeer followed the retreating ice margins. Wolves (*Canis lupus*) pursued the dispersing reindeer herds. Middendorf (1875) reported that in the 19th century in the tundra zone of northern Eurasia the distribution of the wolf was limited by the distribution of the semidomestic and wild reindeer. During the last hundred years, man has become a dominant factor in the population dynamics and distribution of both the wolf and the reindeer.

RECENT HISTORY AND STATUS OF THE WOLF IN NORTHERN EUROPE

The present range of the wolf in northern Europe is very clearly bipartite. Within the area of western culture, the species has been exterminated almost entirely, whereas in eastern Europe the wolf populations have shown the opposite trend during the same time.

In the late 19th century, the wolf still inhabited vast areas in Fennoscandia (Norway, Sweden, and Finland), Russia, and the Baltic countries. In the 1800s, however, great changes took place in the Fennoscandian wolf populations. Official statistics of the wolves killed have been regarded as a fairly reliable index of these changes (Pulliainen, 1965; Johnsen, 1929).

According to the Swedish statistics, the first marked decrease in the number of wolves killed took place in the 1830s and 1840s (Lönnberg, 1934). Twenty years later a second decrease took place. After this, the numbers of wolves killed fluctuated within relatively narrow limits. In Norway, a remarkable decrease parallel to the second Swedish decrease took place in the 1860s (Johnsen, 1929). After this, the situations in the two countries were similar. In Finland the first significant decrease took place in the 1880s. Subsequently, the corresponding numbers of wolves fluctuated fairly evenly from year to year. At times there have been small increases, as in the beginning of the 1960s.

Thus, in Norway and Sweden the last marked decrease in the numbers of wolves killed took place simultaneously, and in Finland some 20 years later. The causes which led to these changes in the wolf populations were common to all these countries. The most important factor was man and his activities. When the populations of the larger predators were dense, these animals killed great numbers of domestic animals yearly. It is natural that in these conditions the state paid a bounty for wolves shot (Lönnberg, 1934). Many decades of keen hunting produced a result; i.e., their numbers decreased strongly and did not increase again remarkably. However, small populations still survived in the vast forests and on the fells and mountains of Fennoscandia. The expansion of settlement and agriculture was also a factor aiding in the extermination of the large predators, especially the wolf.

The last surviving Finnish wolves retreated to the most suitable habitats in Lapland where food resources and snow conditions were favorable. Also in this area, man very effectively regulated the density of the wolf population. The Lapps knew the wolves of the district individually by their tracks. By destroying wolf dens and cubs, they kept the wolf population below a certain limit. They could distinguish the sexes from the size of their tracks, and did not kill the females because they then had lost the easy bounty for the cubs. For this reason the mean age of wolves increased.

In the late 1960s the total amount of wolves in Norway and Sweden did not exceed 15 individuals (Myrberget, 1969). In Finland there were 10–20 wolves which mainly inhabited the frontier zone between Finland and the U.S.S.R. Because the species had already been exterminated in western and northern Finnish Lapland, the above mentioned wolf population of the Scandinavian mountain range had been isolated rather well from the wolf population of the U.S.S.R. (Haglund, 1968).

Before the First World War people in both Eastern and Western Europe took action against wolves. From the west to the east, wolves were gradually exterminated (Kneževič and Kneževič, 1956). Since the First World War, great changes have taken place in the political structure of Europe. The centralized economic system dominates in the east. In the west, traditional free enterprise dominates. When the political system came into being in the Eastern European countries, great structural changes (Kauri, 1957) took place in the society. Factors of biological and ecological importance were the changes in land ownership, agriculture, and hunting, for instance. Hunting as a sport and trade ceased to exist. Hunting under state control took its place. Government hunters were mainly interested in furbearers, and little attention was paid to wolves and other predators. Land ownership and game management were the province of the state. The status of the individual as a sportsman and hunter had changed. It is not surprising that in these conditions game tended to increase. Increases in populations of small game tend to be followed by increases in the populations of their predators. Since the destructive effect of man on predator populations was absent, the changes of such an increase were perhaps optimal. Besides wild animals, the predators could use as their food domestic animals, such as cattle (Kneževič and Kneževič, 1956). As the numbers of predators increased, the population pressure naturally increased and the species widened their ranges. It is to be noted that this took place simultaneously throughout the Eastern bloc (see the review by Pulliainen, 1965). The expansion of the species took place into its former ranges. This expansion was also observed on the eastern frontier of Finland in the late 1950s and early 1960s.

The increase in the numbers of wolves in the U.S.S.R. led to war against wolves in the 1950s and 1960s. The government used helicopters and other machines and soldiers in hunting wolves. By the 1960s this activity had significantly decreased the numbers of wolves at least in the Baltic countries and in the eastern parts of the U.S.S.R. (Harry Ling, personal communication). This trend was also reflected in the decreased movements of wolves on the frontier between Finland and the U.S.S.R., for instance. However, there is still a viable wolf population in the northern parts of the U.S.S.R. in Europe.

EXPANSION PATTERNS (RECOLONIZATION)

In the late 1950s and early 1960s an expansion of wolves from the east could be observed on the frontier between Finland and the U.S.S.R. in North Karelia. Studies carried out by the author on this recolonization gives a good picture of the expansion–movement patterns of the species (Pulliainen, 1965).

Before the Second World War, the wolf bred at Pieninkä, a vast virgin forest in the area of Olonetz and Onega, behind the eastern frontier of Finland. After the Second World War, Finland ceded large areas in Karelia to the U.S.S.R. Most of the ceded land remained neglected, fields and meadows returning to forest. These areas became very suitable for game. In these conditions the wolf population of Pieninkä, as well as the more eastern wolf populations, increased and spread westwards. Before the Second World War, the wolves of Pieninkä extended their winter migrations to the north coast of Lake Ladoga. As the range of wolf population widened, these animals reached the area ceded after the Second World War. There they found a very plentiful game population, and very soon occupied the area. Then (in the beginning of the 1950s) the wolves migrated to Finnish North Karelia in winter. During the years 1956–58, the winter migrations of the species to this area increased very markedly. In 1959, wolves were seen in the neighborhood of the eastern frontier before the summer was over. Perhaps they had been there earlier, but had not been observed. In 1960, the migrations clearly increased. This could be seen from the increase in the numbers of domestic animals killed by wolves, the increase in the number of sightings, etc. In 1961, the expansion was so strong that it could be followed throughout the year and compared with the annual rhythm of the wolf.

During the years 1956–58, the population was still a considerable distance from the frontier and only in winter, the season when the wolf's range is at its widest, were wolves seen on the frontier. In summer, 1959, wolves were seen on this side of the frontier, which was connected with the widening of the range in late summer. In 1960, the numbers of wolves were greater and they appeared earlier than in the previous year. In 1961, the population was so near the frontier that some wolves were observed in June, and in August the females took their cubs with them on the Finnish side of the frontier. During the summer, the range of the wolves continuously expanded in North Karelia. At the same time, there was a great increase in wolf-hunting. Partly owing to man's activities and partly owing to population pressure, wolves migrated northwards from North Karelia. Effective hunting produced a result — in summer 1963, only a few domestic animals were killed by wolves in North Karelia. Likewise, the observations on wolves decreased. Here again man demonstrated that he can regulate the numbers of wolves.

BREEDING BIOLOGY

According to Ognev (1931, 1959), pairing begins in the middle of January when the young females are expelled from the pack. Munsterhjelm (1946) and Ognev (1959) reported that fighting takes place or at least begins in February. In eastern Finland (North Karelia), copulation took place on March 9 and in northern Finland around April 8. Palmgren (1920) mentioned that in southern Finland the copulation of penned wolves took place on the following dates: February 15 and 25, and March 20. The corresponding dates (which were calculated from the birth dates) recorded by the present author are February 22 (1 case), February 28 (1 case), March 13 (2 cases), and March 21 (2 cases). Thus, it seems that copulation mainly takes place at the end of February and in March in southern Finland, but in northern conditions it may even be postponed to the beginning of April. Similar observations were reported by Haglund (1968) in Sweden.

Wolves are mature at the age of 22 months at the earliest (Makridin, 1959). The young females have their first heat before the young males show any overt sexual behavior (Ognev, 1931). The older wolves generally have an earlier heat than the younger ones (Ognev, 1959). Wolves may have very violent rivalry fights, which are generally won by the older males (Munsterhjelm, 1946).

The average number of cubs of European wolves is usually 5–6 (Ognev, 1931, 1959; Kneževié and Kneževié, 1956). The corresponding figure for the wolf population of northern Finland, however, was only 2.8 (Pulliainen, 1965). Individuals of this population were also reared in captivity. The mean number of cubs of 6 progeny was 3.8. It is very difficult to say to what extent the high average age of the wolf population, coupled with a low population density and inbreeding, have led to this lowered fecundity in this wolf population.

The position and structure of the den depend on the nature of the ground in the locality. If possible, the wolf digs a new den every year in sandy ground. The wolf can also use old dens of the red fox (which it enlarges), or rock caverns, and dens may even be under bushes. In the fell region, wolf dens are generally in the valleys, not on the tops of the fells. In Finnish Lapland, wolves mostly dig new dens every year, but some wolves have taken over the old dens of the red fox (Pulliainen, 1965).

According to Ognev (1931, 1959), in Eurasia both parents take care of the young. Pulliainen (1963) pointed out that in certain conditions the female can take care of the young alone. Ognev (1931) reported that for the first few days after the birth of the cubs the female is fed by the male. During suckling, the female remains within a radius of about 2 km from the den (Pulliainen, 1965).

When suckling ceases, the cubs are fed with vomited matter (Siivonen, 1956). The observations made by Pulliainen on penned wolves showed that at the age of 6 weeks the cubs ate fresh fish and milk very eagerly (Pulliainen, 1965). As the cubs grow up, their range naturally continuously expands (Kneževié and Kneževié, 1956; Ognev, 1959). At the age of 3–4 months, the cubs accompany their mother when she is in the surroundings of the den. However, at this age the den still serves as a hiding place. At the age of 5–6 months, cubs feed at the place where the prey is caught. They begin to catch for themselves at the age of 7–8 months (Pulliainen, 1962, 1965).

In September, the female and her cubs keep together and form the natural basis for a pack, the female being the leader of such a unit (Pulliainen, 1965). The autumn howling of wolves serves to link up the members of the pack. This concerns both the female and cubs, and the lone wolves of the previous summer. A remarkable increase in the sociability of the latter is to be noted. Later on, such lone wolves may join a pack earlier formed by a family (Pulliainen, 1965). During the winter months, the sociability of the wolf is so great that even though the heat fights, for instance, break up the packs temporarily, the wolves have a great tendency to join together again.

FOOD AND FEEDING HABITS

The recent studies (Makridin, 1959; Pulliainen, 1965; Haglund, 1968) carried out on the food ecology of the wolf have shown that there are significant differences in the composition of the food between different areas. Both availability and active selection are involved in this respect. Pulliainen (1965) studied the food and feeding behavior of the wolf in both northern Finland (on the range of semidomestic reindeer) and in more southern areas where there were, however, a few wild reindeer available. The following prey animals were identified in the stomach contents of wolves killed in winter in eastern Finland, where no semidomestic reindeer were available: the red fox, moose, dog, small rodents, black grouse, and other tetraonids. In addition, the wolves had fed on different kinds of carcasses and on garbage heaps. Especially in summer, the wolves killed a lot of domestic animals (dogs, cats, horses, cows, and sheep). These animals were even caught by wolves in the yards of houses. In the statistics of 235 prey animals killed by wolves in North Karelia, eastern Finland, the brown bear, moose, snow hare, squirrel, and tetraonids represent wild animals (Pulliainen, 1965). In Ostrobothnia, western Finland, the wolves had killed at least 2 moose, 5 calves, 30 sheep, 7 dogs, 2 hares, and 2 black grouse. During the years 1959–63, the great predators (mainly wolves) killed at least 1092 domestic animals in North Karelia. The bulk of these animals comprised sheep (86.3%). In addition, horses (0.4%), cows (3.6%), heifers and young bulls (2.0%), calves (7.4%), and pigs (0.3%) were represented.

In the area of reindeer husbandry in Finland, semidomestic reindeer comprise 96% of the prey animals killed by wolves in winter (Pulliainen, 1965). Makridin (1959) identified reindeer in 93% of 74 stomach contents of wolves originating from northeastern Europe. Besides reindeer, moose, snow hares, red foxes, small rodents, tetraonids and other birds, and others belong to the winter diet of the wolf in Finnish Lapland (Pulliainen, 1965). Some of the wolves inhabiting the eastern parts of Finland seem to prefer feeding on carcasses (Pulliainen, 1970). Haglund (1968) tracked a pair of wolves and a single wolf for 147 km in Sweden. Those hunts which were controlled by tracking, 6 in all, were brief and ended within a kilometer. Only one of them was successful. Haglund (1968) reported that on 5 different occasions wolves returned to previously-killed quarry.

Pulliainen (1965) found that in most cases the wolves ate wild animals killed by them entirely, while domestic animals were generally only partly eaten. About 20–30% of the domestic animals (sheep, calves, cows and horses) killed by wolves in North Karelia in 1962 were eaten entirely. The small domestic animals like cats and dogs were, however, generally eaten entirely. Haglund (1968) estimated that some hours after slaughtering, two wolves had eaten an adult cow reindeer during a single meal, approximately 15 kg of their prey's flesh. Makridin (1959) mentioned that the wolf digests all its food during a few hours so that only keratin (e.g., hair, hoofs, feathers, and claws) and tendons remain unchanged.

Pulliainen (1965) described 5 cases when the wolves had used runways or hunting routes in Finland. When traveling, those wolves kept to a certain timetable. The mean distance traveled by wolves per day in winter was 23 km. In South and West Finland, wolves have used the same migration routes for decades. The interval between two observations at a certain place might be several decades. However, the wolves observed used exactly the same route when traveling, even to within a few meters.

RELATIONS WITH NONPREY SPECIES

Pulliainen (1965, 1968) showed that when the population density of the wolf was great in Finland, wolves kept the population (density) of the lynx (*Lynx lynx*) low. After the decrease of the wolf population, the population (density) of the lynx increased. The predator–prey relationship between these two species was thus very clear.

Pulliainen (1965) reported a case when wolves had eaten a brown bear (*Ursus arctos*). Besides this predator–prey relationship, general interspecific intolerance was reported to occur between these two species. A similar conclusion was drawn on the relations between the wolf and wolverine (*Gulo gulo*). Later studies carried out by the present author have shown that the wolf, bear, and wolverine may also inhabit the same large area at the same time. However, because of the very low population densities, individuals of the different great predators meet each other only rarely (Mech, 1970).

WOLF–MAN RELATIONS

The wolf and man are competitors at the upper end of natural food chains. The recent history of the wolf in all parts of its range tells about this competition and defeats of the wolf in the fight. In the motivation of this war, a lot of attention has been paid to the attacks of wolves on humans. In most cases there have been evidences that the wolves in question have been rabid (Rutter and Pimlott, 1968). Fennoscandian literature documents two wolves which have killed humans (Godenhjelm, 1891; Marwin, 1959). Suomus (1970) reinvestigated the data published by Godenhjelm (1891) and concluded that during the years 1880–1881 a single wolf killed 22 children in southwestern Finland. This wolf was hardly rabid, because it lived so long. In general it can be said, however, that wolves have never been a real threat to man in northern Europe (Rutter and Pimlott, 1968).

PART

SOCIAL BEHAVIOR

There are many limitations in both field and laboratory studies concerning social behavior and these are clearly spelled out in this section; in fact, field and laboratory studies are complimentary and it is important for the laboratory worker to be familiar with what is known about the animals he studies under natural conditions.

Many questions that arise from field studies can only be answered by studies of captive animals; extraneous variables can then be better controlled, and social groups can be manipulated in various ways (e.g., by removal of an animal of known rank or introduction of a stranger). Many field workers would benefit from observing wild animals at close quarters in captivity; often in the field, distance alone makes interpreting behavior extremely difficult and some prior knowledge at close proximity would help. One field worker tells me that there is no aggression in the wolf pack, a conclusion drawn from direct observation from an airplane — a vantage point from which a low threat growl and a stare could certainly be missed!

In this section, a variety of different field and laboratory studies are described. They provide a good survey of the various methods and techniques of recording and analysis of selected aspects of social behavior in wild and captive animals. Although the main focus of this book is on wild canids, two field studies of domesticated dogs (*C. familiaris*) that are now feral and/or free-roaming are included; such animals are becoming a social and ecological problem in the United States and in other countries. Their study may not only lead to improved control measures but also to an understanding of behavioral adaptation.

22 A LONGITUDINAL FIELD STUDY OF THE BEHAVIOR OF A PAIR OF GOLDEN JACKALS

I. GOLANI and A. KELLER
Department of Zoology
University of Tel-Aviv
Tel-Aviv, Israel

This paper presents a longitudinal field study of the behavior of one pair of golden jackals. It describes their behavior in some detail during a period of 2 years and compares it to the behavior of other pairs and individuals observed simultaneously in the same population.

This manner of presentation was chosen because any premature attempt at a generalization masks the variability and the differences between interactions, individuals, and pairs. Since individuality and variability are of some importance in biological systems, one way to present them is to proceed carefully from lower to higher orders of descritpion, i.e., from the desciption of specific interaction sequences to generalizations on the behavior of specific individuals during specific periods, and only then to generalizations on the behavior of individuals, pairs, and populations. Since we are concentrating on the subtle intricacies of one pair's interactions, some readers might find the presentation too detailed. We are also not in a position to write about jackal social behavior in general terms at the present state of our study.

Our observations were carried out on 17 identified individuals during the years 1968–1970 on the sand dunes of the Israeli coastal plains.

Six out of the 17 individuals formed three pairs which we observed longitudinally for at least 1 year. These three pairs, which formed during our study, were our main subject. We saw how they were formed of individuals whom we knew separately; we recorded their interactions with other individuals and pairs, and among themselves; we saw them court, copulate, and in two pairs out of three, bring up their offspring.

Jackal pairs were the most durable social entity observed during our observations. Specific additional individuals often followed two of the pairs for a period of days, weeks or months, but subsequently disappeared and a new individual or individuals soon joined the pair. We never observed a durable group which did not include a pair of adult jackals. All these facts make the pair and the animals that followed it a natural social group.

The study area was a small area of dunes south of Tel Aviv of about 250 acres (100 hectares), bordered by the Mediterranean on the west, the Soreq River on the north, and Kibbutz Palmachim on the south. Our observation point was on the western side, on a high hill which views most of the area (Figure 22-1).

Jackals' hiding places and dens were situated in the thick vegetation along the river. The main feeding sources were the garbage dump of the kibbutz which was situated about 1 km away from the study area, edible refuse deposited on the beach, and grapevines which grew during the summer on the sand dunes.

SUBJECTS AND METHOD

The subject of the present paper is *Canis aureus syriacus* Hemprich and Ehrenberg 1833 which inhabits Lebanon, Syria, Israel, parts of Sinai,

Figure 22-1 The area where field studies were carried out. **A** = first resting place of Gingi and Belly. Later occupied by Handsome and Elongated. **B** = second resting place of Gingi and Belly. **C** = area most often visited by Timid and her male. **R.D.** = river

and Jordan. This animal looks completely different from the animals which were the subjects of field studies carried out in East Africa (Wyman, 1967; van Lawick-Goodall, 1970). The East African golden jackals (*Canis aureus bea,* Heller) are much more slender and have longer legs, tails, and ears and have wolflike faces. No doubt the taxonomic status of the jackals united by Ellerman and Morrison-Scott (1951) under *Canis aureus* should be reevaluated (H. Mendelssohn, personal communication).

Our observations were made on the following 17 individuals (Table 22-1) and an unspecified number of unidentified individuals.

Observations were made during late afternoons and early evenings with the aid of 20 × 120 Kowa binoculars and a tape recorder from a conspicuous observation point on top of the above-mentioned hill. They were carried out from December, 1967 to March, 1971 with a gap from May to November, 1968. An average of 2–3 observations per week were made with a total of 192 observations. Each observation lasted 2–3 hours. Recently, observations were again started on a regular basis.

At the beginning of our study, the animals either emerged when it was almost dark or avoided, at least before dark, the bare sand dunes. Later they got used to us and emerged as early as 2 hours before sunset. This was probably due to the fact that we introduced into the area a few loudly peeping chicks before each observation. When the jackals heard the peeps they came running from as far as 600 m away, swallowed the chicks, and stayed in the area. In a few months time, the animals got used to us and to the area and emerged early, also in the absence of chicks. As far as we could judge, they almost ignored our presence. The introduction of the chicks changed the

delta, situated 200 m to the left. **T.D.** = Timid's den, situated 80 m to the left. **H.D.** = Handsome's den. Picture was taken from observation point.

TABLE 22-1 List of Observed Individuals and Period of Observation

Name	Sex	Period of Observation	Reference to Other Individuals
1. Gingi	Male	December 1967–July 1970	Adult pair
2. Belly	Female		
3. Handsome	Male	November 1969–February 1971	Adult pair
4. Elongated	Female		
5. Timid	Male	February 1970–August 1972	Adult pair (were not observed during 1971 and the beginning of 1972). Female was a follower of Gingi and Belly until she got engaged.
6. Timid	Female	December 1969–August 1972	
7. Red	Male	February–March 1969	Young pair. Interacted with Gingi and Belly.
8. Light	Female		
9. Ear Cut	Male	May 1970–February 1971	Young pair. Male offspring of Handsome and Elongated.
10. His Female	Female	February 1971	
11. Tail Cut's Male	Male	February 1971	Young pair. Possibly one of them or both offspring of Handsome and Elongated.
12. Tail Cut	Female		
13. The Urinating Male	Male	December 1969	Young pair. Interacted with Gingi and Belly.
14. The Urinating Female	Female		
15. Spot	Male	June–September 1970	Adult pair. Male was follower and feeder of Handsome's offspring.
16. His Female	Female	August–September 1970	
17. The Dark Female	Female	February–March 1969	Follower of Gingi and Belly.

spatiotemporal habits of the animals but jackals are nevertheless attached to human settlements and a large part of their diet is based on food expelled to garbage dumps. Our 4–5 chicks did not affect their diet appreciably but were still an attractive factor. Also, we introduced the chicks because our other alternative was to observe jackals in the dark or in dense vegetation which was practically impossible.

For the meaning of behavioral terms used in the present paper, readers are referred to a previous publication (Golani and Mendelssohn, 1971).

The following descriptions are verbal transcriptions, recorded in field conditions, of interactions which sometimes occurred between as many as five jackals simultaneously. It should be kept in mind that the results are essentially an oversimplification of tremendously complex interactions. Even the most trained observer will miss much more than he records when confronted with so many events occurring simultaneously. How much more so when the tools for description are words which are essentially incompetent for the description of movement. Suffice it to say that during a previous study (Golani, 1973) as many as 10–21 relevant events were simultaneously recorded for each second from films of two pairmate jackals performing precopulatory behavior. In another study, which used the Eshkol–Wachman movement notation for the description of the same interactions, as many as 22–57 relevant events were recorded simultaneously for each 1/6 second (Golani, Eshkol, and Zeidel, 1969; Golani, in press).

The following descriptions are, therefore, only clues as to what really happened.

GINGI AND BELLY

Pair Formation

At the end of 1967, after a few observations on an unidentified 12.25.67
group of four young jackals, two specimens of the group attracted our attention. One was a red male with short fur and the other, a light-colored female. They seemed to form a pair since they groomed each others' fur and urinated once in succession at the same spot, two behavioral interaction patterns which are typical of pairmates.

A fifth young female with a large belly approached the group and stayed somewhat away. The others ignored her, thus indicating to us that they knew her from before.

When we saw the group again a month later it included 2.3.68
the red male, the female with the belly, and two 1 year old males. This time after the male urinated, the two young males approached, sniffed the ground in succession, and one of them lifted a hindleg without urinating. The female urinated by flexing a hindleg. Afterwards, the group split and the two young males took a different common route.

2.23.68 Three weeks later the group included the red male, his supposed mate — the light-colored female, the female with the belly, and a young male. The fully-developed fur of the light-colored female with its contrasting shades of black and white strengthened our belief that she was the red male's pairmate, as usually only paired animals have such a fur.* However, when the red male passed near the female with the belly, she suddenly became rigid and performed tail-flagging,** a pattern performed by pairmates during courtship. This, for us, was the beginning of a 2½ year acquaintance with the pair later known as Gingi and Belly.

3.1.68 The next time they were seen together Belly was in the lead, every now and then pausing, looking back and proceeding when Gingi approached. At that time of the year, pairmates would usually urinate extensively, often at the same spot. However, during the common trot, only Gingi urinated by fully extending a hindleg whereas Belly only sniffed the ground, a fact which was in concert with her dull fur and the recentness of her pair bond to Gingi. When Belly stopped and sat down, Gingi approached her while raising his tail, groomed her neck with his teeth, and sniffed gently the corners of her mouth.

Pair Followers

4.3,8,10.68 The trot by the two of them did not last very long since during the following month a female with a crouching walk was usually observed to follow the pair. On approaching Belly, she would wag her tail, turn her ears backwards, crouch, or sit. Then Gingi and Belly would groom her. Sometimes when both females approached the male, they both wagged their tails. Gingi bared his teeth and approached the second female who, by that time, lay on her side while extensively wagging her tail.

Later, it became very common to see the pair followed by a specific individual or individuals but Belly's response to these followers changed: she did not wag her tail anymore upon meeting them and very often performed a spinning jump on her forelegs and hit them with her hindquarters (see Fig. 11, in Golani and Mendelssohn, 1971).

At that point, observations were stopped for 7 months and Belly
12.19.68 was identified only 8 months later in a group of six jackals.
The group formed during the observation out of two separate pairs

*Both in the field and in captivity there was a marked difference in the appearance of paired versus unpaired jackals. Paired animals have thicker, glossier fur with greater color contrast (see Fig. 44 in Golani and Mendelssohn, 1971), whereas unpaired animals have a dull pelage.

**Previously termed tail-waving.

and two separate individuals. Entrance into the group was accompanied by extensive tail-wagging and body undulations and later, all the jackals of the group except Belly participated in mutual grooming. Because of the great distance, we could not tell whether Gingi was there too.

Only 2 months later, we saw the pair together with a 2.3.69
third individual lying under a tree near the river. Gingi groomed Belly extensively. He started above a front leg, shifted to her waist, then to her throat, and when she turned over on her back, continued to groom her belly. When he stopped grooming they started to trot slowly, each leading alternatingly. The third jackal followed them for a few minutes and then trotted away. They urinated in succession at the same spot when suddenly Belly noticed a pair of young jackals, later recognized as Red and Light. Belly bristled, lowered her head, turned her ears forward and galloped in their direction, her tail wagging passively in a way characteristic only to her. Gingi followed her. When the young pair disappeared behind a hill, both pairmates stopped and urinated, an event which often followed chases and fights and which sometimes was accompanied by scraping.

It soon turned out that the young pair had been lying behind a hillside together with a third dark-looking female who stood and stared at Belly and Gingi. Belly noticed the female first and darted in her direction, Gingi following. The young pair who saw the chase joined in by closely following the dark female, who ran away, and finally lay on their sides a few meters away from each other, their tails between their legs. When Belly arrived, the dark female was engaged in sniffing the ground, unbristled with drooped tail. Belly walked among the three jackals, bristled and with puckered mouth, and then followed Gingi who proceeded trotting, completely ignoring all four animals. A few meters away they stopped and performed the usual after-chase behavior: i.e., they both urinated and Gingi also scraped. Upon continuing the trot, the dark female joined them and all three sat down about 50 m away from the young pair.

Howling

Without standing up, Gingi produced one long howl. There was no response. He stood up, urinated and walked away, the other two following. Again they sat. This time Belly started a howl, going through the long, continuous, monotoned howl, the alternating rising and falling howl, and the series of short, staccato howls. Gingi joined in during the first phase. Gingi and Belly stood up during the howl, whereas the other three, who joined in during the third phase with high-pitched voices, sat while howling. This is how young animals and followers usually howl.

At the end of the howl, Gingi and Belly approached each other and performed a short interaction sequence previously described (Golani and Mendelssohn, 1971) as related to precopulatory behavior — the so-called T-sequence. Upon approaching each other, they both raised their tails, Belly to vertical and Gingi to upward on a diagonal, and Belly formed the top of the T while tail-flagging. The T-sequence was very short and Gingi started to trot, Belly following and urinating every now and then at the same places where Gingi did. The dark female followed them.

Adult pairs often performed a T-sequence after a howl (for the details of after-howl behavior, see Golani and Mendelssohn, 1971, pp. 189). The T-sequence which was performed by Gingi and Belly was similar to other T-sequences performed by them during that season: relatively short, with tail-flagging by either one or both pairmates, and with Belly at the top of the "T".

Interaction Sequences

The beginning and end of interaction sequences between the adult pair and other jackals often occurred completely unexpectedly. This can also be demonstrated in the above sequence: the dark female who ran away from Belly with drawn back lips and ears stood and sniffed the ground with drooped tail when Belly reached her; Belly passed by her without interacting with her and Gingi passed by all four without interacting with them. Similar instances were also observed in captivity when, during a charge, the attacked jackal did not assume any so-called "submissive posture" but sniffed the ground or looked in a different direction. The attacking jackal passed near it without interacting.

Leaving or joining the adult pair was usually accompanied by various behavioral events such as tail-wagging, bristling, chasing, or hitting with hindquarters. Sometimes, however, it happened that the change in social structure occurred without any overt change in behavior. A jackal arrived from over a hill and joined a pair which had been by itself for a long while, and the pairmates did not even look back. Later on, the follower often disappeared for a moment behind a sand dune and when it reappeared, a fight, hitting with hindquarters, or biting followed. The following observation, which was recorded a day after the
2.4.69 previous one, might serve as an example.

After roaming the area for an hour or so, Gingi and Belly all of a sudden found themselves about 30 m away from Red and Light. One of the young jackals followed our adult pair and the other wandered in a different direction. No overt behavioral event was observed. Later, another individual, possibly the dark female, joined the three, again with no overt change in behavior. Gingi and Belly then left the other two and later the dark individual joined them again. A few minutes later, when Belly urinated, she bared her teeth and lowered her head in its direction. The jackal wagged its tail and continued to follow. When the three jackals trotted to the seashore, the young pair was already there. As soon as the three saw the young pair, they all darted in their direction and reached them after a short chase. Whines and barks were heard but because of the darkness and the thick vegetation, we could not obtain any details of the interaction. Since, at that time, both pairs were often seen on the seashore, the sudden attack could not be explained in terms of territorial interaction.

During the next month similar interactions between the two pairs became a part of the usual evening routine. Seven more interaction sequences were observed and then the young pair disappeared.

Since all interactions occurred between the same animals within a relatively short period, it would be worthwhile to examine the degree of their similarity and variability. As a reference, we shall use the following description, which by all means *does not* represent a characteristic interaction.

The 6th interaction between the two pairs started when Gingi, Belly, and the dark female arrived at the area. Red and Light were already there and started to trot in their direction, Red sniffing Light's inguinal region during the trot. In a matter of seconds, the group of the three jackals started to gallop in the young pair's direction. Red and Light sat down immediately. When the three arrived, the young animals were already lying on their sides. They proceeded in the direction of Gingi and Belly while dragging the side of their belly on the ground and vigorously wagging their tails. Belly stood over one of them, Gingi urinated and scraped, and the dark female ran around with drawn back ears and raised tail. Suddenly, the dark female joined the young pair and lay on her side adjacent to them. Belly and Gingi, highly bristled with lowered neck and head, urinated in succession at the same spot and then with their hindquarters, they both hit one of the young pairmates who responded by biting Gingi's neck. All of a sudden, Gingi left the young jackal and approached Belly rigidly while raising his tail, as if initiating a T-sequence. Belly did not respond to his approach. She was baring her teeth to one of the young jackals who, at that stage, stood up and walked away with drooped tail. Again Gingi and Belly stood near the three jackals who were lying on their sides. Gingi started to groom Belly extensively and she responded by gently touching his head with her snout. When the young pair and the dark female started to move around, Gingi and Belly ran toward them with lowered head and puckered faces and the three lay down again. This time, the adult pairmates stood over them for 10 minutes and then performed a T-sequence. Gingi, who formed the base of the T, shifted to grooming. The two young pairmates started to groom each other as well. Again Gingo left Belly, approached one of the young animals and performed a peculiar rhythmic activity with his head,* moving it swiftly forward and backward, each time touching the young jackal's head with his closed, puckered mouth. The dark female stood up and walked away. When Gingi came back to Belly, she was already sitting. He joined her and all four jackals sat together for a quarter of an hour. Meanwhile, the dark female came back and sat down as well. Ten minutes later, Gingi stood up and put his forelegs on Belly's back. Then the four animals walked away. The dark female continued to sit there, looking for a long while in their direction with drawn back ears and lips.

The only component which was characteristic of all the other interactions and was also present in the above-described sequence is the phase in which the adult pair walked around the young pair who were lying or sitting. In all cases during this phase, Gingi and Belly were bristled, lowered their heads, drew back their ears, and puckered their faces. In all cases the young animals tended to keep as close as possible to each other, moving in each others' direction in a crouching walk or dragging their bodies on the ground. In all other respects, the content of each out of the 10 interaction sequences was variable.

Only 3 sequences included a violent physical contact, i.e., hitting with hindquarters or biting. These contacts were of no predictive value as to the form of the termination of the sequence. During the interactions, the young pair was either sitting or lying on their bellies, sides, or backs. Sometimes they wagged their tails vigorously, whereas at other times they kept their tails between their legs. Drawing back of their ears and lips occurred often but not always.

The form of the beginning and end of the various sequences was also variable. Some sequences were started by Belly and Gingi, and others by Red and Light. When the distance was great, the sequences started, if at all, by either pair running in the other pair's direction. When the animals were close, the sequence began sometimes when Red and Light suddenly sat or lay on the ground, and at other times when Gingi and Belly bristled, lowered their heads, puckered their mouths, and withdrew their ears. In one case it started when Gingi trotted toward the young pair in what was formerly described as a

*Probably the same as nose-stabs or muzzle "punches" in coyote — Ed.

stylized trot (see Fig. 9 in Golani and Mendelssohn, 1971). Previously this gait had been observed only during T-sequences.

Termination of sequences occurred, in some cases, when Gingi and Belly were engaged in grooming or in a T-sequence. Then the young pair stood up and walked away. In other cases it was Gingi and Belly who went away, leaving the young pair behind. In still other cases, both pairs stayed together without any overt interaction between them or followed each other in a single file while leaving the area where a violent interaction had just taken place. Once after an interaction, Red followed the adult pair and his female was left behind. She lay there for a long while until it was dark and we could not see her anymore.

The dark female, who was at that time a persistent follower of Gingi and Belly, was often involved in the encounters with the young pair. During these encounters, and during encounters with other young individuals later on, she alternated between running bristled around the attacked individuals or even hitting them with her hindquarters, and sitting or lying on her side near them, while the adult pair stood bristled near the three of them. In one case, an interaction sequence started when she suddenly hit a young pairmate while the elder and younger pairs lay relaxed together. In two other interactions, Belly hit the dark female with her hindquarters and right away they both together hit another individual.

It is impossible to show that the distance between the two pairs became greater or smaller during the 10 interaction sequences nor is it possible to demonstrate a regular change in any physical or behavioral variable. It is, therefore, impossible to state anything about a regular change in the relationships between the two pairs. Rather, the interactions between the two pairs took various forms within the above-specified range of possible combinations.

The behavior described for Gingi and Belly was not an exclusive 2.17,19.69
response reserved for interaction with Red and Light. Similar interac- 3.10,12,16,
tion sequences were observed with other individuals. These se- 17,24.69
quences usually started when Belly hit other individuals with her hindquarters and Gingi, who was usually drawn into the interaction by Belly, followed her, bit these individuals, or stood bristled near them and then groomed them extensively. Sometimes he groomed them right away. From the few cases in which we could tell the sex of the attacked individuals, it seemed that both Belly and the dark female charged both males and females. We do not know if Gingi differentiated between sexes by his response.

Interwoven in these interactions were T-sequences and grooming sequences between Gingi and Belly, and sometimes between the attacked individuals. When such a sequence was performed by Gingi and Belly, the other individuals sometimes started to lick or groom each other, or stood up and walked away. During the charge, the attacked individuals tended to stay as close as possible to each other and to the attacking jackal. It was a common picture to see two young individuals who had just been attacked run ahead of Belly and Gingi who just left them behind and then lie down on their sides in front of them, very close to each other like two inert corpses.

In many of the above-described interactions we could not identify the attacked jackals. Nevertheless, the similarity to those interactions in which we knew the jackals' identity seems to indicate that in all the above cases the jackals were relatively young (under 2 years of age) and were acquainted with our pairmates before the encounter took place.

2.23.69 One unusual interaction took place during the same period. Its participants were adult pairmates. At first, the new pair noticed Gingi, Belly, and the dark female and sneaked away. Later that evening we simulated a howl and the two groups, which were separated by a hill, joined the howl and darted in each others' direction. During the encounter that followed, the two adult males grabbed each others' necks and shook vigorously. Then Gingi hit the other male with his hindquarters. The three females observed the fight from over a hill while bristling, baring teeth, and arching their backs (we could not obtain more details because of the bushes around them). After that short fight, the males joined their females and trotted away, highly bristled.

We continued to observe Gingi and Belly after the encounter described above for another year and a half, but we never observed a similar interaction. Somehow Gingi and Belly never met other adult pairs, although such pairs were present in the area. In other observed fights between adult pairs, the females often stayed behind as well. Nevertheless, when an adult paired male met a paired female in the field, he attacked her in a similar way, first by grabbing her neck and shaking and then hitting her with his hindquarters. But the first phase of shaking was very short.

T-Sequences

T-sequences occurred during a common trot, after lying or sitting together, after a howl, or immediately after a charge on other jackals (see Table 22-2). Of 28 sequences observed during the seasons of 1969–1970 (2 which followed an attack were not included in the table because of lack of other information), 7 occurred during or after a so-called agonistic encounter. For example, "two young individuals lay inert on their sides, their heads on the ground, while Gingi walked highly bristled around them. Suddenly he left them and approached Belly who lay on her stomach nearby. He raised his tail and she stood up and performed tail-flagging while forming the top of the 'T.' He then sniffed her head, stuck his snout into her ear, and walked back to the two young jackals. His tail dropped while approaching them and he returned to Belly, raising his tail again. Forming the base of the 'T,' he touched her ear and groomed her face. Their tails dropped. The two young jackals went away." Another example was presented previously during an encounter with Red and Light.

The common occurrence of both agonistic interaction sequences and T-sequences is not exclusively related to the above situation. Fights or T-sequences also occurred after a howl, depending on the social context (Golani and Mendelssohn, 1971, p. 189).

Most T-sequences started when Gingi approached Belly rigidly while raising his tail. In the few cases in which a sequence was started when Belly was active, she froze, stared at Gingi, and performed tail-flagging. He responded by running toward her and performed a T-sequence with her. Sometimes when they were close, they just turned to each other and started a sequence together. It was never started by Belly running toward Gingi, an event that was very common in other pairs.

During 1969, Gingi and Belly had their own special style of performance of T-sequences. Unlike the T-sequences of other pairs which were long, rich in events, and tremendously complex, our pair's sequences were, except for one case, relatively simple and short. While they moved around each other or stopped, Belly's side was almost always directed at Gingi's head. Gingi formed the base of the "T." While in this position, either one or both of them often performed tail flagging. This behavioral event was never observed as often in other pairs. After freezing for a few seconds at the base of the "T," Gingi sometimes placed his forelegs on Belly's back and sometimes shifted to grooming. He would sniff Belly's ear and start grooming her, at the same time dissolving his rigidity and dropping his tail. In a way, this was similar to what happened between Gingi and the young jackals that he attacked; the tension and the rigidity often dissolved and he shifted to grooming.

Copulation

In the middle of one day before the week of copulations started, 3.15.69
Gingi and Belly performed a T-sequence which, in contrast to their former short and poor T-sequences, was long and elaborate.

Copulations started the next day, during which they were still observed to perform the last three T-sequences for that year. The fact that

TABLE 22-2 T-Sequences Performed by Gingi and Belly during 1960–1970

Performance Date	Preceding Event	Start of Sequence	Description	End of Sequence
2.3.69	Howling.	Both approached each other.	B formed top, tail-flagged, raised tail to vertical. G raised tail to diagonal upward.	G trotted away.
2.15.69	B attacked a follower.	G trotted toward B.	B formed top, G raised tail to vertical and shifted backward to lick vulva.	G licked vulva.
2.17.69a	Trotting.	B raised tail and stared at G. He ran toward her.	B tail-flagged. G raised tail to diagonal downward.	B charged a follower.
2.17.69b	G stood bristled near 2 followers who were lying on their sides.	G trotted to B.	B formed top, tail-flagged. G inserted snout into her ear. Her tail dropped to horizontal.	G walked back bristled to followers.
2.17.69c	G stood bristled near followers (continuation of previous sequence).		G formed based, raised tail to horizontal, touched her ears, groomed her face. Both tails dropped.	G groomed B.
2.18.69a	G and B stood bristled near followers.	G approached B.	B formed top, tail-flagged. G sniffed her ear, groomed her ear.	G groomed B.
2.18.69b	Lying near each other.	G approached B.	G raised tail, formed base, placed forelegs on B's back.	G trotted away.
2.18.69c	Trotting.	G trotted to B.	B formed top. G tail-flagged, sniffed her vulva. B tail-flagged.	G trotted away.
2.19.69	G groomed young jackal.	G trotted to B.	G formed base. B stood up, both raised tails to vertical. G groomed B.	G groomed B.
2.26.69	Trotting.	G trotted to B.	G and B raised tails to vertical. G formed based, both tail-flagged.	G licked vulva.

3.15.69a	Howling.	B stood up. (G was already standing.)	Both tail-flagged, antiparallel position, B formed top, G turned around B; they both turned around each other, G placed forelegs on her back, came off, sniffed vulva, again placed forelegs on the side of her back, shifted to her back, clasped her, got off, formed top. When he formed top, she turned, and he was again at base. They formed antiparallel position, both tails raised to diagonal upward. B tail-flagged. G formed base and placed forelegs on her back, shifted to clasping, got off, sniffed vulva, formed base, his tail horizontal, hers diagonal upward, both slightly bristled. G formed base in parallel position.	G trotted away.
3.15.69b	Walking.	G mounted B.	G got off, pressed chest to vulva with raised tail, dropped tail to vertical, tail-flagged.	G trotted away
3.15.69c	Walking.	G sniffed B's vulva, raised tail.	G formed base, placed forelegs on her back, got off, tail-flagged, stood at right angles to B's pelvis, tail-flagged.	B walked away.
3.16.69a	G urinated.	G trotted to B.	G formed base. B bared teeth to G. Both raised tails to vertical.	G licked vulva. B urinated and trotted away.
3.16.69b	Trotting.	G trotted to B.	B averted tail. G licked vulva.	Both trotted away.
3.16.69c	Howling.		B averted tail, G licked vulva, pressed chest to vulva. B formed top, G raised tail to horizontal. They faced each other and tail-flagged	
3.16.69d	4 successive mountings with licking.	B became rigid.	B formed top. G mounted. They both raised on hindlegs in a two-footed clinch, baring teeth to each other.	Successive mountings with final copulation.
2.7.70a	G urinated.	G trotted to B.	G formed top, shifted to base, placed a foreleg on her back. B trotted away, G followed, placed forelegs on her back, B moved away.	Both trotted away.

TABLE 22-2 Continued

Performance Date	Preceding Event	Start of Sequence	Description	End of Sequence
2.7.70b	After sniffing Timid's urination, B stared at her.	B approached G.	Both tail-flagged. G licked vulva, raised tail to vertical while B tail flagged.	G groomed B. B trotted away.
2.20.70	B chased Timid away. Both G and B bristled and urinated.	G approached B.	G formed base, groomed B, licked vulva. B averted her tail	B walked away.
2.22.70	Trotting.	G trotted to B.	G formed base, placed forelegs on her back. B moved away.	G urinated and watched B.
2.23.70a	Trotting, B leading.	B paused until G reached her.	B raised tail to horizontal. G licked vulva. G formed base.	B urinated, G urinated at same spot.
2.23.70b	Both urinated (continuation of previous sequence).	G puckered mouth and approached B.	G formed base, placed forelegs on back, B moved away.	Both trotted with horizontal tails.
2.23.70c	Trotting (continuation of previous sequence).	G approached B.	G formed base, licked vulva.	G trotted away.
3.5.70a	Howling.	Both turned to each other.	G formed top, licked vulva.	G urinated and scraped.
3.5.70b	G urinated and scraped (continutation of previous sequence).	Both turned to each other.	G formed top. B sniffed G's penis with vertical tail.	B urinated and scraped, then G urinated and scraped.
3.9.70	Both sat. B stood up and urinated.	G stood up, both turned to each other.	Both faced each other, one of them tail-flagged.	The one who didn't tail flag walked away.

T-sequences occurred right before copulations was quite a surprise to us, since in captivity all pairs ceased to perform T-sequences a few days before copulations started. In general, this pair did not show in its T-sequences the richness of events and the complexity in the changing structure of their grouping that were observed in captivity.

During the first day of heat Belly once bared her teeth in Gingi's 3.16.69
direction when he mounted her. Once again they both stood on their hindlegs in a two-footed clinch and bared their teeth to each other. Similar behavior was observed more often in captivity and its occurrence in nature indicates that it is, at least, not an exclusive artifact of confinement.

During the following week they walked very little; when they walked, Gingi followed Belly closely. A large part of the time was spent on licking and mounting, with and without contact of genitalia, and with or without thrusting movements (see Table 22-3). In one case when Gingi mounted Belly, he averted his tail in a manner typical for females during licking or mounting. Similar behavior was also observed in captivity when a male was mounted by a female (see Fig. 68 in Golani and Mendelssohn, 1971). The two observed copulations during the first and second day terminated in locking which lasted 4 minutes 30 seconds and 4 minutes 25 seconds, respectively. On the seventh day, Belly's vulva started to lose its tumescence and Gingi started to lead, occasionally stopping, following Belly, and licking her vulva.

The dark female followed our pair consistently throughout the week of Belly's heat. Another unidentified female joined her twice. Sometimes the dark female bared her teeth toward the second follower and sometimes hit her with her hindquarters. Gingi and Belly hit both of them and chased them away without obvious results. Once locking took place, the followers approached and lay on their sides near the locked pair.

Courtship Feeding

In the middle of Belly's heat, Gingi once grabbed and ate 3.18.69
a chick from Belly's mouth. He often tried to do it but usually he was unsuccessful. A week later, when they both finished eating the few chicks that were presented to them, Belly approached Gingi, wagged her tail, lowered and slightly rotated her head, and pressed her snout to his lips. Gingi at once regurgitated a chick and Belly ate it. Henceforth, this behavior was regularly observed in almost each observation during April, May, and June. In July, when Belly pressed Gingi's lips as usual, he did not respond. She was observed to perform this behavior twice more during the summer, but Gingi did not respond.

TABLE 22-3 Behavioral sequence of events during copulations week of 1969*

Mountings (Day of Heat)	Thrusting Movements	Licking of Vulva	Urinations Belly	Urinations Gingi	Other Events	Behavior of Follower	Copulations
3.16.69 (1st day)			+	+		Dark female present.	
			+ss		Scraping.	Dark female bares teeth to 2nd follower who joined the group.	
				+			
				+	T-sequence.		
				+			
		+					
			+	+			
			+				
		+		+	G chases follower. Howling. Pressing chest to B's vulva. T-sequence.		
				+			
		+					
+	+	+					
+							
+	+						
+	+				B and G rise on hindlegs in a two-footed clinch.		
+	+						
+		Sniffing.					
+	+						
+	+						+

						Lies near locked pair	Duration of locking 4 minutes and 30 seconds.
							pairs
3.17.69 (2nd day)					G and B bristle.	Dark female present. 2nd follower joins.	
				+			
	+		+	+			
+					B sits.		
+		+			B stands up.		
+		+					
+	+	+					
+					B and G bare to followers.	Dark female lies down.	
		+	+	+ss			
		+					
		+					
		+					
			+	+ss			
		+					
+					G bares to dark female.		
		+			G bares to dark female.		
+		+					
+		+					
+		+			B sits.		
+							
+		+					
+	+				Howling.		
+	+						
+	+						

TABLE 22-3 Continued

Mountings (Day of Heat)	Thrusting Movements	Licking of vulva	Urinations Belly	Urinations Gingi	Other Events	Behavior of Follower	Copulations
+					B Sits.		
+	+						
+							
+	+						
+	+						
+	+						
						Dark female lies near locked pair. 2nd follower approaches dark female, bumps into her. Dark female present.	+ Duration of locking, 4 minutes 25 seconds.
3.18.69 (3rd day)							
+					B sits.		
				+			
			+	+ss	B hits and shakes		
		+	+	+ss	the dark female.		
			+	+ss			
				+	Howling.		
		+			G grabs a chick from B's mouth.		
3.21.69 (6th day)			+	+		Dark female present.	
		+					
+				+			

				+ss	+	G and B hit follower with hindquarters.
	+					
	+					
	+		+			
	+	+	G sniffs vulva.			
	+	+				B bares to a follower.
	+	+				
	+	+	+			
	+	+				
	+	+				
	+	+				
	+					
	+	+				
	+					G averts his tail while mounting.
	+					
	+					
	+	+				Stays on female's rump. Then gets off and they trot away.
3.22.69 (7th day)			+			B's vulva loses tumescence.
			+			
			+			

*Temporal sequence is represented by successive lines; + indicates that the event occurred; ss is an abbreviation for "urinating at the same spot."

Failure to Give Birth

Golden jackals' pregnancy lasts 63 days so we were expecting Gingi to trot by himself during the second half of May. But Gingi and Belly continued to go together, indicating thereby to us that Belly had failed to give birth. They continued to go together for another year when the same story repeated itself. We did not observe Gingi and Belly longitudinally throughout the first year of their pair bond in 1968, so we shall never be sure whether she gave birth that year. However, the recency of their pair bond which started in February, 1968, the fact that they were still young at that time, and the low frequency of urinations performed by them during that season all lower the chances that she gave birth. Possibly her distended belly was somehow connected with the fact that she failed to produce offspring.

More about Followers

If indeed Belly never had her own pups, then the jackals that followed our pair were not their offspring from previous years. Thus, neither the dark female, who later disappeared, nor a new female named Timid, who later joined the pair and followed it for a few months until she herself became pair-bonded, were our pair's offspring. Handsome and Elongated, another pair whom we observed longitudinally, were also consistently followed by specific animals who were not their offspring. We know this because we observed the pair formation of that pair and the followers joined them during their first season.

Spring is a short season in Israel. It ends in March or sometimes the first days of April and then the long summer sets in. Throughout the lingering dusk hours of the long summer evenings from April to October, our pairs' social behavior was relatively unchanged. At the beginning of the period, jackals change their winter fur into summer fur; therefore, we could no longer identify the dark female and we do not know if she was still one of the three followers.

Since the behavior of jackals changes during the beginning of this period, we were not able to use behavior as a clue for identification. Two of the followers were females and the third was a young specimen of undetermined sex. During more than half of the occasions in which our pair was observed, three individuals, possibly the same ones, followed. On two other occasions, two more individuals joined our pair, and on one occasion our pairmates joined a group of ten jackals which consisted of four adult animals, one of whom was a suckling female, and six pups. Once, three of the pups were seen following Gingi and Belly.

Throughout this period the jackals were usually seen to lie resting on a hillside, every now and then capturing and eating insects or feeding on grapes which were abundant during that time of the year. Sometimes one or two of them vainly initiated a chase after a hare which occasionally passed by. Often during these long "sit-ins," interaction sequences occurred in which the followers bristled, bared teeth, chased, and stood bristled with arched backs near each other. Usually when a follower joined the group, it was he, if anyone at all, who was attacked by one of the followers that was already present. Gingi and Belly ignored these interactions and kept lying leisurely. In only two such cases, Belly gaped and bared her teeth toward one of the followers.

In most cases, it was the followers who came running and lay in front of Belly or Gingi while wagging their tails. Only then our pairmates bristled, gaped their mouths, and bared their teeth. Throughout this period Belly was never observed to hit a jackal with her hindquarters and Gingi bit only once, after grooming a follower. Chasing was started only by Gingi. Although Belly chased and once even hit two females later on, she never reverted to her old agonistic manners. Henceforth, she resembled much more the other observed adult paired females.

The fact that Gingi and Belly were consistently seen together during this period was no doubt partly due to the fact that Belly kept following Gingi. Gingi showed a considerable degree of independence in his movements during these months. In 5 observations, he left the group and disappeared. In still others, he was in the lead, never looking back or waiting for the others to join him. Belly used to look every now and then in his direction and in a few cases in which she was in the lead, she looked back and waited for him to join her. His freedom of movement fits well with the fact that during that period jackal pairmates who brought up offspring were often seen to forage by themselves.

The beginning of November was associated with a change in social structure. During 2 observations, Gingi and Belly were seen to go unfollowed and Belly started to scrape after almost every urination. Then a third female joined them. This female, who was later called Timid, followed them consistently for 3 months until she left them to form her own pair.

Other Jackal Pairs and Their Relations to Gingi and Belly

During our first observations on Timid, Gingi, and Belly, we saw 12.1.69
for the first time a new pair of young jackals. They roamed the area together, urinating and scraping at an exceptionally high frequency. As a result, we called them the Urinating Pair. They looked very young and their fur was still not developed as in adult jackals.

They both knew us well since they moved freely, ignoring our conspicuous presence and simulated howls. We could not tell whether
12.5.69 we had seen them previously. During our next observation they suddenly saw Gingi, Belly, and Timid who also noticed them. The distance between them was only 30 m. The urinating Male lowered his head, withdrew his lips and ears, and put his tail between his legs. Gingi bristled, raised his tail to a vertical position, urinated, and scraped a few times. Belly bristled and raised her tail to a horizontal position. The Urinating Female was completely relaxed. When our pair and Timid started to walk away, the Urinating Pair followed them, the Urinating Female in the lead. Later she joined the three jackals while her male stayed behind, hiding in the bushes. Possibly she knew our pair well. Belly, who later followed Timid and the Urinating Female, suddenly hit both females in succession. Nevertheless, they both continued to follow Gingi and Belly. The Urinating Pair was seen two more times together with Gingi and Belly and then, like Red and Light during the previous year, disappeared. Timid, like the dark female, continued to follow.

The most peculiar feature of Timid's behavior was the fact that she sometimes urinated in sequence with Gingi and Belly. Whereas our pairmates often urinated in succession in the same spot, she often subsequently urinated 10–20 m away. Young unpaired females were previously seen following a pair and urinating in relatively high frequencies. However, what was exclusive for Timid was the temporal distribution of some of her urinations and the fact that some of them were visually triggered from a distance of 20–30 m (visual triggering of urinations often occurred in pairmates too). Of course, Timid also urinated and scraped by herself without any obvious external stimulus.

When Timid started to follow our pair, she was once hit by Belly who charged both her and the Urinating Female. But this occurred only once. Later, Timid always avoided violent physical contacts with the pair. During almost every observation, Gingi swung around on his forelegs with his hindquarters in the air, almost, but never hitting Timid. She always moved away at the very last moment. Note that in 1969 when the dark female and other followers were charged by Gingi or Belly, they either proceeded in the attackers' direction or sat on the ground, waiting for the hit. In 1970 when Belly chased Timid, bristled with puckered mouth, she never established physical contact with her. This difference in behavior between the two years was partly due to Gingi's and Belly's own behavior. This was indicated by the fact that the same behavior also occurred with other individuals who approached the pair.

When Timid was charged, she often left the pair and rejoined it a few minutes later. In some cases, Timid was in the lead and Gingi and Belly, who had been completely bristled when they first saw her approaching, followed her unbristled. Once she even started a howl, a behavior that was never observed in other followers. When she howled, no one joined her and she soon stopped. In some cases, she was seen to follow Gingi alone, ignoring his occasional puckered mouth and bared teeth, and moving away when he swung toward her with his hindquarters. When Belly arrived, she always chased Timid away and the follower left them only to rejoin them later.

The performance of T-sequences in the new year started as late as
the beginning of February, 1970 and the last T-sequence of the season 2.7–3.9.70
was observed a month later. Then Gingi and Belly disappeared for 2 weeks and when we saw them again no precopulatory behavior was performed. During that period we observed them to perform ten T-sequences (see Table 22-2). This year, Gingi again formed either the base or, less often, the top of the "T." He often placed his forelegs on Belly's back and she typically jumped away sideways, thus avoiding the position. Tail-flagging was performed by either or by both. Licking of the vulva was performed throughout the period and terminated many sequences. As in the majority of last year's interactions, sequences were short and poor in events. Only one T-sequence followed agonistic behavior. However, during that season agonistic behavior was relatively rare.

One day in the second part of February, 1970, a strange male arrived
at the area. Gingi, Belly, and Timid chased him away and one of them 2.19.70
(we could not tell who) hit him with his hindquarters. A day later when the same male lay on his belly on a dune, Timid approached him and urinated and scraped near him twice. The male, who did not look in her direction, stood up, scratched his shoulder with a hindleg, urinated, and started to trot. Timid followed him. A few minutes later when he urinated, she also urinated at the same spot. Two days later the male was seen to follow her. Timid was still seen to urinate at the same spot where the male had just urinated. When they both saw Gingi and Belly, Timid proceeded in their direction and her male followed her, somewhat behind. Our pairmates ignored her and Gingi lowered his head and bared his teeth at her male. Later that day, Timid and her male were both seen chasing away a third individual. The next day, Timid was still seen with Gingi and Belly, but henceforth she was either seen alone or with her male. They had offspring about 2½ months later. We never observed them to meet Gingi and Belly again, although their den was only about 100 m away from Gingi's and Belly's resting place.

Since our observations were restricted to the late afternoons and early evenings, we do not know to what degree the animals left the area where they were seen to trot, forage, court, and fight during observation time. We do know, however, that their dens were situated in the area (see Figure 22-1) and that their hiding places during daytime were also situated there. After dark, their motility always increased tremendously and they were seen trotting in the direction of the Kibbutz poultry house, the seashore, or their common feeding ground on the garbage dump where they gathered in large numbers at night.

Although Gingi and Belly were seen over a large part of the area covered by Figure 22-1, their resting place and the area where they were seen most frequently was around the small hills along the vegetation of the river (designated by **A** in Figure 22-1). Toward the end of 1969, they shifted westward and were mainly observed on the slope of the hill (designated by **B**). This change occurred simultaneously with the appearance and settling in Area **A** of a new male, later called Handsome. At the end of November, 1969, a fully-grown adult male that was never observed before came trotting into the area, climbed a hill, looked around as if watching the horizon, and produced a series of short, peculiar vocalizations which were between a bark and a howl. As a result, four females came trotting from all directions. One of these females formed the top of a "T" in front of him. Henceforth, he stayed in the area, established a pair, and brought up offspring who were later also observed to establish pairs. Handsome was mainly seen in area **A** where Gingi and Belly were formerly seen. At about that time, Timid pair-bonded and was mainly observed with her male in the area designated by **C.** Handsome's pair and Timid's pair had numerous agonistic interactions. Gingi and Belly, however, never met any of the two pairs during our observations. In March, 1970 they shifted about 300 m westward to the delta of the Soreq River, where it bifurcates and flows into the sea.

After they settled in the new place, we saw them very seldom. Once in June, they attracted our attention by joining in a howl which progressed along the river. We noticed that they were followed by a third jackal who approached them during the howl with drawn back ears and wagging tail. He crossed Belly's field of vision but she did not respond. Three weeks later we saw Gingi foraging along the beach by himself. Then, one day in July, 1970 we saw them meeting each other in the middle of the delta. Belly wagged her tail extensively, lowered her head, rotated it slightly, and gently touched Gingi's snout. Possibly they had not seen each other for awhile. This was the last time we saw them.

BEHAVIORAL REGULARITIES OF GINGI AND BELLY IN COMPARISON TO OTHER PAIRS

We shall first summarize the long-term regularities which were observed in the behavior of Gingi and Belly during the entire period of observation. Then we shall describe some short-term regularities which were observed during a part of that period. Only then shall we look for similar regularities in the behavior of the other two observed pairs.

Gingi and Belly were seen together most of the time. In only 8 out of 92 observations did we see one of them without the other (Table 22-4). In only 5 more cases, Gingi was seen to leave Belly during observation time. Belly was never observed to leave Gingi.

Although they spent a large part of their time unaccompanied by other jackals, during almost every observation they were sooner or later joined by one or more individuals.The pair was observed solely by itself for a maximum period of 3 successive observations (Table 22-4). The animals that followed them were not their offspring, in all cases their fur was still not as developed as in adult jackals, and they looked under 2 years of age. During interactions with these followers, Gingi and Belly performed chasing, standing near or over the follower, lowering of the head, bristling, puckering of the mouth, and baring of teeth. Other behavioral events were not performed regularly. After these interactions, the attacked individuals often joined our pair.

Throughout the period of observation, except for one case, Gingi and Belly did not interact with other adult pairs, although such pairs were present in the area. The one interaction with adults was different from all the other observed interactions since it occurred only between the males, including biting and shaking, and was terminated by separation.

While trotting together, Gingi and Belly urinated in relatively high frequencies, Gingi urinating more often than Belly (see Table 22-5). About 40% of their urinations were carried out in succession, out of which more than half were performed at the same spot. Urination in succession occurred as a result of either olfactory or visual stimulation. Sometimes the jackal who followed his mate's path did not see his mate urinate, but still urinated at the same spot after sniffing the ground. In other cases, one of the pairmates observed his mate urinating from a distance of 10–20 m and either urinated immediately or trotted directly to the same spot where urine had just been sprayed and urinated there. This sequence of events should have resulted in the fact that the pairmate who followed his mate should have also

TABLE 22-4 Maximal Number and Identity of the Animals that Joined or Followed Gingi and Belly during an Observation

Date	Number of followers	Identity of followers	Date	Number of followers	Identity of followers
1968					
2.3	+2	A light colored female.	7.12	+1+1	(Gingi is absent)?
2.19	+2	A light colored female.	7.16	+1	(Gingi is absent), young
2.23	+2	A light colored female.	7.22	+6+4	adult female, 3 adult
3.1	0				jackals and pups.
3.3	+1	Female.	8.6	+3	Pups.
3.8	+1		8.13	0	
3.18	0		10.23	+1+1	Young individual.
4.8	0		10.25	1	
4.10	+1		10.30	0	
12.19	+2+1+1+1	(Gingi is absent?)	11.1	0	
1969			11.18	+1	Young individual.
2.3	+2+1	Red and Light, Dark Female.	11.28	0	
2.4	+2+1	Red and Light, Dark Female.	12.1	+1+2	Timid, the Urinating Pair.
2.6	0		12.5	+1	Timid, The Urinating Pair.
2.10	+2	Red and Light.	12.7	+1+1	Timid, the Urinating Pair.
2.14	+2	Red and Light.	12.12	0	
2.15	+2+1	Red and Light, Dark Female.	12.14	+2	The Urinating Pair.
2.17	+2+1+1	Red and Light, Dark Female.	12.17	0	
2.18	+2+1	Red and Light, Dark Female.	12.18	+1+2	Timid, The Urinating Pair.
2.19	+2+1+1	Dark Female.	12.19	+1	Timid, The Urinating Pair.
2.23	+2+1	Adult pair.	1970		
2.24	+2	Red and Light.	1.9	+1	Timid (Belly is absent).
2.26	+1	Dark Female.	1.11	+1	Timid.
3.2	0		1.18	+1	Timid.
3.3	0		1.29	0	

3.5	+2+1	Red and Light, Dark Female.	2.1	0	
3.10	+2		2.2	+1	Timid.
3.12	+2		2.6	+1	Timid.
3.15	0		2.7	+1+1	Timid.
3.16	+1+1	Dark Female.	2.9	+1	Timid.
3.17	+1+1	Dark Female.	2.17	0	
3.18	+1	Dark Female.	2.19	+1+1	Timid and her male.
3.21	+1	Dark Female.	2.20	+1+1	Timid and her male.
3.22	+1	Dark Female.	2.22	+1+1	Timid and her male.
3.24	+3		2.23	+1	Timid.
4.1	0		3.5	0	
4.10	+1		3.7	0	
4.14	+3		3.9	+1	Female.
4.17	+4	(Gingi is absent.)	3.24	0	
4.24	0		3.27	0	(Gingi is absent.)
4.29	+1+1		3.30	0	(Gingi is absent.)
5.15	+1		4.4	0	
5.18	+1+1+1+1		5.11	0	
5.25	+2+3	Young individual, male.	6.10	+1	
6.19	+2+2+1	Young female, 2 pairs.	6.11	0	
6.21	+1+1+1	3 young individuals.	6.30	0	(Belly is absent.)
6.29	+3	3 females.	7.4	0	
7.6	+2+1	Young pair, male.			

followed in urinations. Nevertheless, regardless of who was in the lead, Belly urinated after Gingi in succession in 95% of the cases. Followers often sniffed the fresh urine of the pair but did not urinate at the same spot. They did urinate, however, at places where they themselves had urinated previously.

Gingi and Belly often urinated and scraped in the presence of strangers or immediately after their disappearance. A similar phenomenon has also been observed in scent marking behavior of other mammals (Ralls, 1971).

Another group of interaction sequences in which Belly and Gingi showed some regularity were the T-sequences, which were relatively short and poor in events. Precopulatory behavior, i.e., urination in succession in high frequencies, T-sequences, licking of vulva, and mountings terminated in March when copulations took place. These behavioral interactions were never observed between either of the pairmates and a stranger.

Short-term regularities were already mentioned in the description section of Gingi and Belly. We shall mention only a few here.

During the first year of observations, Gingi and Belly quite often established violent physical contacts with other jackals. During the second year, such contacts were usually avoided. This change coincided with a change in the behavior of Belly. Whereas during the first year she was the first to chase and hit other jackals and Gingi would follow her, during the second year she stopped hitting other jackals with her hindquarters and chases were started by Gingi.

The number and the identity of the followers were also regular within short periods of a few months. In summer they were accompanied by one to five unidentified individuals. In the months which preceded copulations, they were followed during each breeding season by a specific female and, for a shorter period, by an additional specific young pair. The behavior which was performed during interactions with these individuals was also regular; that is, specific in some way to that individual or pair: for example, the fact that they did not hit or bite Timid or the fact that they differentiated in their behavior between the Urinating Male and Female.

Short-term regularities were also observed in their use of space. Except for the fact that they twice changed the area where they used to rest and trot most of the time, they had specific preferred routes during specific periods. For months they were observed using the same path and lying and sitting under the same bush. Then they changed and new routes and places were preferred for awhile.

The question is whether similar regularities are also demonstrable in the behavior of the other two observed pairs. Timid and her male, and

TABLE 22-5 Urination Frequencies of Three Observed Pairs*

Identity of Pair	Period of Observation	Number of Observations	Total Number of Urinations	Number of Urinations in Succession	Number of Urinations at the Same Spot
Gingi			92		
	11.30.69–5.11.70	30		42	20
Belly			51		
Handsome			26		
	11.30.69–5.11.70	15		5	4
Elongated			8		
Timid's male			15		
	11.30.69–5.11.70	30		6	4
Timid			48		
Gingi			100		
	12.19.68–5.18.69	30		45	32
Belly			76		

*Gingi and Belly are represented by two parallel periods. Absolute time of observation was not taken into account.

Handsome and Elongated were seen by themselves, unaccompanied by their respective pairmates, in as many as half of the observations. Possibly, a part of this relative independence was due to the fact that they, unlike Gingi and Belly, had offspring and were seen to forage singly and carry food back to the den.

Like Gingi and Belly, Handsome and Elongated were often followed by young jackals. Since we knew Handsome before he was paired, we concluded that, as in the case of Gingi and Belly, these followers were not the pair's offspring. During the months that preceded and followed copulations, Handsome and Elongated were consistently followed by one to three individuals, often identified as females. In May, 1970 these females disappeared and one month later a male called Spot joined the pair. This male fed their pups persistently throughout the summer. He was usually seen foraging either by himself or together with Elongated or with both pairmates. During the second half of August, 1970 he was joined by a new female, they started to go together, and he left the pair. Since Gingi and Belly had no offspring, we could never tell whether pups were regularly fed by followers.

Unlike Gingi and Belly, Handsome and Elongated attacked followers quite seldom by running in their direction and baring their teeth. Elongated also hit a follower once or twice with her hindquarters, and Handsome charged a follower a few times. They were never observed to attack or bare teeth to Spot, the follower who fed their young. This may have been due to the fact that he followed them in summer during which Gingi and Belly did not attack followers either. Timid and her male were never joined by a follower.

In contrast to Gingi and Belly, the other two pairs were often observed fighting with each other. These fights usually took place between the males, and if females were present, they kept behind. Timid was also attacked once by Handsome who stood bristled over her while she lay on her back. The fights took place on both the area where Handsome and Elongated were usually seen and near Timid's den. During fights, both males took turns at puckering their mouth while the other was on his belly withdrawing his lips and turning his ears backwards. Near Timid's den, Handsome puckered his mouth much less often than in other places. Violent physical contacts were rare. Whereas Timid's male usually avoided the area where Handsome was seen (designated by **A** in Figure 22-1), Handsome often visited the area near Timid's den.

Timid and her male, and Handsome and Elongated urinated much less often than Gingi and Belly (see Table 22-5). They also urinated much less often in succession. Whereas Handsome urinated 3 times as much as Elongated, in Timid's pair this proportion was reversed and she urinated 3 times as much as the male. These figures are partly misleading because they depend upon how long each individual was observed. Yet they do give an idea of the differences in the frequency of urinations as performed by the various individuals.

The first interaction between a specific male and a specific female that were later observed to form a pair was initiated by the females in all 3 observed cases. When Gingi and Belly were observed in a group of four jackals, it was Belly who was first observed to perform tail-flagging when Gingi passed in front of her. At that time, Gingi showed no overt preference for Belly. When Handsome produced a series of short vocalizations, four females came trotting and one of them (unidentified at the time) approached him while raising her tail and formed the top of a "T" in front of him. It was Timid who first approached her male, urinated and scraped in front of him, followed him, and urinated at the same places where he did. Since this regularity is based on only 3 observations, it has to be looked for in future studies.

Once a pair was formed, it was never observed to separate. Belly and Gingi were observed together for 2½ years; Handsome and Elongated were observed for an entire year, which included two breeding seasons; and, Timid and her male, who were observed together for 2½ years, were still together at the time of preparation of this paper in the summer of 1972. With one exception, paired jackals were never observed to perform precopulatory behavior with any jackal except their own mate. A violation of this observed regularity was performed during the winter (March) of 1970 by Handsome and Timid. In the presence of Timid's male, Handsome licked Timid's vulva and then urinated in succession at the same spot where Timid's male and then Timid had just urinated.

In conclusion, behavioral regularity is much more obvious when taken on the level of each pair separately than on all three pairs together. This demonstrates the character of individuality in golden jackals. Because of the great individual differences between various pairs of jackals, we should first observe regularities in the behavior of many more pairs and only then draw conclusions about regularities in jackal behavior in general.

Acknowledgments This work was supported by a research grant of the General Federation of Labor in Eretz Israel. Special thanks are due to Prof. H. Mendelssohn for his encouragement and interest in this project. We would also like to thank the following people for their assistance: Drs. A. Segre-Terkel and D. Kleiman, Mr. Z. Havkin and Miss M. Stavi for the critical reading of the manuscript; Miss M. Stavi, Mr. A. Pelleg, and Mr. O. Habani for assistance in the technical work, and Mrs. I. Watson for preparation of the final manuscript. Some members of Kibbutz Palmachim assisted us during the field studies.

23
SOCIAL DYNAMICS OF THE WOLF PACK

ERIK ZIMEN
National Park
Bayerischer Wald
West Germany

The social behavior within a wolf pack in the wild or in captivity is not the same to each member of the pack. Rather, the usual interactions between different wolves vary due to factors such as age, sex, rank order, and so on. Also, the behavior between two wolves is liable to slow or sudden changes. Such alterations in social behavior between wolves can (a) be correlated to the season of the year, (b) be due to an ontogenetic change in one or both wolves, and (c) appear apparently independent of such endogenous factors and, at our present level of understanding, can be interpreted only in terms of the interaction effects of the social relationship of one animal with one or more others. Frequently the changes observed in the behavior of wolves can be traced back to the combined influence of two or more of these factors. The dynamics of such social relationships between wolves will be the theme of this contribution.

MATERIALS AND METHODS

All observations were made on wolves in captivity. Only in this way is it possible to observe a pack of wolves over an extended period of time without interruption. Altogether, 19 Euro-Asian wolves were kept and observed. From April, 1967 to January, 1970, the wolves lived in a 500 m^2 enclosure at an outer location of the Institut für Haustierkunde, University of Kiel, Germany. After a year of intermediate stay in the zoological garden of the Institute, 4 3-year old wolves of the original pack were taken in February, 1971 to the National Park Bayerischer Wald, Germany. Here the animals live in a 6.5 hectare (65,000 m^2) enclosure. In May, 1971 the pack was enlarged by 5, and in May, 1972, by 4 cubs.

With the exception of 6 animals, all the wolves were taken away from their mothers at an age of 6 days to 3 weeks and artificially reared. By this method, 10 of the 13 artificially reared wolves were tamed. This had the advantage that the tame wolves could be led out of the narrow enclosure every day and be taken on long walks. The wolves thereby run partly tied and partly fully free; they could hunt within restricted bounds.

These excursions had the greatest significance in the development of the wolves. While the untame wolves showed stereotyped movements after 1 year, the tame wolves developed a great deal more naturally. They are also more active as adults, while the untame wolves give one a rather apathetic impression.

The disadvantage of this method is the inclusion of the observer in the social interactions of the pack. As a consequence, certain problems arise; for example, aggressive behavior of the animals toward the observer. Also, the somewhat fake social situation needs to be considered. In spite of that, this method appears to be the best among the accepted conditions of observation on captive living wolves, although it can never entirely substitute for observations in the wild. Comparisons with observations on wild living wolves would for this reason be very interesting.

There are several quantitative methods for elucidating the social rank order in a group of animals. Probably the most frequently employed method is the comparison of the number of attacks of an animal towards other animals of the group and the number of attacks upon him by the others. Another method developed for dogs by Scott and Fuller (1965) utilizes a food morsel as the test object. Two animals of a group are placed in a test cage. Between them lies a bone. That animal which by fighting obtains and keeps the bone becomes designated, after repeated trials, the dominant animal.

Both methods cannot be used to evaluate the Social Rank Order (S.R.O.) of a wolf pack. The first method doesn't work because, as we will see, in some social situations subdominant wolves more frequently attack a dominant wolf than they themselves are attacked by the dominant animal. The second method of the test bone does not work with wolves, because this registers only ordinary food aggression while the S.R.O. is not always identical with a possibly existing food rank order (Zimen, 1971).

A third method, which could have qualified itself as a special method for wolves, uses the frequency of occurrence of active submission (Schenkel, 1947, 1967) as the criterion for the position in the S.R.O. It has also, however, proved a failure, since too many factors are involved.

For the evaluation of the S.R.O. in a wolf pack, I proceeded, therefore, from qualitative characteristics of behavior. Social display (expressive behavior, Schenkel, 1947, 1967) of wolves is of the most subtle distinction whereby one can come very quickly to an evaluation of the social hierarchy in the pack (Figure 23-1). Interobserver agreement in all tests was over 95%.

Figure 23-1 Expressive behavior in wolves is of the most subtle discriminability. The picture shows a social interaction between 4 ♂♂, 1 ♀ and 2 cubs. The α-♂ (walking from right to left) keeps his tail high and legs straight, the ears point forward. A subdominant ♀ (behind α-♂, mostly covered) shows submissive behavior toward α-♂. The B-♂(behind α-♂) keeps his head low and the ears are pulled downward when α-♂ comes close. A low-ranking ♂ (in front of α-♂) keeps his head low and the ears are pulled down and backwards facing α-♂. The 2 cubs show active submission towards the second low-ranking α-♂, who himself expresses submission towards the α-♂. Reprinted with permission from Erik Zimen, *Wölfe und Königspudel,* R. Piper & Co. Velerlag, München, 1971.

A MODEL FOR THE SOCIAL RANK ORDER OF A WOLF PACK

The S.R.O. in a wolf pack is not constant. Therefore it would not be worthwhile to describe all rank situations that occurred in the pack observed by me. I would prefer to set out a generalized model for the S.R.O. in a wolf pack. There are some regularities of the social structure, which I have, in spite of all the transformations, constantly met in my pack and also in other packs maintained in captivity (Hamburg, München, Rotterdam, and Salzburg zoos).

As Schenkel (1947) has already pointed out, there are two rank orders (R.O.) in the pack, one for the males, one for the females. At the top of the two sexual R.O.'s stand the two highest ranking wolves, the so-called alpha-male and the so-called alpha-female. They form the center of the pack.

Beneath the two top ranking wolves in the R.O. is the group of subdominant wolves. Among the males within this group usually no or only minimal rank differences are involved. They can, however, become suppressed with different intensity by the alpha-male; i.e., they have, contrary to the alpha-female, a variable extent of "social freedom." In this way wolf C, endowed with fewer social rights, can stand lower in the S.R.O. than B, although no rank difference exists between B and C. Among the generally somewhat more aggressive females, on the other hand, we observe between the subdominants mostly a so-called linear R.O., which means, for example, the alpha-female is dominant over B and C, and B is dominant over C.

Following these higher ranking wolves, we find a group of essentially younger, lower ranking animals. Also within this group we observe among the males mostly slight rank differences, and among the females, a more linear R.O.

Aggressive interactions between adult wolves of different sex are of low intensity and rarely arise. In spite of this, the alpha-female is usually dominant over the subdominant males, these in turn being dominant over the low ranking females, and the subdominant females are dominant over the low ranking males, and so forth. In other words, the two sex rank orders are not totally independent. No R.O. between wolves of different sexes is found, only between wolves on about the same level in the S.R.O. Between wolves of different aged subgroups, there is a rather fixed classification independent of sex. Severe rank disputes are invariably restricted to wolves of the same sex.

Within each subgroup there is, if there are enough wolves in the group, a dominant male and female. The subgroup dominant wolves often behave within the group much like the two alpha-wolves behave within the whole pack. They show the social display of alpha-wolves (tail high, straight legs, etc.) when they are together with other members of their group. Just as the alpha-wolves tend to keep the whole pack together, they also do the same within their group. Also, like the alpha-wolves, they are more aggressive towards pack strangers than the other pack members.

At the bottom of the S.R.O. stands, when existing, the group of cubs. They stand somewhat outside the R.O. of the adults since, on the one hand, they never participate in the rank disputes of the adults and, on the other hand, they enjoy a certain "accentric freedom." They can show a freedom of behavior that older subdominants never demonstrate facing dominants, as, for example, encroaching upon the individual distance of sleeping adults.

Likewise, the littermates of all litters observed had no S.R.O. with one another. To be sure, an imbalance of strength could be repeatedly observed but this was, however, usually only temporary, being caused, for instance, by the possession of a food morsel or play object or by mutual assaults in the course of play. This, however, is not an established R.O. of any sort.* The first real rank disputes between the cubs were not observed before the cubs were about 9 months old, at the beginning of their first winter. A very slight and often changing R.O. is established at this time, at first more or less independent of sex. Intensive rank altercations do not exist until the cubs reach their second year of life, then being limited to animals of the same sex.

At the bottom in the R.O. sometimes stands a so-called "scapegoat." The wolf, depending upon which position he occupied earlier in the R.O., is assailed by several or all other pack members, with the exception of the cubs. If the "scapegoat" had been at one time a high ranking or even alpha-wolf of his R.O., he will be assaulted according to how "unpopular" he had been previously; that is to say, according to how strongly he suppressed his previous rank subordinates, he will be assaulted either by all pack members or, if he generally behaved peacefully beforehand, only by those animals of the same sex that had fought with him over his rank position. If the "scapegoat" previously belonged to a group of subdominants, he occupies this position usually only within this group. The older, higher ranking wolves do not attack him. They sometimes even defend him. Joint attacks towards "scapegoats" occur quite seldom in the pack. When they occur, such animals, if living in the wild, might presumably withdraw temporarily or entirely from the pack. The way a wolf becomes a "scapegoat" will be described later.

SOCIAL RANK ORDER AND SOCIAL BEHAVIOR

There is, as we have seen, a complex S.R.O. in a wolf pack. In order to describe this in the wolf pack according to the rank situation, we may

*A battery of tests has recently been developed to determine temperament and social rank of 6–8 week old wolf cubs and has shown that it is possible to identify dominant, middle, and low ranking cubs at such an early age (Fox, 1972). Sex differences were not, however, observed at this time.

divide the pack members (in a somewhat simplified way) into five classes:

1. Alpha-male.
2. Alpha-female.
3. Group of subdominant males.
4. Group of subdominant females.
5. Cubs up to an age of 1 year.

Further, we will divide the social behavior of the wolf into three categories:

1. Friendly behavior. This includes all behaviors of a neutral character, for example, sniffing and muzzle contact, as well as play behavior and other friendly behaviors.
2. Aggressive behavior. To this category belongs all behaviors of threat, attack, combat, imposition (subordination), and defense.
3. Submissive behavior. To this category belongs active and passive submission (Schenkel, 1947, 1967).

Proceeding from this classification of social groups and of social behavior, a highly simplified sociogram (who does what to whom) for the wolf pack can now be drawn (Figure 23-2).

Accordingly, the behavior is mainly friendly between the following groups:

1. The two alpha-wolves to each other.
2. All adult subdominant wolves of different sex to each other.
3. The cubs with one another.
4. The adult wolves to the cubs.

Of course aggressive behavior can also occur between the wolves of these groups. Nevertheless, such interactions are seldom and are usually only of slight intensity. They do not have the character of rank altercations and are restricted to local-rank conflicts (Figure 23-3).

Friendly, submissive behavior is to be found among the following groups:

1. All cubs to all adult wolves.
2. All subdominants to the alpha-wolves of the opposite sex.

Here too, the behavior remains the same over a long period of time. Rank altercations are not to be found.

The following groups usually display friendly and submissive, but sometimes also strongly aggressive behavior:

1. All subdominant wolves to the alpha-wolf of the same sex and vice versa.
2. All subdominants of the same sex with one another.

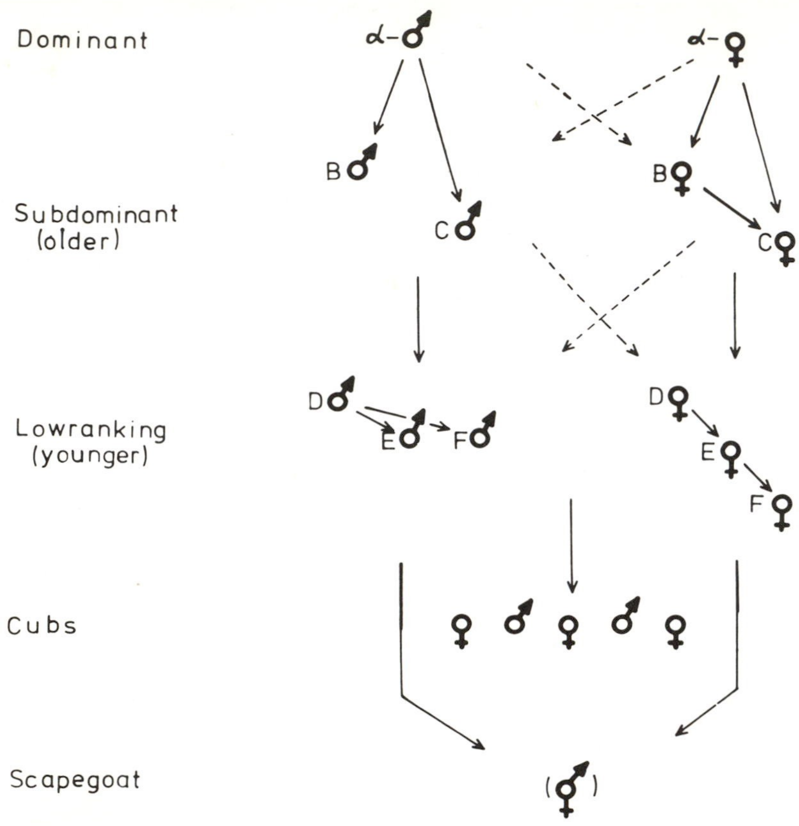

Figure 23-2 The social rank order in the wolf pack — a model.

Although as a rule friendly and submissive behavior is observed here, aggressive altercations can often occur between two or more wolves of these groups. These cannot be classified as momentary disputes, but as rank fights in their open settling phase (see later). Changes in the S.R.O. can arise out of such fights.

Social Dynamics in the Wolf Pack

In dealing with the social dynamics of the wolf pack we need to concentrate our interest on that behavior which possibly leads to a change in the social hierarchy of the pack; this is real aggressive behavior which is restricted to the relationship between adult wolves of the same sex. The frequency and intensity of aggressive behavior between two wolves is dependent upon the social relationship between the

	α — ♀	α — ♂	**Subdom.**	**Subdom.**	**Cubs**
α — ♀	—	Friendly	Friendly Aggressive	Friendly	Friendly
α — ♂	Friendly	—	Friendly	Friendly Aggressive	Friendly
Subdom.	Friendly Submissive Aggressive	Friendly Submissive	Friendly Submissive Aggressive	Friendly	Friendly
Subdom.	Friendly Submissive	Friendly Submissive Aggressive	Friendly	Friendly Submissive Aggressive	Friendly
Cubs	Submissive Friendly	Submissive Friendly	Submissive Friendly	Submissive Friendly	Friendly

Figure 23-3 A simplified sociogram for the wolf pack. Slight aggressive behavior is observed between all wolves. Severe aggressive behavior however, occurs only between adult pack members of the same sex.

two, but as emphasized earlier and enumerated below, it is also influenced by both seasonal and ontogenetic changes.

Seasonal Changes in Social Behavior

The mating season of wolves is in late winter. All adult females come into heat, although not all copulate and have a litter. Usually only one female of the pack bears cubs each year. How this "mother selection" works will be shown later.

Beside the marked increase in sexual behavior in those wolves which have a "sexual right," we also observe an increase in other social activities (Figure 23-4). General friendly pack ceremonies like chorus howling and what Murie (1944) called the "friendly general get-together" are seen much more often. Also, the frequency of neutral forms of social behavior like muzzle contact or fur sniffing is more common. Play activity is also high.

In general, it seems that free-living packs also stay closer together in winter (Jordan, Shelton, and Allen, 1967; Mech, 1970). It is not known whether this is due to the mating season or the necessity to form bigger hunting units in winter. In any case, it is interesting that the frequency of pack integrating forms of behavior is also much higher in a captive-living wolf pack during winter (Figure 23-5). This indicates a seasonal change in the tendency for social aggregation in wolves. Social bonds seem to be stronger in winter. Beside this increase in social interactions in winter, the aggressiveness of wolves also rises. This is of

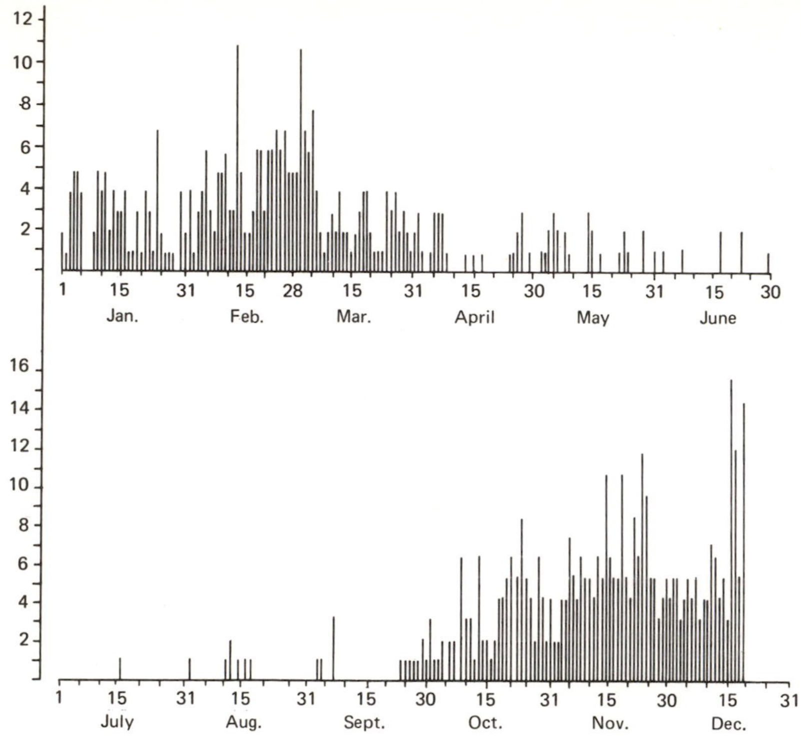

Figure 23-4 Frequency of chorus howling in the wolf pack in 1969. Reprinted with permission from Erik Zimen, *Wölfe and Königspudel,* R. Piper & Co. Verlag, München, 1971.

special interest for us, because changes in the S.R.O. are mostly due to some form of aggressive behavior at this time.

The increase in aggressive behavior starts in early fall, reaches its peak in late winter before, during, and after the mating season, and is back to "normal" again at the beginning of the summer. However, this is just a general tendency. Severe aggressive altercations between two or more pack members can be observed in the summer as well. In winter, the high rate of aggression is not necessarily displayed by all members of the pack. For example, the alpha-male, Wölfchen, showed only a slight increase in aggressive behavior in the winter 1971–72 towards the other pack members, while the two other adult subdominant males and the alpha-female were much more aggressive. The reason for this is found in the social situation of the pack in the winter 1971–72. The alpha-position of Wölfchen was not threatened in any way by the subdominant males. They showed only friendly and submissive behavior towards Wölfchen. They were, however, aggressive

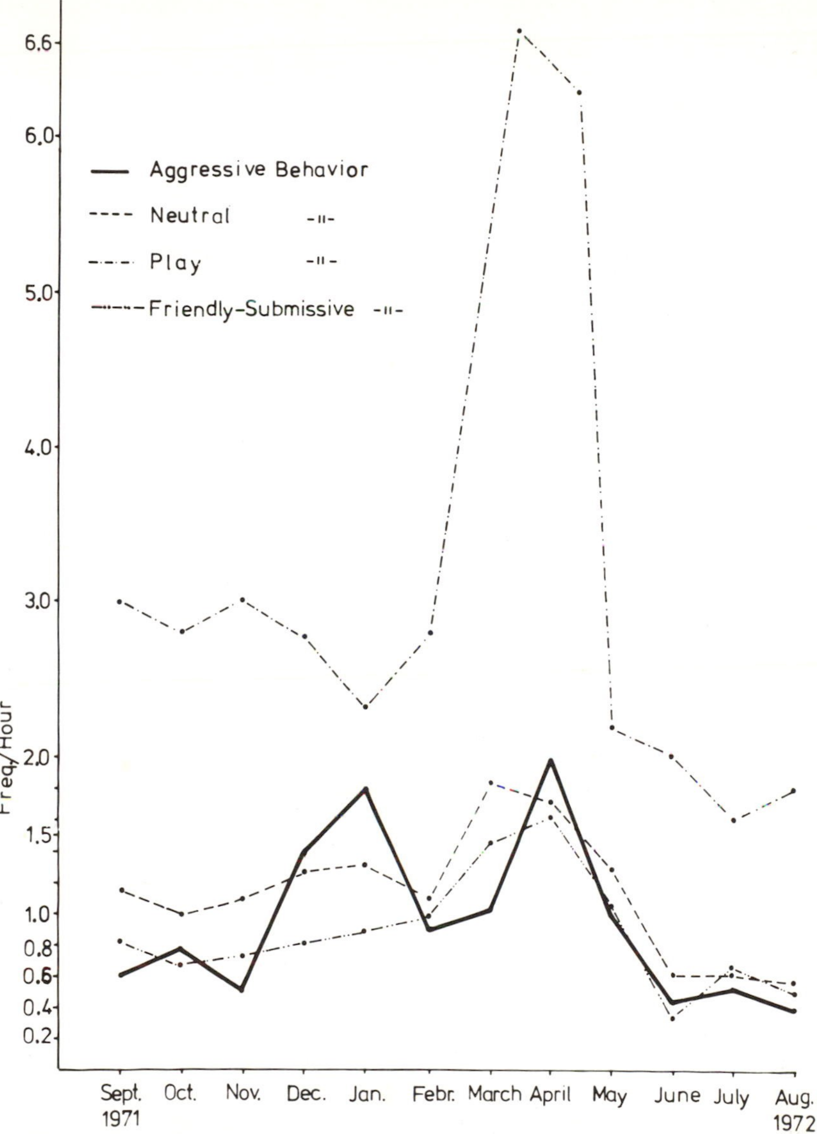

Figure 23-5 Average frequency (freq./hour and wolf) of social behavior in the pack from September, 1971 until August, 1972.

towards the low ranking and now almost 1 year old cubs. This shows, then, that the relationship between the individual wolves of the pack does not necessarily have to be more aggressive in winter. In a stable rank situation the behavior stays mostly friendly. This is at least true for the males. If there is more than one adult female in the pack, the alpha-female tends to suppress most social activities of the subdominant females by means of severe aggressive behavior. The males seem to be more friendly in this respect.

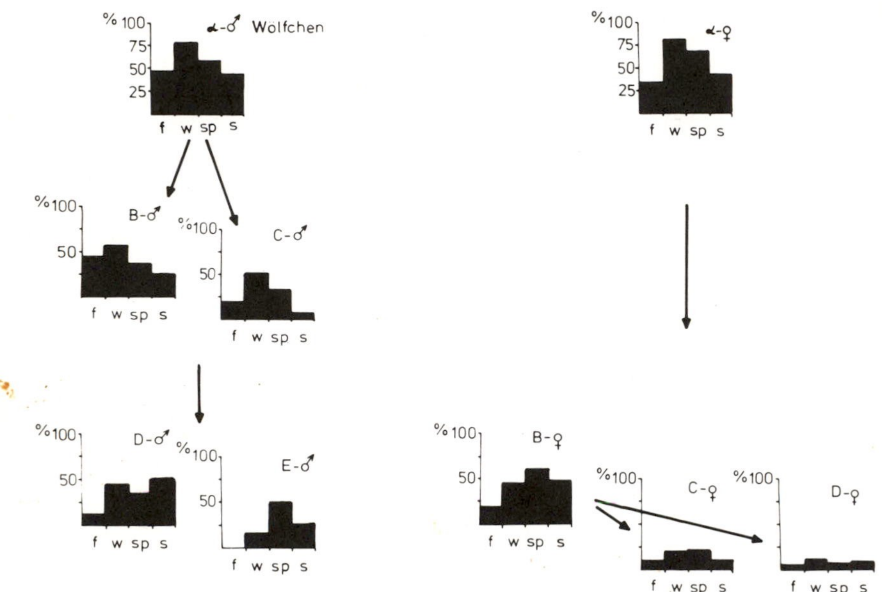

Figure 23-6 Aggressive behavior towards pack strangers by the pack members in correlation to the S.R.O. in the pack (aggressive behavior in % of all observations of strange dogs coming to the fence of the enclosure).

Number of observations:

Fall 1971 (f=Sept., Oct., Nov.)	23
Winter 1971–2 (w=Dec., Jan., Feb., Mar.)	21
Spring 1972 (sp=Apr., May)	11
Summer 1972 (s=June, July, Aug.)	48
Total	103

When we observe the behavior of the wolves towards pack strangers (Figure 23-6), the heightened aggressiveness in winter is obvious. Every once in a while domesticated dogs come to the long fence of the enclosure. All wolves react to these dogs more aggressively in winter than in summer, the most aggressive being the two alpha-wolves. Not only do they react aggressively more often, but also with a much higher intensity. This strong aggressiveness of the two alpha-wolves is due to their *role* as alpha-animals and is not primarily a matter of individual differences. When the C male (Alexander) of the pack in this winter 1971–72 was alpha-male in 1970, he was just as aggressive as

Wölfchen was now. Wölfchen had been a very gentle wolf when subdominant.

It is also significant to record that aggressive behavior towards pack strangers was intense in the two highest ranking wolves of that subgroup which were the cubs of last year. In many respects they behaved like "sub-alphas." Clearly, therefore, the aggressiveness in wolves is not only a matter of season and social relationship but is also dependent upon the rank (and role) of each animal. The higher a wolf ranks, both within the whole pack and within a subgroup, the more aggressive he seems to be (Figure 23-7).

Development of Aggressive Behavior

Aggression is also a function of age. Young wolves are very friendly, but their aggression constantly increases until they reach maturity.

Figure 23-8 shows the average frequency of aggressive behavior in different litters up to the age of 2 years. In spite of the different condi-

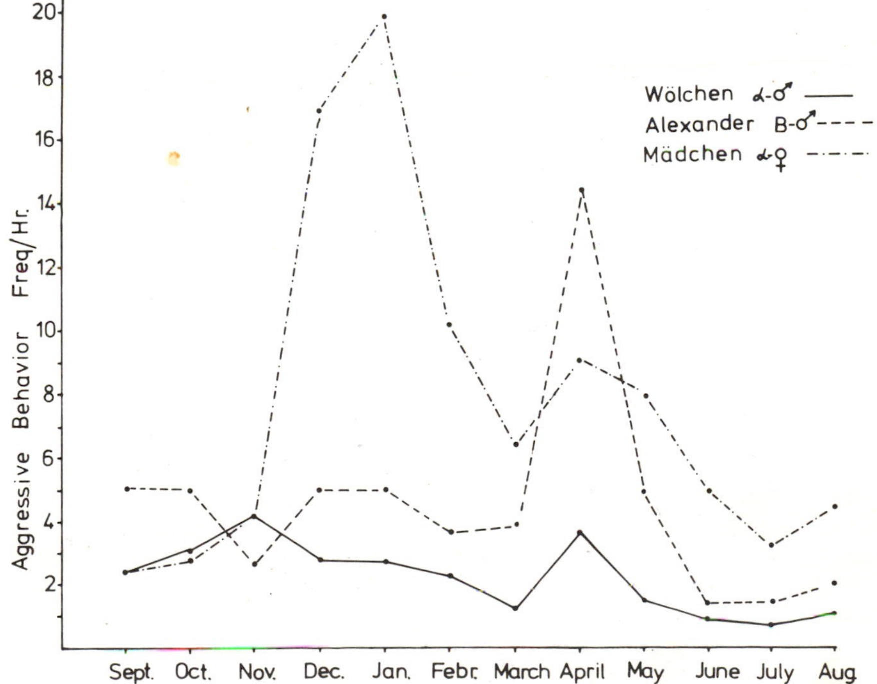

Figure 23-7 Average frequency (freq./hour) of aggressive behavior in 3 adult wolves from September, 1971–August, 1972.

tions under which the cubs grew up, the development of aggressive behavior was more or less alike in all litters.

In their first month of life the cubs show quite a bit of aggressive behavior. This is followed by a period of few aggressive disputes until an increase is noticed at the end of their first year.

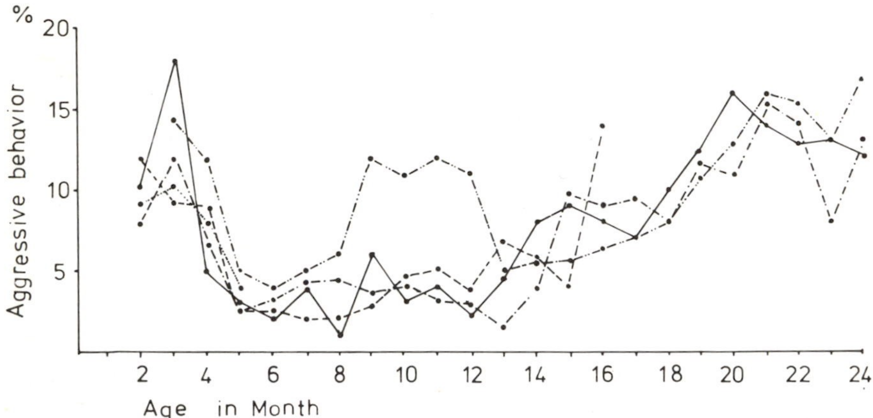

Figure 23-8 Ontogenetic development of frequency of aggressive behavior in 5 different litters. (1)(· ·—· ·—· ·): Anfa (0/1), born March, 1967, (—·—·—):litter (2/1), born May, 1967. (——): litter (4/1), born May, 1968. (— — — — — —): litter (2/4), born May, 1971. (· · · · · ·) (litter (1/2) born May, 1972.

The very low aggressiveness of the cubs was most striking each summer and this is also true for the young cubs. The high frequency of aggressive behavior at this age occurs mostly during feeding. The fast-growing cubs have a very high need for meat and eat a lot. There is much fighting over food morsels since, because it has not yet developed, individual distance does not block aggressive altercations of low intensity as it does in the old wolves. But in spite of this high frequency of aggressive behavior, the young cubs cannot be called aggressive. A few seconds after a fight the cubs may play with each other or sleep closely together. Normally, aggressive behavior at this age has nothing to do with rank disputes but is restricted to local conflicts. Strange wolves, dogs, and humans (should the cubs be socialized to humans) are greeted in a most friendly, submissive way.

Longer-lasting aggressive disputes between the cubs do not occur until the beginning of winter. At this time, the first slight aggressive reactions to pack strangers entering the pack territory are noticed. Real aggressive behavior, however, is not observed before the young wolves reach their second year of age.

Rank Altercations in the Wolf Pack

Three typical interactions in the social dynamics of the wolf pack will now be described. First, the behavior of three females before, during and after the mating season; secondly, the behavior of three males after the alpha-male was removed from the pack; and finally, the behavior of a group of young wolves towards the adult wolves of the pack will be described.

The Behavior of Three Females. In October, 1968 there were 8 (5/3) wolves in the pack, 4 almost adult 1 year olds and 4 6-month old wolves. Anfa, the oldest female, is slightly dominant over Andra. Mädchen, the 6 month old female, is low-ranking.

The behavior between the three females is generally friendly until October, 1968. All three females play with each other. In November, the two adult females become more and more aggressive. Andra shows a strong expansion tendency towards the alpha position of Anfa. She attacks here, threatens, and shows imposition behavior. This happens at first during play, and Andra's activity increases, while Anfa and especially Mädchen, who by now often has to take hard attacks by Andra, play less and less (see Figure 23-9).

In January and February, 1969, no rank difference exists between the two adult females. Aggressive interactions become more and more frequent. Mostly the initiative comes from Andra. Play activity decreases further now in all three females.

At the beginning of March, Anfa, and then about 2 weeks later, Andra come into heat. Mädchen, too, shows vaginal bleeding. Without a real fight, as far as I know, Andra is now the dominant female. She suppresses not only all social activities of Mädchen but also of Anfa. Sometimes Anfa tries to get closer to one of the males but immediately Andra gets between them and drives Anfa away, displaying her alpha-position dramatically. None of the three play anymore. Every play intention of Anfa and Mädchen is at once blocked off by Andra. Andra herself is totally occupied observing and suppressing Anfa and Mädchen. None of the three females mate.

After the mating season in April and May, the social suppression of Anfa and Mädchen slowly decreases. Play activities by the females are again observed. Sometimes Andra still tries to block the social activities of the subdominant females, but has less success with Anfa since she now protests strongly. The aggressiveness of the females has clearly decreased.

In June, the situation between Andra and Anfa again becomes more

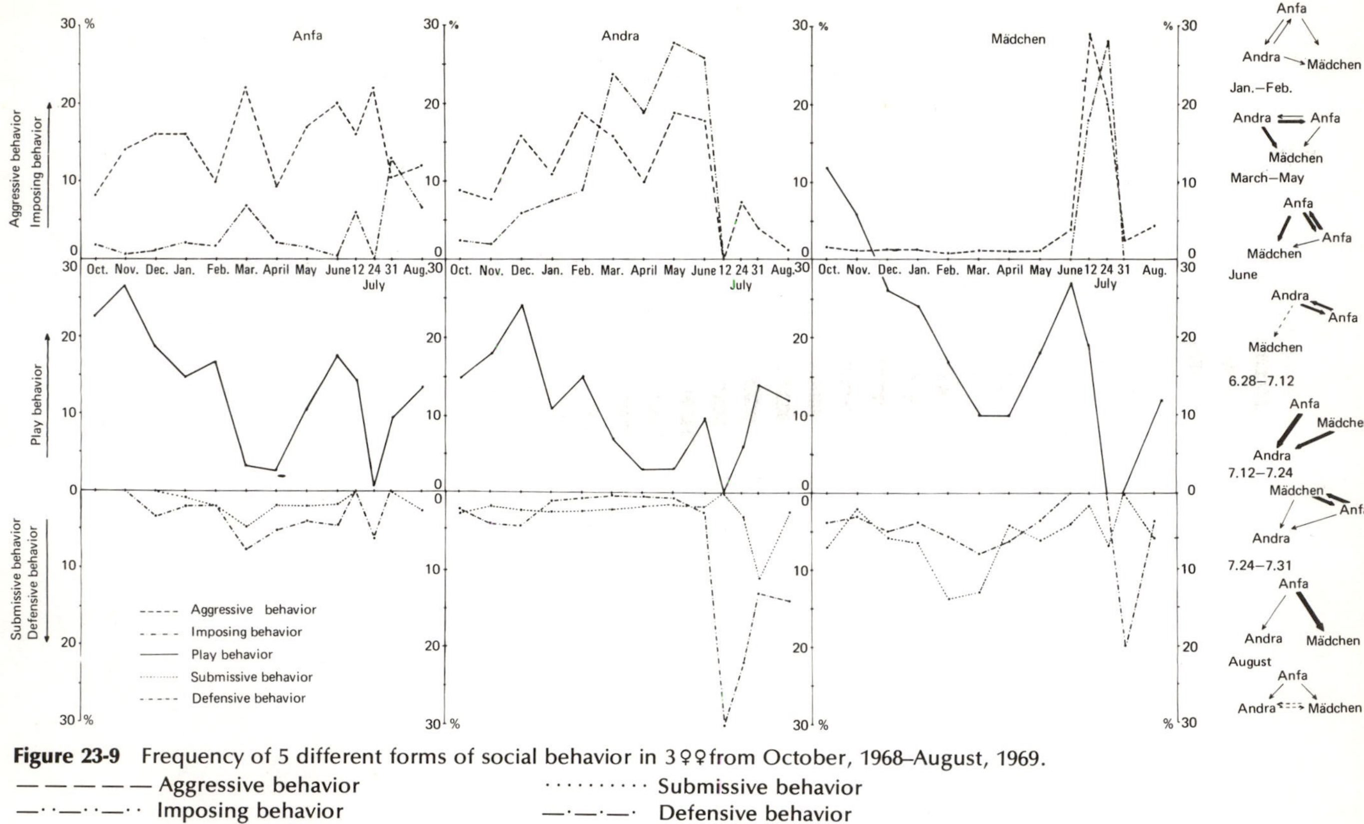

Figure 23-9 Frequency of 5 different forms of social behavior in 3 ♀♀ from October, 1968–August, 1969.

— — — — — Aggressive behavior
—·—·—·—· Imposing behavior
———— Play behavior
·········· Submissive behavior
—·—·— Defensive behavior

critical. Now Anfa shows the initiative. In the posture of the subdominant wolf (tail between the legs, head low) she now attacks Andra. Although Andra rarely attacks Anfa, she still shows the imposition behavior of an alpha-wolf. Mädchen at this time has recovered all her social freedom and her play activity is back to normal again. Thus, the high frequency of aggressive behavior in June is not like the late winter social suppression but is rather a rank altercation between the two adult females. Anfa shows a strong expansion tendency towards the alpha-position of Andra.

In the late afternoon of June 29, the decisive fight between the two adult females occurs. The whole day Anfa attacks Andra. Andra reacts only with imposition behavior. The situation is extremely tense, although hard biting has not occurred. The two females still show a bite inhibition, normal for all aggressive altercations with the exception of the decisive fight, which now faces the two females. The fight starts suddenly. As usual, Anfa attacks Andra first, but this time she does not just threaten and snap at Andra, but bites hard. Andra bites back at once. A real fight starts. Mädchen joins in and fights the alpha-female Andra together with Anfa. The main fight, however, is carried out by Anfa and Andra. They bite and shake their heads vigorously. The fight is silent and there is no display of any sort. After a few minutes of intense fighting, it is obvious that Andra is losing. In a very defensive posture she withdraws into a corner, but the other two females continue the attack with full intensity. The five males stand close by showing much interest in the event but they do not join in. Again and again Anfa and Mädchen attack their old alpha-female. All three are injured and a lot of blood is spread around. Slowly, the fighting weakens although Anfa and Mädchen continue to run around Andra who, for hours, has positioned herslef into a hole in the ground. Only her head sticks out and she shows intense threat display every time either of them comes close.

On the next day, the attacks without bite inhibition on Andra continue. Anfa is now alpha-female. She runs with legs stiffened and tail high, urinates with a leg lifted and shows all the other displays typical for an alpha-wolf. Andra, on the other hand, moves around in an extreme defensive posture. Sometimes she is even attacked by the males. Andra has become the "scapegoat" of the pack. The most intensive attacks, however, are caused by the females, and especially by Anfa.

With time, the attacks on Andra weaken. Then suddenly and most unexpectedly, Mädchen shows an expansion tendency towards Anfa, and the relationship between the two gets critical within a few days. Mädchen attacks Anfa who most vigorously protests against this new behavior of Mädchen. But Mädchen goes on attacking. On the evening of the July 12, a bib fight occurs. This fight is just as intense as the one of Anfa and Mädchen against Andra. After about 15 minutes of fighting, the two females separate. Both females are badly injured. The winner cannot be determined.

The next day, the fight is suddenly resumed. Anfa receives a bad injury on her nose; Mädchen seems to be somewhat superior, but Anfa shows no escape behavior. The females separate and now Mädchen shows the alpha-posture. However, Anfa, unlike Andra, has not lost completely. Now, in the posture of a subdominant, she threatens the imposing Mädchen and sometimes even attacks her (Figure 23-10). These attacks on Mädchen become more vigorous and on July 2 a severe fight again starts. This time Andra joins and fights on the side of Anfa against Mädchen. The altercation lasts about 10 minutes. Then Mädchen cannot stand the united attacks any longer and flees. Anfa pursues her, once again in the posture of an alpha-wolf.

On the following days Anfa continuously attacks Mädchen, who in return has to defend herself very hard. Andra has got most of her social freedom back. She plays again and shows intense submissive behavior towards Anfa.

Eventually the excitement calms down and Anfa is now the unchallenged alpha-female, followed by Andra and Mädchen who is now last in the R.O. About the middle of August, Mädchen shows submissive behavior towards Anfa for the first time. At first Anfa bites, but later on she leaves Mädchen alone. All three play again. The S.R.O. between females is stabilized.

In the fall, all three enjoy a great deal of social freedom. Anfa is a "tolerant" alpha-female. In winter, however, and most of all during mating season, Anfa, just as Andra did in the previous year, suppressed most social and sexual activities of the other females. In the spring, social freedom increased and in the summer everything was back to "normal" again. The three females interacted more or less in a friendly and friendly–submissive manner. An expansion tendency towards the position of Anfa is now nonexistent.

The Behavior of Three Males in the Absence of an Alpha-Male. In October, 1969 Grosskopf was the alpha-male in the pack. Between the three subdominant males, Alexander, Näschen and Wölfchen (all 1½ years old), no S.R.O. could be observed. Grosskopf played with all three and showed no form of suppression. The three subdominant males showed friendly–submissive behavior towards Grosskopf, and also among themselves interactions were friendly. Aggressive interactions were rare and of low intensity.

Figure 23-10 Active protest and attack of the subdominant Anfa against Mädchen between July 12 and 21.

At the end of October, Grosskopf was removed from the pack. At once, aggressive interactions among the three subdominant males increased considerably (Figure 23-11). Näschen was especially aggressive and soon seemed to occupy the alpha position. The other two animals moved out of his way but showed intense protest behavior when Näschen got too rough; the situation was not yet clear. Hardly any play was observed between the males.

In January, 1970, Näschen's aggression was directed only at Wölfchen. Toward Alexander he displayed only imposition behavior, while Alexander grew more aggressive towards Näschen. On January 23, the anticipated fight starts. Wölfchen joins in at once and together with Alexander, fights Nächen. Although no severe fighting occurs, Näschen gives up quickly. Without injury, he withdraws in order to only defend himself.

Just as fast as this change of power occurred, the excitement of the three males decreased, but Näschen was not totally defeated. Although he had lost the alpha-position to Alexander, he was still superior to Wölfchen. Since he was tolerated by Alexander, a rather

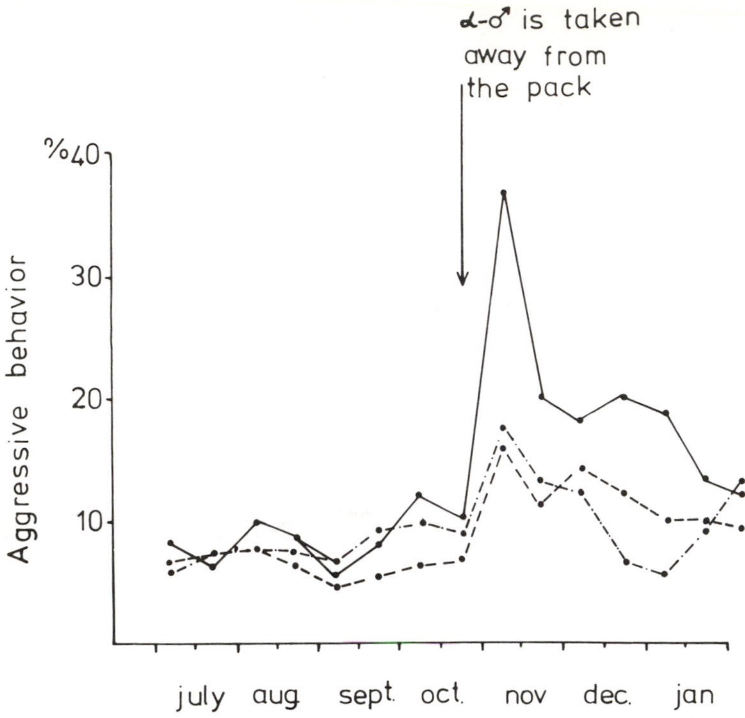

Figure 23-11 Frequency of aggressive behavior in 3 subdominant ♂♂ before and after the α - ♂ was taken away from the pack.

stable and calm situation existed in the pack for a while, but not for long. Soon Wölfchen, "Number 3" in the males R.O., started to be aggressive to "Number 2," Näschen. This was the beginning of Wölfchen's "long march to the top." Alexander supported Wölfchen's attacks on Näschen. This was too much for Näschen and the S.R.O. changed to : Alexander, Wölfchen, Näschen. However, aggressive interactions between Wölfchen and Näschen continued. On March 23, an intense fight erupted between the two, which Näschen won (Alexander had been locked up in another kennel and could not join the fight). This put Näschen again ahead of Wölfchen in the S.R.O., but due to Alexander's support, the R.O. changed once again in favor of Wölfchen.

These alterations in the R.O. among the subdominants went on for quite a while, but no further serious fights took place; Alexander's alpha-position was unthreatened. On his way to the top, Wölfchen first settled things with Näschen.

During the summer, Alexander was quite friendly to the two subdominants. Since he became alpha-male, however, he grew more

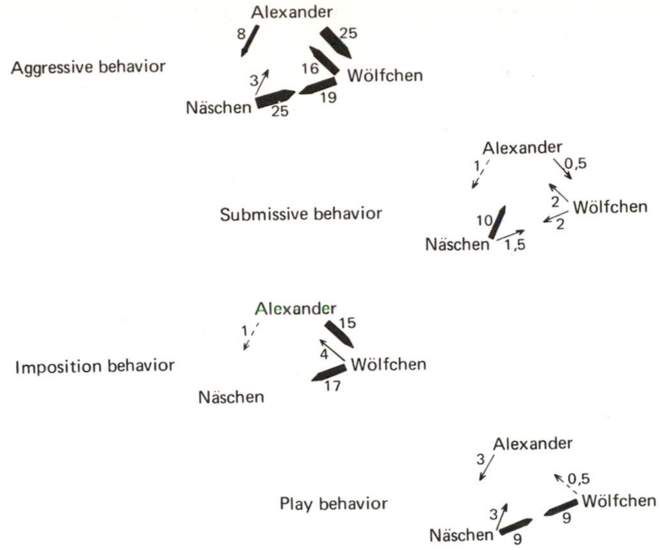

Figure 23-12 Social behavior between the 3 ♂♂ wolves of the pack in September, 1970. The number beside the arrow indicates the frequency (% of all observations this form of behavior was shown in one ♂ towards one of the other two ♂♂ .

aggressive towards me. When I entered the enclosure he greeted me holding his wagging tail high and stiff. When I approached him he assumed a threatening posture with the hair on his back raised. Other people who came to the enclosure fence were threatened even more, and when he saw strange dogs he turned really wild. The few times I managed to take him out of the enclosure on a leash, he did not threaten me, but he was highly aroused, urinated at every tree and would, if I had not kept him back, have attacked anybody. Under these circumstances I preferred to keep him in the enclosure. The other two males were as friendly as usual.

At the end of summer, Wölfchen, now clearly superior to Näschen, showed the first expansion tendencies towards the alpha-position of Alexander. This is the last act of the drama of Wölfchen's rise to the top.

Figure 23-12 shows the interactions of the male wolves of the pack for the month of September, 1970. The frequency of aggressive behavior is very high. In 25% of all observations, Alexander, as well as Näschen, show some kind of aggressive behavior towards Wölfchen, who himself behaves very aggressively towards the other two. The relationship of Alexander and Näschen, on the other hand, is friendly, Näschen showing submissive behavior towards Alexander. This indicates a stable rank situation between these two. The rank relationship is contested betweeen Alexander and Wölfchen and between Wölfchen and Näschen. Imposition behavior was shown by Alexander towards Wölfchen and by Wölfchen towards Näschen, just like the S.R.O. indicated. In 4% of all observations, Wölfchen also tried to impose on Alexander. This clearly indicated Wölfchen's expansion tendency. Also interesting is the lack of imposing behavior by Alexander towards Näschen. Social display of this kind normally occurs between a dominant towards a subdominant wolf and is more intense if the rank situation is unstable. In a stable rank situation, as between Alexander and Näschen, the dominant animal hardly imposes at all on the subdominant. Further, it is interesting to note that Wölfchen tried to impose on Alexander, but Näschen did not impose on Wölfchen. This again indicates that rank altercations are much more intense for the alpha-position than for the subdominant positions. This becomes even more evident when we look at their play behavior. The two wolves with a stable rank relationship play with each other although not very often (Alexander was occupied observing Wölfchen, and played very little at all). The opponents Alexander and Wölfchen did not play at all with each other, while the opponents Wölfchen and Näschen, however, played almost as often as normal.

Let us summarize the behavior of the three wolves in September. The stable relationship between the alpha-male and a subdominant (Alexander to Näschen) was characterized by little aggressiveness and hardly any imposition, by friendly–submissive behavior, and mutual play. The unstable relationship of the alpha-male and a subdominant who has strong expansion tendencies towards the alpha-position (Alexander–Wölfchen) was characterized by repeated intensive aggressive interactions, frequent imposition behavior by the alpha-animal, and the lack of mutual play. The unstable relationship of two subdominants (Wölfchen–Näschen) was characterized by numerous but not intensive aggressive interactions, frequent imposition behavior to the lower ranking males, nearly no submissive behavior, and almost normal play behavior.

In the last days of September the aggressive behavior of Alexander and Wölfchen reached its peak. The initiative was taken by Wölfchen who stayed around Alexander all the time, using every opportunity to attack. Alexander showed imposition behavior but did not take the initiative to attack by himself any more.

On October 1, the long-expected final fight broke out. The silent, but intense, fight lasted 15 minutes; Näschen joined in too and fought along with Alexander against Wölfchen. Nevertheless, Wölfchen managed to injure Alexander seriously and Alexander retreated to a kennel where he was forced to defend himself for many hours. In the days that followed, aggressiveness was still high; Alexander stayed in the kennel until October 20.

Following every decisive altercation for the alpha-position, it was interesting to observe the sudden changes of behavior in the new alpha-wolf. Within minutes, Wölfchen took up the traits of an alpha-wolf, e.g., high tail, legs stiffened, etc. Outside the enclosure he soon acted like Alexander did before (excited, urinates everywhere, and tries to attack strange dogs, and especially inside the enclosure, he becomes more and more aggressive towards me). From the day that Alexander lost the alpha-position, he behaved in a freindly way. On the leash he was calm and very friendly to strange dogs and humans. Even the previously "hated" keeper was now greeted in a most friendly, submissive way.

Wölfchen soon became the unchallenged alpha-male of the pack. He still occupies this position today. Especially in winter, Alexander is somewhat more suppressed by Wölfchen than is Näschen. But in winter none of the subdominant males are so hard suppressed by the alpha-male as the subdominant females are by the alpha-female.

Among the two subdominant males no R.O. has been established. After the lost fight, Alexander, who was a rather "tolerant" alpha-male, was not attacked by the other pack members. Näschen even takes care of him. He often sleeps close to him, licks his wounds, and defends him against the attacks of Wölfchen. This friendly, rankless relationship still exists. It seems to be typical for subdominants of the same age, if there is a strong, unchallenged alpha-male in the pack.

The Integration of Five Cubs into the S.R.O. of the Adults. Typical for the actions of cubs towards adult wolves is their submissive behavior. This is displayed either spontaneously or in response to any sort of aggressive behavior by the adults. Spontaneous submissive behavior occurs especially after a separation when the cubs meet up with adults again, or at the beginning of an activity period (the "general, friendly get-together," Murie, 1944).

In the spring of 1971 we had a group of five cubs. They were kept together with Wölfchen (alpha-male), Mädchen (alpha-female), Näschen, and Alexander (subdominant males). At first the cubs showed submissive behavior towards all four adult wolves alike (Figure 23-13). Soon however, we observed an increasing preference for Wölfchen and Mädchen, the alpha-wolves. Somehow the cubs realized the rank differences between the adults.

At the beginning of winter we also observed a slight preference of the male cubs toward Wölfchen and the female cubs toward Mädchen. At this time the cubs also had their first rank disputes. A slight and

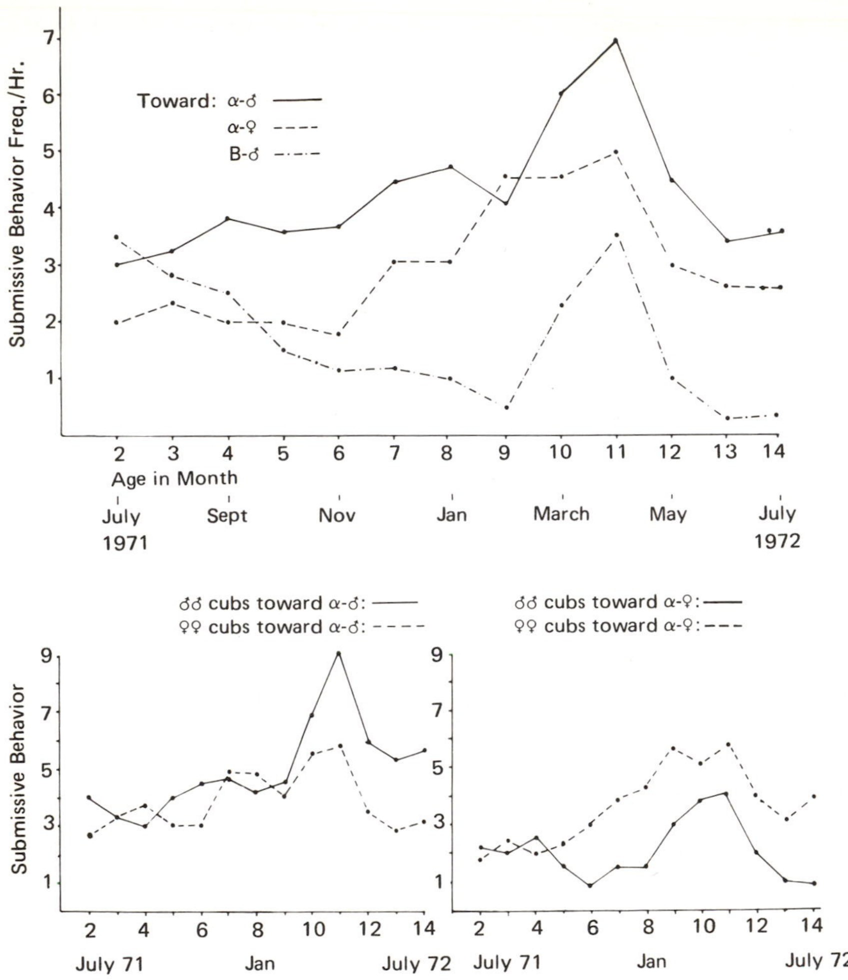

Figure 23-13 At top: submissive behavior (freq./hour and wolf) in the cubs (2/4) of 1971 toward 3 adult wolves of the pack. Below: submissive behavior (freq./hour and wolf) in the ♂ and the ♀ cubs toward the α - ♂ and the α - ♀ .

often changing R.O. was established, and as in adults, an R.O. existed for females and for males. Aggressive interactions between cubs of opposite sex, however, were much more common at this age than among adults. This resulted in a dominance of the two larger male cubs over the smaller females, but severe aggressive behavior was not observed.

In March, social activities of the pack reached a peak. There was much playing among the cubs and between the cubs and the lower-ranking adult male, Alexander. (Näschen had run away and was thus not in the pack anymore.) When playing, the almost year old cubs sometimes showed slight aggressive behavior towards Alexander, especially during more or less playful group attacks on him. To this Alexander reacted aggressively. When playing individually with the cubs, every once in a while he pinned them to the ground in a half-aggressive, half-playful manner or showed them his superiority in another way; play began to get rougher. The young wolves reacted to this slightly increasing aggressive relationship with Alexander with a much stronger tendency to display active and passive submission towards him. Thus, Alexander had for the time being, anyway, successfully blocked off all expansion tendencies by the young wolves.

SOCIAL DYNAMICS AND SOCIAL RANK ORDER: A MODEL AND SUMMARY

These three events, like many others that have been observed, allow us to supplement the sociogram drawn up in Figure 23-2. The sociogram illustrates how severe aggressive behavior is restricted to altercations between adult wolves of the same sex. We are now acquainted with those social situations in which aggressive behavior occurs and also in which adults behave in a more friendly way towards each other.

In Figure 23-14, the behavior of two adult wolves of the same sex is correlated with their rank situation in which A is dominant over B. Although rank altercations for the alpha-position are far more intense than among subdominant wolves, the model is applicable to the behavior of all adult wolves of the same sex.

Four possible rank situations among adult wolves are found:

1. A firm and mutually accepted rank situation in which A is clearly dominant over B.
2. The subdominant B shows an expansion tendency towards the dominant A.
3. The dominant A suppresses the subdominant B.
4. No rank difference.

If the rank situation between A and B is firm and accepted by both, as well as if no rank difference exists, neutral and friendly social interactions are normal. In the first case, if A sometimes displays his superiority, B reacts to this with submission. He also shows spontaneous submissive behavior to A. Should A impose too much or should he otherwise behave too aggressively, B protests. These aggressive reactions indicate that subdominants in such a rank situation are not at all without social rights. Rights have to be maintained by the subdominant as

	Expansion tendency of B towards A ⟵	**Stable rank relationship between A and B**	**A suppresses B** ⟶
Behavior of A	Imposing behavior Attacks	Friendly behavior Slight imposing and aggressive behavior	Imposing behavior Attacks
Behavior of B	Spontaneous attacks Imposing behavior	Friendly and submissive behavior Protest when aggression of A too strong	Flight and defensive behavior
Behavior of A+B	Real (decisive) fight with no bite inhibition	Play	—

Figure 23-14 Forms of social behavior between a dominant wolf (A) and a subdominant wolf (B) in relation to their rank situation.

well as by the dominant. The rank difference between the two is not really absolute and stable but is characterized by a somewhat dynamic social equilibrium.

Spontaneous aggressive behavior of B towards A indicates an expansion tendency of B. The first signs of this are normally observed in usual friendly interactions, like play or the "friendly get-together." B bites a little bit harder than usual or displays aggressive and imposing behavior somewhat out of his normal behavior. A immediately reacts aggressively, either in a more playful manner or directly by means of threats, attacks, and imposition behavior.

Generally, most expansion tendencies in subdominants are blocked off at this early stage. This probably happens quite often but is not always noticed by the observer. The situation becomes evident, however, if B does not stop attacking. Friendly behavior and play decreases and finally disappears. The initiative is slowly taken over by B. He initiates most of the attacks, while the behavior of A is more and more restricted to imposition behavior. Such a tense situation soon causes a decisive fight. The wolves now bite as hard as possible. They show no aggressive displays and the fight is silent. Other wolves may join the fight, to either help a "friend" or fight an "enemy." These participants may fight just as intensely as the main opponents or may only snap occasionally. The fight can end: (a) before a decision has been reached, (b) with one of the fighters in a slight disadvantage, and (c) with the total defeat of one fighter. In case of a draw, the rank situation does not change. Soon there will be another fight. The tense situation also does not change if one opponent has a slight advantage at the end of the fight. However, a change of dominance may have occurred. The former subdominant animal now displays the expressive behavior of a dominant and the former dominant wolf now shows all the traits of a subdominant with an expansion tendency. Again the probability of a new severe fight is high.

The rank situation remains unchanged until one of the opponents has totally lost the struggle. The winner (former or present A) will continue attacking the loser B as hard as possible. The loser has to defend him or herself or flee. Submissive display in this situation is of no use; it will not inhibit further severe aggression.

If other wolves join the attack, B will become a "scapegoat." This, however, is not a general rule. Whether other wolves will join the attacks on B depends mostly on their relationship to B before the fight. If the relationship was tense and aggressive, they are likely to join; otherwise, it is not usual. A good "friend" may even defend B against the attacks of A. Whether B gets to become a "scapegoat" or not, he will, just as other heavily suppressed wolves, tend to keep away from the pack for the first days and even weeks after the fight. In the wild, the so-called "trailing wolves" (Jordan, Shelton, and Allen, 1967) probably are those wolves who were extremely suppressed by some or all members of the pack. It is possible that they leave the pack totally and live by themselves.

If B stays close to the pack (in an enclosure, of course, he must), he will not try to join the pack again before his wounds are almost healed. At a time of "general friendly behavior" in the pack, he will try to get closer. At first he will show friendly, submissive, or playful behavior to less aggressive pack members. If A comes close, he will run away. But as the attacks of A weaken, B will show, at a distance first, signs of submissive behavior. If A does not react with attack, B will then come closer and soon the first very submissive direct contacts will take place. Slowly, the dynamic equilibrium of a normal social relationship between a dominant and a subdominant wolf then takes place. This "resocialization" or reintegration of extremely suppressed "outcast" wolves lasts — as I have observed it — from 2–3 weeks up to a couple of months. None of our wolves were extremely suppressed for a longer time.

The suppression of a subdominant by a dominant wolf is not always due to a fight for dominance. In winter, especially during the mating season, the alpha-wolves often suppress subdominants of the same sex. This is especially true for the alpha-female. Intense attacks, threats, and imposition behavior block off most social and sexual activities of the subdominants. Social suppression of this kind probably prevents the sexual synchronization of the subdominants which is necessary for successful mating. This may be the reason for the frequently observed phenomenon that only one female of the pack, normally the alpha-female, mates and has a litter. This has been observed in many wolf packs held in captivity (Rabb, Woolpy, and Ginsburg, 1967; Zimen, 1971). Also, the big pack studied on Isle Royale most likely did not produce the possible maximum of litters (Mech, 1966; Jordon, Shelton and Allen, 1967). However, 89% of all adult females caught by Rausch (1967) in Alaska during late winter were pregnant. This high percentage is probably due to the small size of Alaskan wolf packs. It is also possible that too few wolves exist to form big packs and therefore the reproduction rate is high. Also, those few big packs that still exist in unhunted areas, as in Mt. McKinley National Park, sometimes produce more than one litter (Murie, 1944; Haber, personal communication). This has especially been observed in packs where, probably due to the abundance and behavior of the main prey species, pack splitting often occurs. In contrast, usually only one litter a year has been observed in those packs which stay closer together. Nevertheless, this form of potential population control is a most interesting phenomenon in the wolf pack.

24
HUNTING BEHAVIOR IN TWO SIMILAR SPECIES OF SOCIAL CANIDS

L. DAVID MECH
Wildlife Research Biologist
U.S. Bureau of Sport Fisheries and Wildlife
Twin Cities, Minnesota

The Canidae as a family of carnivores has been remarkably successful in colonizing most of the land area of the earth. With some 14 living genera, family representatives thrive on every continent except Antarctica. Individual member species occupy remarkably diverse ecological niches, with some species functioning basically as hunters and others as scavengers. Some are almost tiny, and others much larger. Several are solitary-living, whereas a few are social.

Two of the social species, the gray wolf (*Canis lupus*) and the African or Cape hunting dog (*Lycaon pictus*), stand out in occupying similar ecological niches but on opposite sides of the globe. The wolf originally inhabited most of the North Temperate and North Frigid Zones in North America and Eurasia, whereas the hunting dog lives in the Torrid and South Temperate Zones of Africa. Thus, the climate, soils, flora, and fauna differ substantially in the ranges of these two canids. Nevertheless both the wolf and the hunting dog face the same problem: securing a livelihood by preying on large ungulates.

Because of the similarity of the ecological problem but the differences in the environmental circumstances faced by the wolf and the hunting dog, it should be enlightening to examine the behavioral and ecological solutions that the two species have evolved. Therefore, this paper will compare the wolf and the hunting dog in terms of their traits and habits most related to the problem of hunting, killing, and feeding upon large ungulates.

PHYSICAL ATTRIBUTES

Physically, the wolf and the hunting dog are quite similar, being long-legged and larger than most of the other canids. Adult wolves weigh from 60 to 120 pounds (27 to 54 kg), with 175 pounds (80 kg) being the heaviest on record (Mech, 1970). Adult hunting dogs range from 40 to 60 pounds (18 to 27 kg) (Estes and Goddard, 1967). The faces of each species are typically canid, but hunting dogs have much larger ears, while wolves possess longer noses.

These differences in the external portions of the sensory organs may reflect differences in the relative importance of the various senses of these two species. Wolves in wooded areas rely considerably on their sense of smell to locate prey (Mech, 1970), although in open areas vision sometimes becomes important.

Hunting dogs, coursing the open plains of Africa, depend primarily on vision and presumably hearing to locate and follow their prey, although in bush country they may also use their olfactory sense (Estes and Goddard, 1967). Probably their heavy reliance on vision is the cause of their basically diurnal hunting. Wolves hunt day or night (Mech, 1970).

SOCIAL BEHAVIOR

Both wolves and hunting dogs hunt in packs, which allows them to locate, chase, kill, and consume large animals most efficiently. Most packs of wolves range in size from 2 to 12 (Mech, 1970), and the largest pack reliably reported contained 36 (Rausch, 1967). For hunting dogs, Estes and Goddard (1967) concluded that 4 to 6 would be the minimum effective pack size, and packs of 21 or more are known (Estes and Goddard, 1967).

In wolves, packs are basically family groups, with several successive litters or occasionally concurrent litters making up the larger packs (Murie, 1944), and this probably is also true of hunting dogs (van Lawick-Goodall, 1971). Thus, in both species the skills of stalking and killing prey are easily learned as the pups and yearlings accompany the adults on their hunting forays.

Within packs both of wolves (Schenkel, 1947) and of hunting dogs (van Lawick-Goodall, 1971) a dominance hierarchy not only helps maintain order but also aids in hunting. Mech (1966) has observed the lead wolf in a pack of 15 being the most highly motivated in pursuit of moose (*Alces alces*), and Estes and Goddard (1967) noted that it was the lead hunting dogs that singled out certain members of herds of prey. By concentrating on an individual, a pack can focus all its attention on that individual and thus increase the chances of success. This requires that some pack member must be commonly accepted by the others as the one to single out the prey animal.

As part of their social natures, wolves and hunting dogs communicate within their packs through a rich repertoire of vocalizations, and these also add to the species' hunting success. Howling among wolves helps separated pack members to get together quickly (Mech, 1970), and the same appears to be true of the "contact call" or "hooing" of hunting dogs (Estes and Goddard, 1967). "Twittering" by hunting dogs just before or during a chase is thought to foster concerted pack action (Estes and Goddard, 1967). Similarly, it is widely believed by laymen that wolves "yip" when pursuing prey, although this contention has not yet been documented.

Both wolves and hunting dogs also communicate through scent-marking by urine and feces (Mech, 1970; van Lawick-Goodall, 1971). Scent-marking is thought to help advertise and reinforce the dominant pack members and thus strengthen the social order, and also to mark areas used by each pack and perhaps help exclude neighboring packs.

The exact nature of the spacing of adjacent packs is unknown for both wolves and hunting dogs. However, an increasing amount of evidence is accumulating that wolf packs are basically territorial (Mech, 1966; Jordan, *et al.*, 1967; Pimlott, *et al.*, 1969; Mech and Frenzel, 1971), with pack territories ranging in size from approximately 64 to 384 km^2 (25 to 150 $mile^2$) in Minnesota, for instance (Mech, unpublished). In Alaska, where the degree to which wolf packs are territorial has not yet been established, one pack traveled an area of some 12,800 km^2 (5000 $mile^2$) (Burkholder, 1959).

Much less is known about the spacing and home range sizes of hunting dog packs. The van Lawick-Goodalls (1971, p. 89) believe that these packs are not strictly territorial but that several packs may wander over a large area. They give range sizes of 1280 to 3840 km^2 (500 to 1500 $mile^2$) for individual packs, but do imply that when one pack is occupying a certain area temporarily other packs will not intrude there. Whatever the case, it is apparent that both the spatial territories of wolves and the postulated spatiotemporal territories of hunting dogs would be of considerable advantage in reducing competition for prey.

FOOD UTILIZATION

Because hunting dogs are somewhat smaller than wolves, on the average it would be expected that they would require somewhat less prey to sustain themselves, and this actually seems to be the case. Estimated food consumption rates for wolves vary from 2.6 kg (5.6 pounds)

(Mech and Frenzel, 1971) to 6.3 kg (13.9 pounds) (Mech, 1966) per wolf per day. For hunting dogs, the estimated rates are 2.0 to 4.0 kg (4.5 to 9.0 pounds) per dog per day (Estes and Goddard, 1967). When put on a common base, however, the estimated consumption rates for both wolves and hunting dogs compare very favorably (0.09 to 0.19 kg food per kg of wolf and 0.11 to 0.15 kg food per kg of dog).

It should be obvious from the traits discussed above of both the wolf and the African hunting dog that they are very similar animals. Therefore, it should not be surprising that the actual hunting behavior of these two canids is also very much alike.

HUNTING BEHAVIOR

Wolves and hunting dogs engage in a similar type of social behavior before beginning a hunt. Murie (1944, p. 31) described the ritual in wolves as follows: "Considerable ceremony often precedes the departure for the hunt. Usually there is a general get-together and much tail-wagging," and this ceremony sometimes terminates in a group howl. For hunting dogs, Estes and Goddard (1967, p. 57) described similar behavior: "Play and chasing tended to become progressively wilder and reached a climax when the whole pack milled together in a circle and gave the twittering call in unison."

After this group ceremony, which is very similar to the food-begging ceremony of pups and which may serve to motivate the leaders to the hunt (Mech, 1970), the pack members trot off in search of prey. At this time, individuals of either species may snatch small prey and devour it immediately. However, the packs are clearly programmed to concentrate on large animals.

Mech (1970) classified the hunting behavior of wolves into several stages, and it is evident from the accounts of Estes and Goddard (1967) that comparable stages exist in the hunts of the African dogs. The first stage is *prey location*. As discussed earlier, in open areas both species use their vision, but in wooded areas they resort to olfaction, and perhaps to hearing. Little has been reported about the distance that hunting dogs can smell prey, but wolves can detect certain prey at distances of a kilometer or more (Mech, 1966).

After sensing prey, the packs of both canids begin the *stalk*, during which the individuals approach deliberately, with their attention focused on the prey. "The dogs appeared to be attempting to get as close as possible without alarming the game, and certainly the flight distances were much less than when the pack appeared running" (Estes and Goddard, 1967, p. 58). "The wolves sneak as close as they can to the prey without making it flee" (Mech, 1970, p. 200).

The next stage of the hunt is the *encounter*, when the prey and predator confront each other, often at a distance. With larger prey of either the wolf or the hunting dog, for instance, moose and wildebeest (*Connochaetes taurinus*), respectively, the prey may stand their ground as the predators approach and may aggressively defend themselves. Smaller prey species, however, usually flee. The moment they do, this activity triggers the urge in the hunters to bolt after them, in the *rush* stage of the hunt.

The rush gives the hunters their best head start toward the prey and may determine the ultimate outcome. Usually this stage continues into the *chase*. During the chase, hunting dogs and wolves perform quite comparably, running at speeds of 35 to 40 mph (56.3 to 64.4 kmph) (Mech, 1970; Estes and Goddard, 1967). The lead animals direct the pursuit, with most of the other members following. If in a group of fleeing prey certain individuals falter or fall behind, they are the ones likely to be attacked. Murie (1944) documented this with wolves chasing caribou (*Rangifer tarandus*), and van Lawick-Goodall (1971, p. 62) concluded from watching numerous hunts by hunting dogs that they "select an individual from the herd that is, in some way, weaker and slower than his fellows."

The chases by either wolves or hunting dogs do not usually last long or cover great distances. Generally wolves give up if not successful within about 3 km (2 miles), and only rarely do they persist for more than about 5 km (3 miles) (Mech, 1970). Hunting dog chases cover an average of 1.6 to 3.2 km (1 to 2 miles) (Estes and Goddard, 1967).

Of course, neither wolves nor hunting dogs succeed during every hunt. Wolves hunting the solitary moose on Isle Royale in Lake Superior during winter had a success rate of about 8% (Mech, 1966). Wolves preying on deer (*Odocoileus virginianus*) in Ontario succeeded on an estimated 25% of their hunts during one winter and on 63% during another, although these estimates may be higher than the actual success rates (Kolenosky, 1972). Hunting success figures reported for the hunting dog range from about 34% (van Lawick-Goodall, 1971) to 85% (Estes and Goddard, 1967). It must be stressed that success rates reported for wolves are usually from hunts involving only single prey animals or those in groups of up to four, whereas the figures from hunting dogs usually involved attacks upon herds of prey, where there were many more prey individuals that might be vulnerable.

When packs of hunting dogs or wolves do catch up with prey, they attack the animals in similar fashion. Both generally throw down smaller prey such as deer or gazelles (*Gazella thompsonii*) and tear at them anywhere. Usually they attack larger animals such as moose or wildebeest from behind, in the rump, flank, or hindlegs.

Parallels even exist in a further, very specific tactic used by both wolves and hunting dogs when attacking larger prey. One wolf will often grab a moose by the nose and hold it, while the rest of the pack works on its rump (Mech, 1966). Compare this with the following description of hunting dogs attacking a zebra (*Equus burchelli*) from van Lawick-Goodall (1971, p. 65): "One dog seizes the upper lip and pulls hard whilst the rest of the pack disembowels the prey."

In feeding, both wolves and hunting dogs bite into the rump and the abdomen of their prey. They arrange themselves side-by-side around the carcass and tear at it in every direction. With large packs, even the largest prey may be devoured in minutes.

At this point, a further similarity in the ecological niches of these two large social canids becomes apparent: both species have their attendant mammalian and avian scavengers that share their prey with them. In keeping with the greater diversity of the tropical ecosystems, more species of scavengers are associated with hunting dogs, but the basic parallel remains. Hyenas (*Crocuta crocuta*) and jackals (*Canis* spp.) dart in and out among the dogs, snatching pieces of the carcass, and various vulture descend upon it from the sky (van Lawick-Goodall, 1971). With wolves, such species as foxes (*Vulpes* spp.), fishers (*Martes pennanti*), and wildcats (*Lynx* spp.) share the spoils, although they usually wait until the wolves are resting at a distance (Mech, 1970). Ravens (*Corvus corax*) and eagles (*Haliaeetus leucocephalus*) float above the kills of the wolves, and I once watched a lone wolf, an eagle, and several ravens all sharing a fresh carcass.

CONCLUSIONS

The similarities outlined above in the ecological niches of the wolf and the hunting dog, in the social behavior of these two species, in their basic physical attributes, and in their hunting behavior are striking. There is no question that each species is the direct ecological counterpart of the other in distant geographic areas. Despite the widely different environmental circumstances faced by both, they have solved the same ecological problem in remarkably similar ways.

Indeed, the similarities in the behavior of the wolf and the hunting dog even call into question the taxonomic status of the two species. Although placed in separate genera, they seem to be so similar that one wonders whether they should both be considered *Canis*.

25

DINGO SOCIETY AND ITS MAINTENANCE: A PRELIMINARY ANALYSIS

L. CORBETT and A. NEWSOME
Division of Wildlife Research
Commonwealth Scientific and
Industrial Research Organization
Canberra, Australia

A 10 year study of the dingo (*Canis familiaris dingo*) began in 1966. It is being conducted by a small team of investigators in two contrasting regions: arid central Australia, and the wet, heavily-wooded mountainous terrain of southeastern Australia. The study in the former area is well-advanced because dingoes there are relatively common and accessible compared with southeastern Australia. The investigation of the dingo's social behavior forms an integrated part of the overall program which encompasses the following aspects:

1. Sampling wild populations for data on diet, breeding, population structure, phenotype, parasites, and diseases.
2. Maintenance of colonies of dingoes to study behavior, growth rates, reproduction, water physiology, genetics of coat color, and hybridization with domestic dogs.
3. Distribution, numbers, and movements in the wild.
4. Social behavior in the wild.
5. Water physiology in the wild.
6. The effects of instituted control measures.

Items 3 to 5 are heavily interdependent. They all rely on extensive observations made at watering points on moonlight nights over the years, on capturing and individually collaring animals, and, more recently, on telemetry. So although the bulk of the data on social behavior of dingoes has been collected in central Australia by one of us (L.C.), other members of the team have contributed considerably. It must be emphasized that conclusions presented here are tentative, constituting hypotheses that will be tested in the next few years now that we have a telemetering system working. For example, it now seems certain that an important aspect of social behavior, the care of young, can vary according to the availability of food (see below).

CLIMATE AND HABITATS IN CENTRAL AUSTRALIA

To place this study in context, the environment needs brief mention. The climate of that part of central Australia where we work (within a radius of 321–482 km of the township of Alice Springs) is semiarid or arid. Rainfall in the study area averages 15 to 25 cm per year and potential evaporation 250 to 300 cm, illustrating the dryness of the climate. Though most rain falls in summer, there is no real rainy season. Droughts can be long and severe.

The driest regions are deserts but extensive grasslands of *Astrebla, Aristida, Eragrostis*, etc., and shrublands of *Acacia* grow in the better watered regions, forming rangelands grazed by cattle. Water for the cattle comes mostly from subartesian bores and small surface dams scattered 4.8 to 8.0 km apart. Dingoes also use these waters.

A spectacular series of ancient mountain ranges, notably the MacDonnell and Harts Ranges, extend for 400 km across central Australia from east to west, centered on Alice Springs. The ranges are broken into valleys and gorges, some with permanent water holes. Dingoes are relatively common in these ranges, and so the main area for this study of social behavior is located there, on The Gardens, a cattle station about 64 km northeast of Alice Springs.

The other important study area in the arid region was chosen for comparison, being a drier habitat. This area was the extensive, sparsely-treed gibber (rocky) plains on the cattle station, Mt. Dare, fringing the Simpson Desert.

SOME BACKGROUND BIOLOGY OF THE DINGO

A single breeding season occurs in the autumn. Pups are whelped in late winter and spring, and become independent at 3 to 4 months of age in early or midsummer. Because the weather can be very hot and dry then, pups may often be found very close to water. The heat of day forces dingoes to be mostly nocturnal. Such a habit makes the dingo difficult enough to study, but also it is secretive, intelligent, and rarely vocalizes.

In good seasons, dingoes eat the commonest small animals available. The introduced European rabbit (*Oryctolagus cuniculus*), which can become extremely abundant in good seasons, is the mainstay of the diet. Even in drought when rabbits are rare, they are commonly eaten. Good seasons can also induce an abundance of rodents (*Rattus villosissimus, Notomys* spp., *Mus musculus*) which dingoes also eat. Lizards, particularly *Amphibolurus inermis,* can be important dietary items at times. In the absence of these larger prey, grasshoppers and other insects may be eaten, especially by pups.

During drought especially, some kangaroos (*Megaleia rufa*) are killed as are cattle, mostly calves. Cattle killed by the drought can comprise over half the items eaten then. The occurrences of cattle carcasses and native game in the diet are, in general, inversely related.

The cattle waters provide dingoes with assured, though widely separated supplies of water. Dingoes may use these waters at least once a day in summer, less frequently in winter. The waters, therefore, have been the foci of much of our work in almost all aspects of the overall study, including direct observations of behavior. Though such observations may be biased (for example, little hunting behavior is seen around the waters), the reaction of dingoes to one another around this important communal resource has provided one of the main keys to our present understanding of the organization of dingo society.

SOCIAL BEHAVIOR OF DINGOES

Social Postures

The social postures of the dingo are very similar to the domestic dog (Scott and Fuller, 1965). Some patterns, however, are more strongly developed and stereotyped than in the domestic dog, such as the rituals associated with dominance and submission. In the captive colonies, the foremost social postures seen are agonistic. They arise in pups at 4 weeks of age in captivity, and at a slightly older age, about 5 to 6 weeks, in the wild.

Social Units

A notable fact is that most dingoes seen in the study have been solitary (see Table 25-1). Thus of 1000 animals seen, 73% were by themselves, 16.2% were in pairs, 5.1% were in trios, 2.8% in quartets, and there

TABLE 25-1 The numbers of dingoes in various groups.

Season	1	2	3–7	Totals
Non-breeding (Jan–Mar)	251	54	30	335
Breeding (Apr–June)	198	49	26	273
Whelping and Weaning (July–Sept)	176	33	29	238
Pups Independent (Oct–Dec)	105	26	23	154
Totals	730	162	108	1000

were 2 groups each of 5, 6, and 7. When these sightings are split between various seasons of the dingoes' calendar, as in Table 25-1, slightly larger groups were seen in the last quarter of the year than at other times (reducing the table to two by two, $p < .01$). This is due to pups which sometimes hang together in groups after they become independent, or attach themselves for a while to an adult, invariably a male. Groups have also been seen when dingoes were attacking kangaroos (Stephens, personal communication) and cattle.

That most dingoes are seen alone tells nothing, however, about any temporary nonaggressive associations that they may form at other times. The few results we now have from telemetry indicate that many short-term and loose associations are formed between apparently "solitary" animals. Perhaps the instances where groups have been seen (Table 25-1) were examples of such short-term associations.

A good example of the kind of social structuring suggested was found in a group of 3 animals, 2 males and an estrous bitch, followed and watched in their activities for 3 hours one cool morning. At first they were hunting as a trio for rabbits and lizards. The larger male was obviously dominant, and the other male's chasing the bitch off a rabbit carcass showed their relationship. After about 30 minutes, they separated, but all appeared to be purposely headed in the one direction. The female was seen to catch a rabbit before we lost her, but the large male was then sighted about 400 m away, and he was then joined by the other male. They trotted off another kilometer or so where the large male lay down to rest behind rocks. The subordinate male wandered off out of sight still hunting for lizards and was not seen again. After about 30 minutes, the large male gave a howl and then shifted position to a vantage point overlooking the route he had come. He kept watch in that direction and eventually, very alert, sprang up and ran back in that direction for the female was coming in. A fine greeting display followed and both trotted off to another rock ledge where the bitch rested. The male kept on going over a low ridge nearby, presumably to another resting place close by, but we could not find it. So here was an example of animals acting independently but within apparently loose social bonds. Recently, 7 apparently "solitary" animals were seen together or inferred to be so from telemetric results, in one association or another at different times in different localities, with no agonistic behavior observed. Thus, it looks as though dingoes, while apparently operating mostly independently, belong to loose but amicably social associations of many animals.

There is evidence of two positive processes that operate to keep social units and individuals apart throughout the year. The dingoes caught each year in our sampling operations exhibit a steady low incidence of fresh scars and wounds mostly about the head (see Table 25-2). Despite other evidence of heightened sociality during the breeding season (see below), there is no seasonal trend in the incidence of wounds. So it appears that aggression exists in dingo society throughout the year at sufficient intensity for fighting to result occasionally. The second set of evidence, the reaction of different individuals or groups to one another at waters, is most noteworthy. Should an individual or group be watering and another individual or group arrive, either those present depart and the others come in, or the latter hang back until those drinking have finished. Precedence does not seem to be determined by sex, size, or numbers, but by which animal(s) detect the other(s) first. There is even a particular howl sometimes uttered by dingoes coming in to water that seems to be a warning signal (see below). There appears to be, therefore, a strong tendency to avoid strangers; a reaction that is apparently reinforced sometimes by fighting.

TABLE 25-2 The Incidence of Fresh Wounds, 1966–1971.

Month	No. Dingoes Examined (n = 1121)	Percentage with Fresh Wounds
Dec–Jan	161	3.4
Feb–Mar	164	4.3
Apr–May	283	2.8
June–July	286	3.5
Aug–Sept	139	2.9
Oct–Nov	88	3.4

Vocal Communication

Dingoes are not highly vocal animals. Their howling has a highly seasonal trend linked to the breeding cycle (see Figure 25-1). Three kinds of vocalizing have been recognized, which we describe as howls, "bark-howls," and moans. The messages of the latter two seem clear. Howling on the other hand may have considerable subtlety.

Howling is the vocalization most often heard. Three basic kinds of howls have been recognized with at least 12 variations in all, whose different meanings, if any, are not known. The three basic kinds are: holding the one note, inflected, or abrupt. They may have an even rise and fall, an abrupt rise and protracted fall, the reverse, or a rise to be held as a high note. Sometimes individuals are heard howling alone and at other times in groups.

During the height of the breeding season, howling is most intense (Figure 25-1). Howls can be heard off and on throughout the night, one dingo often being answered by others several kilometers away. Sometimes a chorus will be struck up by dingoes from several points of the compass, continuing for half an hour or more. Pups can be distinguished by their higher pitch, and are largely responsible for the summer peak in groups howling (Figure 25-1).

The hypothesis is that howling has one basic function, and that is location. But it may serve two purposes: to attract congeners in a unifying process, and avoid strangers to reduce social friction. The abundance of howls in the breeding season may initially be to help find a mate and to maintain contact with congeners during hunting. An example was mentioned above, the male calling to his mate. Pups have been heard howling excitedly near a calf freshly killed by adults. We did not hear any earlier calls, but suspect that, rather than drag the carcass to the pups, they had been called in. An example of the use of howling to avoid others could have been the howls heard each morning around 7 am during the breeding season emanating from two pairs of dingoes, one on one end of a small mountain and the other at the other end, presumably their daytime camps. We have witnessed hurried departures of dingoes from watering places when howls have come from nearby, which are more certain examples. Spectrographic analysis and telemetry will be used to test this hypothesis.

The "bark-howl" is an agitated cry, starting with one or more barks like those of a domestic dog, followed by a howl. It has been heard only in situations of extreme alarm, for example, when a dingo suddenly detects our presence on its line of travel or in towers watching waters at night, or a bitch locates us near her pups. The "bark-howl" may act to warn the pups of danger at the time because pups have

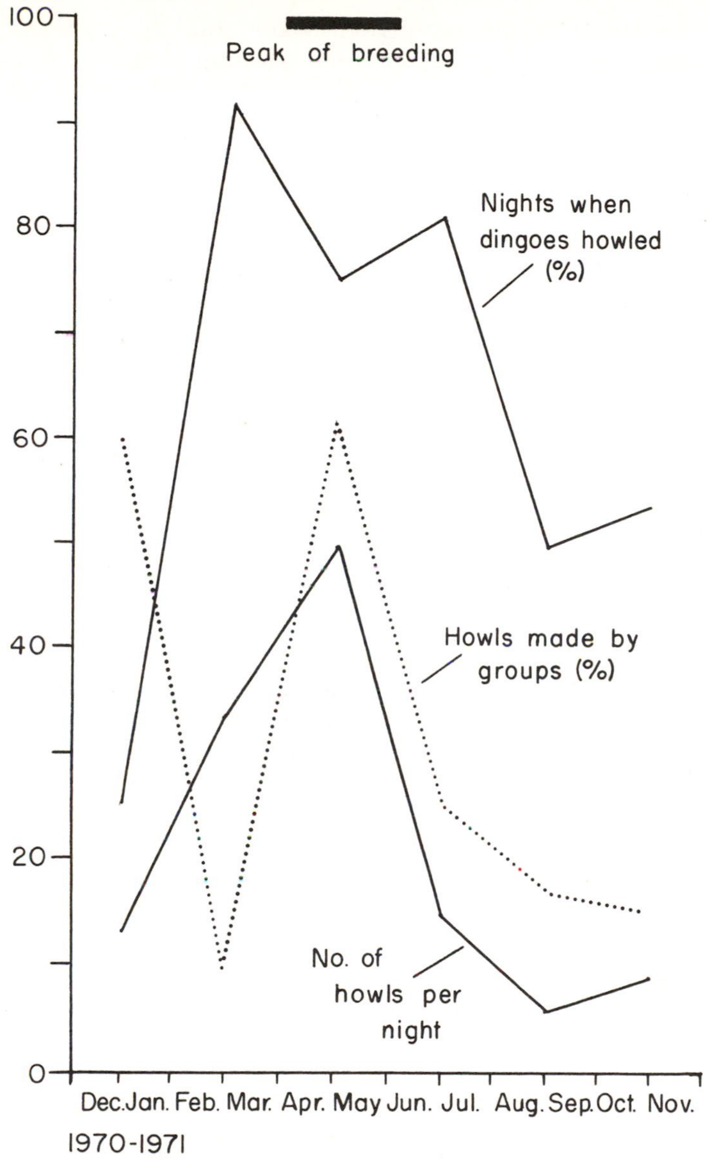

Figure 25-1 The frequency of howling.

been seen to head immediately for cover, only answering the mother once. Another adult nearby may also be heard to answer just once on that occasion despite a performance lasting half an hour by the bitch.

The moan is a low-pitched soft howl that does not carry far. It sometimes precedes dingoes coming in to water and has only been heard at such times. We regard it as a signal warning others of intention to come to drink.

In summary, therefore, vocalizing of all kinds is thought to serve the following purposes:

1. To locate mating partners.
2. To locate congeners for regrouping.
3. To locate and avoid strangers.
4. To signal intention to come in to water.
5. To warn pups and probably others of danger.
6. To express alarm.

Pheromones

The role of chemical messages in maintaining dingo society is not known. Work begun a few years ago during the breeding season to isolate soluble fractions from urine, dung, and secretions from the anal gland was at first extremely promising. After a few months, however, the captive dingoes used for the experiments responded vaguely and equivocally to fractions originally of great interest to them. It was known that the females' visible reactions were not so intense as that of males (e.g., in urination), and, of course, that they were anestrous for much of the year and presumably poor test subjects anyhow. It came as a great surprise to discover some time later, however, that males too have what appears to be an endogenous reproductive cycle, becoming almost aspermous during summer. They show scant interest in estrous domestic dogs which they readily mount in autumn and winter.

Work on pheromones is to be recommended in the near future, but data on sites and abundance of scent posts in the wild are being gathered, and their importance are being experimentally examined.

Whelping

Bitches usually whelp in excavated rabbit warrens, caves, hollow logs, etc., mostly within 2 or 3 km of water. Litters average about 4 to 5 pups. Extensive data have been gathered from four dens, one of which was watched 24 hours a day in all moonlit periods for 3 months. The 5 pups remained at the den site for 3 months. They were attended, to our surprise, by two pairs of animals, one pair being nonbreeding yearlings. The adult male regularly visited the den and was greeted most affectionately by the pups though neither male was ever seen to retrieve food for them.

During the day the pups were left alone though at least one adult was probably camped nearby. When small, they stayed down the burrow, but eventually made separate camps above ground, along a dry river bed. At sundown, the pups always gathered expectantly at the den. Eventually, one or more of the attendants would arrive to be greeted excitedly by the pups. After about half an hour, the larger dingoes present would head off together, presumably to hunt. It is certain that the mother always brought back a rabbit for the pups, and possibly so did the other female.

Rabbits were abundant everywhere at the time, even about the den, but the dingoes ignored those close by and always brought back a rabbit for the pups from elsewhere. Eventually, the pups began to stalk the local rabbits and the virgin bitch was seen to coach them. But the local rabbits seem to have been definitely left for the pups.

During hot weather after dark, one attendant or another was often seen at a water hole 2.5 km away and soon after at the den. We wondered how the weaned pups obtained water until one day we saw the mother return from the direction of the watering point. The pups quietly came up to her, not agonistic toward one another as if she had food, and licked at her mouth whilst she was seen to regurgitate. She was watering the pups.

The mother abandoned the pups when they were between 3 and 4 months old. The pups then left the den, but not before we had caught and marked them. The adult male had been marked sometime previously, and the virgin female was also caught and marked with an individual collar. Two of the pups were seen separately near the adjacent watering point 4 months later, one of them accompanying the adult male. Other evidence from marked animals indicate localized movements (perhaps within a range of 10 to 15 km, sometimes more), but this is so far the only evidence that young may maintain associations formed with adults. Current work this breeding season with telemetry is aimed at understanding mechanisms of forging and maintaining social relationships.

Telemetry this year has allowed us to study closely the three other dens mentioned above. In contrast to the first den, food was scarce near these three dens, rabbits and mice being rare or absent, and the pups being fed mostly on insects. As soon as the pups were mobile (about 4 weeks old), they were shifted about the countryside over distances of at least 3.5 km. When they were about 5 to 6 weeks old, it became apparent that two of the mothers had split their litters into smaller groups, planting them in different localities. One group was seen in three such temporary den sites, one of which was only 20 m from a cave containing another bitch's litter. Litters even became intermingled, one bitch having some of her own pups plus two others. Isolated groups of pups were probably being fed by both bitches, possibly aided by at least one male. One of the bitches, three of her pups, and the male were seen traveling together.

As mentioned above, pups are found in the company of one or more males after independence. Pups from two litters have been seen with the one male. We think that such relationships arise passively. Pups live near water after the mother leaves, because the weather is hot. They probably attach themselves to males that come in to drink, the male tolerating their presence at least for some time, and the pups mimicking the adult male's every move.

ADAPTIVE SIGNIFICANCE

If dingoes are basically loners with social ties loose for most of the year, what is the adaptive significance of such a social system? On present evidence, the answer appears to lie with the food supply, as the following hypothesis suggests.

Weighing about 20 kg and standing about 50 cm high at the shoulders, dingoes are not built to kill large animals like kangaroos unless they cooperate. Before the introduction of the European rabbit, it is likely that the most abundant prey would have been the medium-sized bandicoots *(Chaeropus, Perameles,* and *Macrotis),* rat-kangaroos *(Bettongia)*, and hare-wallabies (*Lagorchestes*). Such game is now much less common, even rare or extinct, compared with earlier days, whereas the red kangaroo (*Megaleia*) was uncommon then (Newsome, 1971). Moreover, the dingo may have arrived relatively recently in Australia, after the giant marsupials became extinct. The oldest fossil dingoes so far found are about 3000 years old (McIntosh, 1969) and between 7000 to 8000 years old (Mulvaney, 1969).

The rabbit of today is about the size of the marsupial game mentioned above. When pups are small, one rabbit will feed them all. But since one rabbit will eventually be able to feed only one dingo, the close social ties of puppyhood would be strained and animals forced to hunt alone. It is hard to imagine any native game in early times being more abundantly available than rabbits today; so the same problems should have arisen then. Dingoes might be expected to display egocentric behavior early in life, which they do, when about a month old, about one third of the way through puppyhood.* Such behavior is well-developed when pups become independent.

*See also Chapter 30 for further details on this phenomenon in other canids.

Strong ties, however, need to be formed during the breeding season. Waters are communal resources and so provide places where dingoes may meet or leave sign of their presence. Another possible mechanism is vocalization which is most manifest before and during breeding. That males appear to have an endogenous reproductive cycle with heightened social behavior during breeding is consistent with this hypothesis.

SUMMARY

The investigation of the social behavior of the dingo is an integrated part of a 10 year study of its ecology in the wild. Conclusions as yet are tentative, but results indicate that dingoes are basically solitary animals which are nonetheless loosely bonded in small amicable groups. Social ties strengthen in the breeding season when non-breeding yearlings of as yet unknown origin help parents rear young. Howling is heard most during the breeding season. Its main function is thought to be location, keeping contact with congeners out hunting, but also helping to avoid strangers.

Strong avoidance behavior is seen at watering points, precedence being given to animals first detected. One particular kind of howl heard near waters is interpreted as a warning signal. Another kind of howl, the "bark-howl," is an expression of great alarm that acts to warn pups. Egocentric behavior emerges early in pups, at about 1 month of age, and pups behave independently after about 6 months.

The evolutionary forces acting to produce this type of canine society are thought to involve the small size of the most abundant prey. While a medium-sized mammal will feed all pups when they are young, it will only feed one animal later on, forcing dingoes to hunt alone. Dingoes display egocentricity early in life when about 1 month old.

Acknowledgments It is a pleasure to acknowledge the cooperation and help of other members of the team studying dingoes. In particular, we would like to thank Mr. L. Best and Dr. B. Green for permission to use their data, and Mr. P. Hanesch and Mr. A. Wakefield among others whose willing efforts in the field have made the study possible. We are also grateful to Mr. G. Christman for drawing the figure.

The dingo study has been financed mainly by the Australian Meat Research Committee and we gratefully acknowledge their generous support.

26

THE ECOLOGY OF "FERAL" AND FREE-ROVING DOGS IN BALTIMORE

ALAN M. BECK*
Department of Mental Hygiene
School of Hygiene & Public Health
The Johns Hopkins University
Baltimore, Maryland

"In 1966 in the United States there were an estimated 24.7 million owned dogs, in addition to an undetermined number of ownerless strays. Almost one half of the dogs in the United States are found in the cities; between 20 and 40% of American city-dwelling families own at least one dog" (Schwabe, 1969). Djerassi, *et al.* (1973) using marketing data for animal products, estimate the owned dog population is now over 33 million with 38% of the households owning an average of 1.4 dogs. A survey in urban and rural Davis, California revealed the actual percent of ownership to be even greater, 42.3 and 58.6, respectively (Franti and Kraus, 1974). Despite this far from coincidental concentration and merging of dog and human populations, the scientific and educational value of urban dogs has been sadly neglected. Urban studies are very much in need at this time.

Dogs, being large, social, territorial, individually recognizable, mostly diurnal, and tolerant of human proximity make excellent educational tools for the teaching of ecological methods in the urban-centered school (Beck, 1974a). Studying the readily available dog population could prepare students of behavior and ecology before they

*Present address: The City of New York, Department of Health, 125 Worth St., New York.

embark to study more illusive animals, like wolves and coyotes. Also, there are many public health implications since dogs share and at times compete with man in the urban environment (Beck, 1974b; Feldmann and Carding, 1973).

ABUNDANCE

As pound and dog license records give little indication of true abundance, wildlife estimating techniques were applied to the populations of the study areas.

The first method is merely a modification of Schnabel's variation for multiple recapture of the Petersen-Lincoln Index (Schnabel, 1938) by photographing every dog observed within one half block of my car as I rode the same route for 9 consecutive days. The natural variation of the dogs facilitated individual recognition in determining whether or not a dog had been previously photographed, i.e., recaptured. Trapping the animal to put on additional markings has never been necessary. As the dogs are habituated to passing automobiles, they ignore me and do not know they are being photographed. There is no trap bias since the animal is neither trap prone nor trap shy. Estimates of 600 dogs/2.59 km^2 were generated by this method.

The second method involved plotting the location of every dog photographed on a map of the 0.65 km^2 area for each day, then superimposing a grid of 100 squares (6475 m^2) over each day's plot. In this way small sections of the total study area that contained dog sightings could be mathematically removed as formulated by Hanson (1968). This removal is analogous to actual removal, and yielded an estimate of 800 dogs/2.59 km^2.

The third method was a reevaluation of the recapture data using Darroch's (1958) method which placed the estimate for the same area at 612 dogs/2.59 km^2.

Finally, my own proportion method is now being tested and so far has given me the impression that dogs in similar sociological neighborhoods have similar densities.

By extrapolation for the whole city, Baltimore's population would be around 80,000 to 100,000 animals, which agrees nicely with city (Berzon, 1971) and state (Crawford, 1970) health official estimates. In 1960, Baltimore's dog population was estimated to be 75,000 (Crawford, 1964). Therefore, in the last 10 years, while the human population actually decreased slightly, the dog population increased by 25,000 animals. There is 1 free-roving dog for every 9 humans in Baltimore (see Beck, 1973 for details).

ACTIVITY

While making continuous hourly automobile runs through a study area so as to include a full 24 hour period, I recorded every dog and human being observed. During the hot summer months, dog activity is fairly restricted to two periods: 5–8 am and 7–10 pm, specifically avoiding the heat of the day. This early morning and early evening pattern is generated by natural crepuscular rhythms and the common habit of people releasing their pets to the street before and after usual work hours. Human street activity is more unimodal going from 6 am to 1 am (19 hours). Fall periods are not so clear as midday activity is more common. These patterns apply to the population generally, but individuals may be found on the streets at almost any time of day except during summer midday heat.

More detailed information about dog activity was gained by following chosen individuals for extended periods of time to eventually include several 24 hour days (Beck, 1971, 1973). Two such animals, both large, 16 kg dogs, lived in some shrubbery and followed a bimodal summer activity regime very similar to the one previously mentioned. It is of interest to note that even during the so-called "activity" periods nearly half the time was spent resting or sleeping. Rural free-running dogs also are active in the early morning hours (Perry and Giles, 1970).

HOME RANGE

Following these two animals continuously revealed a home range of nearly 0.26 km^2. The range is irregularly shaped with extensions to open fields and feeding sites, which are alleys containing garbage. The animals spent most of their time near the center of the range. Feral dogs in rural areas have larger ranges (Scott and Causey, 1973).

FOOD

Almost all dogs observed for any period of time find some edible food by going through garbage cans. Dogs can tip over garbage cans and boxes, open paper and plastic bags containing garbage, and are often observed carrying such bags back to protected areas where careful rummaging takes place. Stray dog populations must be considered by any municipality planning to convert to a paper or plastic bag containerization of garbage in lieu of the traditional noisy and heavy metal cans. However, metal cans that can be knocked over afford little protection aganst dog damage. This damage, together with an insufficient number of containers, almost always provides for some litter on the ground, even after garbage collection. Thus, garbage collection causes only slight shifts in dog usage of an alley and on the whole, trash collection is not a major ecological catastrophe for the dog (Figure 26-1).

Garbage is readily available because of : (1) the high density of people, (2) only two weekly pick-ups of refuse, (3) the concentration of garbage cans in narrow alleys, (4) no provisions for preventing cans from being knocked over or lids removed by dogs, (5) inadequate number of cans, (6) the difficulty in opening dumpster doors, so garbage is left at base, and (7) no garbage disposal units in use in high density areas.

Figure 26-1 Typical "foraging area" in Baltimore.

Food is also available to strays through the kindness of area residents via handouts. Food is either left for any dog or is given to specific dogs. In my low-income study area, I interviewed every head of household (usually female) that was home along a block that cut through a very low income area and 20% of the people observed people putting food out for dogs or had done so themselves. This phenomenon was uncommon in my second area (with comparable dog density) that was of middle class residence.

The two individual dogs previously mentioned spent no more than 11% of their activity periods feeding and were also the recipients of a regular handout — food was dropped to them from a second floor window somewhat regularly in the mornings. These dogs' short feeding times, small home range, extensive rest periods, and apparent good health all indicate adequate food for the stray.

SHELTER

The urban environment contains numerous areas where dogs can find cover and protection against adverse weather conditions, other dogs, and people. "Den sites" offering complete cover include: (1) dense shrubbery around buildings and in woodlots, (2) vacant buildings and garages that were previously occupied and those under construction (urban renewal supplies many buildings in this category), (3) hallways of occupied buildings, and (4) under porch steps. There are also topographic features in the environment that, while not providing total cover, do offer some protection against extreme weather conditions. Dogs, who are sensitive to extreme heat, show various forms of heat avoidance behavior, like sleeping under cars during the day, and insulating themselves from the hot or wet ground by resting on the tops of discarded mattresses, grave stones, and cars. In one area, the complex topography of a land fill area which turned into a dump was used extensively by dogs for, shelter and socializing.

SOCIAL ORGANIZATION

Few groups appear stable though small groups, usually 2 in number — either both males or one male and one female — are common. Larger groups often appear to vary regarding size and membership. These groups form and dissolve within minutes to days, the latter true when forming around a bitch in estrus. Leadership in the group is usually discernible. Groups reform after being dispersed by people and traffic.

Rapid observations of dogs active in the morning revealed the following group patterns: 50.6% of all dogs were seen alone, 25.9% in groups of 2, 16.3% in groups of 3, 5.3% in groups of 4, 1.9% in groups of 5, and isolated groups on other occasions up to 17 animals.

The dog is truly a social animal, as about half are in the company of other dogs. This type of social structure appears to be significant in the urban environment. I have observed large dogs knocking over garbage cans making food available for smaller dogs. Dogs vary considerably in their wariness of people and new situations. For example, some individuals of a pack will feed while others will not, thus lessening individual competition. People even prefer to feed some individuals over others. For a social animal like the dog, morphological and behavioral variations are either real biological phenomena serving some adaptive function at the population level or a fortuitous result of man's capricious manipulation of inherent canine variability. In any case, the variations are conducive to greater utilization of the habitat's resources.

ORIGIN

From observation and interviews, it appears that free-roving dogs come from : (1) pets released in early mornings and evenings for daily unsupervised runs (interviews in my areas revealed from 37 to 51% of the families owned at least one dog and *at least* a third of those dogs were permitted freedom of the streets), (2) pet escapes, (3) pets abandoned to streets when families move out of area, (4) pets that run away or are released after being stolen by children, and (5) birth (under porch steps and in woodlots). The order listed is approximately in order of magnitude.

Stray dogs are not intimidated by people as I have observed mating with "tie" in alleys, sidewalks, and streets with heavy traffic. Ownerless strays have been observed nursing pups under porch steps, in alleys, and in woodlots; however, survival of pups appears almost nil.

MORTALITY

I have observed numerous car kills and injuries. Dogs appear to act submissively (cowering while looking up) when faced with a rapidly approaching car, possibly explaining why Baltimore's veterinarians consider the car the most common cause of injury and seconded only by disease as an agent of mortality (Large, 1971). From July 1, 1968 to June 30, 1969 the city's Municipal Shelter collected 15,264 dead dogs from the streets and gassed 6565 dogs that were collected alive and not adopted or retrieved. Obviously, not all dead dogs are found, as they may be discarded with garbage, go unnoticed on vacant lots and private property, or be scattered by traffic, so this figure is an underestimate of true mortality. Even so, the total kill represents nearly 20% of the total estimated population, indicating a rapid turnover of individuals and probably an overall lowering of mean age. Such a situation has many epidemiological implications, as a young population of new individuals would be more susceptible to diseases and *Toxocara* infection (Beck, 1973).

The Dog Pound

In addition to the automobile, the Municipal Shelter or pound is a predator of the dog, as its function is to collect strays as well as sick, injured, and unwanted animals. During the year period previously mentioned, the pound collected a total of 11,444 live dogs (only 495 of which were retrieved by owners) and received some 14,000 phone calls requesting service (Harmon, 1971). Each day, any one of four trucks rides for 7 hours, covering 70–100 miles (112–160 km) of streets and

Figure 26-2 "Predation" by dog-catchers.

impounds from 20 to 40 dogs (Figure 26-2); mostly sick, injured, and unwanted ones. In my travels with the pound truck I have often observed people chasing dogs away to interfere with pound function and, in general, I sensed a negative feeling for the service. I believe efficiency would be greatly improved if information were readily available explaining the philosophy and techniques of municipal dog control.

Animals that are not retrieved, adopted, or sold to medical institutions are gassed and brought to an animal rendering plant where meat and bone scrap (with a crude protein content of not less than 50%) is prepared and sold as chicken and hog food supplement. The grease is used in the manufacture of low phosphate soaps. This is one of the few examples of recycling of natural components in the urban ecosystem.

The dog population is also influenced by the activities of the local anti-vivisection society and the S.P.C.A., but their records were not made available to me.

PUBLIC HEALTH IMPLICATIONS

There are significant interactions between man and dog as dogs are large, well-armed, very vocal, capable of transmitting disease, and larger dogs occur in ever-growing numbers in the urban setting. Some public health problems are unique to strays, whether owned or ownerless, while others are related to all dogs.

Firstly, garbage disruption is a problem caused by unsupervised dogs. Rummaging through and overturning garbage cans and bags creates a poor appearance, decreases trash collection efficiency, and makes food available for pest species like rats, mice, roaches, and flies (Figure 26-3). After many hours of nighttime observation I never saw dogs make any attempt to capture rats, though I have seen them chase cats. Cats have been observed to stalk rats though a capture was never witnessed. The dogs' scattering of contained trash and chasing cats

Figure 26-3 The rat, "commensal" of the urban dog.

may be important to the rats. Also, rats benefit from food and water left out for pet dogs who are often fed in backyards.

All other health hazards involve stray and nonstray dogs alike. Dog bite is probably the most significant public health problem with 6227 reported bites in 1969 and it is expected there will be an excess of 7000 in 1970. It should be noted that only about 50% of all bites in Baltimore are reported (Crawford, 1964). Baltimore's bite rate of 74/10,000 is about twice the U.S. average, probably because of the large straying pet and stray population. About 25% of the bites are from strays that are not captured and held for examination (Berzon, *et al.*, 1972). Children under age 15 receive 60% of the bites even though this age group constitutes only 30% of the population (Berzon, *et al.*, 1972), indicating that children are the most victimized, which is consistent with findings in other cities (Parrish, *et al.*, 1959). The bite rate is more than double during the summer months (Berzon, *et al.*, 1972), undoubtedly related to the increased street activity of both children and dogs. People avoid larger groups of dogs and a few pack attacks on children and zoo animals have occurred. Cyclists encounter severe dog harassment (Freedman, 1971). The problem will surely become more serious as large, bite-prone dogs are secured for protection in cities with increasing crime rates. Future urban wildlife managers may have to decide whether or not it is advisable to permit people to buy attack-trained dogs when protection is best served by warning barks (Caras, 1971).

Dog feces is another aspect related to all dogs. Feces in the streets and parks offends most people and is a potential health hazard.

Firstly, there is the possibility of visceral larva migrans (VLM) caused by the larval form of *Toxocara canis* (common dog worms). Symptoms include fever, coughing, enlargement of liver and spleen, eosinophilia and more rarely, convulsions and blindness (Beaver, 1969; Brown, 1970, 1974; Snyder, 1961; Zinkham, 1968). It affects mostly children, usually those with a history of pica, which is widespread in Baltimore (Cooper, 1957). Patients with *Toxocara* invasion of the eye have been seen over the years at The Johns Hopkins Hospital (Wilkinson and Welch, 1971). VLM is rarely serious, but since there is no definite diagnosis, its effects and even its frequency of occurrence is unknown. Baltimore's potential for VLM is very great as children and dogs compete for soil in backyards and under street trees.** More detailed study in this area is definitely indicated.

Secondly, *Salmonella* may be transmitted from dog to man via flies feeding on dog feces (Quarterman, Baker, and Jenson, 1949; Wolff,

**Various species of flies may also carry the parasite eggs and so aid dissemination — Ed.

Henderson, and McCallum, 1948). Incidently, dog feces is second only to open garbage cans (another canine product) in the breeding of house flies (Quarterman, *et al.*, 1949). In wealthier residential areas where exposed garbage is less available, feces are even more important for fly breeding.

Thirdly, feces may be a source of food and moisture for rats (Carroll, 1971). Also, pet and stray dogs chase and detain rat eradication personnel as they go through alleys during poisoning campaigns. As you can see, the rat benefits from dogs in many ways. In turn, the rat creates a nuisance that demands attention, attracting personnel and funds away from potential dog control. Rats and dogs exhibit a sort of urban-style symbiotic relationship.

Over the last 5 years pet dogs have been implicated more and more in the spread of leptospirosis to people (CDC, 1974). Leptospirosis is passed in dog urine, even in pets that have been vaccinated (Feigin, *et al.*, 1973).

Miscellaneous public health problems include: (1) the noise of barking, especially the chain-reaction kind heard in high density areas, (2) the killing of zoo animals, (3) the hindering of traffic, (4) the killing of street trees (Pivone, 1969), and (5) their role as hosts or vectors for diseases that are on the increase in the area, like rabies and tick-borne typhus (Rocky Mountain Spotted Fever) (Peters, 1971). Rabies in dogs is still rare in Baltimore City and the typhus is vectored by the American dog tick *Dermacentor variabilis* which is more common on dogs that run through woodlots than on those that run only on the streets, who are almost always parasitized by the brown dog tick, *Rhipicephalus sanguineus,* during the warmer months.

Very much part of the public health aspect are the positive roles of the urban dog. Dogs as pets, not strays, are obviously appreciated as reflected by over 33% dog ownership (by family unit) in my areas. There is the widespread belief that dogs afford protection against city crime. However, many people are of the opinion that burglars live in the areas they burglarize and know the dogs, who therefore do not bark at them. Attack-trained dogs in the hands of untrained owners is rapidly becoming a threat to public health.

Many pet owners do not extend their affection for strays and, in general, dogs are not appreciated as wildlife, except possibly for a small but vocal "anti-vivisectionist" faction. Several assorted groups, who begrudgingly concede the need for stray control, periodically capture newspaper headlines with charges of animal cruelty by the city. Counter charges have also been levied.

It is important for anyone involved in any sort of management program to realize that feelings for dogs extend through the entire spectrum from dog hate (involving extreme cruelty) to dog love (where the person substitutes dogs for humans as love objects).

For some people, dogs are companions and protectors while for others they are additional sources of anxiety in a crowded city. In any event, dogs play a significant role in the lives of the urban dweller — a role which is very much part of the ecology of man and dog.

27

ECOLOGY OF A FERAL DOG PACK ON A WILDLIFE REFUGE

WILLIAM H. NESBITT *
Department of Zoology
Southern Illinois University
Carbondale, Illinois

The distinction between a feral, stray, or free-ranging dog is sometimes a matter of degree. A free-ranging pet might become stray and perhaps finally feral. A feral dog is a wild animal. If owned at one time, the dog no longer will freely approach humans and usually shows strong fear of them. Such a dog is fully capable of surviving in nature and reproducing without aid from man. McKnight (1964) mentioned German shepherds, Doberman pinschers, and collies as dog types most often becoming feral. He further noted that feral dogs may have been present on the North American continent before white men since the early Indian tribes had dogs.

Feral and free-ranging dogs have long been condemned in a variety of publications. These inflammatory articles suggest the destructive potential of such dogs but offer little insight into their total ecology (examples: Banks, 1957; Giles, 1960; Gilsvik, 1970; Hunter, 1971; McKnight, 1964; Morrison, 1968; Ward, 1954). Often these articles are exaggerations. Cochran (1967) repeats a tale of a "35 pound, wolflike mongrel that hunted alone, never barked while on trail, and killed like a big cat, leaping on its victim's back breaking a deer's neck or biting deeply behind its ears."

*Present address: Manager, Hunting Activities Department
National Rifle Association of America
1600 Rhode Island Ave. N.W.
Washington, D.C. 20036

HUNTING ACTIVITIES

Some recent studies have attempted to go beyond single incidents in assessing the complex situation of free-ranging pets, strays, and truly feral dogs. Progulske and Baskett (1958), although noting hounds harassing white-tailed deer on their study area during all seasons, concluded that dogs were seemingly negligible causes of direct mortality under their study conditions. Sweeney, *et al.* (1971) recorded 65 experimental chases of 6 radio-telemetered deer by 1 to 9 hounds. Average chase time was 33 minutes (3 to 155) and the average chase length was 3.9 km (0.3 to 22). No deer was caught by the dogs or died due to the interaction. A variety of escape patterns was exhibited by the deer. Perry and Giles (1971), studying free-ranging pets in Virginia, found "roaming" activity to be sporadic in a particular dog's routine. They also noted that dog control projects seldom maintain adequate records or evaluations of such programs.

Scott (1971), studying feral dog packs in east-central Alabama, found them feeding on garbage, carrion, and small mammals. He observed no predation on deer or cattle during his 20 month study. He concluded that field observations and theoretical considerations indicate that dogs probably do not regulate deer populations.

In January, 1969, I began a 5-year study of feral dogs in Crab Orchard National Wildlife Refuge, Carterville, Illinois. Dogs were present in this area, including feral residents, free-ranging pets, and strays. Hawkins *et al.* (1970) recorded the removal of over 100 dogs from the refuge during 1964 through 1966. These investigators, studying long-term population dynamics of the refuge herd of white-tailed deer, attributed 7% of the total known mortality of 687 marked deer during 1962 to 1968 to dogs. During the same period, hunter harvest accounted for 49% and highway accidents for 28%. The 9 marked deer killed by dogs during this period included 6 fawns, 2 adult bucks, and 1 adult doe. These authors noted that most of the deer known to be attacked by dogs were in corral-type deer traps or were injured; dogs did not appear to them to be an important mortality factor of healthy, free-ranging deer. However, they felt that if dog control had not been practiced in 1964–1966, dogs might have become an important mortality factor of even healthy deer.

FIELD STUDY OF SOCIAL ECOLOGY

In January, 1969, I began observing the activities of 4 collie-type feral pups, apparently past weaning stage. They were essentially independent of their very wild mother. Later in the study, I was able to verify that these pups were from a litter of 5 born in the area in September, 1968. This information was obtained from a refuge employee who captured the fifth pup at a few weeks of age. This dog never became fully domesticated and was finally donated to a laboratory. One pup was shot and apparently died in the spring of 1969. The remaining pups (1 female and 2 males) formed the nucleus of a pack that I closely followed for the next 5 years, using sightings, tracks and other indirect evidence, and limited radio-telemetry.

These three pups showed a definite collie influence. The muzzles were narrow and elongated, the ears erect, and one pup had long hair. Coat colors were red-brown and white on one male and the female, and black, brown, and white on the third. (The fourth pup was also black and white, with a nearly completely black face and upper neck.) The short hair and black in the coat was apparently due to a mongrelizing influence.

The pack bonds developed before the end of the pups' first year. The pack as first formed included the three pups, an adult male collie with more white in the coat than usual, an adult female black and white mongrel, and a dropper (setter–pointer cross) of known origin and the same approximate age as the pups. In the early stage of pack development, the female pup showed aggressive tendencies, but the tricolored male pup assumed definite leadership in pack activities during the second year of the study. This dog was found dead in October, 1970, apparently shot, at age 2 years. The female pup assumed leadership following this dog's death. Some time before this, the other male pup ceased pack participation to run with a "new" dog in the area (a pack of 2).

I was able to trap several pack members during 1970 and 1971 to examine physical characteristics and attach radio-telemetry collars to aid in pack location. When examined, both the adult male collie and the adult female mongrel were aged by tooth wear as over 10 years old. Weights of adult dogs examined (1 year or older) varied from 17 to 25 kg. One uncaptured dog was estimated as 30 kg and was the largest pack member at any time during the study. All dogs examined were in apparently excellent health with occasional infestations of ticks as the only discernible ectoparasites. This was true even during times of the year when strays in the area were observed with "ribs showing" from apparent lack of ability to obtain food.

Females left the pack to whelp in heavy cover but did not dig a den. As the pups neared weaning age, the mother left them for increasing time periods while she ran with the pack. In one case, the apparent father was observed "guarding" the general area of the young family. In early activities, pups were often vocal. After the first year of life, this lessened noticeably. Adult pack members clearly discerned human voices and responded by retreat. Scott (1971) found the feral dogs of his study to be essentially nonvocal.

Pups born to pack members were accepted into the pack at about 4 months of age. However, due to low pup survival (example: 2 observed litters of 8 pups total had only 3 entering the pack), pack numbers showed little fluctuation. The pack size "usually" noted year-round was 5 to 6. Perhaps this is an ecologically efficient size. Scott (1971) reported feral dog pack sizes in his study as 2 to 5 animals.

There was a definite leadership hierarchy in the pack with single-file formation being exercised as the pack moved. The lowest ranking individuals were at or near the rear of the column. Pups of pack members were accepted at the rear of the column. The pack appeared "closed" to other dogs, although strays and free-ranging pets were constantly in the study area. It was not possible to determine if a dominance hierarchy was present. The definite leadership ranking suggests such.

The female pack leader often "scouted" ahead before moving the pack, with the pack awaiting her return on such occasions. I never observed the tricolored male doing this although both leaders often "patrolled" the general area at frequent intervals when the pack was at rest. The pack moved as a concerted unit when a vocal signal or movement by the leader was given.

Pack movements appeared to be influenced by human activity in the area, food opportunities, wooded cover preference, interactions with other wildlife of the area, and similar factors. Pack movements occurred at all times of day and night. Nocturnal and crepuscular times appeared favored, apparently as a reflection of the relatively heavy level of human activity in the area during daylight hours. The easiest travel lanes such as roads, deer trails, and crop rows were used when possible. If frightened in the open, the dogs immediately sought cover in woods or tall herbaceous growth.

Pack range included nearly all of a 28.5 km^2 portion of the inviolate area of the refuge. This portion is defined by fencing and fire lanes on two sides, by a two-lane highway on the third, and the margin of Crab Orchard Lake on the fourth. Within this portion is a mixture of cropland and fallow rotation, pine plantations, hardwood forest, and several small ponds. Small game and white-tailed deer are abundant. The refuge is a major waterfowl wintering area. The pack evidenced a reluctance to leave this portion that was demonstrated on several occasions by a reversal of travel direction to prevent crossing a boundary. This was perhaps a response to the great increase in human usage found outside the inviolate area boundaries.

In food gathering activities, the pack was opportunistic. During season, they utilized crippled waterfowl, road-killed animals and other carrion, small game, crippled and young deer, persimmon fruits and some green vegetation, and occasionally garbage from the refuge dump. The pack range included several pastures in which cattle were grazed (both calves and adults) from April to September of each year of this study. I was unable to document a single case of livestock depredation by any of these dogs. (Free-ranging pets with leather neck collars killed 3 calves on another portion of the inviolate area during 1970.)

Movements of this pack constantly put them in contact with other wildlife of the area. Interspecific recognition of dog intent appeared evident on several occasions. I have observed deer running at a determined pace long before the dogs appeared on days when the pack was apparently seeking food. On another occasion, 3 deer merely raised their heads from feeding to watch a pack member pass within 22 m with no aggression evidenced by either the deer or the dog.

In summary, this pack of truly feral dogs did not appear to be a regulatory force of any magnitude on animals of this refuge; they are a valuable part of the fauna by their sanitary predation activities.

ADDENDUM

I stated in the paper that the 5 original pups forming the pack nucleus were born in September, 1968. I did not include dates for litters whelped by pack members. As can be expected from the nonseasonal whelping of pet dogs, there was not a discernible breeding or whelping season. This leads into Fox's (the editor's) comment concerning Mengel's paper on hybrids of a captive coyote and a pet mongrel (*J. Mammal.*, **52**(2), 316–336). Certainly I cannot question the results he reported for his study animals but I would hesitate to extend his conclusions to wild populations. The dysynchrony of estrus periods discussed by Mengel might have an effect upon hybrid stabilization, but probably not on closure of the pack to strangers. I would suggest behavior of the pack as a group as being much more influential on acceptance of new members. I have in my own limited research seen strong evidence of exclusion of strange dogs by the pack, once the pack formative period has passed. This has appeared to be true of even strange dogs in estrus. In one case, operation of a large, corral-type trap for 59 days with 3 separate females in estrus as bait (dog chained in back of trap) produced no captures of the 4 sexually mature males of the pack. One might readily contrast this with the use of females in estrus by city dog wardens with excellent results. Baiting of this same trap with carrion immediately after the last dog-bait days gave pack entrance into the trap within 1 week. (Dirt entrance of trap was swept clean daily to record tracks; no dogs entered with female in estrus as bait.)

Editor's Note: Pack formation and subsequent closure with regard to strangers may therefore further limit the chances of hybridization of feral dogs with wild canids (wolf, coyote) in addition to the dysynchrony of estrus proposed by Mengel. "Loner" feral or free-roaming dogs with no allegiance to a pack as such, especially male dogs that constantly rather than seasonally produce sperm, and more rarely a female dog whose estrus was in phase with the breeding season of the wolf and coyote, could result in hybrid matings. Subsequent dysynchrony of heat periods in hybrids, as proposed by Mengel, could isolate them reproductively from the wolf or coyote, as well as increasing mortality of offspring born outside of the optimal spring period. Further field studies are needed in order to answer these questions.

PART

IV

GENETICS AND PHYSIOLOGY

Few studies have been undertaken on genetic and physiological processes underlying the behavior of wild canids. The papers in this section comprise the beginnings of exploration into this neglected area. Much more remains to be done; more biochemical parameters need to be investigated, as well as continuous monitoring of physiological and biochemical events underlying changes in behavior with minimal disturbance of the animal being essential. The relationships between coat color and behavior, and between temperament ("wildness") and pituitary–adrenal and gonadal activity are discussed in this section. The findings are unconclusive but stimulating and provide considerable impetus for what promises to be a fruitful area of research.

28
GENETICS OF BEHAVIOR VARIATIONS IN COLOR PHASES OF THE RED FOX

CLYDE KEELER
Department of Medical Genetics
Milledgeville State Hospital, Georgia

Mr. Ed Fromm, a fox fur rancher, unlocked the high, iron-framed gate to an 80 acre (32.4 hectare) priming range where his foxes develop the winter coat. It was feeding time and numbers of foxes could be seen at a distance gradually moving toward the food trough. As I walked across the range enclosure, I noted that six or seven curious foxes were trailing me at a distance of about 4–5 meters. They were all ambers, although far beyond them were many pearls, silvers, and platinums. Why were they not distributed at random? Over a number of years Mr. Fromm had noticed this curious segregation in distance from people assumed by these particular color phases.

When I arrived at the feeding station, I made myself small by getting down on one knee, and this seemed to relieve the animals. I held my wrist to my mouth and produced a series of mousy squeaks. At this, an interested crowd closed in on me from every direction. All the foxes near me were ambers with an occasional glacier. Farther away I could recognize a pearl or a silver and in the far distance a few platinums. Red foxes never show themselves like this, although they sometimes find their way into the range — nobody knows how. "In fact, we never see them until the final roundup," explained Mr. Fromm. It will be our purpose in this paper to examine the phenomenon of color phase-associated trends in temperament and behavior and place it on a physiological basis for better understanding.

The color phases mentioned above are due to gene mutations derived almost without exception from the red fox. These phases have a fairly uniform residual hereditary background of genes due to crosses within the herd. In order to set up a standard with which to compare the behavior of the color phases, it will be necessary to describe that of the red fox.

RED FOX BEHAVIOR

Vulpes vulpes fulva is an extremely nervous species. We had observed that when first brought into captivity as an adult, the red fox displays a number of symptoms that are in many ways similar to those observed

Figure 28-1 Male amber fox exhibiting fear. His eyes are wide open and staring with enlarged pupils, ears lowered, mouth slightly open, back arched, hindquarters crouched with hindlegs bent, tail bushed and held between the hindlegs, body rigid and shaking (especially the legs). These features are all manifestations of fear. The female, shown by contrast, exhibits no evidence of fear. Reprinted with permission from the March–April 1970 (Vol. 61, pp. 81–88, 1970) issue of the *Journal of Heredity*. Copyright 1970 by the American Genetic Association.

in psychosis (Keeler and Mellinger, 1966). They resemble a wide variety of phobias, especially fear of open spaces, movement, white objects, sounds, eyes or lenses, large objects and man, and they exhibit panic, anxiety, fear, apprehension, and a deep distrust of the environment (Figure 28-1). There are: (1) catalepsy-like frozen positions accompanied by blank stares, (2) fear of sitting down, (3) withdrawal, (4) run-away flight reactions, and (5) aggressiveness. When one approaches the door of a high, ranch "puppy cage" 4.9 m (16 feet) deep, the red fox mutants squat, defecate, and then panic. The frenzied animals seek refuge at the far end of the cage. They scramble about frantically trying to hide under each other, the seething mass settling down quickly into a shapeless, motionless, catatonic layer of fur upon the floor, two or three foxes deep, plastered together in impossible positions. Such fearful, catatonic foxes in withdrawal resist all attempts to move them with a stick or otherwise.

Sometimes the stress of captivity makes them deeply disturbed and confused, or may produce a depression-like state. Extreme excitation and restlessness may also be observed in some individuals in response to many changes in the physical environment.

The fox's maladaptive psychotic-like behavior in captivity yields to treatment with various psychomimetic drugs (Keeler, MacKinnon, and Fromm, 1966), especially Oxazepam, the effects being quantitatively proportional to the dosage as measured on our 20 point fear–courage scale. At step 1 of this scale the fox is hiding in the back of his kennel. At step 20 the fox is standing on his hindlegs upon the knees of the seated observer trying to take meat from the latter's hand. In a few hours the drug effect wears off and his fear returns.

Foxes may also show extreme aggression at times, maiming and killing each other; they may be more susceptible to crowding stress than the mutant forms.

BEHAVIOR OF AMBER FOXES

In comparison with our wild red foxes, we can detail the behavior of an experimental pair of ambers (Keeler and Fromm, 1965) as follows: (1) they are relatively odorless or employ their musk glands very little, (2) they have greatly reduced activity, but may occasionally run in an arc on side of the wall like the red fox, (3) they are less disturbed by noises or movements in the environment, (4) they have not been heard to bark, (5) they seldom hiss or bite, and the tail is less often highly bushed, (6) they are not much disturbed by colors such as white (red foxes prefer black, fear white), (7) they do not fight back or try to bite when properly handled, (8) they retain marked fear manifested in low crouching or arched stance with ears drooped and catatonic immobility, (9) they have depressed mood in captivity and they lack affection like red foxes, (10) they remain at a distance of 1.2 to 1.8 m (4 to 6 feet) from an observer and move about in arcs at this distance, constantly regarding the observer, (11) such ataractics as Mellaril reduce fear in these foxes without affecting activity (Keeler, MacKinnon, and Fromm, 1966), (12) such mood elevators as Tofranil and Elavil change crouching to sitting position and eliminate the eye-shifting "guilt" reaction, and (13) the relative degree of nervousness in both red and amber foxes is indicated by the frequency of urination and defecation.

Red foxes have more fear than silvers, silvers have more fear than pearls, and pearls have more fear than ambers, as measured by the average distance from an observer these grown color phase animals maintain on the priming range (Keeler, 1970). Actually, wild reds that have occasionally succeeded in entering the range stayed much more than 180 m from the observer, not having been seen at all before the roundup. Silvers tended to remain at an average of 180 m from the observer. Pearls were sighted at 130–150 m and ambers came as close as 4–5 m to strangers. It appears that the natural curiosity of all foxes draws them toward the observer, but their relative quantity of fear deters them, so that a balance is struck, setting the distance to which they will approach an investigator. The silvers are much tamer than the reds, but wilder than the pearls, ambers, and glaciers.

THEORY

Although Freud (1933) generally ignored the distinction between fear and anxiety, fear is usually considered to be one of the symptoms found in schizophrenic psychosis, especially of the pseudoneurotic type (Polatin, 1964), and closely-related anxiety is recognized as a basic element of the neuroses (Lief, 1967). The gradational removal of fear by means of increased drug dosage (Keeler, *et al.*, 1965, 1966) is a major advance of modern medicine, but a stepwise reduction of fear and anxiety response associated with the adding together of Mendelian genes has been little explored.

It was found that in anger the secretion of noradrenalin increases (Lief, 1967) and that fear is accompanied by increased secretion of adrenalin. In acute fear (Funkenstein, 1958), discharge of the entire autonomic nervous system is probably followed by a dominant parasympathetic cholinergic (trophotropic) response and then a secondary sympathetic adrenergic (ergotropic) reaction.

"More commonly, fear is less acute, and a shift to the sympathetic side of the balance in the hypothalamic system occurs that may produce such symptoms as sweating, tachycardia, pupillary dilation, and perhaps tremor due to increasing muscle tonus. The balance may also shift in adrenalin–noradrenalin output" (Lief, 1967).

Hoffer (1957) reported that in schizophrenia there is "an increase in the concentration and activity of acetylcholine within the brain, and an abnormal metabolic diversion of adrenalin to some aberrant indole compound."

It is well-known that tyrosine is transformed into dopa (3,4-dihydroxyphenylalanine) (editorial note, 1969), from which adrenalin is derived by one series of chemical reactions, and melanin pigment is produced by another series of which the first step is dopa quinone. Hence, we may expect that a mutation affecting the chemical structure of dopa may conceivably affect the constitution of behavior-modifying adrenalin and noradrenalin as well as of melanin. This gives a rationale for the pleiotropic association of pigment gene mutations with variations in adrenalin output as well as relative quantities of fear.

Physical Variations. In searching for physical variations that could help to account for differences in quantity of fear, we found histologically that the iris of animals containing the chocolate gene carry chocolate pigment only, and this is restricted in distribution in such a manner that amber foxes seem to be somewhat disoriented visually when in the sunlight (as when they run to find cover). A study was made of the lateral surface of the cribriform plate (Keeler, *et al.*, 1964). In spite of the fact that the amber's skull is larger than that of the red, the aperture areas in the lamina cribriformis of 6 ambers are inferior to those of 6 reds in cross section by an amount 1.3 times the standard error of the difference. Significance would probably be obtained with greater numbers.

Color movies demonstrate that the silvers of 1942 were more active, have greater speed, and are more vicious than ambers of 1965.

Silvers challenge each other at the food trough. Ambers eat quietly, and as a result, ambers fatten more readily. It was found that larger doses of tranquilizing drugs were necessary to control wild red foxes than ambers.

PROBLEM AND METHOD

A psychopharmacologist is likely to ask two questions: (1) do the wilder and more active red foxes function on relatively more adrenalin per unit of body weight than do their color phases? and (2) are their differences associated with the relative size of the pituitary?

We weighed foxes before pelting, and dissected out the adrenals and pituitary for weighing (Keeler, *et al.*, 1968). Under the assumption of unaltered proportions of adrenal cortex to medulla, and of unaltered capacity of secretion by each medullary cell, relative adrenal size might account for much of the hereditary activity and behavior differences observed between red foxes and their several derivative color phases.

In the study, 374 foxes were dissected. (Keeler, 1970). Their adrenals were weighed. The data on relative adrenal and pituitary size were separated into old and young, male and female, as well as specfic color phase.

RESULTS

Adrenal Weight. Ratios of adrenal weight divided by body weight for red foxes were compared with similar ratios for silver, pearl, amber, and glacier animals (See Table 28-1).

TABLE 28-1 Ratio of Adrenal Weight to Body Weight for Various Color Phases[a]

Females	
16 red .1343 ± .01175 > 15 silver	.0654 ± .00194.
	dm = .0689. s = 5.8
16 red .1343 ± .01175 > 36 pearl	.0557 ± .00195.
	dm = .0786. s = 6.6
16 red .1343 ± .01175 > 8 glacier	.0478 ± .00262.
	dm = .0865. s = 7.2
16 red .1343 ± .01175 > 59 amber	.0521 ± .00142.
	dm = .0822. s = 6.9
Males	
15 red .1540 ± .01477 > 9 silver	.0644 ± .00467.
	dm = .0896. s = 5.8
15 red .1540 ± .01477 > 32 pearl	.0513 ± .00148.
	dm = .1027. s = 6.9
15 red .1540 ± .01477 > 5 glacier	.0462 ± .00510.
	dm = .1078. s = 6.9
15 red .1540 ± .01477 > 88 amber	.0486 ± .00113.
	dm = .1054. s = 7.1

[a]dm = the difference between the means; s = significance, or the difference between the means divided by the standard error of the difference.

In all comparisons, the ratio of adrenal weight to body weight for reds is significantly greater than that of silver, pearl, glacier, and amber animals for both males and females. When the adrenal weight/body weight ratio of pearl animals was compared with that of amber and glacier, the pearl ratio was found to be greater than that for the other two color phases in 10 out of 12 comparisons. It would appear that with large numbers of animals, and with laboratory refinements in the technique of raising foxes, the pearl adrenal weight/body weight ratio will tend to be significantly greater than that of both amber and glacier.

When the adrenal weight/body weight ratio for groups of ambers was compared with that of glaciers, the amber ratio was found to be greater than the glacier ratio in 5 out of 6 comparisons.

Pituitary Weight. Silver exceeded pearl in pituitary weight relative to body weight ratios in 5 out of 6 silver–pearl comparisons. Silver males were greater than pearl males, with statistical significance in one case and almost significance in a second case. Young silver males were greater than young pearl males with signficance.

There were 6 silver–amber pituitary comparisons. In 4 out of 6 comparisons silver exceeded amber. Other comparisons were nonconclusive due to limited numbers. In 2 comparisons of females, silver exceeded amber significantly. In 3 out of 6 pearl–amber comparisons for relative pituitary size, pearl was greater than amber. When 6 amber–glacier comparisons were made, amber exceeded glacier in 3.

Body Size — Adrenal Function. Average body size in foxes increases progressively with the addition of the coat color genes for black, pearl, amber, and possibly glacier (Keeler, 1970). Taking the average body weight of red foxes as 100%, silvers are 49% larger than reds, pearls are 19% larger than silvers, glaciers are 8% larger than pearls, and ambers are 3% larger than glaciers, in keeping with our findings on rats (Keeler, 1942, 1947) and mink (Keeler and Moore, 1961).

Actual adrenal weight tends to decrease with the addition of the above named genes except that the pearl combination may possibly produce a smaller adrenal than does the amber.

The reduction of activity and wildness found in the fox color phases, silver (bearing the black gene homozygously), pearl (bearing the black and blue genes homozygously), amber (bearing the black, the blue, and the chocolate genes homozygously), and the glacier (bearing probably the black, the blue, the chocolate, and the white gene homozygously), in comparison with the wild red fox show somewhat graded behavior tendencies that appear as pleiotropic manifestations of the mutant coat color genes listed above. These graded levels of reduced activity and tameness could be partially explained if a parallel series of adrenal functional levels were found among the color phases.

Histological examinations were made of cross sections of adrenals of red foxes and ambers (Keeler, *et al.*, 1968). From these studies we must conclude that relative adrenal size also means relative amount of medullary secretory tissue, and that the medullary cells of coat color mutants produce their normal quantity and quality of adrenalin.

From the comparisons described, it would appear that the homozygous fox strains studied can be arranged in order of their adrenal weight relative to body weight, in descending order as follows:

Wild red (no mutant genes)
Silver (black mutation + silver modifiers)
Pearl (black and blue mutations)
Amber (black, blue, and chocolate mutations)
Glacier (black, blue, chocolate, and white mutations)

It would appear that adrenal size reduction is correlated with both the number of hereditary coat color mutations possessed, and with a pleiotropic reduction in activity and wildness, much as has been described for mink (Keeler and Moore, 1961) and rats (Keeler, 1942, 1947).

It appears that relative pituitary size also tends to decrease with the addition of coat color mutations. A similar and statistically significant reduction of the pituitary was described in mink (Keeler and Moore, 1961).

The work of Christian (1955) and Levine (1957) would suggest that possibly the problem is even more complicated than we have indicated, and may involve also a tendency toward emotionally determined hypertrophy of the adrenals in animals under great stress from their environment. If this is true, then the size of adrenal is influenced both by pleiotropic manifestations of the coat color genes and also according to the fear, frustrations, and resultant flight activity that the individual has recently undergone. How extensive this effect of recent experience can be we do not know, but Christian and Levine stated that it could be as great as 9% in mice. However, a possible variation of this magnitude due to stress would not interfere with our conclusions, especially in view of the fact that there was no long period of stress in any of our animals prior to killing for dissection.

RATING THE FEAR OF EXPERIMENTAL FOXES

During this study, 34 foxes were produced by crossing tame amber to wild red, and then crossing the wild red hybrids back to tame amber. It will be noted that these 34 animals are of 8 distinguishable phenotypic coat color classes. They are heterozygous for any recessive genes (black, blue, chocolate) for which they are not homozygous.

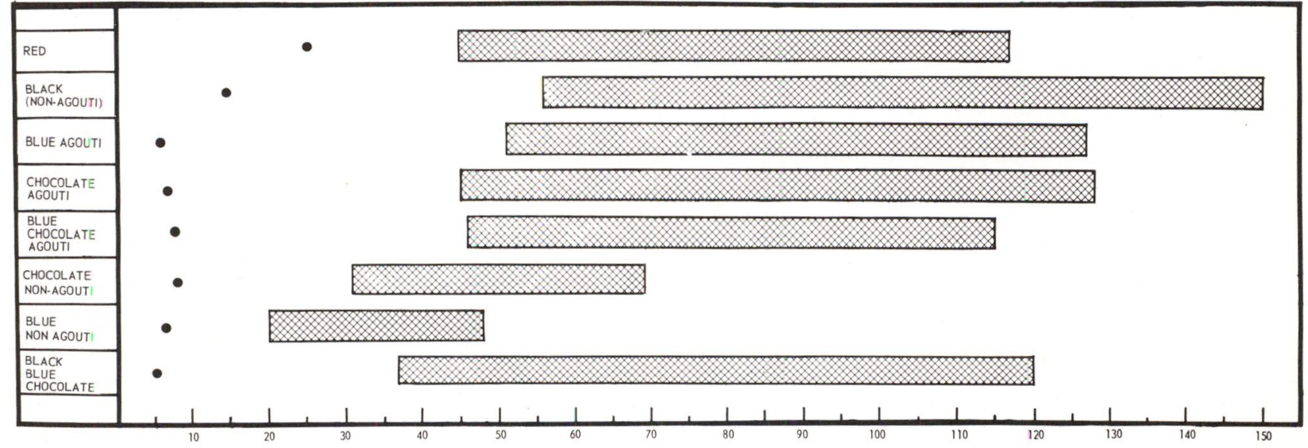

Figure 28-2 Dots represent average startle distance (in feet) from the observer on first day of testing. Left ends of cross-hatched strips represent average startle distances during subsequent 4 days of testing; lengths of strips show average distances run after startle. The column on the left names the genes for which each group of foxes is homozygous, with the exception of red, which are homozygous for no receissive mutations. (1) red, (2) black, (3) blue nonagouti (pearl), and (4) black, blue, chocolate (amber) are the same phenotypic colors. Animals exhibiting the coat color effects of (5) chocolate and agouti genes, (6) blue, chocolate, and agouti, (7) chocolate and nonagouti, and (8) blue and agouti are new combination color phases. Reprinted with permission from the March–April 1970 (Vol. 61, pp. 81–88, 1970) issue of the *Journal of Heredity*. Copyright 1970 by the American Genetic Association.

They were turned loose in an enclosure about 30 × 45 m (Keeler, 1970). Clumps of grass about half a meter high served as natural cover in which foxes may hide.

When three observers approached the enclosure not more than one or two foxes appeared to be present in the enclosure, because nearly all of them were lying down, hidden by clumps of grass. When the observers entered the enclosure slowly and quietly, animals all over the area became curious, they sat up and faced the intruders, shaking their heads sharply, flicking and focusing their ears on them. One observer walked slowly and quietly the length of the field and recorded distances at which individual foxes startled and ran, as well as their color phase.

TABLE 28-2 List of Data for Comparison (these Data Are Compared in Table 28-3)

No. encounters *	Color description	Startle distance	No. mutations	Gene constitution		
15	red (agouti)	45±1.2	0	*Aa*	*Bb*	*Cc*
14	black (nonagouti)	56±5.4	1	*aa*	*Bb*	*Cc*
23	blue agouti	51±3.1	1	*Aa*	*bb*	*Cc*
22	chocolate agouti	45±3.4	1	*Aa*	*Bb*	*cc*
33	blue nonagouti	20±2.3	2	*aa*	*bb*	*Cc*
28	chocolate nonagouti	31±2.6	2	*aa*	*Bb*	*cc*
6†	blue chocolate agouti	46±8.8	2	*Aa*	*bb*	*cc*
8†	blue chocolate' nonagouti	37±4.3	3	*aa*	*bb*	*cc*

**aa,* nonagouti; *AA or Aa,* agouti; *bb,* blue dilute; *BB* or *Bb,* intense pigment; *cc,* chocolate; *CC* or *Cc,* black.

†Unfortunately, we had not noticed that we had only 6 records for blue chocolate agouti; and 8 records for blue chocolate nonagouti; rendering the standard error of the mean too high for comparison in time to increase the number. Part of the reason appears to be that on occasion these relatively tame animals were passed by and did not startle at all but lay hiding in the grass. However, there are significant differences between the startle differences of animals having 0 mutations or 1 mutation, and those bearing 2 mutations, as shown in Table 28-3.

The distances run after startle during the next week (indicated in Figure 28-2 by the length of the cross-hatched strips) did not differ much from the first observations for most of the colors, with the notable exception of individuals homozygous for both the nonagouti (black) and blue genes, as well as animals homozygous for both the nonagouti and chocolate genes.

It will be seen that the greater startle distances after the first day (indicated by the left ends of the cross-hatched strips in Figure 28-2) are

TABLE 28-3 Comparisons Between Color Phases*

15 red agouti	45±1.2 > 33 blue nonagouti	20±2.3=25±2.6	s=9.6
15 red agouti	45±1.2 > 28 chocolate nonagouti	31±2.6=14±2.8	s=5.0
15 red agouti	45±1.2 > 8 blue chocolate nonagouti	37±4.3= 8±4.4	s=1.8
14 black (nonagouti)	56±5.4 > 8 blue chocolate nonagouti	37±4.3=19±6.8	s =2.7
14 black (nonagouti)	56±5.4 > 22 chocolate agouti	45±3.4=11±6.3	s=1.7
14 black (nonagouti)	56±5.4 > 33 blue nonagouti	20±2.3=36±5.8	s=6.2
23 blue agouti	51±3.1 > 33 blue nonagouti	20±2.3=31±3.8	s=8.1
23 blue agouti	51±3.1 > 28 chocolate nonagouti	31±2.6=20±4.0	s=5.0
22 chocolate agouti	45±3.4 > 33 blue nonagouti	20±2.3 =25±4.1	s=6.1
22 chocolate agouti	45±3.4 > 28 chocolate nonagouti	31±2.6=14±4.3	s=3.2
28 chocolate nonagouti	31±2.6 > 33 blue nonagouti	20±2.3=11±3.4	s=3.4

*The mean startle distance for red agouti (0 mutations), black nonagouti (1 mutation), blue agouti (1 mutation), chocolate agouti (1 mutation) were compared with blue nonagouti (2 mutations), chocolate nonagouti (2 mutations), blue chocolate agouti (2 mutations) and blue chocolate nonagouti (3 mutations). *s* = significance, calculated as the difference between the means divided by the standard error of their difference. When this quotient is 2 or more, the difference between the means is considered to be statistically significant. Thus, in 8 comparisons, color phases with 0 or 1 mutation on the average have statistically significantly greater startle distances than have color phases with 2 mutations. In 2 other cases the differences approach significance.

still not nearly so great for reds, blacks (silver), and blue (pearls) as observed on the priming range after a much longer time at liberty. Ambers (homozygous for black, blue, and chocolate) have an average startle distance less than all others except the extremely calm chocolate-colored animals (homozygous for chocolate and nonagouti) and blue-colored animals (homozygous for blue and nonagouti genes) that not only had a shorter startle distance, but also an extremely short running distance. Many startle distance comparisons are statistically significant. See Tables 28-2 and 28-3.

It should be pointed out that of these backcross foxes, each one carries a recessive gene for the 3 color mutations (black, blue, and chocolate) that might dampen the fear and activity effect of any dominant alleles that they may carry.

PRELIMINARY BIOCHEMICAL TESTS

Blood drawn from the hearts of wild red, black (silver), blue (pearl), and amber animals were tested for protein-bound iodine, which was measured in micrograms per milliliter (mg/ml). Tests were made on 4 animals of each color phase. The values were averaged for each color phase and are represented on curves in Figure 28-3.

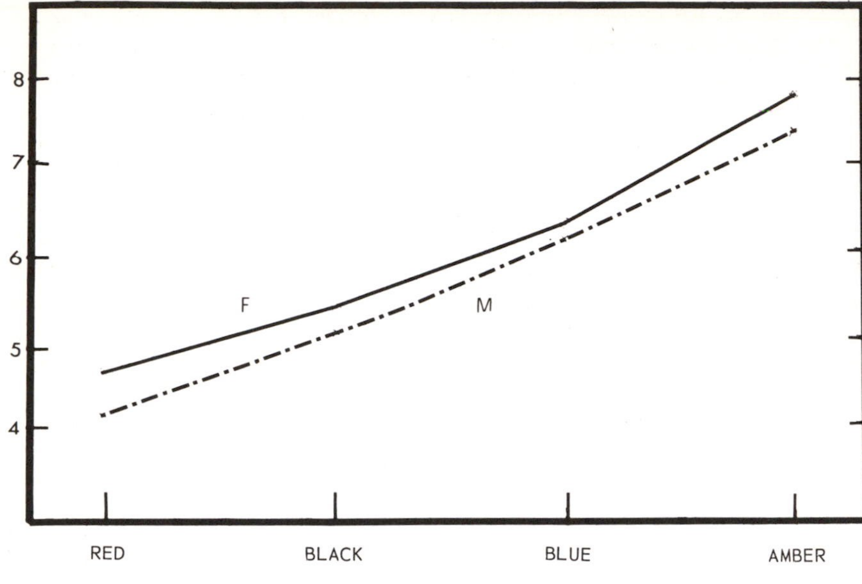

Figure 28-3 Male and female average values of protein-bound iodine (μg/ml) for 4 animals of each color phase. Reprinted with permission from the March–April 1970 (Vol. 61, pp. 81–88, 1970) issue of the *Journal of Heredity.* Copyright 1970 by the American Genetic Association.

Because iodine bound into a globulin protein in the thyroid breaks down into thyroxin, the measurement of this protein-bound iodine may be taken as an assessment of thyroid function (Table 28-4).

TABLE 28-4 Protein-bound iodine tests (measured in μg/ml)

A. Males			
4 red	4.1±.71 < 4 amber	7.4± .53=3.3±.88	s=3.7
4 silver	5.1±.64 < 4 amber	7.4± .53=2.3±.83	s=2.7
B. Females			
4 red	5.1±.75 < 4 amber	7.8± .43=2.7±.86	s=3.1
4 silver	5.4±.35 < 4 amber	7.8± .43=2.4±.55	s=4.3
C. Combined sexes			
8 red	4.6±0.46 < 8 silver	5.3±0.19=0.7±.49	s=1.4
8 red	4.6±0.46 < 8 amber	7.6±0.20=3.0±.50	s=6.0
8 red	4.6±0.46 < 8 pearl	6.2±0.48=1.6±.66	s=2.4
8 silver	5.3±0.19 < 8 pearl	6.2±0.48=0.9±.51	s=1.8
8 silver	5.3±0.19 < 8 amber	7.6±0.20=2.3±.28	s=8.2
8 pearl	6.2±0.48 < 8 amber	7.6±0.20=1.4±.52	s=2.6

The values for males were consistently lower than for females, although this sex difference is statistically insignificant because so few animals were tested. It will be noted that with the addition of 1 mutation (black), 2 genes (blue), and 3 genes (amber), the PBI is raised in a stepwise fashion, so that the values for PBI for animals having 3 gene mutations are about 2 times as high as those for wild reds.

Although there are but 4 animals in each group represented by a point in Figure 28-3, the differences in PBI are significant in both sexes when the mean for reds is compared with that of ambers, and when the mean of silvers is compared with that of ambers (see Table 28-3).

If we neglect the slight sex differences and simply compare the means of color phases, we get significant differences in all comparisons except red *vs* silver and silver *vs* pearl. It appears highly probable

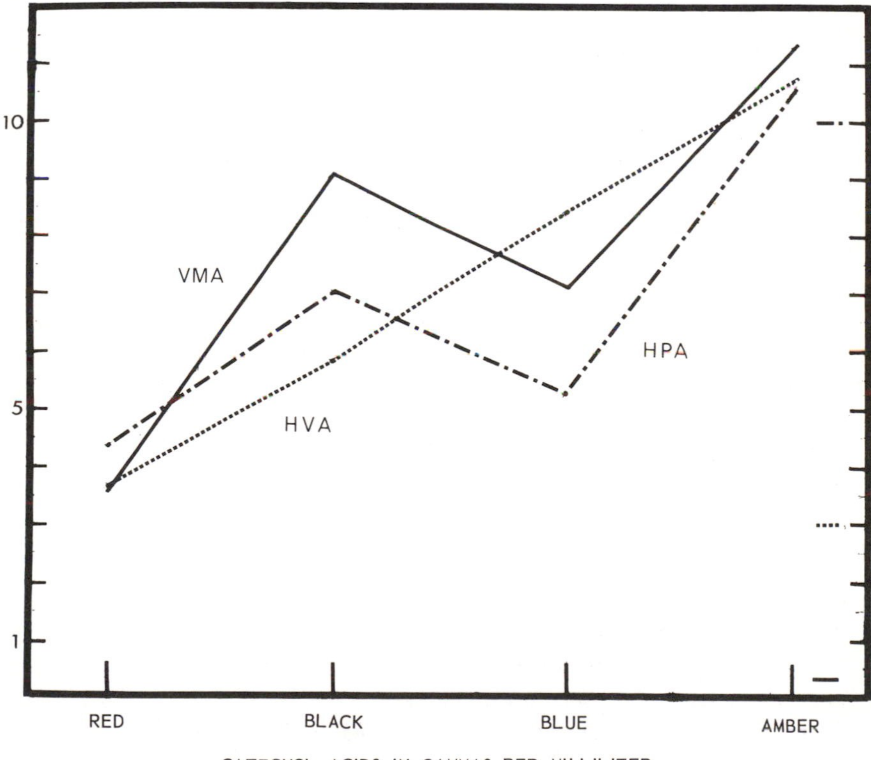

Figure 28-4 Average values of 3 catechol acids (mg/ml) for 4 color phases, (VMA = vanylmandelic acid: HVA = homovanillic acid: HPA = dihydroxyphenylacetic acid). Reprinted with permission from the March–April 1970 (Vol. 61, pp. 81–88, 1970) issue of the *Journal of Heredity*. Copyright 1970 by the American Genetic Association.

that when larger numbers are tested, all differences will become statistically significant.

Urine samples taken by catheter were tested using the photo fluorometric method by Mellinger (1968) for derivatives of adrenalin measured in milligrams per milliliter (mg/ml): homovanillic acid, dihydroxyphenylacetic acid, and vanylmandelic acid. Since males and females did not differ in range of values, they were averaged to provide the points on which the curves in Figure 28-4 are based. Although catechol acid tests generally vary more than PBI tests, it is noted that the values for all 3 acids in ambers are about 3 times as great as those for wild reds, which suggests that the introduction of 1 mutation (black) and 2 mutations (blue) produce animals with intermediate values (Figure 28-4).

Tests for the catechol acids (homovanillic, dihydroxyphenylacetic, and vanylmandelic) were more variable than those for PBI. For homovanillic acid, the difference between the means divided by the standard error of the difference between 5 reds, 7 silvers, 8 pearls, and 8 ambers was less than 2. However, the difference between 5 reds and 8 pearls was significant, as was the difference between 5 reds and 8 ambers (Table 28-5A).

TABLE 28-5 Catechol Acid Tests (Measured in mg/ml)

A. Homovanillic acid			
5 red	3.7±0.6 < 8 pearl	8.4±1.4=4.7±1.5	s=3.1
5 red	3.7±0.6 < 8 amber	10.7±1.2=7.0±1.3	s=5.3
B. Dihydroxyphenylacetic acid			
5 red	4.3±1.1 < 5 silver	7.0±0.9=2.7±1.4	s=1.9
5 pearl	5.2±1.0 < 7 amber	10.6±2.3=.54±2.5	s=2.1
5 red	4.3±1.1 < 7 amber	10.6±2.3=6.3±2.5	s=2.5
C. Vanylmandelic acid			
1 red	3.6±0.0 < 6 amber	11.4±2.2=7.8±2.2	s=3.5

Differences between tests for dihydroxyphenylacetic acid were not of statistical significance in 3 comparisons, although it approached significance when 5 reds were compared with 5 silvers. Two other comparisons were significant: 5 pearls to 7 ambers, and 5 reds to 7 ambers (Table 28-5B). In the tests for vanylmandelic acid, only 1 red fox gave a satisfactory response. The results for this animal were different from tests on amber (Table 28-5C).

A possible explanation of the PBI results could be that all foxes take in a similar amount of iodine in their food. This provides a reservoir of protein-bound iodine in the blood stream, and the thyroid of the hyperactive, wild red fox is efficient in drawing from that reservoir. The thyroids of the mutants black, blue, and amber are less efficient in this respect. Males may draw upon their reservoir slightly more than females do, and it is well-known that females tend to be tamer than males.

Although our catechol acid determinations for black and blue appear erratic, still those determinations for red and for amber are orderly and quantitatively so different that an explanation should be sought. One possibility supposes that these metabolites of adrenalin in the urine represent the greater ability of red foxes to retain the catechols within the body, or the greater tendency of animals bearing the mutations studied to excrete the catechols.

It appears certain from the present studies that adrenal and thyroid functions are altered by the presence of coat color genes, and that through these altered functions, fear is modified. That the pituitary is also implicated has been shown previously. However, further investigation is needed to determine if our explanation of the chemical mechanisms involved is correct.

SUMMARY

We have considered in detail the unique personality of the red fox and its maladaptive behavior under captivity: aggression, withdrawal, catatonia, panic, flight, and other symptoms based on a very large quantity of fear, and have shown that the genetic constitution affects the quantity of fear response on the priming range. When the genes for black (nonagouti), blue, and chocolate are combined in a single individual, their effects are somewhat additive, and the decrease in distance from the observer is roughly correlated with the ratio of adrenal weight to body weight. When a trihybrid cross of wild red to triple recessive amber was made, and the hybrids were backcrossed to the ambers, the 8 expected color phases were produced.

These backcross animals exhibited roughly the same pleiotropic association of different quantities of fear in the startle tests that we had observed on the priming range, increasing in quantity with the time at liberty. Protein-bound iodine of the blood, and the catechol acids of the urine (homovanillic, dihydroxyphenylacetic, and vanylmandelic) increase with the addition of black, blue, and chocolate genes to the genetic constitution, but the chemical mechanisms involved have not yet been clearly determined.

ADDENDUM

An Hypothesis Of Density-Adapted Morphs Among Northern Canids

R. D. GUTHRIE
Department of Biosciences
University of Alaska
College, Alaska

Dr. Fox has asked me to write a very brief summary of a hypothesis that I am now in process of testing which relates somewhat to Dr. Keeler's work. Data collecting is still at an early stage and much of what we have is still unprocessed. But the hypothesis alone is perhaps worthwhile introducing even though it may prove incomplete or incorrect.

I am proposing that the silver phase common in the northern parts of both Eurasia and North America may be an outward manifestation of a syndrome of behavioral, physiological, and morphological adaptations to unstable population density situation. Population numbers of voles and lemmings in the far north and snowshoe hares in the taiga fluctuate in relatively predictable periodic cycles. The numbers of animals preying on these species cycle in phase with their food sources. Both predator and prey reach plague levels only to fall to virtual extinction during the low of the cycle.

Keeler in this volume and elsewhere has shown that the phases of the red fox are differentially suited to varying densities — the blacks (silvers) doing well at artificially high population levels in captivity but apparently not so well in lower densities as they occur only rarely throughout the central and southern portions of the fox range. Also, our preliminary information indicates that the frequencies of the phases change with the stage of the cycle, the black reaching between 3–10% in some areas.

The hypothesis, then, is that the black (or silver) and possibly to some degree its heterozygotes (the cross) are selected for at very high densities when a relatively nonaggressive, less hyper social profile is more suitable. Its coat is a neotenic expression of the newborn chocolate–black and perhaps it capitalizes on the protection of a younger image (Keeler also mentions that foxes are less wary of dark colors and more intimidated by light ones). The red phase would be selected for at the lower population levels and at increasing densities. This results in a balanced polymorphism.

Our analysis is taking several forms. We are collecting carcasses from trappers to be autopsied. (This is our first year with a moderately high population and just over half of the individuals in our sample have ulcers — some quite acute.) A phase correlation with prey and fox density during the rise, peak, crash, and crash aftermath is being conducted using trapper information and a widespread sighting-record project. Also, we plan on rearing about 40 young of various color phases taken as pups from the wild in a large enclosure and studying individual and phase behavior. We will sacrifice them at the end of the study period for autopsies. If the hypothesis is founded in fact, it may be that the polymorphisms so common among many other northern animals (e.g., the Arctic fox, *Alopex*) is a related phenomenon.

29

SOME GENETIC AND ENDOCRINE EFFECTS OF SELECTION FOR DOMESTICATION IN SILVER FOXES

D. K. BELYAEV and L. N. TRUT
Institute of Cytology and Genetics
Siberian Department of the
U.S.S.R. Academy of Sciences, U.S.S.R.

It is becoming increasingly apparent that in the evolution and selective breeding of animals, the adaptive importance of modified behavioral responses must be çonsidered. Also, behavioral correlations may be made with a number of morphological and physiological characters (Wilcock, 1969; Price, 1967). Of particular importance is the correlated variability in the evolution of characters endowed with high adaptive values; those which are formed under the effect of a stabilizing form of natural selection are characterized by low heritability.

Seasonal reproductive periodicity is a good example and, although inherent to wild animals, it has been lost in the course of domestication. The adaptive significance of this character is quite obvious, since the bearing and rearing of offspring is limited to the most favorable season of the year (usually in terms of availability of food). The mechanisms underlying the loss of the seasonal pattern of reproduction and the acquisition of the capacity to mate more than once a year, the feature of the domestication process of wild animals, remain unclear. Thus, silver foxes *(Vulpes fulvus* Desm.), for instance, have been bred for commercial purposes in fur farms where food is always plentiful for about 80 years. Throughout these years they were selected for early mating periods within the breeding season which begins at the end of January and continues up to the end of March. However, this selection has had no effect on reproduction and silver foxes do not display any tendency towards mating outside of the natural spring breeding season.

The concept was formulated that as a consequence of artificial selection for a certain behavior trait (docility) (Belyaev, 1962), reorganization of reproduction from a rigid seasonal pattern with a monestrous cycle to the loss of this pattern and the adoption of a diestrous cycle had occurred. This would be a logical correlation between behavior and reproductive patterns wrought by domestication. This suggestion is supported by the correlation between the type of defensive behavior and such reproductive characters as fertility and time of sexual activity which we have established in farm-bred silver foxes (Belyaev and Trut, 1964). Although the animals exhibited in general wild type behavior, there was diversity in their response to man. Analysis has shown that the diversity is hereditarily determined. It was found that foxes which do not show aggressive and fear responses when coming into contact with man mated earlier during the breeding season and had larger litters. These observations substantiate the suggestion that the reorganization of the genetic basis affecting the reproductive function, its seasonal pattern, for example, might have evolved through selection for certain behavioral responses which may be especially characteristic of the early stages of domestication of animals.

The purpose of the present study was to produce, in the course of systematic selection for behavior, a type of domestic fox in some measure resembling the domesticated dog in its behavior (i.e., tractability and docility) and to trace the effect of such trait selection on the reproductive activity. It should be emphasized that in the process of domesticating these foxes, profound and complex changes occurred in behavior and in the organismic neuroendocrine state. Of particular relevance is the pituitary–adrenal system which plays an important role in the processes of adaptation. Evidence accumulated so far indicates that there is a relation between this system and different forms of behavior, and also between the pituitary–adrenal system and the pituitary–gonad system (Lissak and Endrosci, 1967; Thiessen and Nealey, 1962; Pare, 1966). These neuroendocrine interrelationships give good reason to believe that the process of the domestication of foxes must have been accompanied by shifts in the pituitary–adrenal system, as well as with functional changes in the state of the reproductive system. Hence, it seemed expedient to study the endocrine effects of selection for domestication.

MATERIAL AND METHODS

The study was carried out on a colony of animals managed on the same basis as a commercial farm at the experimental fur farm of the Institute of Cytology and Genetics of the Siberian Department of the USSR Academy of Sciences in the region of Novosobirsk, Akademgorodok. The selection of foxes for domestic type of behavior was started in 1956. Pups from 1½ to 2 months old were chosen as initial material for selection; these pups, just as their parents, showed no aggressive nor fear responses towards man. The proclivity towards domestication was tested on the basis of contacts with man, which were graded in time, and then assessed quantitatively by the acceptance of food from the hand of man, and response to fondling, handling, and to call. Homogeneous pairs of unrelated animals with similar behavior phenotypes were set up with the highest scores for amenability to domestication. The estimates of the functional state of the reproductive system were regularly made from the gross appearance of the genitals, microscopic studies of vaginal smears, and sexual behavior of the animals.

The total content of free 11-oxycorticosteroids in peripheral plasma (determined by a fluorometric method) served as measure of the func-

Figure 29-1 Foxes from the population selected for domestic behavior.

Figure 29-2 Aggressive fox from nonselected for behavior population.

tion of the pituitary–adrenal system (Stahl and Dörner, 1966; Shorin, 1967). Blood for analysis was collected from the hindleg in the morning before feeding. Since the experimental animals were not trained to the procedures connected with the blood sampling, this alone could be the cause of some stimulation of the pituitary–adrenal system. However, not more than 5 minutes elapsed from the time the experimenter came up to the cage to the completion of blood collection. In such conditions, we determined that these corticosteroid levels were close to initial or baseline levels.

Corticosteroid levels were determined in the experimental animals 2 times a month during the year to scan for possible seasonal fluctuations. Two groups of animals were used in the experiment: one group (5 males and 4 females) was derived from lines selected for docile behavior and characterized by domestic behavior formed on the basis of a definite genotype (Figure 29-1); the control group was composed of animals distributed similarly in number, age, and sex, but obtained from populations which were not selected for domestication and showed an aggressive response to man (Figure 29-2).

In order to determine the response of the adrenal cortex and the pituitary–adrenal system during different seasons of the year, two conditions of stress were produced: exogenous ACTH was administered in a dose of 3 units per kg of body weight and the animals were stressed by immobilization in a portable cage corresponding to their size. Corticosteroid levels were determined prior to and 1 hour after stress.

All the experimental animals and fur farm animals were caged separately and kept on standard rations.

RESULTS

The results obtained in the course of the selection of foxes for docile behavior have shown that the amenability to domestication is hereditarily determined and the degree to which the offspring are domesticated increases with the number of domestic ancestors in their pedigree. In other words, domestication for behavior is a profound process of selection. By means of systematic selection of animals not exhibiting any aggressive or fear responses to man and most easily amenable to taming, genotypes were produced in foxes on the basis of which behavioral traits of the domestic dog were evident. Domestic foxes respond when called, come up to man, and permit themselves to be petted and picked up: the most domesticated animals wag their tails in greeting and emit typical barks at the sight of man.

The specific correlated response to the specific selection for behavior which we have carried out is the reorganization of the behaviorally correlated reproductive characters. This specific response expressed itself in a number of females of the selected population by an activation of the reproductive system outside the season characteristic of these animals (in autumn, September and October and in spring, April and May).

The hereditary nature of this reorganization was confirmed by the analysis of the pedigrees of foxes showing such sexual activation outside of the season. Throughout the years, of all the 310 females so far subjected to selection, the seasonal pattern of reproduction was altered in 48 females in only 5 inbred lines. An illustrative case in this respect is the genealogical line of the female "Laska." Of 60 females belonging to this line (i.e., descendants of "Laska"), 20 exhibited the tendency towards transgressing the seasonal limits of reproduction.

The effect of selection for this behaviorally-correlated character resulted in an increase of the number of animals with an altered pattern of reproduction. Foxes born in 1962 and showing the sexual activation outside the season made up only 6% of the total number of females of the experimental farm, whereas in foxes born in 1969, the figure rose to 40– after only 7 years of selective breeding.

Although complete development of estrus was not observed in most females, signs of sexual activity were clear-cut and the sexual cycle was shifted far beyond proestrus. The reorganization towards 2 annual estrus cycles in foxes were paralleled by decreased capacity to reproduce during the breeding season. Thirty percent of the foxes failed to produce litters: either they did not mate and, in the case when they did mate, they did not produce litters or the litters succumbed due to inhibited lactation or cannibalism. (This circumstance may decrease substantially the actual efficiency of selection.)

The decrease of the reproductive performance of animals from the experimental farm is regarded as a specific response of the reproductive system to the specific form of selecting for docile behavior. This is supported by the fact that the decrease in reproductive efficiency was observed mainly in females displaying a tendency towards diestrous cycles. Thus, in these females, the behavioral and reproductive patterns which had been created earlier by stabilizing natural selection were breaking down (i.e., destabilization) and new reorganized patterns had not as yet evolved.

Studies of the pituitary–adrenal system have shown that animals with different types of behavior and temperament differ in plasma content of the adrenocortical hormones. Thus, in wild type females from such a population which had not been artificially selected, the level of

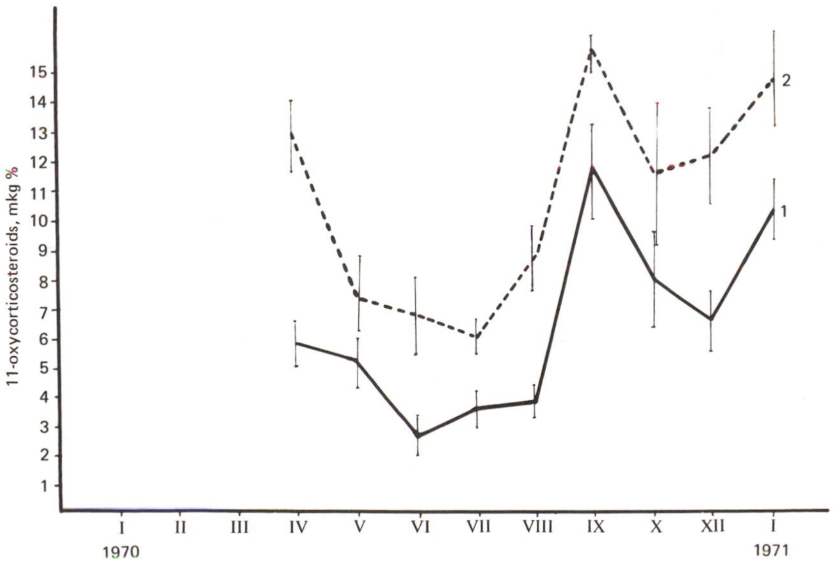

Figure 29-3 Seasonal dynamics of corticosteroid level (M±m) in peripheral blood of tamed and wild silver fox females.

——— tamed animals

- - - - wild animals

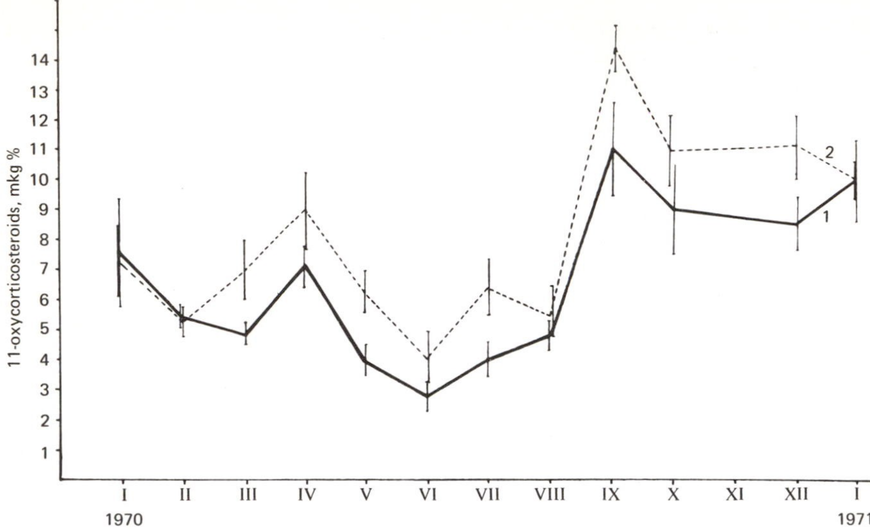

Figure 29-4 Seasonal dynamics of corticosteroid level (M±m) in peripheral blood of tamed and wild silver fox males.
——— tamed animals
- - - - wild animals

plasma corticosteroids waa significantly higher than in "domestic" type females (Figure 29-3). However, the differences in plasma 11-oxycorticosteroids between wild and domestic type male foxes were not as marked as in the females and the differences were absent during the breeding season at the end of January and in February (Figure 29-4). The seasonal fluctuations in corticosteroid level had a similar trend in both tame and wild males and females.

The responsiveness of the pituitary–adrenal system characteristic of domestic and wild animals not only determines the initial level of plasma hormones but may also influence the reactivity to ACTH and to psychological stress. Thus, the administration of the adrenocorticotrophic hormone to "domestic" male foxes was accompanied by a distinct elevation of plasma corticosteroid level during all seasons of the year studied (Table 29-1). The elevation was highest in spring, in May, and lowest in autumn, in October.

In "wild" males, the changes were most pronounced at the end of February, while in May their adrenal response showed a twofold decrease as compared with domestic foxes.

ACTH administration to female silver foxes evoked a similar trend in the adrenocortical response. However, during all the months studied, ACTH response in domestic females was about 2–4 times more intense than in wild females (Table 29-1). As a rule, the level of plasma cortico-

TABLE 29-1 Seasonal dynamics of adrenal reactivity to ACTH in silver foxes (resting level and σ (sigma) % increase in corticosteroids from baseline after ACTH treatment shown)

Animal Groups	Cortico-steroid Level (M±m)	February 1970	May	September	October	December	January 1971
Domestic males	M±m	89.8±7.25	158.6±58.15	85.1±15.5	65.8±17.6	95.8±26.7	79.15±22.95
	σ	14.5	116.3	34.7	39.3	59.58	45.9
Wild males	M±m	142.1±39.1	70.5±42.0	64.9±11.4	83.5±33.7	57.1±12.0	98.3±22.0
	σ	78.2	84.0	25.4	75.3	27.8	51.17
Domestic females	M±m	—	207.8±94.7	122.8±28.15	209.2±119.6	140.8±33.0	81.65±22.0
	σ	—	189.5	56.3	239.2	57.2	44.01
Wild females	M±m	—	131.2±40.9	33.8±8.25	47.4±14.25	60.8±37.2	58.7±4.8
	σ	—	81.9	16.5	28.7	64.4	8.32

Table 29-2 Seasonal dynamics of adrenal reactivity to psychological stress in silver foxes (% increase (σ) from initial level)

Animal Groups	Cortico-steroid level (M±m)	February 1970	March	May	July	August	October	December	January 1971
Domestic males	M±m	49.2±16.85	25.5±12.9	—	152.8±50.8	—	57.8±15.0	74.4±13.4	114.8±29.5
	σ	33.5	22.3	—	113.3	—	33.5	26.8	65.9
Wild males	M±m	72.7±34.9	112.05±20.0	—	162.1±50.7	—	23.6±8.7	38.6±11.3	40.7±14.08
	σ	69.8	40.0		113.2		19.4	19.55	28.16
Domestic females	M±m	—	—	98.4±51.15	116.0±36.0	190.7±135.0	93.6±54.5	141.0±29.35	118.6±37.8
	σ			102.3	72.0	233.5	94.5	58.7	65.4
Wild females	M±m	—	—	48.1±6.07	116.1±31.85	62.9±20.45	86.9±21.6	45.1±26.3	46.98±22.14
	σ			10.5	63.7	40.9	43.2	37.2	44.28

steroids against the background of ACTH administration was lower both in wild and domestic females in autumn and winter than in spring. The only exception was October when an average increase of about 209% in the 11-oxycorticosteroid level occurred in domestic females after ACTH administration, that is, reaching the level attained in May.

These experiments show that domestic and wild foxes differ in the degree to which they respond to the same dose of ACTH, and one must also consider that wild and domestic foxes differ in the functional state of the adrenal cortex.

These experiments using ACTH give some insight into the functional state of only the terminal link of the hypothalamo–pituitary–adrenal system, the adrenals. Since the main purpose of the present study was to elucidate the relation between the type of animal behavior (or trait/temperament) and the function of the entire hypothalamo–adrenal system, it seemed expedient to analyze changes in the corticosteroid level in response to factors affecting the whole system. Experiments introducing psychological (emotional) stress (exposure of animals to a novel situation, confinement in a carrying cage) have shown that during almost all the seasons studied, the hypothalamo–pituitary–adrenal system is activated to a lesser degree in wild foxes than in domestic foxes of the same sex (Table 29-2). July was an exception, when there were no differences in the response of domestic and wild animals of either sex. February and March were also exceptions in that wild males responded more than domestic males.

Thus, experiments using psychological stress have pointed out a feature which had also been noted in experiments with ACTH administration: animals of differing temperaments (wild or docile and tame), vary in their responsiveness to the same stress (see also Fox and Andrews, 1973). The data suggest that wild and domestic foxes differ in the functional state of the entire hypothalmo–pituitary–adrenal system.

CONCLUSIONS

The present study has shown that the amenability of silver foxes to domestication is hereditarily determined and that domestication is a profound process of selection. The reorganization of animal behavior in the process of selection for domestication evokes as a major correlated response the reorganization, or to be more precise, the *destabilization* of a highly stabilized feature as exemplified by the seasonal pattern of reproduction.

It was found that the corticosteroid level is lower in domestic foxes than in wild foxes during all the seasons of the year. The differences in corticosteroid levels were greatest in females. Since in our experiments it was the females that showed sexual activation outside of the season, it may be thought that the differences in plasma corticosteroids in foxes with different hereditarily-determined behavior represent a correlated response to selection for behavior. The selection leads to the destabilization of the reproductive function and to a decrease in the corticosteroid level. This is in compliance with available literature indicating that the effect of domestication enhances the functional activity of the gonads and attenuates that of the adrenals (Richter, 1954; Eiknes, 1959). The interrelation between the adrenals and the gonads, on one hand, and between behavior, estrogens, and corticosteroids on the other hand, are intricate. From the data of this study, it is difficult to discuss the physiological nature of these interrelationships. The indisputable fact remains, however, that the artificial selection of silver foxes for docility induces destabilization (i.e., the reorganization of the seasonal pattern of reproduction, and a decrease in plasma corticosteroids).

Analysis of the responses of domestic and wild foxes to exogenous ACTH and to psychological stress favors the view that the decrease in corticosteroid level in the course of domestication is a reflection of the altered state of not only the adrenal cortex, but also of the hypothalamo–pituitary system as a whole. If only one component of the system, the adrenals, were involved in changes of functional activity, then the responses of foxes with different types of behavior to psychological stress would correspond in magnitude to the response to ACTH. However, the responses to ACTH and to psychological stress in animals of the same sex and behavioral characteristics varied and occasionally either the response to ACTH or to psychological stress predominated during the same season. It should be noted that the response of domestic females to both stressful exposures were much more variable during all seasons than in the wild females.

In the course of this selection experiment, a number of morphological, physiological and karyological characters were also modified. Their discussion is beyond the scope of the paper. Selection for docile, tractable behavior leads to the dramatic emergence of new forms and to the destabilization of ontogenesis manifested by the breakdown of correlated systems created under stabilizing selection. It is for this reason that the concept of destabilizing selection is formulated and referred to a specific form of selection which is expressed vividly in the process of domestication of animals.

PART V

BEHAVIORAL EVOLUTION

30
EVOLUTION OF SOCIAL BEHAVIOR IN CANIDS

M. W. FOX
Department of Psychology
Washington University
St. Louis, Missouri

The comparative ethology in canids, with emphasis on communication, social organization, and ontogeny of behavior, has been described by Fox (1971b). These studies form the basis of the present review, which will correlate what is known of the predation patterns of the various Canidae with their social behavior and methods of communication.

Most of the known species of canids are listed in Figure 30-1, which shows the possible behavior–taxonomic relationships between species. These relationships, which are highly speculative, are based upon preliminary behavioral studies on captive specimens. Much more research is needed before any conclusions can be made, but it will be obvious to anyone familiar with the behavior of some of the canids that earlier classifications, based upon tooth and skull measurements, are inadequate when, for example, the bush dog (*Speothos venaticus*) and the Cape hunting dog (*Lycaon pictus*) are placed in the same subfamily (*Simocyoninae*).

The comparative method as exemplified by the studies of Lorenz (1941) and Johnsgard (1965) is one way of tracing evolutionary changes in behavior and social organization in related species that have adapted to particular ecological conditions. Thus, a knowledge of the life style and habits of related species, and also variations in the same species adapted to different environments, may show evolutionary trends which, in the absence of comparative analysis, might not otherwise be revealed. The intention of this review, therefore, is to construct a general picture of evolutionary trends in the Canidae,

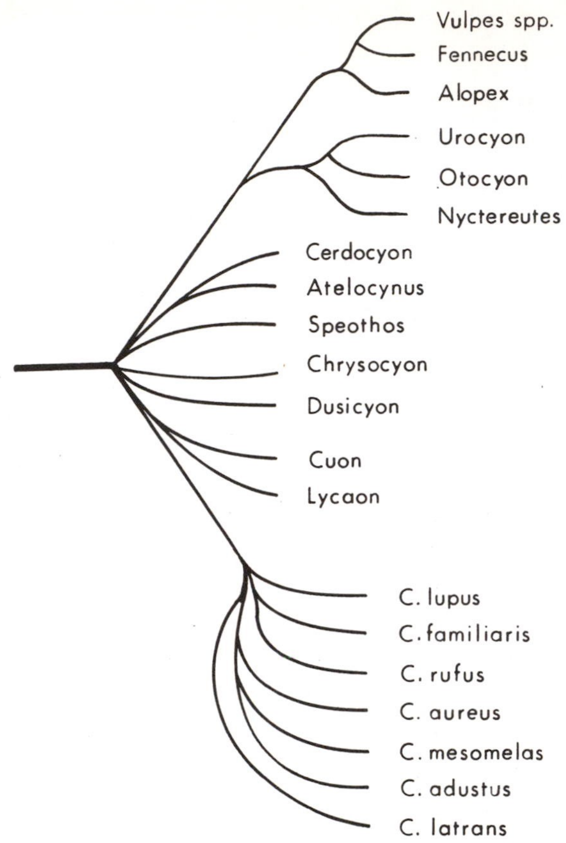

Figure 30-1 Hypothetical reconstruction of evolutionary relationships between various canid species (common names, reading down vertically: red fox and related subspecies, fennec fox, Arctic fox, gray fox, bat-eared fox, raccoon-like dog; from South America, crab-eating dog, small-eared dog, bush dog, maned wolf and culpeo; Indian wild dog or dhole, Cape hunting dog. The subfamily *Canis* includes the wolf, domestic dog, red wolf, golden jackal, black-backed and side-striped jackals, and the coyote.

drawing principally upon the data derived from laboratory and field studies of three highly representative canids; namely, the red fox (*Vulpes vulpes fulva*),the coyote (*Canis latrans*), and the wolf (*C. lupus*).

PREDATION PATTERNS

The Canidae show three types of hunting patterns which are based upon varying degrees of mutual cooperation and interdependence: the method of hunting and also the size of the prey in relation to the size of the predator influence the type of hunting pattern that a particular species most often adopts.

Type I: Solitary Hunter

Most members of the cat family hunt by stealth, ambush, and surprise attack from cover; this method of hunting, by virtue of the difficulties in two or more animals communicating with each other and so coordinating their actions, tends to limit the activity to a single individual. In the Canidae, such hunting patterns have been observed, where one wolf or coyote, for example, will remain concealed while one or more conspecifics drive the prey towards it. This is an example of cooperative hunting and will be discussed subsequently. Canids usually hunt in the open and run down their prey, the cheetah being the only member of the cat family to have evolved this method. Solitary hunters in the Canidae, exemplified by the red fox (Burrows, 1968), are invariably small in size and hunt proportionately small prey in the open, notably rodents, but like other canids they are also omnivorous scavengers, eating fruit, bird's eggs, and carrion. Because of the small size of the prey and its distribution, the fox is by necessity a solitary hunter; the expenditure of energy of two or more foxes cooperating to kill one mouse would be maladaptive. When prey is abundant, uneaten food may be cached by the solitary hunter as well as by the cooperative hunters. Also, at such times when prey is plentiful and concentrated in one area, several solitary hunters may be attracted to the same place, such as an open meadow, and give the impression of coordinating their activities; doing the same thing at the same time and place may be regarded as "parallel" hunting.

Type II: Solitary-Social Hunters (Transitional Type)

Many of the Canidae are efficient solitary hunters of small prey, but are capable of hunting in pairs or as a family unit, and consequently may secure larger prey. The coyote (Murie, 1940) and the golden jackal (H. and J. van Lawick-Goodall, 1971) fit this category, for in these species a permanent pair bond may be formed so that outside of the season when young are to be provided for, the pair continues to hunt together. It is not known whether cooperative hunting in pairs is greater in the Type II hunter during the period when the young must be provided for. Cooperative hunting must be influenced by the size, abundance, and distribution of prey; if prey is small in size and scarce, say during the winter months, solitary hunting may be more efficient, but if the prey is larger in size and scarce, cooperative hunting patterns may be utilized. Muckenhirn and Eisenberg (1973) have observed cooperative activity in a pair of golden jackals hunting axis deer; one partner chased the deer while its mate lay in ambush. Similar observations have been made on coyotes hunting in pairs. In open terrain, a cooperating pair can more efficiently bring down prey where one individual is driving the prey and the other moves across to cut it off. Offspring, prior to moving out of the parental territory, may hunt cooperatively with one or both parents. Again, in this type of hunter the individual's size is closely matched with the size of the larger prey species utilized, which in the absence of social cooperation tends to set an upper limit of prey size for the predator. As noted earlier in the Type I hunters, Type II hunters may be drawn to the same place where there is an abundance of prey or carrion — rodents, jack rabbits, over-wintering elk — and such concentrations of food may lead to a family of coyotes or several individual coyotes to engage in the same activities at the same time and place.* Such coincidence may lead to cooperation between pairs or within family units or to "parallel" hunting as noted earlier in the red fox. The Type II hunter is necessarily an efficient solitary hunter and not wholly dependent upon a partner for success. This is because the partner may become injured or die, and because after leaving the parental territory, solitary hunting and scavenging are essential activities prior to the young canid establishing a territory and pair bond. In the cat family, Type II hunting may occur while the pair is temporarily together and providing food for their offspring.

Type III: Social Hunters (Pack Type)

The truly social hunters that cooperate as a unit include the Cape hunting dog (Estes and Goddard, 1967), the wolf (Mech, 1970) and the dhole or Indian wild dog (Burton, 1941; Schaller, 1967). All these species show some flexibility in that they are capable of effectively engaging in Type I and II hunting patterns. But so far as the author is aware, Type I and II patterns are relatively rare in the Cape hunting dog and dhole, while the wolf in some habitats may engage in the Type II pattern (as a breeding pair) or in Type I pattern as a "loner" wolf. There is also regional variation in pack size and in body size in the wolf, such racial differences being correlated with the major prey species utilized, be it a large moose or a small deer (Standfield, personal communica-

*Coyote, an American Indian and field representative of the Sierra Club, tells me that in the Sierras, where deer form large herds, groups of 10–15 coyotes hunting together are not uncommon.

tion). Pack size may also vary, possibly on the basis of filial relationships or niche occupancy by neighboring packs. Thus, in some regions the pack may be a family unit of some 6–8 individuals, and does not usually mix with neighboring packs although the degree of contact and rate of gene flow has yet to be determined, or it may be an "aggregated" pack of 20–30 which shares the same hunting range as in Mount McKinley Park, and which breaks up into hunting units of 7 or 8 (Haber, personal communication).

Estes and Goddard (1967) also point out regional differences in body size in Cape hunting dogs; the darker, 18.1 kg East African plains type that specializes on diminutive Thomson's gazelle and the 27.2 kg central and southern types that hunt larger prey such as impala.

The only comparable member of the cat family that shows some cooperative hunting is the lion, and in this species there is a division of labor. The females are most often the hunters, while the males mainly defend the territory of the pride from rival prides. The influence of ecological factors, notably the size, distribution, and abundance of prey, on group size alluded to earlier in relation to the wolf is also evident in regional variations in the lion. For example, in South Africa, some prides contain over 40 individuals, while in the Sudan where food is less plentiful, the prides are much smaller (Eloff, 1973). In an even more severe mountainous region in southwest Africa, lions are rarely seen in prides — a socio-ecological phenomenon reminiscent of the now extinct solitary Barbary lion of North Africa. Such ecological variables similarly limit the size of the wolf pack, which may (like baboon troops in seasonally arid regions of Africa) break up, and individuals hunt singly or in pairs or trios. The wolf therefore demonstrates a high degree of flexibility in hunting patterns (ranging from Type I to Type III) and is therefore not limited or niche-specific as is the case in the red fox. The more highly developed sociability of the wolf allows this flexibility which is adaptive to both regional and seasonal variations in abundance and distribution of food.

The pack hunter is also able to transcend the prey–predator body size relationship evident in the Type I hunter in that the individual, supported by cooperative conspecifics, is able to obtain prey many times larger than its own weight. But here too a limiting factor still operates. If the major prey is large, as alluded to earlier, the pack may be larger and individual wolves are larger than other races that specialize on smaller prey in a different region. Also, if the major prey is small but abundant, a large cooperative pack could maintain itself, as in the case of the dhole and the Cape hunting dog.

Estes and Goddard (1967), in discussing effective pack size in Cape hunting dogs note that a large pack (of 20) utilizes prey resources more effectively than a small pack (of 4–6), the latter being close to the minimum effective unit. Competition from hyenas where they are numerous also exerts a strong selective pressure in favor of large packs and for close cooperation at kills.

During the dry season, when the prey are concentrated at water holes, their hunting range is far less than during the rainy season, during which time the packs will range over a great area. Thus, the social organization of the dhole and Cape hunting dog may differ significantly from the wolf, for in the latter species the pack may not remain together throughout the year because of fluctuations in availability of food.

The birth of young in all canids occurs with maximum availability of prey, namely when the latter are rearing their young (see Figure 30-3). Consequently, the greatest mortalities of young canids is during their first winter when prey is scarce. In the Type I hunter, the young leave the parental territory prior to this time and the parents also separate. In the Type II hunter, the offspring may emigrate at this time or remain with the parents until the next breeding season, when they too are forced to leave their parents' territory. The timing of intraspecific intolerance is therefore closely correlated with the availability of prey and with the reproductive cycle. In the pack hunting (Type III) canids, the offspring do not usually emigrate because other contiguous packs are occupying available hunting ranges. Offspring instead are recruited into the pack, and as in the case of lion cubs (Schaller, 1973), mortalities are generally high. Historical accounts of the prairie wolf in enormous packs of 40 or more are not difficult to envisage when one realizes that these wolves preyed principally upon abundant herds of buffalo. Like the large packs of dholes and Cape hunting dogs that hunt frequently and successfully in regions where there is an abundance of larger prey, so certain races of wolf prior to the advent of human ecocide were socio-ecologically adapted as large packs.

Hunting patterns of the Type III hunters have been well-documented by field observers (Mech, 1970; Estes and Goddard, 1967). The canids may get the herd moving to "test" the herd and weaker members will be picked out. The pack may appear to run in relays when chasing prey, first one and then another taking the lead; it is more likely that as the prey zigzags and circles, it is cut off by others behind the lead animal, who for a time take over the lead until another wolf cuts off the prey. An interesting pattern in the Cape hunting dog consists of individuals or pairs of dogs, each chasing an animal until the most vulnerable and weakest one is identified; then other dogs will converge upon it and cooperate in bringing it down (H. and J. van Lawick-Goodall, 1971). Strategy involving ambush, distraction by decoy, and coordinated group attack on one individual separated from the herd have been reported. In the wolf and Cape hunting dog, a ritualized "greeting" ceremony has been observed prior to the pack going out to hunt. All canids will return from the hunt and regurgitate food for the young; in the wolf, meat may also be regurgitated for an adult that was unable to hunt because of injury (Haber, personal communication), and in the Cape hunting dog, caretakers of the cubs will also be fed by others returning from the kill (Kühme, 1965a). Here some division of labor is evident. The social structure of these canids will be discussed subsequently. Studies of prey killing in 6–8 week old red foxes, coyotes, and wolves — arbitrarily designated Type I, II, and III hunters, respectively — presented with live rats under controlled laboratory conditions have revealed some relevant developmental data (Fox, 1969a). Both the red fox and coyote are highly competitive and noncooperative. With few exceptions, young of both species are able to catch, kill, dissect, and ingest the prey on first encounter, without prior experience. What Leyhausen (1965) has termed relative dominance was evident in red fox litters; when in possession of a kill, the individual was dominant over its conspecifics. In coyotes, it was usually the most dominant cub that possessed the prey (i.e., absolute dominance) and unlike the red fox, a subordinate would often give up its prey when threatened by a more dominant conspecific. Wolf cubs tested with live prey often needed several trials before they killed, dissected, and ate the prey. There is some evidence that the more dominant and exploratory wolf cubs will more readily kill live prey than more timid and subordinate littermates (Fox, 1972). When tested in litter groups, relative and absolute dominance in terms of possession of prey was observed in the wolf cubs and also what was interpreted as social facilitation. The more timid and subordinate cubs were more reactive than when tested alone with live prey under the facilitation of their more reactive littermates. Such social facilitation and the formation of leader–follower relationships may be the basis for coordinated and cooperative pack hunting in the Type III predator (Fox, 1972).

In summary, there appears to be a subtle reciprocal evolution between prey and predator,the predator being adapted physically and socially to a particular class of prey. Selection for coordinated pack hunting reduces the selection pressure for increased body size and strength necessary to overcome larger prey; thus, in the Canidae we do not find solitary individuals of large size hunting large prey whereas in the cats, there is a wide range of solitary species varying in body size from the small spotted cat of South America to the large tiger of India and Asia. It will be of interest to investigate regional differences in prey specialization in the canids in order to determine the subtle processes of socio-ecological adaptation. Regrettably the wolf, with its many racial variations, is virtually extinct in North America. In some regions (such as in Wyoming where elk spend the winter in a reserve) coyotes are concentrated where there is an abundance of prey and carrion. Under such conditions, social behavior may be very different and family groups of coyotes may show some of the characteristics of Type III social organization (see later) and hunting patterns (Camenzind, personal communication). In addition, regional and seasonal variations in availability of food may give rise to racial differences (or "social clines") in the timing of reproduction and birth of young as well as influencing social organization, perseverance of pair bonds, and of the ties between parents and offspring. The relationships between conspecifics occupying adjacent territories or hunting ranges and movement patterns during the various seasons also require further study.

EVOLUTION OF COMMUNICATION

The evolution of communication may be followed in the various canid species using the same Type classification because of the close relationships between social organization, socio-ecological adaptation, and intraspecific communication.

Visual Signals

Of the Type I canids, the displays of the red fox have been most thoroughly documented (Tembrock, 1957; Fox, 1970a). A basic repertoire of tail and body displays and facial expressions which lack subtle intensity gradation and often appear as "all or none" at a typical intensity have been described (Figure 30-2).

Identifiable in the Type I repertoire are signals associated with greeting and submission, play solicitation, and defensive and offensive aggression. These same basic display categories, which serve essentially to decrease, increase, or maintain a certain social distance or proximity, are present in Type II and III canids. In the latter type, however, they show more subtle intensity gradation and also successive and simultaneous combination (Fox, 1970a, 1971b). The Type II canids, as would be predicted, are intermediate in their nonvocal repertoire (Figure 30-3) (Fox, 1970a). This increasing complexity of the display repertoire from the Type I–Type III canid is correlated with increasing sociability and more or less year-round close proximity. More detailed analysis of the displays of the various canid species is necessary to verify these tentative generalizations concerning the qualitative and

Figure 30-2 Expressions of red fox: a) threat gape, b) small-mouth threat face, c) submissive greeting, d) and e) open mouth "play face", f) submissive greeting, g) defensive threat gape, h) low intensity threat gape.

quantitative changes in the evolution of body posture displays and of facial expressions.

The nonvocal repertoire (Figure 30-4) of the wolf contains many action patterns which are "derived" from infantile actions (Fox, 1970a, 1971d). Such behavioral neoteny, or the persistence of infantile actions into maturity, seems to be greatest in the Type III canids. This apparent trend towards increasing neoteny with increasing sociability in the canids warrants further verification.

Vocal Signals

Little comparative work has been done on the ontogeny and socioecological significance of canid vocalizations. The red fox, according to Tembrock (1957), has a large repertoire of vocalizations which may be

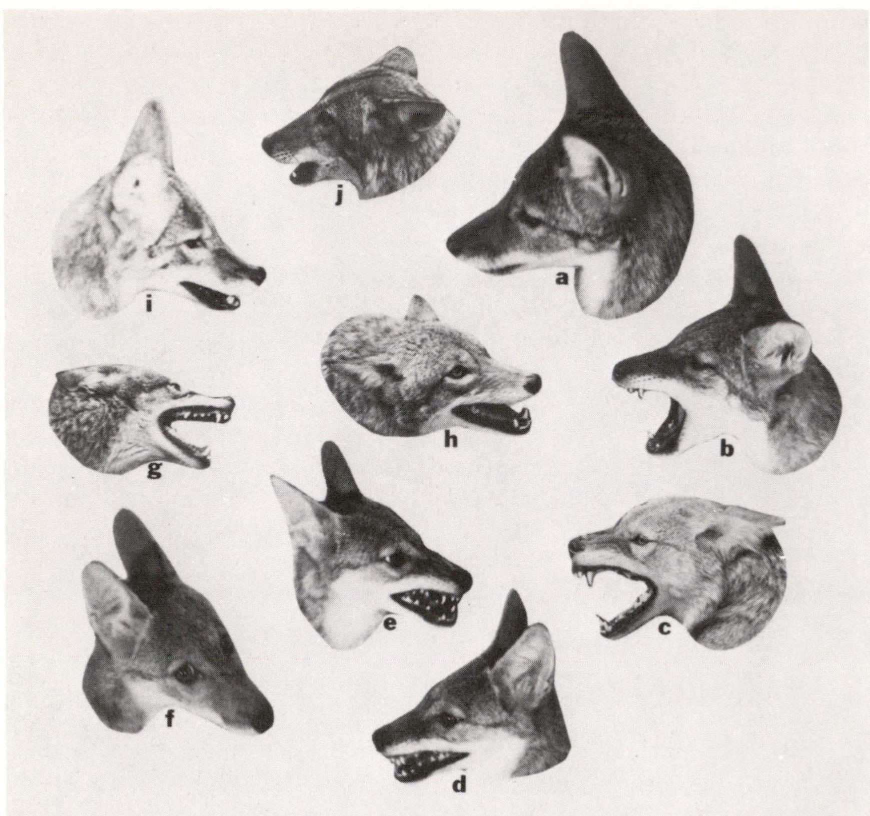

Figure 30-3 Expressions of coyote: a) small mouth threat face, b) aggressive threat gape, c) defensive threat gape, d) and e) defensive threat combined with submissive "grin," f) submissive, avoiding eye contact, g) defensive aggressive threat gape, h) open mouth submissive greeting or "play face", i) "Flehmen" face (upper lip elevated), j) low intensity threat face.

subtly graded in intensity and which may occur in successive or simultaneous combination (Fox and Cohen, 1974). It would appear, at least in the red fox, which has been studied in more detail than any other Type I canid, that the Type I canids may have a greater repertoire of stereotyped vocalizations than the other canid types (by stereotyped, I mean those vocalizations which are emitted with little variation in pitch, frequency, and intensity). Since they are often nocturnal or crepuscular and are concealed much of the time under dense cover where visual signals can only be effective at close proximity, such compensatory adaptation in auditory communication would be expected in a relatively solitary species. In contrast, the Type II and III canids tend to be more diurnal and often frequent open spaces; both these conditions, coupled with close proximity, favor the use of visual signaling. Little work has been done on the low intensity vocalizations that are emitted at close proximity in the Type I, II, and III canids, which include whimpers and whines, panting, grunts, and groans, many of which closely resemble infantile vocalizations.

The Type II and III canids stand out from the Type I canids in that some species engage in collective singing — the dawn and dusk

Figure 30-4 Expression of wolf: a) alert, "neutral" face, b) intimidating stare, c) small mouth threat face, d) passive submission during eye-contact intimidation, e) defensive aggressive gape, f) small mouth threat face with tongue extrusion (signaling ambivalence or friendliness?), g) small mouth threat face, h) active submissive greeting "grin," i) and j) open mouth play face, k) anxious passive submissive face, l) consummatory face (eyes closed during eating), m) passive submissive face.

choruses of coyotes and jackals, and the choirs of wolves. In coyotes, it is often a family that engages in choral singing, and in both coyote and wolf, is accompanied by much reciprocal greeting, face-licking, and muzzle-biting. The function of this collective singing is obscure; it may serve some positive socially cohesive function for the group, and often occurs, for example, prior to the assembly going off to hunt. It may also serve a territorial function, informing other groups of their presence. Solitary howls of wolves may be used for location of (or by) conspecifics, and may also be calls for assembly. More work is needed on the ontogeny and comparative apects of canid vocalizations, a subject which deserves close study in the various canid types which span the spectrum from the solitary to the gregarious and which consequently could be a valuable model group for phylogenetic studies.

Tactile Communication

During social investigation, certain body regions are contacted, and occasionally presented, and these regions often contain scent glands. Social investigation, which is often reciprocal, is frequently initiated and controlled by the more dominant individual. During social investigation, all canids remain passive; the inguinal response which is often evoked during social investigation (Fox, 1971d) is more characteristic of canid Types II and III than of Type I. Social grooming (allogrooming) often follows social investigation; this behavior has received little detailed comparative study, but the author believes that there is a trend towards a decrease in the frequency of social grooming (but an increase in licking and licking intentions) with increasing sociability from Type I to Type III canids. Social grooming (nibbling) especially around the eyes in red foxes and along the back in gray foxes and coyotes has been frequently observed in captive animals, while in captive wolves, face-oriented licking rather than nibbling appears to be the more usual grooming pattern.

Another aspect of tactile communication is lying together in rest and sleep. This tendency decreases in all canids with increasing age; adult red foxes (Type I canid) are rarely seen lying in contact while in gray foxes and coyotes (Type II) and wolves (Type III) this is more common. The spacing of canids in captivity may therefore reflect something about their sociability and proximity tolerance. In the Type I canids there is a phasic shift in proximity tolerance, proximity being lessened during the breeding–rearing season. Also the intraindividual avoidance characteristic of the Type I canid gives way to intraindividual attraction in the Type II and III canids with concomitant intergroup avoidance.

Olfactory Communication

A general trend in the canids is a decrease in the intensity of body odors (notably the supracaudal tail gland) with increasing sociability (Fox, 1971 d). There are always exceptions to any such generalizations; the Cape hunting dog, for example, produces a very strong scent, which Estes and Goddard (1967) propose may permit high speed tracking of the pack by members that have lost visual contact. This is also reflected in their frequency and duration of social investigation. The red and Arctic fox, for example, have higher scores investigating the supracaudal gland of a strange conspecific, wolves having low scores, and coyotes and jackals intermediate scores. There is also a tendency in the more sociable canids to roll in novel odors such as carrion and deer musk, whereas the less sociable canids tend to urinate or defecate on such odors (personal observation). Some social reinforcement may underlie rolling in the Type II and III canids. A great deal of work remains to be done on olfactory communication, on the socio-ecological significance of scent marking and rolling, and on the chemical composition of the various glandular secretions that may contain pheromones which vary according to age, sex, stage in the reproductive cycle, social rank, locale (in relation to type of food) and emotional state.

SOCIAL ORGANIZATION AND SOCIO-ECOLOGICAL ADAPTATIONS

The hunting patterns described earlier in part reflect the social organization, communication, and socio-ecological adaptations manifest by the Canidae, and we may similarly classify the social organization patterns into three "Types."

Type I: Solitary. This type is exemplified by the red fox, except during the mating season and infant rearing period. Males may range further than females and outside of the breeding season may or may not have a hunting range which they mark and defend against intruders (Burrows, 1968). More often, several males occupy a given region and avoid contact by scent marking and barking. Females may stay near to the den site and range outside of the breeding season less often. At the beginning of the breeding season the males set up "rutting trails" which they mark, and vocalize frequently, possibly to repel rival males and at the same time attract one or more females. Several foxes may aggregate in one area at such times, a social phenomenon described by Leyhausen (1965) in the domestic cat. A temporary pair bond is formed, although occasionally the male may mate with more than one female (Burrows, 1968); such cases of polygamy may only be possible when there is an abundance of food permitting the female to be the sole provider for the offspring.

Therefore, in the Type I canid social organization we find a trend towards persistence of the sexual bond through the period during which the young are raised. In the equivalent Type I solitary feline, the sexual bond does not persist and the male leaves the female, who cares for the young herself.

In the Type I canid, the parents either abandon the offspring or drive them off before their first winter (Figure 30-5). Socifugal tendencies such as proximity intolerance and intraspecific aggression, lack of a socially cohesive dominance hierarchy, leader–follower relationships and group coordination most probably contribute to the subsequent dispersion of the litter. Tembrock (1958) has described in detail the changes in proximity tolerance in the red fox before, during, and after the breeding–rearing season; the effects of sex hormones on social behavior await further study (see also Zimen, Chapter 23). One plausible hypothesis is that in the Type I canid there has been a strong selection for one particular temperament which is relatively homogenous within litters. This temperament is characterized by mutual intolerance and "self-reliance" in that all individuals are outgoing and highly exploratory (see further discussion on Type III social organization). Endocrine changes during the breeding–rearing season may effectively lower this intolerance and give rise to socipetal tendencies in the solitary canids, leading to pair formation. Within family groups (Type II) and packs (Type III) endocrine changes during the breeding season may instead have a socifugal effect and male rivalry may lead to dispersal of offspring or pack splitting.

It is clear how the Type I social organization is adapted to the mode of hunting in relation to prey–predator size, distribution, and abundance. As emphasized earlier, an abundance of prey may lead to some temporary social interaction, but within such aggregations, social organization may be minimal, although filial relationships might be evident. A shortage of prey, notably in the Arctic, may lead to migration; Arctic foxes have been observed migrating in large numbers (Ognev, 1962; Sdobnikov, 1967).

Type II: Permanent Pair. This type of social organization typified by the coyote is an extension of Type I in that the sexual–parental bond persists outside of the breeding–rearing season (Figure 30-5). Offspring may remain with the pair several months longer than the Type I

TABLE 30-1 Interrelationships between evolution of family unit pair bond relationships in canids. The shaded areas indicate the absence of a particular type of pair bond.

Pair Bond / Family Bond	Temporary Pair: Sexual	Temporary Pair: Sex and Parental	Permanent Pair	Pairs and/or complex socius
Permanent Family				Type III (Wolf)
Temporary Family			Type II (Coyote)	
Transient Family	Type I (Red fox)			

canid; they over-winter with the parents and survivors emigrate from the natal territory at the onset of the next breeding season. The advantages of remaining with the parents over the winter as a cooperative hunting and scavenging unit may be offset by paucity of food, so that infant mortalities are often high. The permanent pair bond *per se* could, however, increase the hunting success rate and enable the pair to maintain their territory or hunting range against rival intruders. When contiguous territories are occupied, emigrating offspring may have to travel great distances before finding their own territory or pairing up with an established animal that has lost its mate. Mortalities of juveniles are therefore high, and as in the red fox, similar socifugal tendencies may be operating. Captive litters of coyotes do form dominance hierarchies, this most probably being a reflection of what occurs during the first year with their parents. But with increasing maturity, social preferences develop, pair bonds form, and members of such pairs become increasingly intolerant towards others, either of the same or opposite sex. Van der Merwe (1953b) has noted that in litters of black-backed jackals, social pairing often takes places prior to dispersion of the litter, and such a possibility should be entertained for the coyote. A high rate of inbreeding could be offset by mortalities of litter partners and by some individuals pairing with older adults in adjacent territories.

Occasionally, two female coyotes may breed and share the same mate and den. Such occurrences, possibly correlated with an abundance of food, indicate how arbitrary these Type classifications are. Filial and parental relationships which play an important role in the Type III canids (see later) may also significantly influence social pat-

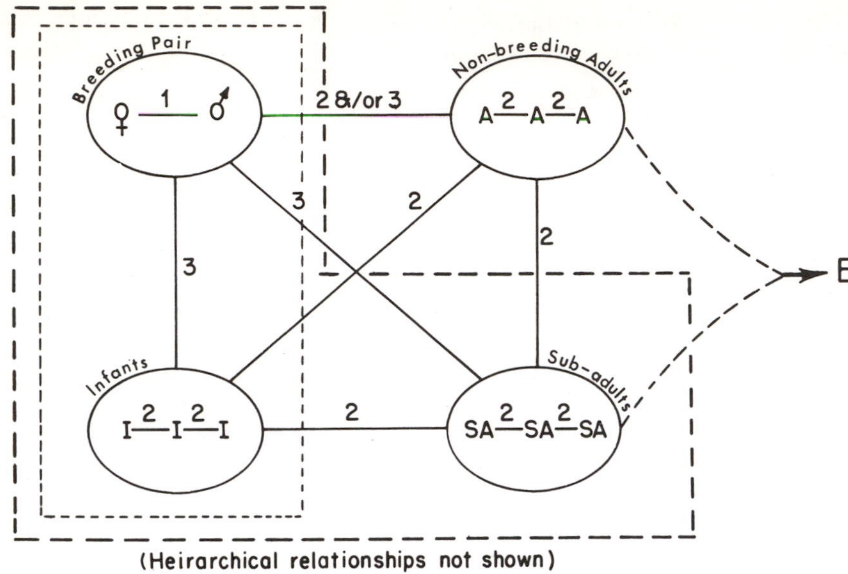

Figure 30-5 Schematic representation of social relationships within a wolf pack; non-breeding adults and subadults may all be familially related to the breeding pair; the pack may divide and one unit emigrates, but more normally the young are "recruited" into the main pack (see text). Also depicted are the social relationships in Type I and Type II canids, red fox and coyote, with small and large broken lines respectively.

terns in the Type I and II canids, but such influences have as yet received little attention. In this example, possibly two sisters or a mother and daughter had remained together and shared the same mate, den site, and hunting range. Under certain ecological conditions, therefore, some flexibility and modification of the usual species Type social organization may be possible. It would be of interest to investigate further the flexibility and modifiability of existing social systems under controlled conditions. Also,field studies of the same species in very different habitats should be encouraged, as exemplified by the socio-ecological differences in baboons, for example (Crook, 1970).

With regard to Table 30-1, it should be noted that the Type II canid may show some of the complex relationships characteristic of the Type III pack canid, when for example, yearling golden jackals are living (and occasionally hunting as a "pack" with the parental pair which may be raising a second litter) (H. and J. van Lawick-Goodall, 1971). Such Type transitions or variations in the usual species Type may be accounted for by a regionally atypical set of ecological conditions such as abundance of moderately large prey and lack of competition from rival predators and conspecies. Young Type II canids, e.g., black-backed jackals (H. and J. van Lawick-Goodall, 1971), that do not possess a territory may be found in loosely-structured roaming packs, where the advantages of cooperative hunting offset the lack of mate, territory, and hunting range which normally exert a sociofugal effect between conspecific pairs. The lack of such sociofugal factors may facilitate the formation of a temporary hunting band. Familial and parental ties which affect sociability and which could influence socifugal factors between conspecies sharing overlapping hunting ranges in the Type I and II canids should also be considered. Proximity tolerance and sociability may also change as a function of the distance from the center of the individual's territory (aggression decreasing with increasing distance from the denning area) and also in relation to the availability of prey and reproductive cycle of the female partner.

*Type III: The Pack.** Type III canids show a further extension of the Type II pattern, and their pack organization is perhaps best described as an extended family unit (Figure 30-5). The bond between a mated pair persists during the rearing period and may persist subsequently in the wolf, but this is not fully verified. Within the wolf pack, there is a clear intrasexual dominance hierarchy, i.e., between males and between females (Rabb, *et al.,* 1967), and there is usually a clearly identifiable leader or alpha individual and frequently a lowest ranking omega or "trailer" wolf. Social relationships within and between the sexes are complicated not only by mate preferences (which may be effectively blocked from breeding by higher-ranking individuals) but also by filial bonds between parents and their offspring and between littermates (Rabb, *et al.,* 1967; Fox, 1973). Not all wolves within a pack are necessarily directly related, although it is not precisely known what the origin of the pack is — be it an extended family or a more heterogeneous unit. The degree of heterogeneity, of pack openness or acceptance of strangers,** and of gene flow between contiguous packs remain to be determined. This may be a function of availability of

*By definition, a "pack" here refers to a social group which hunts together. Asiatic wild dogs (*Cuon alpinus*) may occur in numbers ranging from 30–50, splitting up into smaller groups of 2–15 to hunt. The wild dog exemplifies a further evolution of canid social behavior, where the term "clan" is appropriate for such a large group sharing the same range but rarely hunting together. It therefore constitutes a Type IV canid social class, the clan (analogous to the baboon troop). This socio-ecological pattern is only possible where there is a year-round abundance of prey as in the Nilgiris of S.W. India (personal observation).

**See Fox, *et al.* (1973) for study of "openness" in captive packs where age, sex, social rank, and territory are identified as interdependent determinants of acceptance or rejection by introduced strangers.

food, which serves to limit pack size and to increase socifugal tendencies between packs. Defense of hunting range would be less in regions where prey is abundant and under such circumstances there may be a greater social mobility and acceptance between packs.

Mech (1970) observes that:

> There appear to be four factors that might affect pack size: (1) the smallest number of wolves required to locate and kill prey safely and efficiently, (2) the largest number that could feed effectively on prey (i.e., not all of a pack of sixty wolves could feed on a 150-pound deer), (3) the number of other pack members with which each wolf could form social bonds (the social-attachment factor), and (4) the amount of social competition that each member of a pack could accept (the social-competition factor).

Mech thinks that the first two factors could act only as secondary controls, and that the second two social factors are probably the major regulators of pack size. But as he points out, the larger the pack, the greater the competition would be for food, mates, leadership, or dominance. When such competition reaches a certain level, some members of the pack may be forced to leave in order for the pack to function with maximal efficiency. He concludes that it is this social-competition factor that limits the pack membership to less than the number of wolves that could feed effectively on a prey animal.

Under optimal conditions, a segment of a large pack could emigrate and occupy a contiguous territory, and consequently in some regions many packs may be genetically related. Social ties between contiguous packs would be a function of the frequency of contact and recency of separation, the social bond of familial relationships being broken after the death of the first generation. Thus within one generation, individuals of two packs derived from the same parent pack would no longer have any social (familial) bond. Here the socifugal effects of territorial defense and fear, avoidance, or aggression towards strange conspecifics come into play. Social bonds may be re-established with outsiders during the breeding season, but such instances must be rare and minimized by the socifugal factors which tend to keep packs separated.

A given pack is more or less a closed social group since available hunting ranges and territories are usually occupied by neighboring packs (Eastern Europe being one exception, see Pullianen, Chapter 21). The offspring of a space-limited pack are recruited into its ranks to fill positions vacated by adults that have died or are subordinate and have emigrated to become either "loners" or "trailer" wolves that follow the main pack. Infant mortalities will therefore be high if the turnover rate of adults is low and if the emigration of several adults as a separate pack is prevented by neighboring packs already occupying available range.

One point when a pack is likely to break up under optimal emigration conditions would be during the breeding season, when aggression associated with mate rivalry would be more intense than at any other time. The wolf does not attain sexual maturity until 2 years, and this factor which delays the onset of mate rivalry and conflict within the family nucleus may be of great significance. This delayed sexual maturation would mean that for at least 1 full year the offspring of a breeding pair could cooperate with the pair and form an efficient hunting pack and also assist the pair in raising a second generation. (A comparable pattern is seen occasionally in some of the Type II canids, although some female coyotes do have an estrus period in their first year, and might therefore leave the family group by 1 year.) The second generation would be physically mature and capable of assisting the parents in hunting by 1 year of age, at which time socifugal tendencies between the mated pair and the older and now sexually mature offspring would be appearing. It is not known conclusively what percentage of sexually mature offspring remain with the pack. As pointed out earlier, their recruitment into the pack may be a function of mortality rates, availability of food, and overall reproductive success of mated pairs. In the Type II canids, few if any would remain with the parents — a female, perhaps, that has a sexual bond with her father, for example. In most packs, successful mating is usually restricted to only one pair, and it is rare for two litters to be born the same year in the same pack. The social and ecological restraints on reproductive success remain to be evaluated in the Type III canid. Haber (personal communication) believes that human predation of the wolf may disrupt social organization to the extent that social control of the population breaks down and more offspring are born. This observation awaits further verification.

Another point where the pack may break up is when a mated pair has to dig a den and remain relatively stationary during the rearing period. If food is abundant, the pack may remain with the mated pair and assist in the rearing of the young. But if food is scarce, the main pack may move on. Two alternatives are then possible; either the pair with young becomes a separate pack with some familial "allegiance" with the main pack, or they join up with the main pack the following winter.

Clearly the social ecology of the wolf, a classic example of the Type III canid, is extremely complex, and many questions remain to be answered.

Mech (1970) summarizes the findings of Rabb, Woolpy, and Ginsburg (1967) on the Brookfield Zoo pack, which lucidly illustrates the complexities of social interaction and organization in this Type III canid:

> (1) dominance rivalries increase greatly just before the breeding season, especially among the females, and the dominance order is largely settled at that time, (2) the alpha male is often preferred by most of the females, (3) dominant males often disrupt courtship and mating attempts by subordinate males, (4) the alpha female tries to prevent other females from mating, (5) a dominant animal will often interfere with mating attempts by individuals of the opposite sex, (6) a subordinate male sometimes thwarts a superior's courtship of a mutually preferred female by moving between the two, and (7) the mate preferences seem to be related to the dominance order in the pack when the pups mature, with younger animals preferring older dominant wolves.

The alpha wolf, who acts as a leader in directing, governing, and controlling the behavior of subordinates, also guards the pack and initiates coordinated actions as the most alert and active individual. Both Mech (1970) and Fox (1971b) summarize the leadership pattern as a combination of autocratic and democratic systems, the leader acting often independently of his packmates, who are dependent upon him for direction, and at other times, the leader is influenced by the behavior of other pack members.

Recent studies on captive wolf packs by Lockwood and Fox (in preparation) measuring social motivation (by making an isolated wolf "work" to see members of its pack) show that social motivation is correlated with rank. Alpha wolves show the greatest motivation, while low-ranking littermates have less. A change in rank was also correlated with a change in social motivation in wolves retested 12 months later. In the field then, one would predict from these findings that low-ranking wolves are more likely to leave or be left by the upper echelon of the pack when food is scarce; Haber's (personal communication) field data confirm this. Some wolves that become outcasts and leave the pack may be potential alphas that do not adapt to the social pressures of their pack (Fox, 1973).

The wolf, as noted in the red fox, often displays "relative" dominance in that a subordinate, in possession of food, may effectively threaten a superior who withdraws without asserting its dominance; Mech (1970) aptly terms this the "ownership zone" in that anything within a few centimeters of a wolf's mouth is beyond dispute.

Unlike the Type I and II canids, the young of the Type III canids do not normally disperse, nor are they normally driven away by their parents or by adult pack members. They remain with the pack and are integrated or accommodated into the existing social system which is reorganized once they are assimilated as young adults. But only a few offspring are so recruited into the pack; usually mortalities are high during the first winter, and surviving young essentially replace social niches vacated by expired or aged adults. Recruitment into the pack is therefore limited indirectly by the availability of food, and it is remarkable how constant the pack size is kept over successive generations. In some regions where food is usually scarce, even during the spring breeding season, packs may be limited in size to 2–4 adults, while in other habitats the pack may range from 5–20 or more.

Rather like some Type II canids such as the golden jackal that occasionally shows transition to Type III family-pack organization, the wolf may, like the now extinct Barbary lion, show some transitional Type II patterns under certain ecological conditions. Thus, the Mexican wolf (*C. lupus baileyi*) inhabits sparse mountainous regions where prey is widely dispersed and scarce. Such ecological conditions act as a social constraint, for a large pack would be maladaptive to such conditions. As a result, Mexican wolves are often seen singly or in pairs and may have adapted to a predominantly Type II pattern.

SOCIO-ECOLOGICAL SIGNIFICANCE OF INDIVIDUAL DIFFERENCES IN BEHAVIOR

Studies of individual differences in litters of wolf cubs have revealed that, unlike the red fox litter, there is great heterogeneity of temperament. It has been proposed that such behavioral or temperamental polymorphism facilitates the establishment of stable social relationships within the litter (Fox, 1972). This contrasts the homogencity "monomorphism" of red fox litters and the "oligomorphism" of coyote litters, individuals which are temperamentally equipped for a relatively solitary, indpendent life. Dominance hierarchy, leader–follower and dependency relationships, social facilitation, and coordination of group activities are consistently observed in the polymorphic wolf litters. Such a polymorphic litter is potentially a complete pack unit, but as noted above, it is rare for an entire litter to survive to maturity. But the potential is there, and it is not inconceivable that a litter with, say, its parents and an additional one or two adults, could move into unoccupied or vacated territory and become an established pack. Little is known about such possibilities, although Mech is gaining insight into the territorial and pack size consistencies which is only possible after many years of continued field studies.

A large pack of some 20–30 wolves may be composed of three family units that are interrelated and which, instead of occupying three different (and defended) hunting ranges, share a large hunting range. The large pack breaks up into smaller hunting units, the composition of which may be a function of familial relationships.

Very similar patterns of infant recruitment and consistency of pack size due to limitation of food and of available hunting range because of occupancy by rival packs may be present in the Cape hunting dog and the dhole. Little is known about the social dynamics of this latter canid, but the Cape hunting dog apparently has a hierarchical social organization comparable to the wolf pack, namely an alpha leader and female and male dominance hierarchies (H. and J. van Lawick-Goodall, 1971). Developmental studies are needed to evaluate the socialization process and ontogeny of social relationships in these two canids, as well as additional field work on a continuing basis, especially for the dhole.

No canid simply mates and then leaves the female; the importance of precopulatory pair-maintaining displays (and of social grooming) remain to be evaluated. The Type II and III canids must have something more than purely sexual motivation to maintain the pair bond outside of the breeding–rearing season. Greeting rituals between conspecifics and displays of active submission towards the alpha wolf who in turn reciprocates by displaying dominance may serve to maintain social relationships. Schenkel (1967) speaks of a "polarized social field" in the wolf pack, where the bond is one of mutual attraction and affection based most probably upon early socialization. Pair-maintaining and pack-maintaining and integrating displays, clearly await further analysis.

SOCIO-ECOLOGICAL RELATIONSHIPS BETWEEN PREDATORS

Where there is an abundance of many different prey species ranging greatly in size, a greater diversity of predators would be supported. Thus wolf, coyote, and red fox, or Cape hunting dog, golden jackal, and bat-eared fox may be found in the same habitat, each in its own particular niche and each occasionally scavenging on the remains of a kill made by a larger predator. Indirectly, one predator may assist another in the complex food chain by regulating population growth and preventing starvation. For example, smaller predators that specialize on small herbivorous mammals prevent the latter from becoming too abundant, which would lead to overgrazing, which could cause infertility, starvation, and death of larger herbivores. The larger predators would then suffer.

Rosenzweig (1966) has discussed the community structure in sympatric carnivores. He concludes that the coexistence of predators that use similar hunting methods is afforded by size differences between predator species, which leads each species in a hunting set to take a different group of prey species. In addition to these physical factors, social factors, where some predators hunt in packs or pairs while others are solitary hunters, may also facilitate the coexistence of sympatric predators. The social and ecological dynamics between sympatric species remain to be evaluated, especially in terms of conservation and possible imbalance where one species, such as the coyote or gray fox, may invade the niche of another predator of the same Type, such as the red wolf or red fox, respectively.

DISCUSSION

The preceding review of the evolution of social organization in the Canidae poses many questions which remain to be answered until long-term field studies and comparative, developmental studies of the various species under controlled laboratory conditions are undertaken. In spite of limited data, some general trends and predictable consistencies in the patterns of socio-ecological adaptation in the three Type categories are evident. The interrelationships between past environmental selection factors (such as distribution, abundance, and size of major prey utilized) and these factors acting as determinants of genotypic/phenotypic characteristics are schematized in Figure 30-6. As emphasized earlier, it remains to be shown to what extent these factors have influenced the genotype of a given species; or in other words, to what extent the development of a given behavior is genetically determined or experientially dependent.* Regional variations in

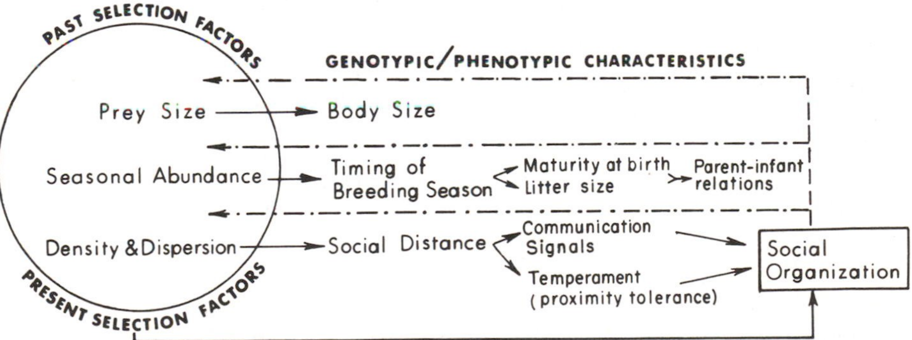

Figure 30-6 Summary schema of how behavioral characteristics and type of social organization are selected by and become adapted to a particular set of ecological factors. With more complex social organization and division of labor (as in the Type III canid) there is increasing independence and freedom from these environmental restraints. A contemporary change in the ecology may or may not lead to some modifications in behavior and social organization; the degree of flexibility and adaptibility to such a change in the environment remains to be determined.

*This statement is not a rephrasing of the nature–nurture problem, but rather poses a very basic developmental question (see Fox, 1970b for further discussion of genotype–environment interaction).

gene pools and the degree of developmental flexibility (or modifiability of the phenotype) under different sets of environmental selection factors await future analysis. Contemporary selection factors may have their initial impact on social organization (Figure 30-6), as, for example, when an abundant supply of prey may permit yearling offspring of Type II canids to remain in the natal territory. The next logical question is to what extent temperament (which is, in part, inherited) works with or against environmentally-derived socifugal or socipetal influences; for example, does the "solitary" nature of the red fox, which normally has a socifugal effect, change when there is a sustained abundance of food? What percentage of a given population of Type I canids have the capacity for sustained social interaction and social organization under socipetally optimal conditions? Data so far would predict that temperament (proximity tolerance) is sufficiently inflexible and essentially monomorphic or of minimal variability within litters to insure that independent of environmental influences, the Type I social patterns will persist. Greater flexibility in social behavior and adaptability to different or changing environmental conditions would be predicted in the Type II and III canids by virtue of the oligomorphy and polymorphy of litter temperaments.

Differences in social behavior and proximity tolerance are apparent in the various canid Types when caged together. Eisenberg (1966) postulates that different species of rodents have different thresholds of sensitivity to experimental grouping, depending on whether they are solitary, semisolitary, or communal.

Temperament changes and a marked change in proximity tolerance may occur in the Type II canids around the time of puberty, which effectively disperses the litter. It is perhaps relevant that aggressive interactions first occur in laboratory mice around puberty (Lagerspetz and Talo, 1967) and in feral mouse populations, the young leave the parental home range at puberty and begin to range some distance (Anderson, 1961).

Butterfield (1970) in his elegant studies of the life-long pair bond in the Zebra finch has brought up some intriguing questions which are relevant to our discussion of the social relationships in the Type II canids which maintain a more or less permanent pair bond. The primary motivation may be sexual, but since the breeding season lasts for less than 1 month over the entire year, other factors must be involved. First we should consider socialization in early infancy, for the pair bond may be between littermates or between parent and offspring. Second, as emphasized by Butterfield, on the question of motivation within the bond, there are two possible explanations. The presence of or contact with the mate "satisfies" purely social tendencies: the mate has acquired secondary reinforcing properties for sexual motivation, with intermittent courtship and consummation as primary reinforcement (Butterfield, 1970). The answer may lie in a subtle combination of both social and sexual motivations. Intermale competition in the Zebra finch at the time of pairing may have contributed to the development of sexual dimorphism; in the canids, however, sexual dimorphism is only slight and this may be correlated with a lower incidence of intermale competition. Crook (1970) similarly proposes that the marked sexual dimorphism in baboons may be a consequence of intermale rivalry.

As in the Zebra finch pair (Butterfield, 1970), so also in the Type I and Type II canid pair we find that the heterosexual relationship involves exclusive behaviors which are not shared with other conspecifics (with the exclusion of the pair's offspring), notably sexual behavior, courtship and play, mutual grooming, feeding (or food-sharing), and contact behavior. Aggression between partners is rare and in Zebra finches and also in captive wolves (Saunders, personal communication) copulation with other females has not been observed when both partners are present. After pairing, sexual competition is consequently reduced and the bond is indirectly maintained via heterosexual isolation.

It is clear that comparative studies on dyadic interactions can lead to some preliminary generalizations across species. Butterfield (1970) concludes that "the formation and maintenance of a dyadic unit within a social system imposes particular problems which appear to have been resolved, in some respects, in parallel ways for the heterosexual adult bond and the parent–young bond. These include: the maintenance of spatial proximity, social facilitation in various activities, behvior after separation, behavior maintaining the attachment, behavior on reunion, and 'protective' or 'defensive' behavior." The factors which lead to disruption of the pair bond in the Type I canid such as the red fox, and those which contribute to the pack-bond or allegiance in the wolf, quite independent of heterosexual affiliations but most probably based upon early socialization, remains to be explored. Pair bond formation may be essential for rearing the young, where both parents cooperate in feeding the offspring. Where food is abundant (so that the female alone can adequately provide for her offspring) polygamy is common.

Leyhausen (1965), in his discussion of social organization in solitary mammals which is concerned primarily with the domestic cat, gives some insights which are relevant to an understanding of social organization in the Type I canids. He notes that territories and home ranges overlap and that territorial behavior involves some form of communication, which is primarily olfactory (marking). He explores some of the socipetal factors operating in the so-called "solitary" species. These include the communal use of paths or runways, in which scent marking gives information as to occupancy or recency of occupancy. Fixed time and place (space) relations tend to keep individuals apart. Leyhausen observed cats occasionally hunting at the same time and place although they were separated by 27.4–60.0 m, and also gathering socially at night, at a point adjacent to or within the fringe of their territories; the cats sit 1.8–4.5 m apart and will occasionally groom each other. Such gatherings may be a forewarning of the mating season. Similar observations have been made on red foxes and are intriguing social phenomena in species which generally show strong socifugal tendencies. He notes that territorial marking in the cat is not intimidating or "to warn off" because another cat will approach and leisurely sniff and then mark the scent site of another. The mark may primarily be to inform who has passed by recently, and who might, should, or should not be encountered.

Leyhausen observed that female cats "remain faithful" to lower ranking males and that the dominant male in a captive colony does not usually interfere. He concludes that this mating system seems to be designed so that the greatest possible number of healthy males have an equal chance for reproduction.

It is clear that the relatively asocial domestic cat has the capacity for social organization (which breaks down if the colony is too large). Relative and absolute hierarchies, as seen in the canids in captive groups, develop. As individual territories shrink and a social group forms, a group territory develops. The faculty for both social grouping and solitary life may be inherent in certain species and warrants further experimentation in captive groups of Type I and II canids; since ecological factors determine what social structure a population will have, artificial ecologies, where variables such as space and food dispersion are experimentally manipulated, could yield valuable data on the degree of flexibility and adaptability of the various canid species.

Interspecies competition and competitive exclusion should also be considered: one species may tend to restrict another to a particular prey and therefore to a particular niche and type of social organization. A wolf pack, for example, would tend to make coyotes more solitary, but in the absence of a wolf pack, coyotes in a given area may hunt larger prey and be more social.

"Solitary" animals show the faculty for developing some form of social life, and the opposite is true for the more social species (e.g., the "loner" wolf and the Mexican wolf which socially may resemble the Type II canid). Is this also true for primates? The wapiti is territorial in some habitats, and nonterritorial in others.

Leyhausen (1965) proposes that although both forms of social hierarchy (relative and absolute) are present in group animals, absolute rank order predominates over relative rank order. He states that "I should predict that absolute rank order would predominate over relative rank order, the bigger the herd, and the less there is a tendency to subdivide into small groups." One should be guarded in extending such notions across species, and such a generalization and an oversimplified division of dominance into "relative" and "absolute" forms does not facilitate the analysis of social relationships in a complex social group such as the wolf pack and baboon troop. Marler (cited by Leyhausen, 1965) found both types of social rank order in his chaffinch study, the identification of the two forms often not being clear cut, and there being an added variable of seasonal incidence in rank-order formation. In the more complex social groups one must take into account filial attachments and allegiances (i.e., special social attachments between individuals) as well as seasonal (breeding) variations and *role* development. The leader wolf or baboon, for example, may "police" and break up fights between subordinates. One must also consider the role played by adult nonbreeding males in the pack or troop, which may be to assist in group hunting or group defense. In some primates, the nonbreeding males form a separate group; in hamadryas bands, there are small one-male harems and a group of peripheral males; in gelada herds, there are several one-male reproductive units and all-male groups (Crook, 1970). In other primate species, there may be several reproductive males in the group (e.g., *Gorilla gorilla)*.

Crook (1970) recognizes five categories of social grouping in primates which are of comparative interest:

1. A small group consisting of one male, one or more females and young; tend to be very territorial, e.g., forest dwelling *Colobus guereza* and gibbon.
2. Larger groups of several males, females and young, e.g., forest and forest fringe dwelling *Allouatta palliata.*
3. Several males, numerically more females with offspring and juveniles; occur in savanna and forest fringe, home ranges overlap and show more avoidance than overt aggression towards other groups, e.g., *Cercopithecus aethiops, Macaca mulatta,* and *Papio spp.* (but not *hamadryas*).
4. One male, several females and young and an all-male group of juveniles and nonbreeding adults, e.g., savanna herds or bands of *Papio hamadryas.*

5. Several males, females and young form temporary gatherings with frequent exchange of individuals between groups, e.g., savanna, forest, and forest fringe dwelling chimpanzee. The forest gorilla band usually has a leader male and there is far less frequent interchange between groups; there tends to be mutual avoidance of other bands.

Crook (1970) goes on to demonstrate the ecological factors which underlie these differences in social organization in primates. Coordinated troop behaviors in baboons is attributed especially as a consequence of predation. Dispersion of food items reduces conflict within the troop, but troop cohesiveness, a large number of males and few females in estrus has resulted in male–male rivalry and selection for great sexual dimorphism (Crook, 1970). A complex male hierarchy is evident with patterned promiscuity, where low ranking males first have the females coming into estrus, but at the height of estrus the high ranking males have them. Crook notes similar structure and function in hamadryas and gelada one-male reproductive units, but very different social dynamics are evident. Gelada units wander independently, entering and leaving a herd at will, and females of a unit are free to wander far from their male. Hamadryas females are not allowed to roam. Hamadryas, but not gelada males "adopt" young females and so form a harem. Hamadryas herds or troops are substructured into bands of one-male units whereas there is no such substructuring in gelada. The all-male group is less well-defined in hamadryas and young males tend to be peripheral to the harem. Two-male teams are seen in hamadryas, one old male and a younger male with the harem, whereas an old male may be a nonbreeding leader and determine the band's movements. In the gelada, old males join up with the all-male group.

From Crook's extensive review, there is little evidence of leadership being based on dominance *per se*, but rather on "role" as a distinct social position within a group structure. More submissive and sexual presentations are seen in subordinate males of the multimale troops in other baboons, compared to the one-male unit in gelada herds and patas. Rowell (1966) proposes that status hierarchy arises not so much because of the dominant animal's behavior, but because of the responsiveness of juvenile and subordinate ones towards their elders, this proposal being supported in Fox's (1972) study of the development of social relationships in wolves. The sum of behaviors towards older individuals, juveniles, and like and opposite sex make up the social repertoire: a set of behavior styles. The sum of these characteristics define the individual's social position (Crook, 1970). Roles change with maturity and with the death of others occupying certain social positions. Role change is also associated with escape and emigration where a solitary male or subordinate male becomes the leader of a new "splinter group." Allegiance affects status and role; changes with estrus lead to "consort" relationships; kinship and matrifocal allegiances are evident; affiliation by subordinates with infants as a means of entering high status groups has been observed. Also, with an increase in size of the group, social tensions tend to increase.

At this point the question arises as to whether dominance and leadership are trait-specific or situation-related; that is, determined by some basic temperament qualities or by the social context. Social experiences during development, interacting with a particular temperament substrate or template, most probably determine role and social rank (Fox, 1972; Fox and Andrews, 1973). What the qualities are that endow individuals with varying degrees of social flexibility and what physiological changes occur when behavior is modified when role or social relationships are altered remain to be ascertained.

Once a sufficient degree of social organization has been attained, where there is a division of labor and individuals cooperate and otherwise support each other (in hunting, care of young, defense of territory, etc.), interindividual synergy would give some freedom or emancipation from environmental constraints (Figure 30-6). For example, animals can hunt certain prey more effectively as a group, and can also secure larger prey; the location of prey may be facilitated by individuals engaging in separate search patterns and communicating with conspecifics once prey has been located. Survival of young may be enhanced by other adults, in addition to the parents, providing food and protection for them. These advantages would tend to be reinforcing and serve to perpetuate social dependency and cooperation through successive generations; (a "social imperative" or drive *per se* might also be operating in this regard).

In the arid savanna, where the dry season reduces the supply of food, subordinate baboon females and males in a large coherent group would suffer, thus one-male and all-male groups have evolved. In the rainy season, when food is plentiful, they come together into big bands. Different timings of reproductive seasons in the primates are associated with the supply of food, being plentiful during pregnancy, lactation, or during the infancy of offspring.

Crook (1970) notes that in primates, social systems are determined primarily by ecological factors, but protocultural processes may yield shifts in social organization independently from environmental processes. In man, cultural control of society has come to mold not only social change, but also our ecology.

Intraspecific Variation in Relation to Ecology (Socio-Ecology)

Crook (1970) uses the baboon as the best example of ecologically-related variations in social behavior. The Type III structure is seen in the savanna *Papio papio*. In *P. anubis* forest troops, males move freely between troops, but in the Ethiopian Rift, they occur only in integrated, closed troops. Clearly, regional (ecological) variations lead to social changes, so that new races (social "clines") or subspecies evolve.

According to Crook (1970), the environment may affect:

1. Variance in group composition and social organization in different species and races.
2. Socio-ecological differences — "protocultural" social traditions, "roles" (or particular activities performed) of age and sex classes, and mortality and reproductive rates (and the number of nonbreeding adults).
3. Socio-ecological selection pressures for modifying the genetic basis of individual temperament and behavior within contrasting groups. (In man, a great heterogeneity of temperament is associated with great niche diversity of cultural roles or opportunities.)
4. Relation between social signaling and the social system in which it occurs.

"Interspecific and intraspecific variations in social organizations suggest that variance is particularly related to the degree of environmental stability shown by a habitat," (Crook, 1970). The abundance or scarcity of food and the presence or absence of seasonal fluctuations are what he terms environmental stability. We may now ask how is social structure (and population size) conserved from one generation to the next? Crook draws upon the Japanese studies of macaques; it has been found that with an increase in numbers (as a result of provisioning), multimale groups of Japanese macaques show an increase in social tension. The group subsequently splits into smaller units of similar composition. Matrifocal kinship relations apparently correlate with dominance rank, because males born of high status females grow up to occupy high social rank.

SUMMARY

Hunting techniques, methods of communication, and patterns of social organization in the Canidae are reviewed. Interspecies comparisons reveal certain consistencies in some species and clear differences

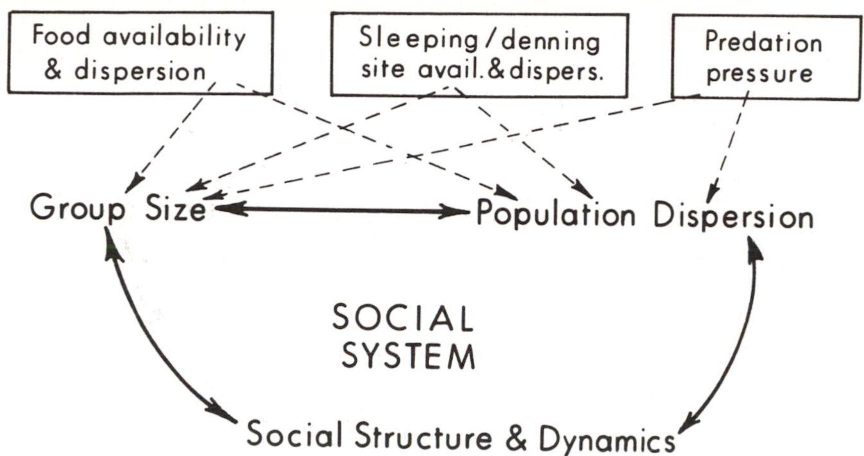

Figure 30-7 Model showing what ecological or environmental factors influence the social system and social dynamics of a group of animals (modified after Crook, 1970).

among other species groups. Three arbitrary Types are identified, which differ markedly in hunting techniques, social organization, and communication. These differences, which are ecologically adaptive, may be related to species differences in temperament and proximity tolerance, which influence sociability. Socipetal or socifugal tendencies either limit or enhance the possibilities for social organization. Variations in social organization under different sets of ecological variables in the Type II and III canids are attributed to greater flexibility in behavior (associated with intralitter oligomorphy and polymorphy).* This capacity to modify social behavior under different ecological conditions is limited in the Type I "solitary" canid (associated with a narrow range of intralitter temperaments, i.e., monomorphy). These ob-

*Kummer (1971a) in his book *Primate Societies* (Aldine, Chicago) also emphasizes that future research should focus upon determining the limitations and capacities for adaptive modification of social behavior under natural and experimental conditions where certain ecological factors are changed. For example, Type I and II canids may appear more social in zoos, especially when they have been raised together as an intact litter and where the confines of the enclosure effectively prevent normal dispersion with maturity. Can Type III dominance hierarchies form, and what other latent capacities might be revealed? Food in one locale and space (territory) in another region (where food is abundant) may be the ecological factor affecting sociability and population density and dispersion. We may then look for behavior–genetic and behavior–experimental differences in the same species from different ecosystems as emphasized and demonstrated by Kummer (1971a) in his studies of baboons.

servations and speculations are compared and contrasted with other mammalian species, notably the primates, in order to construct a general scheme of various patterns of socio-ecological adaptation from the intrinsic complexities of environmental–organismic interaction.

Editor's Conclusions: The prime task of an editor of this kind of book, as I see it, is to find and bring together an outstanding selection of contributors whose experience and knowledge help make a useful text. I feel that what we have here is a prime source of reference material and hard data, coupled with many stimulating and provoking questions. Some of the latter, which at first glance seem to be inadequacies or "holes" in the text, are in fact gaps in our knowledge about the social ecology of canids. The reader will therefore find many areas where further research is needed. We will never know everything and my dissatisfaction with some of these holes in the book is assuaged by the fact that we do need to know more; more in terms of the three divisions of the book: behavioral, ecological, and molecular/genetic/taxonomic. Related problems such as the evolution of the Canidae and their future in man-dominated ecosystems (part of the global human "egosystem"), their conservation, management in the wild and in captivity, and so on are partially answered in this volume. Further research, or rather re-search, is needed for more understanding not only of the canids but also of man's place in nature, the two being intimately related. We must strive to preserve and foster this relationship, this unity and interrelatedness of all things and at the same time reduce man's alienation from nature as well as from his fellow beings. The dynamically balanced intrinsic complexity of life, of ecosystems, of species, of individuals, and of the cells of the body must be appreciated; for man, it must also be understood. Why? In order to predict, control, modify, or manipulate and exploit? I hope simply for the sake of awareness and also for the possible necessity of intervention in times of an ecocatastrophe, as well as to avert such an eventuality. I believe that this book gives us much basic knowledge to help us in these directions and I wish to thank all the contributors for the work that they have done.

REFERENCES

Ables, E. D. (1958). An exceptional fox movement. *Journal of Mammalogy,* **46**, 102.

Ables, E. D. (1968). Ecological studies on red foxes in southern Wisconsin. Ph. D. Thesis, University of Wisconsin, Madison.

Ables, E. D. (1969). Activity studies of red foxes in southern Wisconsin. *Journal of Wildlife Management*, **33**,145–153.

Ables, E. D. (1969a). Home-range studies of red foxes (*Vulpes vulpes*). *Journal of Mammalogy,* **50**, 108–120.

Adams, E. G. Phythian (1949). Jungle memories. *Journal Bombay Natural History Society,* **48**, 645.

Agricultural Statistics (1969–1970). H.M.S.O., London.

Ahmed, I. A. (1941). Cytological analysis of chromosome behaviour in three breeds of dogs. *Proceedings Royal Society, Edinburgh,* **61**, 107–118.

Alekseev, G. (1957). The hunting industry on Taymyr. *Okhota i okhotnich'e khoziaistvo,* **4**, 18–19.

Allen, G. M. (1923). The pampa fox of the Bogota Savanna. *Proceedings Biological Society Washington,* **36**, 55–57.

Allen, G. M. (1939). A checklist of African mammals. *Bulletin Museum of Comparative Zoology, Harvard University,* **83**.

Allen, G. M. (1942). Extinct and vanishing mammals of the Western Hemisphere. *American Committee for International Wildlife Protection, Special Publication No.* **11**.

Allen, J. A. (1904). Report on mammals from the district of Santa Marta, Columbia, collected by Mr. Herbert H. Smith, with field notes by Mr. Smith. *Bulletin American Museum Natural History,* **20**, 407–468.

Allen, J. A. (1911). Mammals from Venezuela collected by Mr. M. A. Carriker, Jr., 1909–1911. *Bulletin American Museum Natural History,* **30**, 239–273.

Allen, P. Z. and Johnson, J. S. (1972). Studies on equine immunoglobulins — III. Antigenic interrelationships among horse and dog G globulins. *Comparative Biochemistry and Physiology,* **41**, 371–383.

Anderson, P. K. (1961). Density, social structure, and non-social environment in house-mouse populations and the implications for regulation of numbers. *Transactions New York Academy of Science Series 11,* **23**, 447–451.

Anderson, R. M. (1937). Mammals and birds of the Western Arctic District, Northwest Territories, Canada. In: *Canada's Western Northland,* W. C. Bethume.

Anderson, R. M. (1946). Catalogue of Canadian recent mammals. *National Museum of Canada Bulletin,* **102**, Biology Series 3.

Andres, A. H. (1938). On the chromosome complex in several Canidae. *Cytologia,* **9**, 35–37.

Andrewartha, H. G. (1971). *Introduction to the Study of Animal Populations*, University of Chicago Press.

Anonymous (1962). Northwest Territories-graphs showing fur take and average prices by species. Government of the Northwest Territories Game Management Services.

Anonymous (1968). Wolves and coyotes in Ontario. Department of Lands and Forests, Ontario, Canada.

Arnold, D. A. (1952). The relationship between ring-necked pheasant and red fox population trends. *Papers Michigan Academy of Science, Arts, and Letters,* **37**, 121–127.

Arnold, D. A. (1956). Red foxes of Michigan. Michigan Conservation Department Bulletin.

Arnold, D. A. and Schofield, R. D.(1)1956. Home range and dispersal of Michigan red foxes. *Papers Michigan Academy of Science, Arts, and Letters,* **41**, 91–97.

Arrighi, F. E. and Hsu, T. C. (1971). Localization of heterochromatin in human chromosomes. *Cytogenetics,* **10**, 81–86.

Asdell, S. A. (1946). *Patterns of Mammalian Reproduction*, Cornell University Press, Ithaca, New York.

Asdell, S. A. (1964). *Patterns of Mammalian Reproduction*, 2nd Ed., Cornell University Press, Ithaca, New York.

Ashbrook, F. G. and Walker, E. P. (1952). Blue fox farming in Alaska. *U.S. Department of Agriculture Bulletin No.* **1350**.

Atkins, D. L. and Dillon, L. S. (1971). Evolution of the cerebellum in the genus *Canis*. *Journal of Mammalogy,* **52**, 96–107.

Augusteyn, R. C., McDowall, M. A., Webb, E. C. and Zerner, B. (1972). Primary structure of cytochrome c from the elephant seal, *Mirounga leonina*. *Biochimica et Biophysica Acta,* **257**, 264–272.

Awa, A., Sasaki, M. and Takayama, S. (1959). An *in vitro* study of the somatic chromosomes in several mammals. *Japanese Journal of Zoology*, **12**, 257–265.

Ayala, J. (1971). Competition between species: frequence dependence. *Science*, **171**, 820–823.

Babero, B. B. and Rausch, R. (1952). Notes on some trematodes parasitic to Alaskan canidae. *Proceedings Helminthological Society Washington,* **19**(1), 15–17.

Bachrach, M. (1947). *Fur: a Practical Treatise,* Prentice-Hall, Inc., New York.

Bailey, A. M. and Hendee, R. W. (1926). Notes on the mammals of northwestern Alaska. *Journal of Mammalogy,* **7**, 9–28.

Bailey, V. (1941). Gray foxes — all Americans. *Nature,* **34,** 493–495, 528.

Banfield, A. W. F. (1954). The role of ice in the distribution of mammals. *Journal of Mammalogy,* **35**, 104–107.

Banks, F. (1957). Outlaws of the Weidenhammer. *Michigan Conservation,* **26**, 22–24.

Barabash-Nikiforov, I (1938). Mammals of the Commander Islands and the surrounding sea. *Journal of Mammalogy,* **19**, 423–429.

Barberis, L., Satri, M. and Sorrentino, R. (1964). I cromosomi di *Canis familiaris:* sistematico ed evolutivo. *Natura,* **55**, 234–240.

Barkalow, F. S., Jr. (1940). Black vulture and red fox found in unusual association. *Wilson Bulletin,* **52**, 278–279.

Barry, T. W. (1967). Geese of the Anderson River Delta, Northwest Territories. Unpublished Ph.D. Thesis, University of Alberta.

Bartlett, M. S. (1947). Multivariate analysis. *Journal Royal Statistical Society Supplement* **9**, 176–197.
Bartlett, M. S. (1965). Multivariate statistics. In: *Theoretical and Mathematical Biology,* T. H. Waterman and H. J. Morowitz, eds., Blaisdell Publishing, New York.
Basrur, P. K. and Gilman, J. P. W. (1966). Chromosome studies in canine lymphosarcoma. *Cornell Vet.,* **56**, 451–469.
Bates, D. N. (1958). *History of the Timber Wolf and Coyote in Ontario,* Ontario Department of Lands and Forests.
Bates, M. (1944). Notes on a captive Icticyon. Journal of Mammalogy, **25**, 152–154.
Baur, E. W. and Schorr, R. T. (1969). Genetic polymorphism of tetrazolium oxidase in dogs. *Science,* **166**, 1524–1525.
Beaver, P. C. (1969). The nature of visceral larva migrans. *Journal of Parasitology,* **55**, 3–12.
Beck, A. M. (1971). The life and times of Shag, a feral dog in Baltimore. *Natural History Magazine,* **80**, 58–65.
Beck, A. M. (1973). *The Ecology of Stray Dogs: A Study of Free-ranging Urban Animals,* York Press, Baltimore.
Beck, A. M. (1974a). Behavior of dogs: canid behavior in a natural setting. In: *Animal Behavior in Laboratory and Field,* E. O. Price and A. W. Stokes, eds., W. H. Freeman and Co., San Francisco, in press.
Beck, A. M. (1974b). The dog: America's sacred cow? *Nations's Cities,* **12**(2), 29–31, 34–35.
Bédard, J. H. (1967). Ecological segregation among plankton-feeding alcidae *(Aethia* and *Cyclorrhynchus).* Unpublished Ph. D. Thesis, University of British Columbia.
Beddard, F. E. (1902). *Mammalia, Macmillan Co., Ltd., London.*
Bednarik, K. F. (1959). The fox: thirteen years of bounty and fur harvest in Ohio 1944–57. *Ohio Department Natural Resources Publication* **177.**
Bee, J. W. and Hall, E. R. (1956). *Mammals of Northern Alaska,* Museum of Natural History, University of Kansas, Lawrence.
Bekoff, M. (1972a) An ethological study of the development of social interaction in the genus Canis: A dyadic analysis. Ph. D. Dissertation, Washington University, St. Louis.
Bekoff, M. (1972b). The development of social interaction, play, and metacommunication in mammals: an ethological perspective. *Quarterly Review of Biology,* **47**, 412–434.
Belyaev, D. K. (1962). On some problems of correlative variability and their significance for the theory of evolution and selection of animals. *Isv. SO AN SSSR,* **N10**.
Belyaev, D. K. and Trut, L. N. (1964). Behaviour and the reproductive function of animals. I. Correlation between the type of behaviour, time of reproduction and fertility. *Bull. MOIP otd. biol.* **t 63**, vyp. 3.
Benirschke, K. and Low, R. (1965). Chromosomes complement of the Coyote (*Canis latrans). Mammalian Chromosomes Newsletter,* **15**, 102.
Bennett, L. J. and English, P. F. (1942). Food habits of the gray fox in Pennsylvania. *Pennsylvania Game News,* **12**, 10–22.
Bertram, G. and Lack, D. (1938). Notes on the animal ecology of Bear Island. *Journal Animal Ecology,* **7**, 27–52.
Berzon, D. R., Farber, R. E., Gordon, J., and Kelly, E. B. (1972). Animal bites in a

large city — A report on Baltimore, Maryland. *American Journal of Public Health,* **62** (3), 422–426.

Besadny, C. D. (1966). Winter food habits of Wisconsin foxes. *Wisconsin Department Natural Resources, Research Report* **20**.

Bigarella, J. J. and de Andrade, G. O. (1965). Contribution to the study of the Brazilian Quaternary. International Studies of the Quaternary. *Geological Society of America Special Papers No.* **84**.

Birdseye, C. (1956). Observations on a domesticated Peruvian desert fox, *Dusicyon. Journal of Mammalogy,* **37,** 284–287.

Birulya, A. (1907). Ocherki iz zhizni ptits polyarnago poberezhya Sibiri. *Zapiski Imp. Akademia Nauk Series* **8, 18**, 1–157.

Bishop, D. W. (1942). Germ cell studies in the male fox *(Vulpes fulva). Anatomical Record,* **84**, 99–115.

Blackith, R. E. (1965). Morphometrics. In: *Theoretical and Mathematical Biology,* T. H. Waterman and H. J. Morowitz, eds., Blaisdell Publishing, New York.

Blackith, R. E. and Reyment, R. A. (1971). *Multivariate Morphometrics,* Academic Press, London.

Blanchet, G. H. (1925). An exploration into the northern plains north and east of Great Slave Lake. *Canadian Field-Naturalist,* **39**, 30–34.

Blomback, B. and Blomback, M. (1968). Primary structure of animal proteins as a guide in taxonomic studies. In: *Chemotaxonomy and Serotaxonomy,* J. G. Hawkes, ed., Academic Press, New York and London.

Boitzov, L. V. (1937). Arctic fox: biology, breeding, feeding. *Transactions Arctic Institute Leningrad,* **65**, 1–144.

Borgaonkar, D. S., Elliot, O. S., Wong, M. and Scott, J. P.(1968). Chromosome study or four breeds of dogs. *Journal of Heredity,* **59**, 157–160.

Bothma, J. du P. (1971a). Reports from the mammal research unit: 2. Food habits of some Carnivora (Mammalia) from southern Africa. *Annals Transvaal Museum,* **27**, 15–26.

Bothma, J. du P. (1971b). Food of *Canis mesomelas* in South Africa. *Zoologica Africana,* **6**, 195–203.

Bothma, J. du P. (1971c). Control and ecology of the black-backed jackal *Canis mesomelas* in the transvaal. *Zoologica Africana,* **6**, 187–193.

Bourliere, F. (1963). Specific feeding habits of African carnivores. *African Wildlife,* **17**, 21–27.

Bowdler, A. J., Bull, R. W., Slating, R. and Swisher, S. N. (1971). Tr: A canine red cell antigen related to the A-antigen of human red cells. *Vox Sanguinis,* **20**, 542–554.

Braend, M. (1966). Serum transferrins of dogs. *Xth European Conference on Animal Blood Groups and Biochemical Polymorphisms,* pp. 319–322.

Braestrup. F. W. (1941). A study on the arctic fox in Greenland (immigrations, fluctuations in numbers based on trading statistics). *Medd. om Grønland,* **131**, 1–101.

Brander, A. Dunbar (1931). *Wild Animals in Central India.*

Bridges, W. (1954). It's the "fearsome Warracaba tiger." *Animal Kingdom,* **57**, 25–28.

Brown, D. H. (1970). Ocular *Toxocara canis. Journal of Pediatric Ophthalmology, 7* (3), 182–*191*.

Brown, D. H. (1974). The geography of ocular *Toxocara canis. Ann. Ophthalmology*, **6** (4), 343–344.

Brown, J. L. and Orians, G. H. (1970). Spacing mechanisms in mobile animals. *Annual Review of Ecology and Systematics,* **1**, 239–262.

Brown, R. C., Castle, W.L.K., Huffines, V. H. and Graham, J. B. (1966). Pattern of DNA replication in chromosomes of the dog. *Cytogenetics,* **5**, 206–222.

Bryant, H. C. (1924). Rabies epidemic among gray foxes. *California Fish and Game,* **10**, 146–147.

Buckton, K. E. and Cunningham, C. (1971). Variations of the chromosome number in the red fox *(Vulpes vulpes). Chromosoma,* **33**, 268.

Bump, G., Darrow, R. W., Edminster, F. C. and Crissey, W. F. (1947). *The Ruffed Grouse: Life History, Propagation, Management,* New York State Conservation Department.

Burkholder, B. L. (1959). Movements and behavior of a wolf pack. *Journal of Wildlife Management,* **23**, 1–11.

Burr, J. B. (1947). Rabid foxes. *Texas Game and Fish,* **5**, 6–7.

Burrows, R. (1968). *Wild Fox,* David and Charles, Newton Abbot, England.

Burt, W. H. (1946). *The Mammals of Michigan,* The University of Michigan Press, Ann Arbor, Michigan.

Burton, M. (1962). *University Dictionary of Mammals of the World,* Apollo, New York.

Burton, R. W. (1941). The Indian wild dog. *Journal of the Bombay Natural History Society,* **41**, 691–715.

Butler, L. (1945). Distribution and genetics of the color phases of the red fox in Canada. *Genetics,* **30**, 39–50.

Butler, L. (1947). The genetics of the colour phases of the red fox in the Mackenzie River locality. *Canadian Journal of Research,* **25**, 190–215.

Butler, L. (1951). Population cycles and color phase genetics of the colored fox in Quebec. *Canadian Journal of Zoology,* **29**, 24–41.

Butterfield, P. A. (1970). The pair bond in the Zebra finch. In: *Social Behavior in Birds and Mammals,* J. H. Crook, ed., Academic Press, New York.

Cabrera, A. (1931). On some South American canine genera. Journal of Mammalogy, **12**, 54–67.

Cabrera, A. and Yepes, J. (1940). *Historia Natural Ediar Mamiferos Sud-Americanos,* Compania Argentina de Editores, Buenos Aires.

Cahalane, V. H. (1947). *Mammals of North America,* The Macmillan Co., New York.

Calhoun, J. B. (1950). Population cycles and gene frequency fluctuations in foxes of the genus *Vulpes,* in Canada. *Canadian Journal of Research,* **28**, 45–157.

Camenzind, F. J. Personal communication. University of Wyoming.

Cameron, A. W. (1950). Arctic fox on Cape Breton Island. *Canadian Field-Naturalist,* **64**, 154.

Caras, R. (1971). Beware of the dog. *Family Circle* (May) **78**, 18, 20.

Carr, W. A. (1945). Gray fox adventures. *Natural History,* **54**, 4–9.

Carroll, C. A., Jr. (1971). Personal communication. Chief, Rat Eradication Program, Baltimore City Health Department.

Caspersson, T. (1970). Identification of human chromosomes by DNA-binding fluorescent agents. *Chromosoma,* **30**, 215–227.

Center for Disease Control (CDC) (1974). Leptospirosis annual summary, 1972, issued Feb., 1974, Atlanta.

Chaddock, T. T. (1939). Report on grey and red fox stomach examinations. *Wisconsin Conservation Bulletin,* **4**, 53–54.

Chance, M.R.A. (1962). Social behaviour and primate evolution. In: *Culture and the Evolution of Man, M. F. Ashley Montagu, ed., Oxford, New York.*

Chance, N. A. (1966). *The Eskimo of North Alaska,* Holt, Rinehart and Winston, New York.

Chesemore, D. L. (1967). Ecology of the arctic fox in northern and western Alaska. Unpublished M.S. Thesis, University of Alaska, College, Alaska.

Chesemore, D. L. (1968). Distribution and movements of white foxes in northern and western Alaska. *Canadian Journal of Zoology,* **46**, 849–854.

Chesemore, D. L. (1968a). Notes on the food habits of arctic foxes in northern Alaska. *Canadian Journal of Zoology,* **46**, 1127–1130.

Chesemore, D. L. (1969). Den ecology of the arctic fox in northern Alaska. *Canadian Journal of Zoology,* **47**, 121–129.

Chesemore, D. L. (1970). Notes on the pelage and priming sequence of arctic foxes in northern Alaska. *Journal of Mammalogy,* **51**, 156–159.

Chiarelli, B. (1965). Interess sistematico ed evolutivo della cariologia dei Canidae (Nota preliminari). *Bollettino di Zoologia,* **32**, 435–444.

Chiarelli, B. (1966). Data on the karyology of different races of *Canis familiaris.Mammalian Chromosomes Newsletter,* **21**, 160.

Chiarelli, B. (1973). A caryological approach on the history of dog domestication. *Bollettino di Zoologia,* **40**.

Chiarelli, B., Sarti, M. and Shafer, D. (1972). Bandeggiamento dei cromosomi mediante tripsina. *Rivista di Antropologia.*

Chiarelli, B., Shafer, D. and Sarti, M. (1973). Chromosome banding with trypsin. *Gentica* (in press).

Chirkova, A. F. (1951). Predvaritel'naya methodika prognosov izmeneniy chislennosti pestosov. *Voprosy Biologii Pushnykh XI, Moscow,* **76**.

Chirkova, A. F. (1953). Dynamics of fox numbers in Voronezh Province and forecasting fox harvests. *Transl. Russian Game Reports,* **3**, 50–69.

Chirkova, A. F. (1955). Opyt massovoy glazomernoy otsenki chislennosti i prognozy "urozhaya" pestosov (1944–1949 gg.). *Voprosy Biologii Pushnykh Zverey XIV, Moscow,* **73**.

Chirkova, A. F. (1967). The relationship between arctic fox and red fox in the far north. *Problems of the North,* **11**, 129–131.

Chitty, D. and Elton, C. (1937). Canadian Arctic wild life enquiry, 1935–36. *Journal Animal Ecology,* **6**, 368–385.

Chitty, H. and Chitty D. (1945). Canadian Arctic wild life enquiry, 1942–43. *Journal Animal Ecology,* **14**, 37–41.

Christian, John J. (1955). Effects of population size on weight of reproductive organs of white mice. *American Journal of Physiology,* **181**, 477–480.

Churcher, C. S. (1959). The specific status of the New World red fox. *Journal of Mammalogy,* **40**, 513–520.

Churcher, C. S. (1960). Cranial variation in the North American red fox. *Journal of Mammalogy,* **41**, 349–360.

Clark, F. W. (1972). Influence of jackrabbit density on coyote population change. *Journal of Wildlife Management,* **36**, 343–356.

Clark, K.R.F. (1971). Food habits and behaviour of the tundra wolf on central Baffin Island. Ph. D. Thesis, University of Toronto, Toronto, Ontario.

Clarke, C.H.D. (1940). A biological investigation of the Thelon Game Sanctuary. *Bulletin National Museum of Canada No.* **96**.

Clutton-Brock, J. (1963). The origins of the dog. In: *Science in Archaeology,* D. Brothwell and E. Higgs, eds., Thames and Hudson, London.

Cochran, B. (1967). Delinquent dogs and dead deer. *Outdoor Oklahoma,* **23**, 12–13, 20.

Coimbra F. A. F. (1966). Notes on the reproduction and diet of Azara's fox *Cerdocyon thous azarae* and the hoary fox *Dusicyon ventulus* at Rio de Janeiro Zoo. *International Zoo Yearbook,* **6**, 168–169.

Collett, R. and Nansen, F. (1900). An account of the birds. In: *The Norwegian North Polar Expedition,* 1893–1897, Vol. I, Part IV., pp. 1–54.

Colson, R. B. and McKeon, W. H. (1952). The wildlife rabies control program in New York. *Proceedings 8th Northeast Fish and Wildlife Conference.*

Coon, C. S. (1951). *Cave Explorations in Iran 1949,* University of Pennsylvania Press Museum Monograph.

Cooper, M. (1957). *Pica,* Chas. C. Thomas, Springfield, Illinois.

Cowan, I. McT. (1947). The timber wolf in the Rocky Mountain National Parks of Canada. *Canadian Journal of Research,* **25**, 139–174.

Cowan, I. McT. (1949). Rabies as a possible population control of arctic canidae. *Journal of Mammalogy,* **30**, 396–398.

Crandall, L. S. (1964). *The Management of Wild Mammals in Captivity,* University of Chicago Press.

Crawford, K. L. (1964). A survey of the animal control activities in Maryland, 1964. Division of Epidemiology, Bureau of Preventive Medicine, Maryland State Department of Health.

Crawford, K. L. (1970). Personal communication. Maryland State Veterinarian.

Creel, G. C. and Thornton, W. A. (1971). A note on the distribution and specific status of the fox genus *Vulpes* in West Texas. *The Southwestern Naturalist,* **15**, 402–404.

Crespo, J. A. (1971). Ecologia de zorro gris, *Dusicyon gymnocercus antiquus* (Ameghino) en la Provincia de La Pampa. *Review Museo Arg. Cien. Natural (Ecology)* **1**, 147–205.

Crespo, J. A. and de Carlo, J. M. (1963). Estudio ecologico de una poblacion de zorros colorados *Dusicyon culpaeus culpaeus* (Molina) en el oeste de la Provincia de Neuquen. *Review Museo Arg. Cien. Natural Bs. As. (Ecology),* **1**, 1–155.

Crisler, L. (1956). Observations of wolves hunting caribou. *Journal of Mammalogy,* **37**, 337–346.

Critchell-Bullock, J. C. (1930). An expedition to sub-arctic Canada, 1924–1925. *Canadian Field-Naturalist,* **44**, 210–212.

Cromwell, H. W., Sweebe, E. E. and Camp, T. C. (1939). Bacteria of the *Listerella* group isolated from foxes. *Science,* **89**, 293.

Crook, J. H. (1970). The socio-ecology of primates. In: *Social Behaviour in Birds and Mammals,* J. H. Crook, ed., Academic Press, New York, pp. 103–166.

Cross, E. C. (1940). Periodic fluctuations in numbers of the red fox in Ontario. *Journal of Mammalogy,* **21**, 294–306.

Cumming, H. G. and Walden, F. A. (1970). The white-tailed deer in Ontario. Ontario Department Lands and Forests.

Dahr, E. (1942). Uber die Variation der Hirnschale bei wilden und zahmen Caniden. *Arkiv für Zoologii Stockholm,* **33A** (16), 1–56.

Danilov, D. N. (1958). Den sites of the arctic fox *(Alopex lagopus)* in the east part of Bol'shezemel'skaya Tundra. *Problems of the North,* **2**, 223–233.

Darroch, J. N. (1958). The multiple-recapture census I. estimating a closed population. *Biometrika,* **45**, 343–359.

Davidar, E. R. C. (1965). Wild dogs and village dogs. An encounter between wild dogs and sambhur. *Journal Bombay Natural History Society,* **62**, 146; **66**, 374 respectively, and other miscellaneous notes in journal.

Davis, D. E. and Wood, J. E. (1959). Ecology of foxes and rabies control. *Public Health Report,* **74**, 115–118.

Davis, W. B. (1960). The mammals of Texas. *Texas Game and Fish Commission, Austin, Bulletin No.* **41**.

Dawson, J. B. (1963). The white-tailed deer in Ontario, past, present and future. *Ontario Fish and Wildlife Review,* **2**, 3–11.

Degerbøl, M. (1961). On a find of a preboreal domestic dog from Starr Carr, Yorkshire, with remarks on other Mesolithic dogs. *Proceedings Prehistorical Society,* **27**, 35–55.

Dement'yev, N. I. (1955). K biologii pestsa Bol'shezemel'skoy tundry. *Voprosy Biologii Pushnykh Zverey XIV, Moscow,* **123**.

Devold, H. (1940). Noen refeksjoner om lemenen og dens forplanting. *Norges Jergerog Fisher-Forbunds Tidsskrift,* **69**, *7–9*.

Dewsbury, D. A. (1972). Patterns of copulatory behavior in male mammals. *Quarterly Review of Biology,* **47**, 1–33.

Dieterlen, F. (1954). Uber den Haarbau des Anderwolfes, *Dasycyon hagenbecki* (Krumbiegel, 1949). Saeugetierk. Mitteil., **2**, 26–31.

Djerassi, C., Israel, A., and Jochle, W. (1973). Planned parenthood for pets? Bulletin Atom. Soc. Jan. 10–19.

Dobie, J. F. (1950). *The voice of the coyote,* Little, Brown and Co., Boston.

Dobryszycka, W., Elwyn, D. H. and Kukral, J. C. (1969). Isolation and chemical composition of canine haptoglobin. *Biochimica et Biophysica Acta* **175** 220–222.

Dolgov, V. A. (1966). Age changes of some structural peculiarities of the skull and baculum in carnivorous mammals: age determination procedure exemplified by polar fox (*Alopex lagopus* L.). *Zoologicheskii Zhurnal* **45**, 1074–1080.

Donaldson, V. H. and Pensky, J. (1970). Some observations on the phylogeny of serum inhibitor of Cl esterase. *Journal of Immunology* **104**, 1388–1395.

Dorst. J. (1970). *A field guide to the larger mammals of Africa,* Houghton-Mifflin, Boston.

Dorst, J. and Dandelot, P. (1970). *A field guide to larger mammals of Africa,* Houghton-Mifflin, Boston.

Dortch, C. E. and Merilees, D. (1971). A salvage excavation in Devil's Lair, Western Australia. *Journal Royal Society, Western Australia,* **54**, 103–113.

Drimme, F., ed. (1954). *The Animal Kingdom*,Vol. I, Doubleday, New York.

Dubinina, M. N. (1951). On the biology and distribution of *Diphylobothrium erinaceieuropaei* (Rud., 1819) Iwata,1933. *Zoologicheskii Zhurnal,* **30**, 421–429.

Dubnitskii, A. A. (1953). *Taenia pisiformis*, a faculative parasite of the polar fox. In: *Akademiia nauk SSSR*. Rabotyu pol gel'mintologii, k-75-letiiu akademika K. I. Skriabina, 234–236.

Dubnitskii, A. A. (1956). Ways of infection with uncinariasis in arctic foxes. *Karakulevodstvo i zverovodstvo*, **9**, 45–46.

Dubrovskii, A. M. (1937). The arctic fox (*Alopex lagopus (L.)*) and arctic fox trapping in Novaya Zemlya. *Transactions Arctic Institute Lenningrad.* (Translation, J. D. Jackson for the Bureau of Animal Populations, Oxford University, England, 1939. Translation 38:F. 1130 A, 1–47.)

Dufresne, F. (1946). *Alaska's Animals and Fishes*, A. S. Barnes and Co., New York.

Echobichon, D. J. (1970). Characterization of the esterases of canine serum. Journal of Bochemistry, **48**, 1359–1367.

Editorial (1969). Peroxidase may mediate melanine formation. *Chemical Engineering News* (Sept. 22).

Egoscue, Harold J. (1956). Preliminary studies of the kit fox in Utah. *Journal of Mammalogy,* **37**, 350–357.

Eiknes, J. (1959). Comments on a paper by J. W. Mason. *Recent Progress in Hormone Research* **15**, 380–387.

Eisenberg, J. F. (1966). The social organization of mammals. *Handbuch der Zoologie Berlin,* **10**, 1–92.

Eisenberg, J. F. and Lockhart, M. (1972). An ecological reconnaisance of Wilpattu National Park, Ceylon. *Smithsonian Contributions to Zoology,* **101**.

Ellerman, Sir John Reeves and Morrison-Scott, T. C. S. (1951). Checklist of Palaearctic and Indian mammals, British Museum Natural History, London.

Eloff, F. C. (1973). In: *The World's Cats, Vol. 1: Ecology and Conservation,* R. Eaton, ed., World Wildlife Safari.

Elton, C. (1931). Epidemics among sledge dogs in the Canadian Arctic and their relation to disease in the arctic fox. *Canadian Journal of Research,* **5**, 673–692.

Elton, C. (1942). *Voles, Mice and Lemmings: Problems in Population Dynamics,* Clarendon Press, Oxford, England.

Elton, C. (1949). Movements of arctic fox populations in the region of Baffin Bay and Smith Sound. *Polar Record,* **5**, 296–305.

Encke, W. (1964). Mähnenwölfe in Krefelder Tierpark. *Freunde Kölner Zoo,* **7**, 33–34.

Encke, W. (1970). Beobachtungen und Erfahrungen bei der Haltung und Zucht von Mähnenwölfen in Krefelder Tierpark. *Freunde Kölner Zoo,* **13**, 69–75.

Erickson, A. B. (1944). Helminths of Minnesota Canidae in relation to food habits, and a host list and key to the species reported from North America. *American Midland Naturalist,* **32**, 358–372.

Errington, P. L. (1933). Bobwhite winter survival in an area heavily populated with gray foxes. *Iowa State College Journal of Science,* **8**, 127–130.

Errington, P. L. (1935). Food habits of Mid-West foxes. *Journal of Mammalogy,* **16**, 192–200.

Errington, P. L. (1937). Food habits of Iowa red foxes during a drought summer. *Ecology,* **18**, 53–61.

Errington, P. L. (1937a). Food habits of the red fox in Iowa. *American Wildlife,* **26**, 5–6, 13.

Errington, P. L. and Berry, R. M. (1937). Tagging studies of red foxes. *Journal of Mammalogy,* **18**, 203–205.

Estes, R. D. and Goddard, J. (1967). Prey selection and hunting behavior of the African wild dog. *Journal of Wildlife Management,* **31**, 52–70.

Etheridge, R. (1916). The warrigal, or dingo, introduced or indigenous? *Memoirs Geological Survey, New South Wales, Ethnological Series* **2**, 43–54.

Ewer, R. F. (1956). The fossil carnivores of the Transvaal caves: Canidae. *Proceedings of Zoological Society of London,* **126**, 97–119.

Ewer, R. F. (1968). *Ethology of mammals*, Plenum Press, New York.

Ewer, R. F. (1973.) *The Carnivores,* Cornell University Press, New York.

Faester, K. (1943). Effect of the climatic amelioration of the past decade on the autumn change of coat in the arctic fox in Greenland. *Medd. om Grønland,* **142**, 1–18.

Failor, P. L. (1969). Calling the gray fox. *Pennsylvania Game News,* **40**, 15–19.

Fay, F. H. and Cade, T. J. (1959). An ecological analysis of the avifauna of St.

Lawrence Island, Alaska. *University of California Publications in Zoology,* **63**, 73–150.

Feigin, R. D., Lobes, L. A. Jr., Anderson, D., and Pickering, L. (1973). Human leptospirosis from immunized dogs. *Ann. Internal Medicine,* **79,** *777–7*85.

Feilden, H. W. (1877). On the mammals of north Greenland and Grinnell Land. *Zoologist,* **1,** 313–321.

Feinstein, R. N., Faulhaber, Joann T. and Howard, J. B. (1968). Acatalasemia and hypocatalasemia in the dog and the duck. *Proceedings Society of Experimental Biology and Medicine,* **127**, 1051–1054.

Feldman, B. M., and Carding, T. H. (1973). Free-roaming urban pets. *Health Service Report,* **88,** 956–962.

Ferris, G. F. and Nuttall, G.H.F. (1918). Anoplura. *Canadian Arctic Expedition,* 1913–1918, **3**, 11–12.

Fetherston, K. (1947). Geographic variation in the incidence of occurrence of the blue phase of the arctic fox in Canada. *Canadian Field-Naturalist, 15-18.* **61**, 66–73.

Fichter, E. (1950).Watching coyotes. Journal of Mammalogy, **31**, 66-73.

Fiennes, R. and Fiennes, A. (1971). *The Natural History of Dogs,* Bonanza Books, Crown Publishing, Inc., New York.

Fisher, H. I. (1951). Notes on the red fox (*Vulpes fulva*) in Missouri. *Journal of Mammalogy,* **32**, 296–299.

Ford, L. (1965). Leucocytes culture and chromosome preparations from adult dog blood. *Stain Technology,* **40**, 317–320.

Ford, L. (1969). Identification and chromomeric interpretation of paphytene bivalents from *Canis familiaris. Canadian Journal of Genetics and Cytology,* **11**, 389–401.

Formozov, A. N. (1946). Snow cover as an integral factor of the environment and its importance in the ecology of mammals and birds. *Materials for Fauna and Flora of the U.S.S.R. New Ser. Zoology,* **5**, 1–152. (Translation Boreal Institute, University of Alberta, Edmonton. Occasional Paper No. 1. W. Prychodoko and W. O. Pruitt, Jr., eds.)

Foster, B. (1955). Arctic fox, *Alopex lagopus,* at Churchill, Manitoba. *Ontario Field Biologist,* **9**, 17–19.

Fox, M. W. (1969a). Ontogeny of prey-killing in *Canidae. Behaviour,* **35**, 259–272.

Fox, M. W. (1969b). The anatomy of aggression and its ritualization in Canidae: A developmental and comparative study. *Behaviour,* **35**, 242–258.

Fox, M. W. (1970a). A comparative study of the development of facial expressions in Canids, wolf, coyote and foxes. *Behaviour,* **36**, 49–73.

Fox M. W. (1970b). Neurobehavioral development and the genotype-environment interaction. *Quarterly Review of Biology,* **45,** 131–147.

Fox, M. W. (1971a.) *Integrative Development of Brain and Behavior in the Dog,* University of Chicago Press, Chicago, Illinois.

Fox, M. W. (1971b). *The Behaviour of Wolves, Dogs and Related Canids,* Jonathan Cape, London; Harper and Row, New York, 1972.

Fox, M. W. (1971c). Socio-infantile and socio-sexual signals in canids: A comparative and ontogenetic study. *Zeitschrift für Tierpsychologie,* **28**, 185–210.

Fox, M. W. (1971d). Ontogeny of socio-infantile and socio-sexual signals in canids. *Zeitschrift für Tierpsychologie,* **28**, 185 —210.

Fox, M. W. (1971e). Possible examples of high order behavior in wolves. *Journal of Mammalogy,* **52**, 640–641.

Fox, M. W. (1972). Socio-ecological implications of individual differences in wolf litters: a developmental and evolutionary perspective. *Behaviour,* **41**, 298–313.

Fox, M. W. (1973). Social dynamics of three captive wolf packs. *Behaviour,* **47**, 290–301.

Fox, M. W. and Andrews, R. V. (1973). Physiological and biochemical correlates of individual differences in behavior of wolf cubs. *Behaviour,* **46**, 129–140.

Fox, M. W., and Cohen, J. R. (1974). Canid communication. In: *How Animals Communicate,* T. A. Sebeok, ed., Indiana Univ. Press, Bloomington.

Fox, M. W., Lockwood, R. and Shideler, R. (1974). Introduction studies in captive wolf packs. *Zeitschrift für Tierpsychologie,* in press.

Frame, G. W. (1970.) On the hunt with the wild dogs of Africa. *Zoology,* **67**, 33–38.

Franti, C. E., and Kraus, J. F. (1974). Aspects of pet ownership in Yolo County, California. *JAVMA,* **164**, 166–171.

Frechkop, Serge (1959). De la position systematique due genre *Nyctereutes. Bulletin Inst. Roy. Sci. Nat. Belg.,* **35**, 1–20.

Fredrickson, L. E. and Thomas, L. (1965). Relationship of fox rabies to caves. *Public Health Reports,* **80**, 495–500.

Freedman, E. (1971). Personal communication. Editor, *Cyclone Magazine*.

Freuchen, P. (1915). Report on the First Thule Expedition. (Scientific work.) *Medd. om Grønland,* **51**.

Freuchen, P. (1935). Field notes and biological observations. Part II. Report of the mammals collected by the Fifth Thule Expedition to Arctic North America. Zoology I; by M. Degerbøl and P. Freuchen. *Fifth Thule Expedition,* 1921–24, **2**, 1–278.

Freud, Sigmund (1933). *New Introductory Lectures on Psycho-Analysis,* Norton, New York.

Friend, M. and Linhart, S. B. (1964). Use of the eye lens as an indicator of age in the red fox. *New York Fish and Game Journal,* **11**, 58–66.

Funakoshi, S. and Deutsch, H. F. (1971). Animal carbonic anhydrase isozymes. *Comparative Biochemistry and Physiology,* **39**, 489–498.

Funkenstein, D. H. (1958). The physiology of fear and anger. In: *Psychopathology,* C. F. Reed, I. E. Alexander and S. S. Tomkins, eds., Harvard University Press, Cambridge, Massachusetts.

Gander, F. F. (1966). Friendly foxes. *Pacific Discovery,* **19**, 28–31.

Gangloff, L. (1972). Breeding fennec foxes *Fennecus zerda* at Strasbourg Zoo. *International Zoo Yearbook,* **12**, 115–116.

Gauthier-Pilters, H. (1962). Beobachtungen an Feneks *(Fennecus zerda* Zimm). *Zeitschrift für Tierpsychologie,* **19**, 440–464.

Gauthier-Pilters, H. (1966). Einige Beobachtungen über das Spielverhalten beim Fenek (*Fennecus zerda Zimm.). Zeitschrift für Säugetierkunde,* **31**, 337–350.

Gauthier-Pilters, H. (1967). The fennec. *African Wildlife,* **21**, 117–125.

Gier, H. T. (1948). Rabies in the wild. *Journal of Wildlife Management,* **12**, 142–153.

Gier, H. T. (1968). Coyotes in Kansas. *Kansas Agricultural Experiment Station Bulletin,* **393**.

Giles, E. (1960). Multivariate analysis of Pleistocene and recent coyotes (*Canis latrans*) from California. *University of California Publications in Geological Science,* **36**, 369–390.

Giles, R. H., Jr. (1960). The free-running dog. *Virginia Wildlife,* **21**, 6–7.

Gills, E.D. (1971). The far-reaching effects of Quaternary sea level changes on the flat continent of Australia. *Royal Society of Victoria,* **84**, 189–205.

Gilmore, R. M. (1946). Mammals in archaeological collections from southwestern Pennsylvania. *Journal of Mammalogy,* **27**, 227–235.

Gilmore, R. M. (1949). The identification and value of mammal bones from archeological excavations. *Journal of Mammalogy,* **30**, 163–169.

Gilsvick, R. (1970). Killer pets that waste our whitetails. *Sports Afield,* **163**, 44–45, 82–83.

Gipson, P. S. (1972). The taxonomy, reproductive biology, food habits and range of wild *Canis* (Canidae) in Arkansas. Ph.D. Thesis, The University of Arkansas.

Godenhjelm, U. (1891). *Minnen från vargåren i Åbo län 1880–1882,* Helsinki.

Golani, I. (1966). Observations on the behaviour of the jackal *Canis aureus L.* in captivity. *Israel Journal of Zoology,* **15**, 28.

Golani, I. (1973). Non-metric analysis of behavioural interaction sequences in captive jackals (*Canis aureus L.). Behaviour,* **44,** 1–2, 89–112.

Golani, I. (in press). Motor homeostatic mechanisms in mammalian display. In: *Perspectives in Ethology,* Vol. II, P. Klopfez and P. P. G. Bateson, eds., Plenum Press, New York.

Golani, I., Zeidel, S. and Eshkol, N. (1969). Eshkol-Wachman movement notation — the golden jackal. The Movement Notation Society Press. Available from the authors, Department of Zoology, Tel-Aviv University, Israel.

Golani, I. and Mendelssohn, H. (1971). Sequences of precopulatory behaviour of the jackal (*Canis aureus L.*). *Behaviour,* **38,** 1–2, 169–192.

Goldman, E. A. (1920). Mammals of Panama. *Smithsonian Miscellaneous Collection,* **69**, 1–309.

Goldman, E. A. (1938). List of the gray foxes of Mexico. *Journal Washington Academy of Science,* **28**, 494–498.

Goldman, E. A. (1944). In: *The Wolves of North America,* S. P. Young and E. A. Goldman, eds., Wildlife Institute, Washington, D.C.

Gould, R. A. (1970). Journey to Pulykara. *Natural History,* **79**, 57–66.

Grafton, R. N. (1965). Food of the black-backed jackal: A preliminary report. *Zoologica Africana,* **1**, 41–54.

Gratzer, W. B. and Allison, A. C. (1960). Multiple haemoglobins. *Biological Review,* **35**, 459–506.

Gray, A. P. (1954.) Mammalian hybrids. A check-list with bibliography. *Technical Communication,* **10**, Comm. Agriculture Bureaux, Farnham Royal, England.

Gray, A. P. (1966). Mammalian hybrids. *Supplemental Bibliography to Technical Communication,* **10**, Comm. Bureaux of Animal Breeding and Genetics, Edinburgh.

Gregory, J. W. (1906).*The Dead Heart of Australia,* London.

Grigor'ev, N. D. and Popov, V. A. (1952). Method for determining the age of the arctic fox. *Akad. Nauk SSSR. Kazanskii filial. Isvestiia. Seria biologicheskik i sel'skolkhoziaistvennyki,* **3**, 207–215.

Grinnell, J., Dixon, J. and Linsdale, J. M. (1937). *Fur-bearing Mammals of California,* 2 vols., University of California Press, Berkeley, California.

Gross, A. O. (1931). Snowy owl migration. *Auk,* **48**, 501–511.

Gross, A. O. (1947). Cyclic invasions of the snowy owl and the migration of 1945–1946. *Auk,* **64**, 584–601.

Grzimek, B. (1970). *Among Animals of Africa,* Stein and Day, New York.
Gubser, N. (1965). *The Nunamiut Eskimos: Hunters of Caribou,* Yale University Press, New Haven and London.
Guggisberg, C.A.W. (1970). *Man and Wildlife,* Arco Publishing, New York.
Guilday, J. E. (1962). Supernumerary molars of *Otocyon. Journal of Mammalogy,* **43**, 455–462.
Gunderson, H. L. (1961). A self-trapped gray fox. *Journal of Mammalogy,* **42**, 270.
Gunderson, H. L., Breckenridge, W. J. and Jarosz, J. A. (1955). Mammal observations at Lower Back River, NWT, Canada. *Journal of Mammalogy,* **36**, 254–259.
Gusdon, J. P., Leake, N. H., Van Dyke, A. H. and Atkins, W. (1970). Immunochemical comparison of human placental lactogen and placental proteins from other species. *American Journal of Obstetrics and Gynecology,* **107**, 441–444.
Gustavsson, I. (1964). The chromosomes of the dog. *Hereditas,* **51**, 187–189.
Gustavsson, I. (1964). Karyotype of the fox. *Nature,* **201**, 950–951.
Gustavsson, I. and Sundt, C. O. (1965). Chromosome complex of the family of Canidae. *Hereditas,* **54**, 249–254.
Gustavsson, I. and Sundt, C. O. (1967). Chromosome elimination in the evolution of the silver fox. *Journal of Heredity,* **58**, 75–78.
Haber, G. Personal communication. Mt. McKinley National Park and Zoology Department, University of British Columbia.
Haberlein, P. J. and Barnhart, M. I. (1968). Canine plasminogen: Purification and a demonstration of multimolecular forms. *Biochimica et Biophysica Acta,* **168**, 195–206.
Haberman, R. F., Herman, C. M. and Williams, F. P., Jr. (1958). Distemper in raccoons and foxes suspected of having rabies. *Journal of the American Veterinary Medical Association,* **132**, 31–35.
Haffer, J. (1969). Speciation in Amazonian forest birds. *Science,* **165**, 131–137.
Haglund, B. (1968). De stora rovdjurens vintervanor II. *Viltrevy,* **5**, 213–361.
Hall, A. M. (1971). Ecology of beaver and selection of prey by wolves in central Ontario. M.Sc. Thesis, University of Toronto, Toronto, Ontario.
Hall, E. R. (1955). *Handbook of Mammals of Kansas,* Museum of Natural History, University of Kansas, Miscellaneous Publication 7.
Hall, E. R. and Kelson, K. R. (1959). *The Mammals of North America,* Vol. II, The Ronald Press, New York.
Haller, F. D. (1951). Fox facts. *Outdoor Indiana,* **18**, 2–19.
Hamilton, W. J., Jr., Hosley, N. W. and MacGregor, A. E. (1937). Late summer and early fall foods of the red fox in central Massachusetts. *Journal of Mammalogy,* **18**, 366–367.
Hanson, W. R. (1968). Estimating the number of animals: A rapid method for unidentified individuals. *Science,* **162**, 675–676.
Hardy, R. (1945). The influence of types of soil upon the local distribution of some mammals in southwestern Utah. *Ecological Monographs,* **15**, 71–108.
Harmon, E., Jr. (1971). Personal communication. Superintendent, Baltimore City Animal Shelter.
Hatfield, D. M. (1939). Winter food habits of foxes in Minnesota. *Journal of Mammalogy,* **20**, 202–206.
Hawkins, R. E., Klimstra, W. D., and Antry, D. C. (1970). Significant mortality factors of deer on Crab Orchard National Wildlife Refuge. *Transactions Illinois State Academy of Science,* **63,** 202–206.

Hediger, H. (1950). *Wild Animals in Captivity,* Butterworths, London; Dover, New York, 1964.
Hediger, H. (1955). *The Psychology and Behaviour of Animals in Zoos and Circuses,* Butterworths, London; Dover, New York, 1968.
Heit, W. S. (1944). Food habits of red foxes of the Maryland marshes. *Journal of Mammalogy,* **25**, 55–58.
Hendrichs, H. (1972). Beobachtungen und Untersuchungen zur Okologie und Ethologie, insbesondere zur sozialen Organisatun ostafrikanisher Säugetiere. *Zeitschrift für Tierpsychologie*, **30**, 146–189.
Hershkovitz, P. (1957). A synopsis of the wild dogs of Columbia. Novedades Colombianas, *Museum History Nat. University del Cauca,* **3**, 157–161.
Hershkovitz, P. (1958). A geographical classification of neotropical mammals. *Fieldiana Zoology,* **36**, 579–620.
Hershkovitz, P. (1969). The recent mammals of the neotropical region: a zoogeographic and ecological review. *Quarterly Review of Biology,* **44**, 1–70.
Hesse, R., Allee, W. C. and Schmidt, K. (1947). *Ecological Animal Geography,* John Wiley, New York.
Hilderbrand, M. (1952). An analysis of body proportions in the Canidae. *American Journal of Anatomy,* **90**, 217–256.
Hilderbrand, M. (1952a). The integument in Canidae. *Journal of Mammalogy,* **33**, 419–428.
Hilderbrand, M. (1954). Comparative morphology of the body skeleton in recent Canidae. University of California Publications in Zoology, **52**, *399–470*.
Hock, R. J. (1952). Golden eagle versus red fox: predation or play? Condor, **54**, **218—319**.
Hoffer, A. (1957). Epinephrine derivatives as potential schizophrenic factors. *Journal of Clinical and Experimental Psychopathology*, **18,** 27.
Hoffman, R. A. and Kirkpatrick, C. M. (1954). Red fox weights and reproduction in Tippecanoe County, Indiana. *Journal of Mammalogy*, **35**, *504–509*.
Holcomb, L. C. (1965). Large litter size of red fox. *Journal of Mammalogy,* **46**, *530*.
Hornadage, B. (1972). *If It Moves, Shoot It*, Review Publ. Pty. Ltd., Dubbo, Australia.
Housse, R. (1949). Las zorras de Chile o Chacales Americanos. *Review Universitaria*, **34**, *33–56* (Anal. Acad. Chilena Cienc. Nat. No. 14).
Hsu, T. C. and Arrighi, F. E. (1966). Karyotypes of 13 carnivores. *Mammalian Chromosomes Newsletter*, **21**, *155–159*.
Hsu, T. C. and Benirschke, K. (1967). *An Atlas of Mammalian Chromosomes*, Vol. 1, Folio 20.
Hsu, T. C. and Benirschke, K. (1967). *An Atlas of Mammalian Chromosomes*, Vol. 1, Folio 21.
Hsu, T. C. and Benirschke, K. (1969). *An Atlas of Mammalian Chromosomes*, Vol. 3, Folio 122.
Hsu, T. C. and Benirschke, K. (1970). *An Atlas of Mammalian Chromosomes*, Vol. 4, Folio 178.
Hsu, T. C. and Benirschke, K. (1970). *An Atlas of Mammalian Chromosomes*, Vol. 4, Folio 179.
Hsu, T. C. and Benirschke, K. (1970). *An Atlas of Mammalian Chromosomes*, Vol.4, Folio 180.
Huey, R. B. (1969). Winter diet of the Peruvian desert fox. *Ecology,* **50**, 1089–1091.

Hungerford, D. A. and Snyder, R. L. (1966). Chromosomes of a European wolf (*Canis lupus*) and of a bactrian camel (*Camelus bactrianus*). Mammalian Chromosomes Newsletter, **20**, 72.

Hunter, G. R. (1971). Hunting Georgia's killer devil dogs. *Guns and Ammo,* **15**, 32–35.

Ihrig, J., Kleinerman, J. and Rynbrandt, D. J. (1971). Serum antitrypsins in animals. *American Review of Respiratory Diseases,* **103**, 377–389.

Ingles, L. G. (1965). *Mammals of the Pacific States,* Stanford University Press, Stanford, California.

Iredale, T. (1947). The scientific name of the dingo. *Proceedings Royal Zoological Society, N.S.W.,* **35–6.**

Irving, L. (1953). The naming of birds by Nunamiut Eskimo. *Arctic,* **6**, 35–43.

Irving, L. (1958). On the naming of birds by Eskimos. *Anthropological Papers of the University of Alaska,* **6**, 61–77.

Irving, L. (1960). Birds of Anaktuvuk Pass, Kobuk, and Old Crow. *U.S. National Museum Bulletin,* **217**.

Jenness, D. (1957). *Dawn in Arctic Alaska,* University of Minnesota Press, Minneapolis, Minnesota.

Jenness, R., Erickson, A. W. and Craighead, J. J. (1972). Some comparative aspects of milk from four species of bears. *Journal of Mammalogy,* **53**, 34–47.

Jennings, J. N. (1959). The submarine topography of Bass Strait. *Proceedings Royal Society of Victoria,* **71**, 49–72.

Jennings, J. N. (1971). Sea level changes and land links. In: *Aboriginal Man and Environment in Australia,* D. J. Mulvaney and J. Golson, eds., Australia National University Press, Canberra.

Jennings, W. L., Schneider, N. J., Lewis, A. L. and Scatterday, J. E. (1960). Fox rabies in Florida. *Journal of Wildlife Management,* **24**, 171–179.

Johansson, I. (1960). Inheritance of the color phases in ranch bred blue foxes. *Hereditas,* **46**, 753–766.

Johnsen, S. (1929). Rovdyr- og rovfugistatistikken i Norge. *Bergens Museums Årbok, 2,* 1–118.

Johnsgard, P. A. (1965). *Handbook of Waterfowl Behavior,* Cornell University Press, Ithaca, New York.

Johnson, D. H., Bryant, M. D. and Miller, A. H. (1948). Vertebrate animals of the Providence Mountains area of California. *University of California Publications in Zoology,* **48**, 221–375.

Johnson, H. N. (1945). Fox rabies. *Journal Medical Association of Alabama,* **14**, 268–271.

Johnson, J. S. and Vaughan, J. H. (1967). Canine immunoglobulins. I. Evidence for six immunoglobulin classes. *Journal of Immunology,* **98**, 923–940.

Johnson, S. T. (1946). Breeding blue foxes for profit. *Black Fox Magazine,* **30**, 24.

Jolicoeur, P. (1959). Multivariate geographical variation in the wolf *Canis lupus* L. *Evolution,* **13**, 283–299.

Jones, F. W. (1921). The status of the dingo. *Transactions Royal Society of South Australia,* **45**, 254–263.

Jones, F. W. (1925). *The Mammals of South Australia,* Pt. 3, Government Printer, Adelaide, Australia.

Jones, R. T., Brimhall, B. and Duerst, M. (1972). In: *Atlas of Protein Sequence and Structure,* Vol. 5, M. Dayhoff, ed., National Biomedical Research Foundation, Washington, D.C.

Jordan, P. A. (1969). Leadership and group response in timber wolves: a socio-ecological analysis. Paper presented at the 1969 annual meeting of the Rural Sociological Society.

Jordan, P. A., Shelton, P. C. and Allen, D. L. (1967). Numbers, turnover and social structure of the Isle Royale wolf population. *American Zoologist,* **7**, 233–252.

Jordan, P. A., Botkin, D. B. and Wolfe, M. L. (1971). Biomass dynamics in a moose population. *Ecology,* **52**, 147–152.

Kahn, P. M., Los, W.R.T., v.d. Does, J. A. and Epstein, R. B. (1973). Isoenzyme markers in dog blood cells. *Transplantation,* **15**, 624–628.

Kalashnikov, M. K. (1963). On "crop service" for the fur industry. *Sovetskaia Arktika,* **11**, 83–84.

Kaminski, M. and Balbierz, H. (1965). Serum proteins in Canidae: species, race and individual differences. In: *Proceedings IXth European Animal Blood Group Conference, Blood Groups of Animals,* Publishing House of the Czechoslovak Academy of Science, Prague, pp. 337–341.

Kantorovich, R. A. (1956). Etiology of a rabies-like disease in arctic animals, 1; biological properties of the virus of this disease. *Voprosy virusologii,* **1**, 32–37.

Kantorovich, R. A. (1957). Etiology of a rabies-like disease in arctic animals, 3; serological and antigenic properties of its virus. *Voprosy virusologii,* **2**, 208–210.

Karpuleon, F. (1957). Food habits of Wisconsin foxes. *Journal of Mammalogy,* **39**, 591–593.

Kauri, H. (1957). Hundid Eestis. *Review Est. Lit. Sci.,* **8**, 310–315.

Keeler, C. E. (1942). The association of the black (nonagouti) gene with behavior. *Journal of Heredity,* **33**, 371–384.

Keeler, C. E. (1947). Modification of brain and endocrine glands, as an explanation of altered behavior trends, in coat-character mutant strains of the Norway rat. *Journal Tennessee Academy Science,* 202–209.

Keeler, C. E. (1970). Melanin, adrenalin and the legacy of fear. *Journal of Heredity,* **61**, 81–88.

Keeler, C. E. and Moore, L. (1961). Psychosomatic synthesis of behavior trends in the taming of mink. *Bulletin Georgia Academy of Science,* **19**, 66–74.

Keeler, C. E., Asteinza, J. and Fromm, E. (1964). Psychosomatics of fear in foxes. *Bulletin Georgia Academy of Science,* **22**, 64–69.

Keeler, C. E. and Fromm, E. (1965). Genes, drugs and behavior in foxes. *Journal of Heredity,* **56**, 288–291.

Keeler, C. E., MacKinnon, I. and Fromm, E. (1966). The effects of the drug Oxazepam upon the red fox. *Bulletin Georgia Academy of Science,* **24**, 125–130.

Keeler, C. E. and Mellinger, T. (1966). Wild foxes for testing psychotic symptoms before and after tranquilizer treatment. *Excerpta Medica,* **117**, 252.

Keeler, C. E., Ridgway, S., Lipscomb, L. and Fromm, E. (1968). The genetics of adrenal size and tameness in color phase foxes. *Journal of Heredity,* **59**, 82–84.

Keith, L. B. (1963). *Wildlife's Ten-Year Cycle,* The University of Wisconsin Press, Madison, Wisconsin.

Kendall, M. G. (1951). *The Advanced Theory of Statistics,* Vol. II, 3rd Ed., Hafner House, New York.

Kilgore, D. L. (1969). An ecological study of the swift fox (*Vulpes velox*) in the Oklahoma Panhandle. *The American Midland Naturalist,* **81**, 512–534.

Kirpichnikov, A. A. (1937). On the biology of the arctic fox on the southwest coast of Taimur. (Translation, J. D. Jackson, 1941, for the Bureau of Animal Populations, Oxford University, England. Translation 101, F1051 A, 1–16.).

Kitchner, S. L. (1971). Observations on the breeding of the bush dog at Lincoln Park Zoo, Chicago. *International Zoo Yearbook,* **11**, 99–101.

Kleiman, D. (1966). Scent marking in Canidae. *Symposia of the Zoological Society of London,* **18**, 167–177.

Kleiman, D. (1967). Some aspects of social behavior in the Canidae. *American Zoologist,* **7**, 365–372.

Kleiman, D. G. and Eisenberg, J. F. (1973). Comparison of canid and felid social systems from an evolutionary perspective. *Animal Behavior,* 21, 637–659.

Knezecić, M. and Knezević, R. (1956). *Vuk Zivot, Stetnost i Tamanjenje,* Sarajevo.

Kneževic, M. and Kneževic, R. (1956). *Vuk, život, štetnost i tamanjenje,* Savajevo.

Koenig, L. (1970). Zur Fortpflanzung und Jugendentwicklung des Wüstenfuchses (*Fennecus zerda* Zimm. 1780). *Zeitschrift für Tierpsychologie,* **27**, 205–246.

Koepcke, H. W. and Koepcke, M. (1952). Sobre el proceso de transformación de la materia orgánica en las playas arenosas marinas del Perú. *Publicaciones Museo de Historia Natural "Javier Prado", Series A (Zoologia),* **8**, 1–24.

Kolenosky, G. B. (1971). Hybridization between wolf and coyote. *Journal of Mammalogy,* **52**, 446–449.

Kolenosky, G. B. (1972). Wolf predation on wintering deer in eastcentral Ontario. *Journal of Wildlife Management,* **36**, 357–369.

Korschgen, L. J. (1959). Food habits of the red fox in Missouri. *Journal of Wildlife Management,* **23**, 168–176.

Kraglievich, L. (1928). Contribución al conocimiento de los grandes cánidos extinguidos de Sud América. *Anales Sociedad Cientifica Argentina,* **106**, 25–66.

Kraglievich, L. (1930). Craneometría y clasificatión de los cánidos sudamericanos especialmente los angentinos actuales y fósiles. *Physis,* **10**, 35–73.

Kraglievich, J. L. (1952). Un cánido del eocuartario de Mar del Plata y sus relaciones con otras formas brasileñas y norteamericanas. *Review Museo Munic. Trad. Mar del Plata,* **1**, 53–70.

Krefft, G. (1865). Notes on the fossil mammals of Australia. *Geological Magazine,* **2**, 572–574.

Krefft, G. (1871). *Mammals of Australia* (unfinished), Government Printers, Sydney. Folio.

Krieg, H. (1948). *Zwischen Anden und Atlantik. Reisen eines Biologen in Sudamerika,* C. Hanser, Munich. (See also *De Zoologische Garten,* **12**, 257–269.)

Krohn, K., Mero, M., Oksanen, A. and Sandholm, M. (1971). Immunologic observations in canine interstitial nephritis. *American Journal of Pathology,* **65**, 157–172.

Kruuk, H. (1971). Review of Fox (1971b). *Nature,* **234**, 288.

Kruuk, H. (1972a). *The Spotted Hyena,* University of Chicago Press, Chicago.

Kruuk, H. (1972b). Surplus killing by carnivores. *Journal of Zoology London,* **166**, 233–244.

Kruuk, H. and Turner, M. (1967). Comparative notes on predation by lion, leopard, cheetah, and wild dog in the Serengeti area, East Africa. *Mammalia,* **31**, 1–27.

Kühme, W. (1965a). Communal food distribution and division of labour in African hunting dogs. *Nature* (London), **205**, 443–444.

Kühme, W. (1965b). Freilandstudien zur Sociologie des Hyänenhundes *Lycaon pictus lupinus* Thomas (1902). *Zeitschrift für Tierpsychologie,* **22**, 495–541.

Kummer, H. (1971a). *Primate Societies,* Aldine-Atherton, Chicago.

Kummer, H. (1971b). Spacing mechanisms in social behavior. In: *Man and Beast: Comparative Social Behavior,* Smithsonian Annual III, J. F. Eisenberg, W. S. Dillon and S. Dillon Ripley, eds., Smithsonian Institution Press, Washington, D.C., pp. 221–234.

Kuyt, E. (1971). Food studies on barren-ground caribou range in the Northwest Territories. *Canadian Wildlife Service Research Progress Notes No.* **23**.

Lagerspetz, K. and Talo, S. (1967). Maturation of aggressive behavior in young mice. *Reports Institute of Psychology, University of Turku,* **28**.

Lande, O. (1958). Chromosome number in the silver fox *(Vulpes vulpes* Des.). *Nature,* **181**, 1353–1354.

Lande, O. (1960). Chromosome number in the blue fox (*Alopex lagopus* L.). *Nature,* **188**, 170.

Lane, H. H. (1948). Survey of the fossil vertebrates of Kansas, Part V: The Mammals. *Transactions Kansas Academy of Science,* **50**, 273–314.

Langguth, A. (1969). Die südamerikanischen Canidae unter besonderer Berücksichtigung des Mähnenwolfes *Chrysocyon brachyurus* Illiger. *Zeitschrift füer Wissenschaftliche Zoologie,* **179**, 1–188.

Large, T. (1971). The plight — and the threat — of man's best friend. *Baltimore Magazine* (August), 19–21, 56.

Larnach, S. L. and Macintosh, N.W.G. (1966). The craniology of the Aborigines of coastal New South Wales. *Oceania Monograph No.* **13**, Sydney.

Larnach, S. L. and Macintosh, N.W.G. (1970). The craniology of the Aborigines of Queensland. *Oceania Monograph No.* **15**, Sydney.

Latham, R. M. (1950). The food of predaceous animals in northeastern United States. *Pennsylvania Game Commission Bulletin.*

Latham, R. M. (1951). The ecology and economics of predator management. *Final Report, P–R Project 36–R,* Pennsylvania Game Commission, Harrisburg, Pa.

Latham, R. M. (1952). The fox as a factor in the control of weasel populations. *Journal of Wildlife Management,* **16**, 516–517.

Laughrin, L. (1971). Preliminary account of the island fox. *P–R Completion Report W–54–R–3,* California Department Fish and Game, Sacramento, California.

Lavrov, N. P. (1932). The arctic fox. (Translation, J. D. Jackson, 1940, for the Bureau of Animal Populations, Oxford University, England. Translation 18.F. 1079 A, 1–92.

van Lawick-Goodall, H. and J. (1971). *The Innocent Killers,* Houghton-Mifflin, Boston.

Lawrence, B. and Bossert, W. H. (1967). Multiple character analysis of *Canis lupus, latrans* and *familiaris*, with a discussion of the relationships of *Canis niger. American Zoologist,* **7**, 223–232.

Lawrence, B. and Bossert, W. H. (1969). The cranial evidence for hybridization in New England *Canis. Breviora, Museum of Comparative Zoology, Harvard University,* **330**, 1–13.

Layne, J. N. (1958). Reproductive characteristics of the gray fox in southern Illinois. *Journal of Wildlife Management,* **22**, 157–163.

Layne, J. N. and McKeon, W. H. (1956). Some notes on the development of the red fox fetus. *New York Fish and Game Journal,* **3**, 120–128.

Layne, J. N. and McKeon, W. H. (1956a). Some aspects of red fox and gray fox reproduction in New York. *New York Fish and Game Journal,* **3**, 44–74.

Layne, J. N. and McKeon, W. H. (1956b). Notes on red fox and gray fox den sites in New York. *New York Fish and Game Journal,* **3**, 248–249.

Leach, H. R. and Fisk, L. O. (1972). At the crossroads: A report on California's endangered and rare fish and wildlife. California Department of Fish and Game, Sacramento, Calif.

LeCrone, C. N. (1970). Absence of special fetal hemoglobin in beagle dogs. *Blood,* **34**, 451–452.

Lembley, W. I. and Lucas, F. A. (1902). Blue fox trapping on the Pribilof Islands. *Science,* **16**, 216–218.

Lemke, C. W. and Thompson, D. R. (1960). Evaluation of fox population index. *Journal of Wildlife Management,* **24**, 406–412.

Lemke, C. W., Ables, E. C. and Gates, J. (1967). Wisconsin fox population considerations. Paper presented at Midwest Wildlife Conference.

Leone, C. A. and Anthony, R. L. (1966). Serum esterases among registered breeds of dogs as revealed by immunoelectrophoretic comparisons. *Comparative Biochemistry and Physiology,* **18**, 359–368.

Leone, C. A. and Wiens, A. L. (1956). Comparative serology of carnivores. *Journal of Mammalogy,* **37**, 11–23.

Leopold, A. (1931). *Game Survey of the North Central States,* Madison, Wisconsin.

Leopold, A. S. (1959). *Wildlife of Mexico: the Game Birds and Mammals,* University of California Press, Berkeley, Calif.

Lever, R.J.A.W. (1959). The diet of the fox since myxomatosis. *Journal Animal Ecology,* **28**, 359–375.

Levine, S. (1957). Infantile experience and the maturation of the pituitary adrenal axis. *Science,* **126**, 1347.

Lewis, H. (1942). Fourth census of non-passerine birds in the bird sanctuaries of the north shore of the Gulf of St. Lawrence. *Canadian Field-Naturalist,* **56**, 5–8.

Leyhausen, P. (1965). The communal organization of solitary mammals. *Symposia of the Zoological Society of London,* **14**, 249–263.

Lief, H. I. (1967). Anxiety reaction. In: *Comprehensive Textbook of Psychiatry,* Freedman and Kaplan, eds., Williams and Wilkins, Baltimore, Maryland.

Lin, C. C. (1972). Personal communication.

Linares, O. J. (1967). El perro de monte *Speothos venaticus* (Lund) en el norte de Venezuela (Canidae). *Sociedad de Ciencias naturales "La Salle", Caracas Memoria,* **27**, 83–86.

Linhart, S. B. (1959). Sex ratios of the red fox and gray fox in New York. *New York Fish and Game Journal,* **6**, 116–117.

Linhart, S. B. (1968). Dentition and pelage in the juvenile red fox (*Vulpes vulpes*). *Journal of Mammalogy,* **49**, 526–528.

Lissak, K. and Endrosci, E. (1967). Neuroendocrine regulation of adaptivity. Academy of Sciences, Budapest.

Lombaard, L. J. (1971). Age determination and growth curves in the black-backed jackal, *Canis mesomelas* Schreber, 1775 (Carnivora: Canidae). *Annals Transvaal Museum,* **27**, 135–1690.

Longley, W. H. (1962). Movements of red fox. *Journal of Mammalogy,* **43**, 107.

Longman, H. A. (1928). Note on the dingo, the Indian wild dog and a Papuan dog. *Memoirs Queensland Museum,* **9**, 151–157.
Lönnberg, E. (1934). Bidrag till vargens historia i Sverige. *K. Svenska Vetenskapsakademiens skrifter i Naturskyddsärenden,* **26**, 1–33.
Lord, R. D. (1961a). A population study of the gray fox. *The American Midland Naturalist,* **66**, 87–109.
Lord, R. D. (1961b). The lens as an indicator of age in the gray fox. *Journal of Mammalogy,* **42**, 109–111.
Lorenz, K. Z. (1941). Vergliechende Bewegungsstudien an Anatinen. *Journal für Ornithologie,* **89**, 194–294.
Lorenz, K. (1954). *Man Meets Dog,* Methuen, London.
Lund, P. W. (1950). Memorias sobre a Palentologia Brasileira. Institute Nac Livro, Rio de Janeiro. (A Portuguese translation of several articles published in Danish in the *Kongl. Dansk Vidensk. Selsk. Naturv. Math. Afhandlinger,* between 1836 and 1846, edited by C. de Paula Couto.)
Luzhkov, A. D. (1960). Studies on helminths of the white arctic fox on the Yamal Peninsula. *Trudy Nauk Issled Inst. Sel'sk. Krain. Severa,* **8**, 52–54.
Luzhkov, A. D. (1961). A case of *Opisthorchis felineus* in the arctic fox on the Yamal Peninsula. *Medicina Parazitol. i. Parazitar Bolezni,* **30**, 361.
Macintosh, N.W.G. (1956). Trail of the dingo. *The Etruscan,* **5**, 8–12.
Macintosh, N.W.G. (1962). In: The archaeology of Mootwingee, N.S.W., F. D. McCarthy and N.W.G. Macintosh, eds., *Records Australian Museum,* **25**, 249–298.
Macintosh, N.W.G. (1964). A 3000 year old dingo from Shelter 6 (Fromm's Landing, South Australia). *Proceedings Royal Society of Victoria,* **77**, 498–507.
Macintosh, N.W.G. (1965a). The physical aspect of man in Australia. In: *Aboriginal Man in Australia,* R. and C. Berndt, eds., Angus and Robertson, Sydney.
Macintosh, N.W.G. (1965b). Dingo and horned anthropomorph in an aboriginal rock shelter. *Oceania,* **36**, 85–101.
Macintosh, N.W.G. (1967a). Fossil man in Australia. *Australian Journal of Science,* **30**, 86–90.
Macintosh, N.W.G. (1967b). Recent discoveries of early Australian Man. *Annals Australian College of Dental Surgeons,* **1**, 104–126.
Macintosh, N.W.G. (1969). A 3000 year old dingo from Shelter 6. *Proceedings Royal Society of Victoria,* **77**, 507–514.
Macintosh, N.W.G. (1971). Analysis of an aboriginal skeleton and a pierced tooth necklace from Lake Nitchie, Australia. *Anthropologie,* **9**, 49–62.
Macpherson, A. H. (1964). A northward range extension of the red fox in the eastern Canadian arctic. *Journal of Mammalogy,* **45**, 138–140.
Macpherson, A. H. (1969). The dynamics of Canadian arctic fox populations. *Canadian Wildlife Service Report Series* **8**, Queens Printer, Ottawa.
Makino, S. (1947). Notes on the chromosomes of four species of small animals (Chromosome studies in domestic animals, V.). *Journal Faculty of Science, Hokkaido University Seriers IV,* **9**, 347–357.
Makino, S. (1949). A review on the chromosomes of domestic mammals. *Japanese Journal Zootechnical Science,* **19**, 5–15.
Makino, S. (1957). *An atlas of the chromosome number in animals,* The Iowa State College Press, Ames, Iowa.
Makridin, V. P. (1959). Materiaĺ po biologii volka v tundrah Nenetskogo natsionaĺnogo okruga. *Zoologicheskii Zhurnal,* **38**, 1719–1728.

Malone, T. H. (1918). Spermatogenesis of the dog. *Transactions American Microscopical Society* **37**, 97–110.

Manery, T. F. Barlow, J. S. and Forbes, J. M. (1966). Electrolytes in tissues, red cells, and plasma of the polar bear and caribou. *Canadian Journal of Zoology,* **44**, 235–240.

Manniche, A. L. (1912). The terrestrial mammals and birds of northeast Greenland; biological observations by A. L. V. Manniche, 1910. *Medd. om Grønland,* **45**, 1–200.

Manning, T. H.(1943). Notes on the mammals of south and central west Baffin Island. *Journal of Mammalogy,* **24**, 47–59.

Mansueti, R. (1955). Case of the displaced fox. *Maryland Naturalist,* **25**, 3–8.

Marsh, D. B. (1938). The influx of the red fox and its color phases into the Barren Lands. *Canadian Field-Naturalist,* **52**, 60–61.

Marsh, H. A. (1962). Notes on gray fox behavior. *Journal of Mammalogy,* **43**, 278.

Marwin, M. (1959). *The Mammals of Karelia,* Petrozavodsk.

Masson, P. L. and Heremans, J. F. (1971). Lactoferrin in milk from different species. *Comparative Biochemistry and Physiology,* **39**, 119–129.

Matthew, W. D. (1930). The phylogeny of dogs. *Journal of Mammalogy,* **11**, 117–138.

Matthey, R. (1954). Chromosomes et systematique des Canides, *Mammalia,* **18**, 225–230.

McBride, G. (1964). A general theory of social organization and behaviour. *Faculty of Veterinary Science Papers, Queensland University, Brisbane,* **1**, 75–110.

McCabe, R. A. and Kozicky, E. L. (1972). A position on predator management. *Journal of Wildlife Management,* **36**, 382–394.

McCarley, H. (1962). The taxonomic status of wild *Canis* (Canidae) in the South Central United States. *The Southwestern Naturalist,* **7**, 227–235.

McCarthy, F. D. (1964). The archaeology of the Capertee Valley, New South Wales. *Records Australian Museum,* **26**, 197–246.

McCoy, F. (1882). *Canis dingo* Blumenbach. In: *Prodromus of the Palaeontology of Victoria,* Decade 7, Geological Survey of Victoria.

McEwen, E. H. (1951). Literature review of the arctic foxes. Unpublished M.A. Thesis, University of Toronto, Canada.

McEwen, E. H. and Scott, A. (1956). Pigmented areas in the uterus of the arctic fox *Alopex lagopus innuitus* Merriam. *Proceedings Zoological Society of London,* **128**, 347–348.

McGeachin, R. L., Pavord, W. M., Widner, D. N., and Prell, P. A. (1966). Comparative inhibition of mammalian amylases by goat antisera to hog pancreatic amylase. *Comparative Biochemistry and Physiology,* **18**, 767–772.

McKnight, T. (1964). Feral livestock in Anglo-America. *University of California Publications in Geography,* Vol. 16, University of California Press, Berkeley and Los Angeles.

Mech, L. D. (1966). The wolves of Isle Royale. *U.S. National Parks Service Fauna Series* **7**, Washington, D.C.

Mech, L. D. (1970). *The Wolf: the Ecology and Behavior of an Endangered Species,* Doubleday and Co., Garden City, New York.

Mech, L. D. and Frenzel, L. D., Jr. (1971). Ecological studies of the timber wolf in northeastern Minnesota. *USDA Forest Service Research Report NC-52.*

Mech, L. D. and Frenzel, L. D., Jr. (1971). The possible occurrence of the Great

Plains wolf in northeastern Minnesota. *USDA Forest Service Research Report NC-52,* 60–62.
Mellinger, T. J. (1968). Spectrofluorometric determination of homovanillic acid in urine. *American Journal of Clinical Pathology,* **49**, 200–206.
Mengel, R. M. (1971). A study of dog–coyote hybrids and implications concerning hybridization in *Canis. Journal of Mammalogy,* **52**, 316–336.
Merrilees, D. (1968). Man the destroyer: late Quaternary changes in the Australia marsupial fauna. *Journal Royal Society Western Australia,* **51**, 1–24.
van der Merwe, N. J. (1953a). The coyote and the black-backed jackal. *Flora and Fauna,* **3**, 45–51.
van der Merwe, N. J. (1953b). The Jackal. *Flora and Fauna,* **4**, 4–80.
Middendorf, A. Th. (1875). *Raise in der äusserstein Norden und Osten Sibiriens während der Jahre 1843 und 1844,* IV, 2, St. Petersburg (not seen, cited after Ognev, 1931).
Miller, G. S. (1912). *Catalogue of the Mammals of Western Europe,* British Museum Natural History, London.
Miller, G. S., Jr., and Kellogg, R. (1955). List of North American recent mammals. *U.S. National Museum Bulletin,* **205**.
Minouchi, O. (1928). The spermatogenesis of the dog, with special reference to meiosis. *Japanese Journal of Zoology,* **1**, 255–268.
Minouchi, O. (1929). On the spermatogenesis of the raccoon dog *(Nyctereutes viverrinus)* with special reference to the sex chromosomes. *Cytologia,* **1**, 88.
Mitchell, B. L., Shenton, J. B. and Uys, J.C.M. (1965). Predation on large mammals in the Kafue National Park, Zambia. *Zoologica Africana,* **1**, 279–318.
Mivart, St. G. (1890). *Dogs, Jackals, Wolves and Foxes: a Monograph of the Canidae,* London.
Montgomery, G. G. (1973). Communication in red fox dyads: a computer simulation study. Ph.D. Thesis, University of Minnesota.
Moore, W. and Lambert, P. D. (1963). The chromosomes of the beagle dog. *Journal of Heredity,* **54**, 273–276.
Moore, W. and Elder, R. L. (1965). Chromosomes of the fox. *Journal of Heredity,* **56**, 142–143.
Morris, R. C. (1937). Mange on wild dog. *Journal Bombay Natural History Society,* **39**, 615 and other miscellaneous notes in journal.
Morrison, D. F. (1967). *Multivariate Statistical Methods,* McGraw-Hill, New York.
Morrison, J. (1968). Hounds of Hell. *Georgia Fish and Game* **3**, 13–19.
Muckenhirn, N. A. and Eisenberg, J. F. (1973). Home ranges and predation of the Ceylon leopard. In: *The World's Cats, Vol. 1: Ecology and Conservation,* R. Eaton, ed., World Wildlife Safari.
Mulvaney, D. J. (1969). *The Prehistory of Australia,* London.
Munsterhjelm, L. (1946). *Pohjolan pedot,* Helsinki.
Murie, A. (1936). Following fox trails. *University of Michigan Museum of Zoology Miscellaneous Publication,* **32**.
Murie, A. (1940). Ecology of the coyote in the Yellowstone. *U.S. Department of Agriculture Fauna Series* **5**, U.S. Government Printing Office, Washington, D. C.
Murie, A. (1944). The wolves of Mount McKinley. *U.S. National Parks Fauna Series* **5**, Washington, D.C.
Murie, O. J. (1936). Notes on the mammals of St. Lawrence Island, Alaska. In: *Archeological Excavations at Kukulik, St. Lawrence Island,* O. W. Geist and F. G. Rainey, eds., *University of Alaska Miscellaneous Publication,* **2**, 1–391.

Murie, O. J. (1959). Fauna of the Aleutian Islands and Alaska Peninsula. *North American Fauna,* **61**, 1–406.

Myrberget, S. (1969). Ulvens status i Fennoskandia. *Naturen,* 160–172.

Nelson, A. L. (1933). A preliminary report on the winter food of Virginia foxes. *Journal of Mammalogy,* **14**, 40–43.

Nelson, A. L. and Handley, C. O. (1938). Behavior of gray foxes in raiding quail nests. *Journal of Wildlife Management,* **2**, 73–78.

Nelson, E. W. (1887). Mammals of northern Alaska. In: *Report upon Natural History Collections Made in Alaska 1877–*1881; *Arctic Series No.* **3**, U.S. Government Printing Office, Washington, D.C. Part 2, 227–293.

Nelson, E. W. (1930). *Wild Animals of North America,* National Geographic Society, Washington, D.C.

Nelson, R. K. (1969). *Hunters of the Northern Ice,* University of Chicago Press, Chicago.

Newnham, R. E. and Davidson, W. M. (1966). Comparative study of the karyotypes of several species of carnivora, including the giant panda *(Ailuropode melanoleuca)*. Cytogenetics, **5**, 152–163.

Newsome, A. E. (1971). Competition between wildlife and domestic stock. *Australian Veterinary Journal,* **47**, 577–586.

New York Conservation Department (1951). A study of fox control as a means of increasing pheasant abundance. *New York Conservation Department Research Series No.* **3**.

Nobrega, F. G., Maia, J.C.C., Colli, W. and Saldanha, P. H. (1970). Heterogeneity of erythrocyte glucose-6-phosphate dehydrogenase (G6PD, E.C.1.1.1.49) activity and electrophoretic patterns among representatives of different classes of vertebrates. *Comparative Biochemistry and Physiology,* **33**, 191–199.

Nordland, O. S. (1955). Occurrence of listeriosis in arctic mammals, with a note on its possible pathogenesis. *Canada Defense Laboratory Report,* **47**, 1–22.

North Dakota State Game and Fish Department (1949). The red fox in North Dakota. North Dakota State Game and Fish Department, Division of Federal Aid.

Novikov, G. A. (1956). *Carnivorous Mammals of the Fauna of the USSR,* Israel Program for Scientific Translations, Jerusalem (1962).

Nowak, R. M. (1970). Report on the red wolf. *Defenders of Wildlife News,* 82–94.

Nunez, E. A., Becker, D. V., Furth, E. D., Belshaw, B. E. and Scott, J. P. (1970). Breed differences and similarities in thyroid function in purebred dogs. *American Journal of Physiology,* **218**, 1337–1341.

Nunez, E. A., Belshaw, B. E. and Gershon, M. D. (1972). A fine structural study of the highly active thyroid follicular cell of the African basenji dog. *American Journal of Anatomy,* **133**, 463–482.

Nuttall, G.H.F. (1904). *Blood Immunity and Blood Relationships,* Cambridge University Press.

Nygren, W. E. (1950). Bolivar geosyncline of Northwestern South America. *Bulletin American Association Petroleum Geologists,* **34**, 1998–2006.

Ognev, S. I. (1931). *Mammals of Eastern Europe and Northern Asia, Vol. II, Carnivora, Fissipedia.* Israel Program for Scientific Translations, Jerusalem (1962).

Ognev, S. I. (1931). *Zveri Vostochnoi Evropy i Severnoi Azii, Vol. II, Carnivora (Fissipedia)*, Moskva-Leningrad.

Ognev, S. I. (1959). *Säugetiere und ihre Welt,* Berlin.

Ognev, S. I. (1962). *Mammals of Eastern Europe and Northern Asia, Vol. II, Carnivora (Fissipedia)*. Publication for the National Science Foundation, Washington, D.C., by the Israel Program for Scientific Translations, Jerusalem.

Ohlsson, K. (1971). Isolation of partial characterization of two related trypsin binding α-macroglobulins of dog plasma. *Biochimica et Biophysica Acta,* **236**, 84–91.

Olive, J. R. and Riley, C. V. (1948). Sarcoptic mange in the red fox in Ohio. *Journal of Mammalogy,* **29**, 73–74.

Olsen, O. W. (1958). Hookworms, *Uncinaria lucasi* Stiles, 1901, in fur seals, *Callorhinus ursinus* (Linn.), on the Pribilof Islands. *Transactions North American Wildlife Conference,* **23**, 152–175.

Omodeo, P. E. and Renzoni, A. (1965). Il cariogramma di aluni carnivori. *Bollettino di Zoologia,* **32**.

Osbahr, A. J., Colman, R. W., Laki, K. and Gladner, J. A. (1964). The nature of the peptides released from canine fibrinogen. *Biochemical and Biophysics Research Communications,* **14**, 555–558.

Osgood, W. H. (1904). A biological reconnaisance of the base of the Alaska Peninsula. *North American Fauna,* **24**.

Osgood, W. H. (1934). The genera and subgenera of South American canids. *Journal of Mammalogy,* **15,** 45–50.

Osgood, W. H., Preble, E. A. and Parker, G. H. (1915). The fur seals and other life of the Pribilof Islands, Alaska, in 1914. *Senate Documents,* Vol. **6**, No. 980, Washington, D.C.

Painter, T. S. (1925). A comparative study of the chromosomes of mammals. *The American Naturalist,* **59**, 385–409.

Palmer, R. S. (1956). Gray fox in the Northeast. *Maine Field Naturalist,* **12**, 62–70.

Palmgren, R. (1920). Högholmens zoologiska trädgard aren 1889–1918. *Acta Soc. F. Fl. Fennica,* **47**, 1–240.

Paradiso, J. (1968). Canids recently collected in east Texas, with comments on the taxonomy of the red wolf. *American Midland Naturalist,* **80**, 529–534.

Paradiso, J. L. and Nowak, R. M. (1971). A report on the taxonomic status and distribution of the red wolf. U.S. Department of the Interior, Fish and Wildlife Service, Washington, D.C.

Pare, W. (1966). Subject emotionality and susceptibility to environmental stress. *The Journal of Genetic Psychology,* **108**, 303–309.

Parker, R. L. (1962). Rabies in skunks in the Northcentral States. *Proceedings U.S. Live Stock Sanitary Association,* **65**, 273–280.

Parker, R. L., Kelly, J. W., Cheatum, E. L. and Dean, D. J. (1957). Fox population densities in relation to rabies. *New York Fish and Game Journal,* **4**, 221–228.

Parnell, I. W. (1934). Animal parasites of northeast Canada. *Canadian Field-Naturalist,* **48**, 111–115.

Parrish, H. M., Clack, F. B., Brobst, D. and Mock, J. F. (1959). Epidemiology of dog bite. *Public Health Report,* **74**, 891–903.

Pascual, R., Ortega Hinojosa, E. J., Gondar, D. and Tonni, E. (1965). Las edades del cenozoico mamalífero de la Argentina con especial attención a aquellas del territorio bonaerense. *Anal. Com. Invest. Cient. Prov. Buenos Aires,* **6**, 165–193.

Patil, S. R., Merrick, S. and Lubs, H. A. (1971). Identification of each human chromosome with modified Glemsa stain. *Science,* **173**, 821–822.

Paulk, L. K. (1962). Systematic serology among the Felidae and other closely related groups. *Bulletin Serology Museum,* **28**, 5–8, Rutgers University, Bureau of Biological Research.

Pauly, L. K. and Wolfe, H. R. (1957–58). Serological relationships among members of the order Carnivora. *Zoologica,* **42–43**, 159–166.

Pearson, O. P. and Enders, R. K. (1943). Ovulation, maturation and fertilization of the fox. *Anatomical Record,* **85**, 69–84.

Pearson, O. P. and Bassett, C. F. (1946). Certain aspects of reproduction in a herd of silver foxes. *The American Naturalist,* **80**, 45–67.

Pederson, A. (1926). Beiträge zur Kentniss der Säugetier -und Vogelfauna der ostküste Grön*lands. Medd. om Grø*nland, **68**, 151–249.

Pederson, A. (1934). *Polardyr,* Kobenhavn, Gyldendal.

Perry, M. C. and Giles, R. H., Jr. (1970). Studies of deer-related dog activity in Virginia. *Proc. 24th Annual Conference Southeastern Association Game and Fish Commissioners.*

Perry, M. C. and Giles, R. H., Jr. (1971). Free running dogs. *Virginia Wildlife,* **32**, 17–19.

Peters, A. H. (1971). Tick-borne Typhus (Rocky Mt. spotted fever). Epidemiologic trends, with particular reference to Vir. *Journal American Medical Association,* **216**, 1003–1007.

Peterson, R. L. (1955). *North American Moose,* University of Toronto Press, Toronto, Ontario.

Peterson, R. L. (1966). *The Mammals of Eastern Canada,* Oxford University Press, Toronto, Ontario.

Peterson, R. L., Standfield, R. O., McEwen, E. H. and Brooks, A.C. (1953). Early records of the red and the gray fox in Ontario. *Journal of Mammalogy,* **34**, 126–127.

Petrides, G. A. (1950). The determination of sex and age ratios in fur animals. *American Midland Naturalist,* **43**, 355–382.

Petroff, I. (1898). Report on the population, industries, and resources of Alaska. In: *Seal and Salmon Fisheries and General Resources of Alaska,* Vol. 4, 55th Congress, Document 92, U.S. House of Representatives, p. 167–450.

Petrov, A. M. and Dubnitskii, A. A. (1946). A contribution to the biology of *Strongyloides vulpis* and an epizootic of the blue arctic foxes. *Akademiia nauk SSSR. Gel'mintologicheskii sbornik,* 202–207.

Pfeffer, P. (1972). Observations sur le comportement social et predateur du Lycaon (*Lycaon pictus*) en Ruplublique Centrafricaine. *Mammalia,* **36**, 1–7.

Phillips, R. L. (1970). Age ratios of Iowa foxes. *Journal of Wildlife Management,* **34**, 52–56.

Phillips, R. L., Andrews, R. D., Storm, G. L. and Bishop, R. A. (1972). Dispersal and mortality of red foxes. *Journal of Wildlife Management,* **36**, 237–248.

Pimlott, D. H. (1967). Wolf predation and ungulate populations. *American Zoologist,* **7**, 267–278.

Pimlott, D. H., Shannon, J. A. and Kolenosky, G. B. (1969). The ecology of the timber wolf in Algonquin Provincial Park. *Ontario Department of Lands and Forests Research Report (Wildlife) No.* **87**.

Pitelka, F. A., Tomich, P. Q. and Treichel, G. W. (1955). Ecological relations of jaegers and owls as lemming predators near Barrow, Alaska. *Ecological Monographs,* **25**, 85–117.

Pivone, P. P. (1969). What causes the demise of city street trees? *New York Times,* Sept. 21, 1969, 41.

Ploog, D. (1970). Social communication among animals. In: *The Neurosciences: Second Study Program,* F. O. Schmitt, ed., Rockefeller University Press.

Plummer, P.J.G. (1947). Preliminary note on arctic dog disease and its relationship to rabies. *Canadian Journal of Comparative Medicine,* **11**, 154–160.

Plummer, P.J.G. (1947a). Further note on arctic dog disease and its relationship to rabies. *Canadian Journal of Comparative Medicine,* **11**, 330–334.

Pocock, R. I. (1912). On the moulting of an arctic fox (*Vulpes lagopus*) in the Society's gardens. *Proceedings of the Zoological Society London, 1912,* 55–60.

Polatin, P. (1964). Diagnosis of schizophrenia: pseudoneurotic and other types. *International Psychiatry Clinics,* **1**.

Porsild, A. E. (1945). Mammals of the Mackenzie Delta. *Canadian Field-Naturalist,* **59**, 4–22.

Prashad, B. (1936). Animal remains from Harappa. *Memoirs Archaeological Survey of India,* **51**, 1–62.

Prater, S. H. (1965). *The Book of Indian Animals,* Bombay Natural History Society, Bombay, India.

Preble, E. A. (1902). A biological investigation of the Hudson Bay region. *North American Fauna,* **22**, 1–140.

Preble, E. A. and McAtee, W. L. (1923). A biological survey of the Pribilof Islands, Alaska. *North American Fauna,* **46**, 1–255.

Price, E. (1967). The effect of reproductive performance on the domestication of the prairie deermouse, *Peromyscus maniculatus bairdii. Evolution,* **27**, 4.

Progulske, D. R. and Baskett, T. S. (1958). Mobility of Missouri deer and their harassment by dog. *Journal of Wildlife Management,* **22**, 184–192.

Pulliainen, E. (1962). The appearance and behaviour of wolves in Finland during the period 1954–62. *Suomen Riista,* **15**, 99–129.

Pulliainen, E. (1963). Food and feeding habits of the wolf (*Canis lupus*) in Finland. *Suomen Riista,* **16**, 136–150.

Pulliainen, E. (1965). On the distribution and migrations of the arctic fox (*Alopex lagopus* L.) in Finland. *Aquilo Ser. Zoologica* **2**, 25–40.

Pulliainen, E. (1965). Studies on the wolf (*Canis lupus* L.) in Finland. *Annals Zool. Fennici,* **2**, 215–259.

Pulliainen, E. (1968). The lynx population in Finland. *Acta sc. nat. Brno,* **2**(5–6), 27–34.

Pulliainen, E. (1970). Kuusamon tappajasusi. *Metassästys ja Kalastus,* **59**, 19–20.

Quarterman, K. D., Baker, W. C. and Jenson, J. A. (1949). The importance of sanitation in municipal fly control. *The American Journal of Tropical Medicine,* **29**, 973–982.

Rabb, G. B., Woolpy, J. H. and Ginsburg, B. E. (1967). Social relationships in a group of captive wolves. *American Zoologist,* **7**, 305–311.

Ralls, K. (1971). Mammalian scent marking. *Science,* **171**, 443–449.

Ranjini, P. V. (1966). The chromosomes of the Indian jackal (*Canis aureus*). *Mammalian Chromosomes Newsletter,* **19**, 5.

Ranjini, P. V. (1966). Chromosomes of *Vulpes bengalensis* (Shaw). *Mammalian Chromosomes Newsletter,* **22**, 216.

Rao, C. R. (1952). *Advanced Statistical Methods in Biometric Research,* John Wiley, New York. Reprinted in 1970 by Hafner Publishing Co., Darien, Connecticut.

Rappazzo, M. E. and Hall, C. A. (1972). Cyanocobalamin transport proteins in canine plasma. *American Journal of Physiology,* **222**, 202–206.

Rath,. O. vom. (1894). Uber die koustanz der chromosomenzahl bei Tieren. *Biol. Zhl.,* **14**.

Rausch, R. A. (1967). Some aspects of the population ecology of wolves, Alaska. *American Zoologist,* **7**, 253–265.

Rausch, R. L. (1951). Notes on the Nunamiut Eskimo and mammals of the Anaktuvuk Pass Region, Brooks Range, Alaska. *Arctic,* **4**, 147–195.

Rausch, R. L. (1953). On the status of some arctic mammals. *Arctic,* **6**, 91–148.

Rausch, R. L. (1953a). On the land mammals of St. Lawrence Island, Alaska. *The Murrelet,* **34**, 18–26.

Rausch, R. L. (1956). Studies on the helminth fauna of Alaska, 30; the occurrence of *Echinococcus multilocularis* Leuckart, 1863, on the mainland of Alaska. *American Journal Tropical Medicine and Hygiene,* **5**, 1086–1092.

Rausch, R. (1958). Some observations on rabies in Alaska, with special reference to wild Canidae. *Journal of Wildlife Management,* **22**, 246–260.

Rausch, R. L. (1967). On the ecology and distribution of *Echinococcus* spp. (Cestoda: Taeniidae) and characteristics of their development in the intermediate host. *Annales de Parasitologie (Paris),* **42**, 19–63.

Reichert, E. T. and Brown, A. P. (1909). The differentiation and specificity of corresponding proteins and other vital substances in relation to biological classification and organic evolution: The crystallography of hemoglobins. *Carnegie Institute of Washington Publication* **116**.

Reilly, J. R. and Curren, W. (1961). Evaluation of certain techniques for judging the age of red foxes (*Vulpes fulva*). *New York Fish and Game Journal,* **8**, 122–129.

Reiter, M. B., Gilmore, V. H. and Jones, T. C. (1963). Karyotype of the dog: *Canis familiaris. Mammalian Chromosomes Newsletter,* **12**, 170.

Rempe, U. and Bühler, P. (1969). Zum Einfluss der geographischen und altersbedingten Variabilität bei der Bestimmung von *Neomys*- Mandibeln mit Hilfe der Diskriminanzanalyse. *Zeitschrift für Säugetierkunde,* **34**, 148–164.

Rengger, J. R. (1830). *Naturgeschichte der Saeugethiere von Paraguay,* Basel.

Reynolds, H. Y. and Johnson, J. S. (1970). Quantitation of canine immunoglobulins. *Journal of Immunology,* **105**, 698–703.

Richards, S. H. and Hine, R. L. (1953). Wisconsin fox populations. *Technical Wildlife Bulletin* **6**, Wisconsin Conservation Department, Madison, Wisconsin.

Richardson, J. (1839). List of mammalia and additional observations. In: *Zoology of Capt. Beechey's Voyage to the Pacific and Behring Strait . . . in His Majesty's Ship Blossom, . . . 1825*–28, London.

Richmond, N. D. (1952). Fluctuations in gray fox population in Pennsylvania and their relationship to precipitation. *Journal of Wildlife Management,* **16**, 198–206.

Richter, C. P. (1954). The effect of domestication on the behaviour of the Norway rat. *Journal National Cancer Institute,* **15**, 3.

Rietz, J. H. (1947). Rabies in foxes. *Journal of the American Veterinary Medical Association,* **111**, 138–139.

Robeson, S. B. (1950). Fox-pheasant relations. *New York Conservation Department P-R Project 27-R, Final Report.*

Robinson, W. B. (1961). Population changes of carnivores in some coyote control areas. *Journal of Mammalogy,* **42**, 510–515.

Robinson, W. B. and Cummings, M. W. (1951). Movements of coyotes from and to Yellowstone National Park. *U.S. Fish and Wildlife Service Special Scientific Report Wildlife No.* **11**, 1–17.

Robinson, W. B. and Grand, E. F. (1958). Comparative movements of bobcats and coyotes as disclosed by tagging. *Journal of Wildlife Management,* **22**, 117–122.

Romer, A. S. (1966). *Vertebrate Paleontology,* 3rd Ed., University of Chicago Press, Chicago.

Rosenzweig, M. L. (1966). Community structure in sympatric carnivora. *Journal of Mammalogy,* **47**, 602–612.

Rossolimo, D. L. and Dolgov, V. A. (1965). Variability of the skull in *Canis lupus* Linnaeus, 1758 from the U.S.S.R. *Acta Theriologica,* **10**, 195–207.

Roth, V. (1941). Notes and observations on animal life in British Guiana. A popular guide to Colonial Mammalia. The Guiana Edition No. 3, *The Daily Chronicle,* Georgetown.

Rowe, J. S. (1959). Forest regions of Canada. *Canadian Department of Northern Affairs and National Resources Bulletin* **123**.

Rowell, T. E. (1966). Hierarchy in the organization of a captive baboon group. *Animal Behavior,* **14**, 430–443.

Rue, L. L. III (1969). *The World of the Red Fox,* J. B. Lippincott, Philadelphia.

Russell, D. N. and Shaw, J. H. (1971). Distribution and relative density of the red wolf in Texas. *Proceedings Annual Conference Southeast Association Game and Fish Commission,* **25**, 131–137.

Russell, D. N. and Shaw, J. H. (1971). Notes on the red wolf (*Canis rufus*) in the coastal marshes and prairies of eastern Texas. *Proceedings Texas Academy of Science,* 5.

Rutter, R. J. and Pimlott, D. H. (1968). *The World of the Wolf,* J. B. Lippincott, Philadelphia.

Samet, A. (1950). *Pictorial Encyclopedia of Furs,* Carey Press Corp., New York.

Sanderson, I. T. (1949). A brief review of the mammals of Suriname (Dutch Guiana), based upon a collection made in 1938. *Proceedings Zoological Society London,* **119**, 755–789.

Santos, E. (1945). *Entre o Gambá e o Macaco, vida e costumes dos maniferos do Brasil,* Briguiet, Rio de Janeiro.

Sargeant, A. B. (1965). Cedar Creek radio tracking project. *U.S. Department of the Interior Fish and Wildlife Service, Branch of Predator and Rodent Control, Minnesota–Wisconsin District Annual Report.*

Sargeant, A. B. (1972). Red fox spatial characteristics in relation to waterfowl predation. *Journal of Wildlife Management,* **36**, 225–236.

Sarich, V. M. (1969a). Pinniped origins and the rate of evolution of carnivore albumins. *Systematic Zoology,* **18**, 286–295.

Sarich, V. M. (1969b). Pinniped phylogeny. *Systematic Zoology,* **18**, 416–422.

Sarich, V. M. (1972). On the nonidentity of several carnivore hemoglobins. *Biochemical Genetics,* **7**, 205–212.

Sauer, C. O. (1950). Geography of South America. Handbook South American Indians, Part 6. *Bulletin Bureau American Ethnol. Smithsonian Institution No.* **143**, 319–344.

Saunders, D. Personal communication. Naval Arctic Research Laboratory, Point Barrow, Alaska.

Savage, D. E. (1958). Evidence from fossil land mammals on the origin and affinities of the western Nearctic fauna. In: *Zoogeography,* C. L. Hubbs, ed., Publication No. 51, American Association for Advancement of Science, Horn-Shafer Co., Baltimore, Maryland, pp. 97–129.

Schaller, G. (1967). *The Deer and the Tiger,* University of Chicago Press, Chicago.

Schaller, G. (1972). *The Serengeti Lion*, University of Chicago Press, Chicago.

Schaller, G. B. (1973). In: *The World's Cats, Vol. 1: Ecology and Conservation,* R. Eaton, ed., World Wildlife Safari.

Schenkel, R. (1947). Expression studies of wolves (Ausdrucksstudien an Wölfen). *Behaviour,* **1**, 81–129. (Translation from German by Agnes Klasson.)

Schenkel, R. (1967). Submission: Its features and functions in the wolf and dog. *American Zoologist,* **7**, 319–330.

Schiller, E. T. (1954). Unusual walrus mortality on St. Lawrence Island, Alaska. *Journal of Mammalogy,* **35**, 203–210.

Schiller, E. T. (1959). Observations on the morphology and life cycle of *Microphallus pirium* (Afans'ev, 1941). *Transactions American Microscopical Society,* **78**, 65–76.

Schmidt-Nielsen, K. (1972). *How Animals Work,* Cambridge University Press, New York.

Schnabel, Z. E. (1938). The estimation of the total fish population in a lake. *American Mathematical Monthly,* **45**, 348–352.

Schnedl, W. (1971). Banding pattern of human chromosomes. *Nature New Biology,* **233**, 93.

Schofield, R. D. (1958). Litter size and age ratios of Michigan red foxes. *Journal of Wildlife Management,* **22**, 313–315.

Schofield, R. D. (1960). A thousand miles of fox trails in Michigan's ruffed grouse range. *Journal of Wildlife Management,* **24**, 432–434.

Scholander, P. F., Hock, R., Walters, V., Johnson, F. and Irving, L. (1950). Heat regulation in some arctic and tropical mammals and birds. *Biological Bulletin,* **99**, 237–258.

Schomburgk, R. (1848). *Reisen in British Guiana,* 2nd Ed., Weber, Leipzig.

Schwabe, C. (1969). *Veterinary Medicine and Human Health,* Williams and Wilkins Co., Baltimore.

Schwartz, C. W. and Schwartz, E. R. (1959). *The Wild Mammals of Missouri,* University of Missouri Press and Missouri Conservation Commission, Columbia, Missouri.

Scott, J. P. and Fuller, J. L. (1965). *Genetics and the Social Behavior of the Dog,* University of Chicago Press, Chicago.

Scott, M. D. (1971). The ecology and ethology of feral dogs in eastcentral Alabama. Ph. D. dissertation, Auburn University, Auburn, Alabama. From abstract of the dissertation, with the author's permission.

Scott, M. D., and Causey, K. (1973). Ecology of feral dogs in Alabama. *Journal of Wildlife Management.,* **37**(3), 253–265.

Scott, T. G. (1943). Some food coactions of the northern plains red fox. *Ecological Monographs,* **13**, 427–479.

Scott, T. G. (1947). Comparative analysis of red fox feeding trends on two central Iowa areas. *Research Bulletin* **353**, Iowa Agricultural Experiment Station, Ames, Iowa, pp. 427–487.

Scott, T. G. (1955a). An evaluation of the red fox. *Biological Notes No.* **35,** Illinois Natural History Survey Division, Urbana.

Scott, T. G. (1955b). Dietary patterns of red and gray foxes. *Ecology,* **36,** 366–367.

Scott, T. G. and Selko, L. F. (1939). A census of red foxes and striped skunks in Clay and Boone Counties, Iowa. *Journal of Wildlife Management,* **3**, 92–98.

Scott, T. G. and Klimstra, W. D. (1955). Red foxes and a declining prey population. *Southern Illinois University Monograph Series No.* **1**.

Sdobnikov, V. M. (1957). Hunting in the far north and tasks for scientific research. *Okhota i okhotnich'e khoziaistvo,* **3**, 8–9.

Sdobnikov, V. M. (1958). The arctic fox in Taimyr. *Problems of the North,* **1**, 229–238.

Sdobnikov, V. M. (1967). Some peculiarities of distribution and abundance of the arctic fox (*Alopex lagopus*) in the Asiatic tundra (in Russian). *Zoologicheskii Zhurnal,* **46**, 1378–1382.

Seagers, C. B. (1944). The fox in New York. *New York State Conservation Department Educational Bulletin.*

Seal, H. L. (1964). *Multivariate Statistical Analysis for Biologists,* Methuen, London.

Seal, U. S. (1969). Carnivora systematics: A study of hemoglobins. *Comparative Biochemistry and Physiology,* **31**, 799–811.

Seal, U. S., Phillips, N. I. and Erickson, A. W. (1970). Carnivora systematics: immunological relationships of bear serum albumins. *Comparative Biochemistry and Physiology,* **32**, 33–48.

Seal, U. S., Erickson, A. W., Siniff, D. and Cline, D. R. (1971). Blood chemistry and protein polymorphisms in three species of antartic seals (*Lobodon carcinophagus, Leptonychotes weddelli,* and *Mirounga leonina*). In: *Antarctic pinnipedia,* W. H. Burt, ed., American Geophysical Union 181–192.

Seal, U. S., Erickson, A. W., Siniff, D. and Hofman, R. (1971). Biochemical, population genetic, phylogenetic and cytological studies of Antarctic seal species. In: *SCAR Symposium #26006, 77–95.*

Seitz, A. (1959a). Ein Bastard Nordafricanischer Goldshakal × Coyote unter besonderer Berucksichtigung des Verhaltens in den ersten 4 Lebensmonaten. *De Zoologische Garten, Leipzig,* **25**, 79–95.

Seitz, A. (1959b). Beobachtungen an Handaufgezogenen Goldschakalen (*Canis aureus algirensis* Wagner 1843). *Zeitschrift für Tierpsychologie,* **16**, 747–771.

Seitz, A. (1965). Fruechtbare Kreuzungen Goldshakol × Coyote und veziprok Coyote × Goldshakal; erste fruchtbare Ruckkreuzung. *De Zoologische Garten, Leipzig,* **31**, 174–183.

Seton, E. T. (1923). The mane of the tail of the gray fox. *Journal of Mammalogy,* **4**, 180–182.

Seton, E. T. (1929). The arctic fox. In: *Lives of Game Animals,* Vol. I, part II, Doubleday, Doran and Co., Inc., New York.

Severinghaus, C. W. and Cheatum, E. L. (1956). Life and times of the white-tailed deer. In: *The Deer of North America,* W. P. Taylor, ed., Stackpole Co., Harrisburg, Pennsylvania, pp. 57–186.

Sheldon, W. G. (1949). Reproductive behavior of foxes in New York State. *Journal of Mammalogy,* **30**, 236–246.

Sheldon, W. G. (1950). Denning habits and home range of red foxes in New York State. *Journal of Wildlife Management,* **14**, 33–42.

Sheldon, W. G. (1953). Returns on banded red and gray foxes in New York State. *Journal of Mammalogy,* **34**, 125.

Shelton, P. C. (1966). Ecological studies of beavers, wolves, and moose in Isle Royale National Park, Michigan. Ph. D. Thesis, Purdue University, Lafayette, Indiana.

Shim, B.-S., Yoon, C.-S., Oh, S.-K., Lee, T.-H. and Kang, Y.-S. (1971). Studies on swine and canine serum haptoglobins. *Biochemica et Biophysica Acta,* **243**, 126–136.

Shorin, J. P. (1967). Fluorometric determination of free 11-oxycorticosteroids in blood plasm. In: *Corticosteroid Regulation of Water–Saline Homeostasis,* Nauka, Novosibirsk.

Shortridge, G. C. (1934). *The Mammals of Southwest Africa,* Heinemann, London.

Siivonen, L. (1956). *Mammalia Fennica,* Helsinki.

Silveira, E.K.P. da (1968). Notes on the care and breeding of the maned wolf. *International Zoo Yearbook,* **8**, 20–27.

Silveira, E.K.P. da (1969). O lôbo-guará (*Chrysocyon brachyurus*) — Possível acão inibidora de certas solanáceas sôbre o nematóide renal. *Vellozia,* **7**.

Silver, H. and Silver, W. T. (1969). Growth and behavior of the coyotelike canid of northern New England with observations on canid hybrids. *Wildlife Monographs,* **17**, 1–41.

Simpson, G. G. (1945). The principles of classification and a classification of mammals. *Bulletin American Museum of Natural History,* **85**, 1–350.

Simpson Vuilleumier, B. (1971). Pleistocene changes in the fauna and flora of South America. *Science,* **173**, 771–780.

Siniff, D. B. and Jessen, C. R. (1969). A simulation model of animal movement patterns. *Advances in Ecological Research,* **6**, 185–219, Academic Press, London.

Sisson, S. (1945). *The Anatomy of the Domestic Animals,* 3rd Ed., 3rd Report, Philadelphia.

Skrobov, V. D. (1960). Interrelationships of the polar fox and fox in the tundras of the Nenatsk National Okrug. *Zoologicheskii Zhurnal,* **39**, 469–471.

Skrobov, V. D. (1960a). Some data on the biology and ecology of polar fox in connection with characteristics of its burrow in Bolshezemelskaja and Malozemelskaja Tundra. *Bulletin Moscow Obshch. Inspit. Prirody (Biology),* **65**, 28–36.

Skrobov, V. D. (1961). The arctic fox. *Okhota i okhotnich'e khozyaystvo,* U.S.S.R. Ministry of Agriculture, 17–20 (January).

Skrobov, V. D. (1961a). Registration and distribution of arctic foxes in Yamal Tundra. In: Organization and methods of censusing terrestial vertebrate faunal resources, 35–36. Summarized reports of a Symposium of the Moscow Naturalists' Society, U.S. National Science Foundation, Israel Program for Scientific Translations, Jerusalem (1963).

Smith, E. L. and McKibbin, J. M. (1972). Separation of dog intestine glycolipids into classes according to sugar content by thin-layer chromatography: Evidence for individual dog intestine specific glycolipids. *Analytical Biochemistry,* **45**, 608–616.

Smith, L. F. (1943). Internal parasites of the red fox in Iowa. *Journal of Wildlife Management,* **7**, 174–178.

Smithers, R.H.N. (1966a). *The Mammals of Rhodesia, Zambia and Malawi,* Collins, London.

Smithers, R.H.N. (1966b). "Matsuana" a Southern bat-eared fox. *Animal Kingdom,* **69**, 163–167.

Snyder, C. H. (1961). Visceral larvae migrans; ten years' experience. *Pediatrics,* **28**, 85–91.

Sokolov, N. N. (1957). Histological analysis of the sexual cycle in the arctic fox of the tundra. *Zoologicheskii Zhurnal,* **36**, 1076–1083.

Soper, J. D. (1928). A faunal investigation of southern Baffin Island. *National Museum of Canada Bulletin,* **53**, Biology Series No. 15, 1–143.

Soper, J. D. (1942). Mammals of Wood Buffalo Park, Northern Alberta and District of Mackensie. *Journal of Mammalogy,* **23**, 119–145.

Soper, J. D. (1944). Mammals of Baffin Island. *Journal of Mammalogy,* **25**, 221–254.

Southern, H. N. and Watson, J. S. (1941). Summer food of the red fox in Great Britain: a preliminary report. *Journal of Animal Ecology,* **15**, 198–202.

Sparks, J. (1967). Allogrooming in primates: A review. In: *Primate Ethology,* D. Morris, ed., Anchor, New York.

Spencer, R. F. (1959). *The North Alaskan Eskimo,* U.S. Government Printing Office, Washington, D.C.

Sperry, C. C. (1941). Food habits of the coyote. *Fish and Wildlife Service Research Bulletin,* **4**.

Srivastava, M.D.L. and Bhatnagar, V. S. (1969). Studies on the somatic chromosomes of a pariah dog at Allahabad. *Mammalian Chromosomes Newsletter,* **10**, 14.

Stahl, F. and Dörner G. (1966). Eine einfache spezifische Routinmethode für fluorometrischen Bestimmung von unkonjugierten 11-hydroxycorticosteroiden in Korperelûssigkeiten. *Acta Endocrinology,* **51,** 175–185.

Stains, H. J. (1967). Carnivores and Pinnipeds. In: *Recent Mammals of the World,* S. Anderson and J. Knox Jones, Jr., eds., Ronald Press, New York.

Standfield, R. O. Personal communication. Department of Wildlife Research, Maple, Ontario.

Stanley, W. C. (1963). Habits of the red fox in Northeastern Kansas. *University of Kansas Museum Natural History Miscellaneous Publication* **34**.

Stefansson, V. (1919). *My Life with the Eskimo,* New York.

Stefansson, V. (1922). *Hunters of the Great North,* New York.

Stephenson, R. O. (1970). A study of the summer food habits of the arctic fox on St. Lawrence Island, Alaska. Unpublished M.S. Thesis, University of Alaska, College, Alaska.

Stevens, R.W.C. and Townsley, M. E. (1970). Canine serum transferrins. *Journal of Heredity,* **61**, 71–72.

Storm, G. L. (1965). Movements and activities of foxes as determined by radiotracking. *Journal of Wildlife Management,* **29**, 1–13.

Storm, G. L. (1972). Population dynamics of red foxes in northcentral United States. Ph. D. Thesis, University of Minnesota.

Storm, G. L. and Dauphin, K. P. (1965). A wire ferret for use in studies of foxes and skunks. *Journal of Wildlife Management,* **29**, 625–626.

Storm, G. L. and Ables, E. D. (1966). Notes on newborn and fullterm wild red foxes. *Journal of Mammalogy,* **47**, 116–118.

Strogov, A. K. (1961). The etiology of the "madness" of arctic and other foxes and of dogs living in the tundra zone of the Yakutsk Autonomous Soviet Socialist Republic. *Nauch. Soobschenija Yakutsk. Filiala Sibir. Otdel. Akad. Nauk. SSSR,* **5**, 101–108.

Studer, T. (1901). Die prahistorischen Hunde in ihrer Beziehung zu den gegenwartig lebenden Rassen. *Abhandlungen Schweiz. palantölogische Geselischaft Zurich,* **28**, 1–137.

Studer, T. (1905). Über einen Hund aus der Palaolithischen zeit Ruslands, *Canis poutiatini. Zoologischer Anzeiger,* **29**(1).

Stutzenbacker, C. D. (1968). Coastal marsh management. Survey Job No. 6. Mottled duck status. *Job Progress Report on Federal Aid Project No. W-96-R-3.*

Sullivan, E. (1955). The fox — friend or foe? *South Carolina Wildlife,* **2**, 2–3, 18–19.

Sullivan, E. G. (1956). Gray fox reproduction, denning range, and weights in Alabama. *Journal of Mammalogy,* **37**, 346–351.

Sullivan, E. G. and Haugen, A. O. (1956). Age determination of foxes by x-ray of forefeet. *Journal of Wildlife Management,* **20**, 210–212.

Suomus, H. (1970). Eräs susimuisto viime vuosisadalta. *Riista ja kala,* 1970–71, 8–14.

Sutton, G. M. (1932). The birds of Southhampton Island. *Carnegie Museum Memoirs,* **12**, 1–275.

Sutton, G. M. and Hamilton, W. J. (1932). The exploration of Southhampton Island, Hudson Bay. *Carnegie Museum Memoirs,* **12**(2), 3–267.

Sweeney, J. R., Marchinton, R. L. and Sweeney, J. M. (1971). Responses of radio-monitored white-tailed deer chased by hunting dogs. *Journal of Wildlife Management,* **35**, 707–716.

Swisher, S. N. and Young, L. E. (1967). The blood grouping systems of dogs. *Physiological Review,* **41**, 495–520.

Swisher, S. N., Young, L. E. and Trabold, N. (1962). *In vitro* and *in vivo* studies of the behavior of canine erythrocyte-isoantibody systems. *Annals New York Academy of Science,* **97**, 15–25.

Switzenberg, D. F. (1950). Breeding productivity in Michigan red foxes. *Journal of Mammalogy,* **31**, 194–195.

Switzenberg, D. F. (1951). Examination of a state fox bounty. *Journal of Wildlife Management,* **15**, 288–299.

Syuzyumova, L. M. (1967). Epizootiology of rabies among arctic foxes on the Yamal Peninsula. *Problems of the North,* **11**, 99–106.

Tate, G.H.H. (1931). Random observations on habits of South American mammals. *Journal of Mammalogy,* **12**, 248–256.

Tate, G.H.H. (1939). The mammals of the Guiana region. *Bulletin American Museum Natural History,* **76**, 151–229.

Tate, G.H.H. (1952). Results of the Archbold Expeditions. *Bulletin American Museum of Natural History,* **98**, 612–614.

Taylor, W. P. (1943). The grey fox in captivity. *Texas Game and Fish,* **1**, 12–13, 19.

Taylor, W. P. (1953). Food habits of the gray fox (*Urocyon cinereoargenteus)* in the Edwards Plateau, Texas. Part of final report project W-31-R-1 for Texas Game and Fish Commission, Austin.

Tembrock, G. (1957). Zur ethologie des Rotfuchses (*Vulpes vulpes* L.) unter besonderer Berucksichtigung der Fortpflanzung. *Der Zoologische Garten,* **23**, 289–532.

Tembrock, G. (1958). Spielverhalten beim Rotfuch. *Zoologische beiträge Berlin,* **3**, 423–496. (English translation by B. Piddack.)

Tembrock, G. (1968). Land mammals. In: *Animal Communication,* T. A. Sebeok, ed., Indiana University Press, Bloomington, Indiana.

Tener, J. S. (1954). A preliminary study of the musk-oxen of Fosheim Peninsula, Ellesmere Island, N.W.T. *Canadian Wildlife Service Wildlife Management Bulletin Series,* **1**, No. 9.

Terres, J. K. (1939). Tree climbing techniques of the gray fox. *Journal of Mammalogy,* **20**, 256.

Terry, R. L. (1970). Primate grooming as a tension reduction mechanism. *Journal of Psychology,* **76**, 129–136.

Thiessen, D. and Nealey, V. (1962). Adrenocortical activity, stress, response and behavioral reactivity of five inbred mouse strains. *Endocrinology,* **71**, 227–232.

Thomas, L. J. and Babero, B. B. (1956). Some helminths of mammals from St. Lawrence Island, Alaska, with a description of the nomenclature of *Echinococcus* in voles. *Journal of Parasitology,* **42**, 500.

Thompson, D. Q. (1952). Travel, range and food habits of timber wolves in Wisconsin. *Journal of Mammalogy,* **33**, 429–442.

Thompson, D. Q. (1955). The 1953 lemming emigration at Point Barrow, Alaska. *Arctic,* **8**, 37–45.

Thornton, W. A., Creel, G. C. and Trimble, R. E. (1971). Hybridization in the fox genus *Vulpes* in West Texas. *The Southwestern Naturalist,* **15**, 473–484.

Tichota, J. (1937). Das Verwandtschaftsverhaltnis des australischen Dingoes zu den prahistorischen Type n des Haushundes. *Zoologischer Anzeiger,* **120** (9/10).

Tinbergen, N. (1957). The functions of territory. *Bird Study,* **4**, 14–27.

Todd, N. B. (1969). The karyotype of the raccoon dog (*Nyctereutes sp.*). *Mammalian Chromosomes Newsletter,* **10**, 21–22.

Trapp, G. R. (1973). Comparative behavioral ecology of two southwest Utah carnivores: *Bassariscus astutus* and *Urocyon cinereoargenteus*. Unpublished doctoral dissertation, University of Wisconsin, Madison, Wisconsin.

Troughton, E. (1941). *Furred Animals of Australia,* Sydney.

Troughton, E. (1971). The early history and relationship of the New Guinea Highland Dog (*Canis hallstromi*). *Proceedings Linnean Society of New South Wales,* **96**, 93–98.

Trouvessart, E. L. (1899). *Catalogus Mammalium,* Vol. 1, R. Friedlander and Sohn, Berlin.

Tsalkin, V. I. (1944). Geographic variability in the skull structure of the Eurasian polar fox. *Zoologischeskii Zhurnal,* **23**, 156–169.

Tsetsevinski, L. M. (1940). Materials on the ecology of the arctic fox in northern Yamal. *Zoologischeskii Zhurnal,* **19**.

Tsushima, T., Irie, M. and Sakuma, M. (1971). Radioimmunoassay for canine growth hormone. *Endocrinology,* **89**, 685–693.

Tuck, L. M. (1960). The murres. *Canadian Wildlife Service Report Series No.* **1**, Queens Printer, Ottawa.

Turner, L. M. (1886). Contributions to the natural history of Alaska; results of investigations made chiefly in the Yukon District and the Aleutian Islands. *Arctic Series of Publications No.* **2**, issued in connection with the Signal Service, U.S. Army, pp. 1–226.

Valenti, C. and Levy, C. (1965). The karyotype of *Canis dingo*. *Mammalian Chromosomes Newsletter,* **18**, 147–148.

Vanzolini, P. E. and Williams, E. E. (1970). South American anoles: the geographic differentiation and evolution of the *Anolis chrysolepis* species group (Sauria, Iguanidae). *Arquivos de Zoologia Sao Paulo,* **19**, 1–124.

Van Valen, L. (1964). Nature of supernumerary molars of *Otocyon*. *Journal of Mammalogy,* **45**, 284–286.

Vasil'ev, V. A. (1938). Experiment in marking and feeding white foxes. *Sovetskaia,* **1**, 89–90.

Venge, O. (1959). Reproduction in the fox and mink. *Animal Breeding Abstracts,* **27**, 129–145.

Verts, B. J. and Storm, G. L. (1966). A local study of prevalence of rabies among foxes and striped skunks. *Journal of Wildlife Management,* **30**, 419–421.

Vibe, C. (1967). Arctic animals in relation to climatic fluctuations. *Medd. om Grønland,* **170**, 7–227.

Von Bloeker, J. (1967). Land mammals of the southern California islands. In: *Proceedings of the Symposium on the Biology of the California Islands,* R. N. Philbrick, ed., Santa Barbara Botanic Garden, Santa Barbara, Calif.

Wagner, F. H. and Stoddard, L. C. (1972). Influence of coyote predation on black-tailed jackrabbit populations in Utah. *Journal of Wildlife Management,* **36**, 329–342.
Walker, E. P. (1964). *Mammals of the World,* Vol. II, The Johns Hopkins Press, Baltimore, Maryland.
Walker, E. P. (1968). *Mammals of the World,* 2nd Ed., The Johns Hopkins Press, Baltimore, Maryland.
Walther, F. R. (1969). Flight behavior and avoidance of predators in Thomson's gazelle (*Gazella thomsonii* Guenther 1884). *Behavior,* **34**, 184–221.
Ward, L. (1954). What's it going to be, deer or dogs in southern West Virginia? *West Virginia Conservation,* **18**, 3–5.
Waters, J. H. (1964). Red fox and gray fox from New England archaeological sites. *Journal of Mammalogy,* **45**, 307–308.
Waters, J. H. (1967). Foxes on Martha's Vineyard, Massachusetts. *Journal of Mammalogy,* **48**, 137–138.
Wetmore, A. (1952). The gray fox attracted by a crow call. *Journal of Mammalogy,* **32**, 244–245.
Wied, M. (1826). *Beiträge zur Naturgeschichte von Brasilien. II Abt. Mammalia,* Vol. 2, Weimar.
Wilcock, J. (1969). Gene action and behaviour: evaluation of major gene pleiotropism. *Physiological Bulletin,* **72**, 1–29.
Wilcomb, M. S. (1956). Fox populations and food habits in relation to game bird survival, Willamette Valley, Oregon. *Oregon State Game Commission Technical Bulletin,* **38**.
Wilkinson, C. P., and Welch, R. B. (1971). Intraocular *Toxocara. American Journal of Ophthalmology,* **71**, 921–930.
Winge, H. (1895). Jordfunde og nulevende Rovdyr (Carnivora) fra Lagoa Santa, Minas Geraes, Brasilien. *E. Museo Lundii,* **2**(2).
Wipf, L. (1942). Chromosomes of the red fox. *Proceedings National Academy of Science,* **28**, 265–268.
Wipf, L. and Schackleford, R. M. (1949). Chromosomes of a fox hybrid (*Alopex–Vulpes*). *Proceedings National Academy of Science,* **35**, 468–472.
Wisconsin Conservation Department (1962). Game and fur harvest reports. Game Management Division, Wisconsin Conservation Department, Madison, Wisconsin.
Wodsedalek, J. E. (1931). Spermatogenesis of the red fox, *Vulpes fulvus. Anatomical Record Supplement,* **51**, 70.
Wolff, A. H., Henderson, N. D. and McCallum, G. L. (1948). Salmonella from dogs and the possible relationship to Salmonellosis in man. *American Journal of Public Health,* **38**, 403–408.
Wolfe, M. L. and Allen, D. L. (in press). Continued studies of the status, socialization and relationship of Isle Royale wolves, 1967–1970.
Wood, J. E. (1954a). Food habits of furbearers of the upland post oak region in Texas. *Journal of Mammalogy,* **35**, 406–415.
Wood, J. E. (1954b). Investigation of fox populations and sylvatic rabies in the Southeast. *Transactions 19th North American Wildlife Conference,* 131–139.
Wood, J. E. (1958). Age structure and productivity of a gray fox population. *Journal of Mammalogy,* **39**, 74–86.
Wood, J. E. (1959). Relative estimates of fox population levels. *Journal of Wildlife Management,* **23**, 53–63.
Wood, J. E. and Davis, D. E. (1959). The prevalence of rabies in populations of

foxes in the southern states. *Journal of the American Veterinary Medical Association,* **135**, 121–124.

Wood, J. E., Davis, D. E. and Komarek, E. V. (1958). The distribution of fox populations in relation to vegetation in Southern Georgia. *Ecology,* **39**, 160–162.

Wood, J. E. and Odum, E. P. (1964). A nine-year history of furbearer populations on the AEC Savannah River Plant Area. *Journal of Mammalogy,* **45**, 540–551.

Wright, B. S. (1960). Predation on big game in East Africa. *Journal of Wildlife Management,* **24**, 1–14.

Wright, T. J. (1949). A study of the fox in Rhode Island. *Rhode Island Department of Agriculture and Conservation, Division Fish and Game Pamphlet No.* **3**.

Wurster, D. H. (1959). Cytogenetics and phylogenetic studies in carnivora. In: *Comparative Mammalian Cytogenetics,* K. Benirschke, ed., Springer Verlag, Berlin, pp. 310–329.

Wurster, D. H. and Benirschke, K. (1968). Comparative cytogenetic studies in the order Carnivora. *Chromosoma,* **24**, 336–382.

Wyman, J. (1967). The jackals of the Serengeti. *Animals* (June), 79–83.

Yakushkin, G. D. (1963). Some data on autumn migration of the arctic fox (*Alopex lagopus*) on the Taimyr Peninsula. *Trudy Vses Sel'skokhoz Inst. Zaochnogo Obrazovaniya,* **15**, 172–180.

Yeager, L. E. (1938). Tree climbing by a gray fox. *Journal of Mammalogy,* **19**, 376.

Young, S. P. (1944). *The Wolves of North America,* Part 1, The American Wildlife Institute, Washington, D.C.

Young, S. P. and Goldman, E. A. (1944). *The Wolves of North America,* The American Wildlife Institute, Washington, D.C.

Young, S. P. and Jackson, H.H.T. (1951). *The Clever Coyote,* The Stackpole Co., Harrisburg, Pennsylvania and The Wildlife Management Institute, Washington, D.C.

Zeuner, F. E. (1963). *A History of Domesticated Animals,* London.

Zimen, E. (1971). *Wölfe und Königspudel,* Vergleichende Verhaltensbeobachtungen, Munich.

Zinkham, W. H. (1968). Visceral larva migrans due to *Toxocara* as a cause of eosinophilia. *Johns Hopkins Medical Journal,* **123**, 41–47.

Zittel, K. A. (1925). *Textbook of Palaeontology,* Macmillan and Co., Ltd., London.

Zuckerman, S. (1953). The breeding seasons of mammals in captivity. *Proceedings of Zoological Society of London,* **122**, 827–950.

Author Index

Subject Index